obal biodiversity. To
appreciate not only what
w they interact with each
wledge in this area, with
esearch conducted in both
tions among taxa at all
re the starting point, but,
nts with other plants, with
nships of human-induced
ying theme of the volume
iversity in tropical regions,
observations in ecological

lant Science at the
tropical forest sites in

the University of Aberdeen.
ia, Sri Lanka, Brazil and

of Sussex, specializing in

Ecological Reviews

Ecological Reviews will publish books at the cutting edge of modern ecology, providing a forum for volumes that discuss topics that are focal points of current activity and likely long-term importance to the progress of the field. The series will be an invaluable source of ideas and inspiration for ecologists at all levels from graduate students to more-established researchers and professionals. The series will be developed jointly by the British Ecological Society and Cambridge University Press and will encompass the Society's Symposia as appropriate.

Biotic Interactions in the Tropics: Their Role in the Maintenance of Species Diversity
Edited by David F. R. P. Burslem, Michelle A. Pinard and Sue E. Hartley

Biological Diversity and Function in Soils
Edited by Richard Bardgett, Michael Usher and David Hopkins

Biotic Interactions in the Tropics

Their Role in the Maintenance of Species Diversity

Edited by

DAVID F. R. P. BURSLEM
School of Biological Sciences, University of Aberdeen

MICHELLE A. PINARD
School of Biological Sciences, University of Aberdeen

SUE E. HARTLEY
School of Life Sciences, University of Sussex

CAMBRIDGE
UNIVERSITY PRESS

CAMBRIDGE UNIVERSITY PRESS
Cambridge, New York, Melbourne, Madrid, Cape Town, Singapore, São Paulo

Cambridge University Press
The Edinburgh Building, Cambridge CB2 2RU, UK

Published in the United States of America by Cambridge University Press, New York

www.cambridge.org
Information on this title: www.cambridge.org/9780521847079

© British Ecological Society 2005

First published 2005

Printed in the United Kingdom at the University Press, Cambridge

A catalogue record for this publication is available from the British Library

ISBN-13 978-0-521-84707-0 hardback
ISBN-10 0-521-84707-9 hardback
ISBN-13 978-0-521-60985-2 paperback
ISBN-10 0-521-60985-2 paperback

Contents

Contributors

Ian J. Alexander School of Biological Sciences, University of Aberdeen, Cruickshank building, Aberdeen AB24 3UU, UK

Anne E. Arnold Duke University, Durham, NC 27708, USA

Joseph Bischoff Department of Plant Biology and Pathology, Rutgers University, Cook College, 369 Foran Hall, 59 Dudley Road, New Brunswick, NJ 08901, USA

Thomas J. Brandeis International Institute of Tropical Forestry, USDA Forest Service, Puerto Rico

Carine Brouat Centre de Biologie et de Gestion des Populations, Campus International de Baillarguet, CS 30016, F-34988 Montferrier sur Lez cedex, France

David F. R. P. Burslem School of Biological Sciences, University of Aberdeen, Cruickshank Building, Aberdeen, AB24 3UU, UK

Charles D. Canham Institute of Ecosystem Studies, Millbrook, New York 12545, USA

Richard Condit Smithsonian Tropical Research Institute, Apartado 2072, Balboa, Republic of Panama

Roberto A. Cordero Smithsonian Tropical Research Institute, Apartado Postal 2072, Balboa, Ancon, Republic of Panama

Ambroise Dalecky Centre de Biologie et de Gestion des Populations, Campus International de Baillarguet, CS 30016, F-34988 Montferrier sur Lez cedex, France

Jim W. Dalling Department of Plant Biology, University of Illinois, 265 Morrill Hall, 505 South Goodwin Avenue, Urbana, IL 61801, USA

H. S. Dattaraja Centre for Ecological Sciences, Indian Institute of Science, Bangalore 560 012, India

Gabriel Debout Centre for Ecology, Evolution and Conservation, School of Biological Sciences, University of East Anglia, Norwich NR4 7TJ, UK

Kleber Del-Claro Universidade Federal de Uberlândia, Av. Engenheiro Diniz, 1178, Cx. Postal 593, CEP 38.400-902, Uberlândia, Minas Gerais, Brazil

Saara J. DeWalt Rice University, 6100 Main Houston, Texas 77005, USA

Bruno di Gíusto Botanique et bioinformatique de l'architecture des plantes, CNRS, UMR 5120, Boulevard de la Lironde TA40 PS2, F-34398 Montpellier cedex 5, France

Lee A. Dyer 6823 St. Charles Avenue, New Orleans, LA 70118, USA

Ahn-Heum Eom Korea National University of Education, 363-791 Korea, Republic of Korea

John J. Ewel Institute of Pacific Islands Forestry, United States Forest Service, Honolulu, Hawaii 96813, USA

José M. V. Fragoso Pacific Cooperative Studies Unit, Department of Botany, University of Hawaii, 3190 Maile Way, Honolulu, Hawaii 96822-2232, USA

Laurence Gaume Botanique et bioinformatique de l'architecture des plantes, CNRS, UMR 5120, Boulevard de la Lironde TA40 PS2, F-34398 Montpellier cedex 5, France

Jaboury Ghazoul Imperial College London, Silwood Park, Ascot, Berkshire, SL5 7PY, UK

Gregory S. Gilbert Environmental Studies Department, 439 I5B, 1156 High St., University of California, Santa Cruz, Santa Cruz, CA 95064, USA

Britta Denise Hardesty University of Georgia, Plant Biology Department, 2502 Miller Plant Sciences, Athens, GA 30602, USA

Sue E. Hartley Department of Ecology and Environmental Sciences, University of Sussex, East Sussex, BN1 9RH, UK

Martin Heil FB-9, Allgemeine Botanik/Pflanzenökologie, Universität Duisburg-Essen, Universitätsstrasse 5, D-45117 Essen, Germany

Edward Allen Herre Smithsonian Tropical Research Institute, Apartado Postal 2072, Balboa, Ancon, Republic of Panama

Ankila Hiremath Ashoka Trust for Research in Ecology and the Environment, 659 5th Avenue Main, Hebbal Bangalore 560 024, India

Stephen P. Hubbell University of Georgia, Plant Biology Department, Athens, GA 30602, USA

Rebecca Husband The University of York, Heslington, York YO10 5DD, UK

Robert John University of Georgia, Plant Biology Department, Athens, GA 30602, USA

Damond Kyllo Smithsonian Tropical Research Institute, Apartado Postal 2072, Balboa, Ancon, Republic of Panama

William F. Laurance Smithsonian Tropical Research Institute, Apartado Postal 2072, Balboa, Ancon, Republic of Panama

Deborah K. Letourneau Department of Environmental Studies, 214 College Eight, University of California, Santa Cruz, CA 95064, USA

Ariel E. Lugo International Institute of Tropical Forestry, USDA Forest Service, Puerto Rico

Scott Mangan Indiana University, 107 S. Indiana Avenue, Bloomington, IN 47405-7000, USA

Robert J. Marquis Department of Biology, University of Missouri-St Louis, One University Boulevard, 99 St Louis, MO 63121-44 USA

Fergus P. Massey Department of Ecology and Environmental Sciences, University of Sussex, East Sussex, BN1 9RH, UK

Zuleyka Maynard Smithsonian Tropical Research Institute, Apartado Postal 2072, Balboa, Ancon, Republic of Panama

Doyle McKey Centre d'Ecologie Fonctionnelle et Evolutive, CNRS, UMR 5175, 1919 route de Mende, 34293 Montpellier cedex 5, France

Luis C. Mejia Smithsonian Tropical Research Institute, Apartado Postal 2072, Balboa, Ancon, Republic of Panama

Helene C. Muller-Landau Department of Ecology, Evolution and Behavior,

University of Minnesota, 1987 Upper Buford Circle, St Paul, MN 55108, USA

C. Nath Centre for Ecological Sciences, Indian Institute of Science, Bangalore 560 012, India

Paulo S. Oliveira Departmento de Zoologia, C.P. 6109, Universidade Estadual de Campinas, 13083-970 Campinas SP, Brazil

Stephen W. Pacala Princeton University, Princeton, New Jersey 08544 USA

Laurence Pascal Centre d'Ecologie Fonctionnelle et Evolutive, CNRS, UMR 5175, 1919 route de Mende, 34293 Montpellier cedex 5, France

Michelle A. Pinard School of Biological Sciences, University of Aberdeen, Cruickshank Building, Aberdeen AB24 3UU, UK

Lourens Poorter Forest Ecology and Forest Management Group, Wageningen University, PO Box 47, 6700 AA Wageningen, The Netherlands

Malcolm C. Press Department of Animal and Plant Sciences, University of Sheffield, Sheffield, S10 2TN, UK

Drew W. Purves Princeton University, Princeton, New Jersey 08544 USA

Nancy Robbins Smithsonian Tropical Research Institute, Apartado Postal 2072, Balboa, Ancon, Republic of Panama

Enith Rojas Smithsonian Tropical Research Institute, Apartado Postal 2072, Balboa, Ancon, Republic of Panama

Lee Su See Forest Research Institute of Malaysia, (52109), Kepong, Kuala Lumpur, Malaysia

S. Srinidhi Centre for Ecological Sciences, Indian Institute of Science, Bangalore 560 012, India

R. Sukumar Centre for Ecological Sciences, Indian Institute of Science, Bangalore 560 012, India

H. S. Suresh Centre for Ecological Sciences, Indian Institute of Science, Bangalore 560 012, India

Maria Uriarte Institute of Ecosystem Studies, Millbrook, New York 12545, USA

Sunshine A. Van Bael Smithsonian Tropical Research Institute, Apartado Postal 2072, Balboa, Ancon, Republic of Panama

Catherine Woodward Smithsonian Tropical Research Institute, Apartado Postal 2072, Balboa, Ancon, Republic of Panama

Editors' Preface

It is well known that tropical forests and savannas house a significant proportion of global biodiversity. However, an appreciation of the diversity of interactions among organisms in tropical ecosystems is only just emerging. Interactions among species are important because they affect the growth, survival and reproduction of individuals, but also because they have a key role in structuring communities and in the functioning of ecosystems. A sound knowledge of these interactions is therefore fundamental to understanding how tropical ecosystems work, as well as informing important practical concerns such as conservation, management and carbon sequestration. The aim of this book, and the meeting from which it derives, is to synthesize the current state of knowledge of biotic interactions in terrestrial communities in the tropics. Each of the 22 chapters of this volume provides a review or a case study of interactions among organisms from tropical ecosystems, with a perspective drawn from the organisms and sites with which the individual authors work. Our aim was to draw on research conducted in both Old and New World tropics and to include biotic interactions among taxa at all trophic levels. Most authors have taken plants (typically trees) as their starting point, but taken together the chapters consider interactions of plants with other plants, with micro-organisms and with animals, and the interrelationships of human-induced disturbance with interactions among species. An underlying theme of the volume is the attempt to explain the maintenance of high diversity in tropical regions, which remains one of the most significant unsolved problems in ecology.

This new synthesis of biotic interactions is particularly timely because current empirical and theoretical advances, as well as technical developments, are yielding important and novel insights into the biology of tropical organisms. These insights result both from a 'scaling up' of empirical studies of tropical communities (e.g. the expansion of long-term censuses of tropical forest plots and recent large-scale field experiments), and from more in-depth experimental manipulations (e.g. detailed case studies of interactions across multiple trophic layers). These exciting new data, together with theoretical developments (see, for example, *The Unified Neutral Theory of Biodiversity and Biogeography*, by S. P. Hubbell;

(Princeton University Press, 2001), have collectively stimulated renewed debate on the relative importance of niche-assembly and dispersal-assembly processes as drivers of species coexistence in tropical forests. Interactions among organisms provide the mechanistic basis for both these sets of processes, and any theoretical reconciliation of alternative views can only emerge from a sound review of the empirical evidence. New evidence is also coming from the application of the techniques of molecular biology to studies of dispersal, population structure and less 'visible' taxa such as fungal symbionts and pathogens.

In July 2003, a symposium of the British Ecological Society was convened at Aberdeen University to review these developments in tropical forest ecology and to summarize the recent advances in our understanding of the role of biotic interactions in tropical ecosystem function. The symposium was a joint meeting with the Association for Tropical Biology and Conservation, and attracted a participation of about 500 delegates who made a total of 420 oral and poster contributions. Some chapters of this volume were presented as oral contributions to the plenary sessions at that symposium, while others were commissioned from specialists working in areas that were under-represented among existing contributions. Authors were asked to review their own research in the context of related research in the field.

The chapters of this book are grouped in four sections of unequal length. The variation in the length of the sections parallels, in part, an imbalance in the distribution of research activity by tropical biologists among different taxa, trophic levels and interactions. Part I concerns interactions among plants and their consequences for ecosystem function and theories of species coexistence in tropical forests. In this section, the assumptions that underpin niche-assembly and dispersal-assembly models of species coexistence are explored, and a theoretical reconciliation of these views is presented in the final chapter. Other chapters of this section consider resource use and partitioning and the extent to which trade-offs among life-history traits contribute to ecosystem function and the maintenance of plant species richness. Part II provides an overview of the diversity of interactions among plants and fungi, both as pathogens and as symbionts. It opens with chapters that explore the ecological importance of fungal pathogens and mycorrhizas, particularly in relation to their roles in the maintenance of plant species diversity. Plant-host specificity and environmental variability emerge as key issues affecting patterns of disease and plant response to mycorrhizal infection. The study of plant–microbe interactions has been constrained by the difficulty of species identifications, and is being revolutionized by developments in molecular techniques. The chapters of this section illustrate the application of these new methods and highlight the opportunities and the pitfalls. Plant–animal interactions receive the greatest emphasis in this volume and form the subject of Part III. They are considered from the perspective of animals as pollinators, predators and dispersers. This section also presents some important reviews and case studies of multi-trophic and complex interactions from the tropics, many of them involving ants, one of the most abundant, diverse and ecologically significant animal groups in the tropics. Finally, Part IV presents a series of case-study reviews of biotic interactions in human-dominated landscapes. In these papers, the impacts of habitat fragmentation, invasive species, human-induced fires and timber management highlight the importance of biotic interactions to the response of communities to human-induced disturbance. The authors consider the evidence for the disruption of biotic interactions and the influence of land-use history and forest history on the outcomes of biotic interactions. The implications of changes in biotic interactions for ecosystem function and management are discussed.

The book emphasizes the richness and diversity of new research on interactions among organisms in the tropics, but, despite this, a small number of unifying themes can be distinguished that cut across groups of related contributions. First, it is evident that no single mechanism will be sufficient to explain the maintenance of tropical forest diversity. There is abundant evidence in this volume for an important role for niche differentiation driven by plant–plant competition, density-dependent recruitment resulting from biotic factors (pathogenic fungi, and seed and seedling predators), dispersal limitation and ecological equivalence. Several contributors thus conclude that the question of whether niche assembly or dispersal-assembly processes are predominant is a quantitative one that is likely to be sensitive to site and species differences. Second, the popular perception that communities of tropical organisms are highly coevolved assemblages with complex trophic structures is supported by studies described in this volume. These chapters provide evidence for direct and indirect interactions across several trophic layers. For example, a case study of *Piper*–ant interactions spans four trophic levels, whilst myrmecophytes in Cameroon are at the centre of an interlocking set of mutualisms involving ants, plants, bacteria and phloem-feeding insects. These sorts of interactions have important implications for the structure and functioning of food webs in tropical communities since their effects cascade well beyond the two immediate mutualistic partners. The high frequency of ants as model species for this research reflects the importance of this group to a wide variety of interactions and processes in the tropics, particularly mutualistic ones, as illustrated by their key roles in seed dispersal, plant nutrition and plant protection. Third, the book highlights the importance of large-scale data-intensive studies to future progress in understanding tropical ecosystems because they bring together practitioners across disciplinary boundaries. The Center for Tropical Forest Science through its network of large-scale forest-dynamics plots has pioneered the consortium approach to collection of large-scale, long-term and spatially explicit data on trees that has stimulated so many developments in tropical forest ecology. A similar scale of research funding and research effort is now required to catalogue other taxa and to describe and interpret the highly complex interactions that link them as ecological communities. Finally, several authors emphasize the importance of temporal and spatial variability to particular interactions, and how the outcomes can vary greatly for different scales of investigation. For example, interactions between plants may shift between competitive and complementary during succession. This change is associated with a shift in the dominance hierarchy within the community and results from the differential effectiveness of species to capture light, nutrients and water. Spatial and temporal variability is also a feature of interactions among animals, for example when the outcome of interactions between peccaries and beetles depends on the stage of the peccary population cycle. These considerations have important consequences for the design and interpretation of studies of biotic interactions.

The book illustrates many of the ongoing debates in tropical forest biology, and highlights the opportunities for future research. The extent to which coexisting trees are ecological equivalents or niche-differentiated habitat specialists is not yet resolved, although evidence supporting both perspectives is provided. In particular, the role of trade-offs in determining acquisition of resources, plant–plant competition and niche differentiation is emphasized, but different authors clearly view different traits as most important. For example, in the case of trees, the role of canopy gaps in the maintenance of species richness is still widely debated, despite the fact that this question has stimulated more research than any other in tropical forest ecology. Debate over the importance of herbivory in controlling growth and survival of tropical forest plants has been reactivated by recent research and this issue is reflected in a number of contributions to this volume. Resolution of this question will be contingent on application of new statistical procedures for deriving robust estimates of tissue loss to herbivores from leaf-census data. New statistical and molecular tools are also helping to resolve old questions concerning the importance of long- versus short-distance dispersal of plants, although practitioners of molecular techniques will need to heed the warning provided here that DNA extracted from 'plants' is very often mixed with that of fungal associates such as mycorrhizas and endophytes.

Although we have attempted to be as wide-ranging as possible, the book nonetheless contains some significant gaps in its coverage. We regret, in particular, the relative lack of studies centred on tropical African ecosystems and by African researchers. There is also a bias towards research in wetter terrestrial (forest) ecosystems rather than arid ecosystems or semi-deserts, and we have made no attempt to consider biotic interactions in freshwater aquatic or marine ecosystems. We believe that the lack of African-authored and arid-ecosystem studies reflects a genuine imbalance in the literature that was available to authors for review, and we hope that future research will fill these gaps, perhaps in part stimulated by the contributions to this volume.

We are deeply grateful to the many individuals and organizations that have contributed to the production of this volume. Most importantly, we thank the contributing authors and the reviewers of their papers for their tremendous commitment of expertise and time. We acknowledge the generous financial and logistical support of the British Ecological Society and the Association for Tropical Biology and Conservation, and we thank the officers and councillors of both societies for supporting our concept for the meeting. We are especially grateful to Hefin Jones, Hazel Norman and Richard English for their help and advice implementing the symposium.

PART I

Plant–plant interactions

CHAPTER ONE

Plant–plant interactions in tropical forests

JOHN J. EWEL
US Forest Service
ANKILA J. HIREMATH
ATREE, Bangalore

Introduction

Some interactions between plants are uniquely conspicuous elements of certain tropical forests; the giant lianas that wend through the canopy and the epiphyte-laden branches of cloud forests are striking examples. Nevertheless, the fundamental processes involved are no different from those in extra-tropical communities, even though diverse, sometimes uniquely tropical, mechanisms may be involved. An individual of one plant species interacting with an individual of a second plant species can lead to any of the same five outcomes at any latitude, and these consist of all combinations of negative, positive and neutral effects (except the non-interaction described by the mutually neutral interaction, 0/0). But interactions among plants in forests seldom involve such simple one-on-one relationships. More commonly, multiple players are involved and the interactions change with time: the liana binds crowns of several trees, the fallen palm frond damages multiple seedlings, and the solum is shared by roots of many species. Furthermore, positive and negative interactions occur simultaneously, so the observer sees only an integrated net effect of multiple interactions (Holmgren *et al.* 1997).

Most symbiotic (mutually positive) interactions in tropical forests involve relationships between plants and animals or between plants and microbes – fungi, bacteria, algae – described elsewhere in this volume. What, if anything, distinguishes plant–plant interactions from plant–microbe, plant–animal or animal–animal interactions? Even though they employ different biotic services (pollination, dispersal; see Ghazoul, Chapter 10, this volume; Muller-Landau & Hardesty, Chapter 11, this volume), higher plants (with a handful of exceptions) all use the same abiotic resources: water, carbon dioxide, photosynthetically active solar radiation and the same suite of 13 mineral elements. Furthermore, the sporophyte is immobile, precluding spatial shifts to accommodate changing

Biotic Interactions in the Tropics: Their Role in the Maintenance of Species Diversity, ed. D. F. R. P. Burslem, M. A. Pinard and S. E. Hartley. Published by Cambridge University Press. © Cambridge University Press 2005.

conditions. Not surprisingly, then, most plant–plant interactions result in negative net impacts on at least one of the players as they compete in place for a common set of resources. Although recent years have witnessed a resurgence of interest in cooperative plant–plant interactions that benefit both participants, most examples concern physically harsh environments and are related to disturbance (Bertness & Callaway 1994).

Identification of interactions is relatively straightforward, but quantitative assessment of their ramifications is much more difficult. The approach we have used to assess consequences of interactions involved construction and 12 years of observation of communities on a fertile soil in the humid lowlands of Costa Rica. The experimental communities contain few species, representative of the life forms that have proven most evolutionarily successful in forests of the humid tropical lowlands. Unlike most studies of interspecific plant interactions, ours involve perennial species having different stature at maturity. Fast growth gives quick results, long-term observations provide an opportunity to assess changes, and perennial plants representative of successful life forms in natural forest give a semblance of real-world relevance. Three tree species (*Hyeronima alchorneoides*, *Cedrela odorata* and *Cordia alliodora*) are grown in monoculture and together with two monocots, one a perennial giant herb (*Heliconia imbricata*) and the other a palm (*Euterpe oleracea*) (Fig. 1.1). We and our colleagues have used this approach to assess invasibility (Gerwing 1995; Hummel 2000; Merwin *et al.* 2003), productivity (Haggar & Ewel 1994; 1995; 1997), stand structure (Ewel & Bigelow 1996; Menalled *et al.* 1998; Kelty 2000), and nutrient use (Ewel & Hiremath 1998; Heneghan *et al.* 1999; Hiremath 2000; Hiremath & Ewel 2001; Hiremath *et al.* 2002; Reich *et al.* 2003; Bigelow *et al.* 2004; Russell *et al.* 2004), in addition to a number of topics involving abiotic factors or intertrophic interactions.

In this chapter we restrict consideration to interactions among established plants – seedlings (or sporlings) through adults. First, we briefly review the mechanisms involved in some prominent, tropical plant–plant interactions, both physical and biogeochemical. Next, drawing primarily on 12 years of research using the experimental communities of low life-form diversity, we turn to manifestations of plant–plant interactions at the stand level, particularly the consequences of using a common resource base. We assess the results of resource use and partitioning in terms of competition, compensation and complementarity, and we consider their implications for ecosystem functioning. We then turn to the often-ignored temporal dynamics of plant interactions, particularly resource use, and discuss time-dependent changes in plant–plant relationships as they affect individuals, communities and systems of the future. Overall findings are then interpreted in terms of implications – positive, neutral, and negative – for the maintenance, restoration and management of tropical ecosystems.

Figure 1.1 Three-life-form community at age 8 years, with tree overstorey (*Cedrela odorata*, in this example), palm mid-storey (*Euterpe oleracea*) and perennial-herb understorey (*Heliconia imbricata*). La Selva Biological Station, in the humid Atlantic lowlands of Costa Rica.

Mechanisms
Physical interactions
Perhaps second only to the striking diversity of tropical forests, physical interactions among plants are one of the most conspicuous features of tropical forests: the fallen tree trunk that leads to a row of seedlings germinating upon it, the right-angle bend in the stem of a treelet temporarily flattened years ago by a fallen palm frond, the tree crowns festooned with epiphytes, the tangles of lianas using one another as paths to the canopy, and crown shyness due to abrasion. At one extreme, a plant can be a passive participant. For example, an important yet easily overlooked interaction among plants is detritus-fall from large individuals that breaks, bends or kills smaller ones. Although this can occur in any forest, it is an especially important phenomenon in fast-turnover tropical forests. The agent may be branches or fronds, a rain of heavy detritus that damages an average of 10 to 20 per cent of a given cohort of seedlings (Aide 1987; Clark &

Clark 1991; Mack 1998; Drake & Pratt 2001) and therefore can affect future composition. At the other extreme, some interactions involve intricate mechanisms and processes, illustrative of the potential of plant evolution. Tropical forests are rich with examples, such as relationships between hemiparasites and their hosts.

In some plant–plant interactions, one member is a passive player in an interaction that is beneficial to the second member. The protection from direct sunlight afforded by a tall plant to a shorter one is an example. Although solar radiation commonly limits plant growth, many plants cannot thrive if exposed to direct sunlight. There are a number of potential causes: a plant may not be able to dissipate the heat accumulated by direct exposure to sunlight; it may lack the shielding that prevents chlorophyll degradation; or it may be unable to sustain an internal water balance when faced with high transpiration demand. Nevertheless, although many species survive and grow in light shade, few of them do better as shade intensity increases; most plants, including tropical trees, respond positively to increasing light.

Some plants are able to detect shifts in radiation wavelengths caused by neighbouring plants and follow that detection with a rapid growth response (Ballaré et al. 1990; Schmitt et al. 1999). Although the phytochrome mechanism involved has not been widely tested, growth responses to shifts in red-to-far-red ratios among tropical seedlings (Lee et al. 1996) indicate that this is probably a common phenomenon.

In most plant–plant interactions, the costs and benefits are less well defined than those just described. Epiphytes, hemiepiphytes, vines and hemiparasites, for example, all involve one plant providing physical substrate, and sometimes nutrition, for another (usually, but not always, of a different life form). The dependence of one life form on another for physical support is a common phenomenon in the tropics, and the wide range of scaffolding, substrates and trellises available, coupled with life forms capable of exploiting them, enrich tropical ecosystems: diversity breeds diversity. The most common forms of some large families are epiphytic (e.g. Orchidaceae, Bromeliaceae), and the local species richness of epiphytes can rise to 200 species or more, constituting up to half the flora (Nadkarni et al. 2001). The consequences of provision of support by one species for another can range from positive to nil to fatal.

Epiphytes that colonize tree limbs benefit from being perched in the canopy (the obligate habitat of many of them), often without apparent harm to the host. This is a reasonably clear case of facilitation (+/0). In some cases, however, heavy epiphyte loads increase the risk of phorophyte breakage (+/−). The relationship may also be mutually beneficial (+/+), for example if epiphytes slow the passage of atmosphere-borne nutrients that would otherwise pass through the system quickly and be lost before being captured by tree roots (Nadkarni 1981).

Many epiphylls, the leaf-colonizing subset of epiphytes, are capable of fixation of atmospheric nitrogen (e.g. Goosem & Lamb 1986), although the degree to which this nitrogen becomes available to the host is unclear. Epiphylls are generally thought to have a negative impact on their hosts because they capture solar radiation that might otherwise energize the leaf they are growing on or because they hold water that fosters the growth of pathogens (Coley & Kursar 1996). They might also filter incoming direct-beam radiation, reducing the likelihood of chlorophyll-bleaching in the host. Nevertheless, epiphyll growth is far more common in the understorey than in the canopy, which argues against such a sun-screening role being widespread. The abundance of epiphylls on shaded leaves may simply be a consequence of the drought vulnerability of epiphylls, most of which are non-vascular plants.

Hemiepiphytes, which are uniquely tropical, shift their growth habit with age. These plants (principally members of the Moraceae, Clusiaceae, and Araliaceae) begin life as epiphytes but become self-supporting and soil-rooted as adults, a feat that requires substantial morphological, anatomical and physiological plasticity (Holbrook & Putz 1996). By starting life in the crowns of trees, hemiepiphytes circumvent severe competition for light as seedlings. During their epiphytic phase they use the nutrient-rich humus that accumulates in the crown of the host tree (Putz & Holbrook 1989). A small subset of hemiepiphytes is commonly dubbed 'stranglers', although they might more accurately be called 'suicide facilitators', for it is the host tree that continues to enlarge when enveloped by such a hemiepiphyte, eventually leading to its death – a distinct disadvantage of having vascular cambium.

Among the most intricate plant–plant interactions are those between hosts and their hemiparasites or parasites. In the case of the hemiparasitic mistletoes (Loranthaceae, Viscaceae, others), both host and parasite photosynthesize, but the mistletoe has invasive haustoria that penetrate the host's branches, tapping its water supply and, in some cases, its photosynthate (Calder 1983; Lamont 1983; Marshall & Ehleringer 1990). Some parasites show little outward evidence of parasitism, as with the widespread (and valuable) sandalwood (*Santalum* spp.) and Australia's *Nuytsia*, both of which look superficially like other trees, but whose roots invade those of non-conspecific hosts (often other trees) on which they depend (Calladine & Pate 2000). In still other cases the parasite has lost its capacity to photosynthesize, thereby becoming entirely dependent upon the host. The most renowned example may be Southeast Asia's *Rafflesia*, notorious because it produces the world's largest (and one of its most foul-smelling) flower.

Vines are a notoriously successful life form in the tropics and, like obligate epiphytes, dependent upon other life forms for support. Seasonal shutdown of water flow leads to high risk of cavitation in plants having large vessels. It is not uncommon for lianas to have vessels with diameters of 200 μm or more (Ewers *et al.* 1991), and diameters as high as 700 μm have been recorded (Tyree & Ewers 1996), so it is not surprising that lianas are a much more species-rich life form in the humid tropics than in cold or dry biomes. The impacts of vines on the host plants that provide them with structural support are invariably negative. These impacts include mass load on tree crowns (Putz 1991), interception of solar radiation (Dillenburg *et al.* 1995) and competition for water (Pérez-Salicrup & Barker 2000; Pérez-Salicrup 2001). In experiments in which trellises of varying diameters were provided or removed at ground level, Putz (1984) concluded that vine growth and survival can be trellis-limited, demonstrating once again the dependence of one life form upon the same or another life form. Twining vines cannot climb large-diameter trellises, so vines themselves become a common path to the canopy for other vines.

Biogeochemical interactions

Although inconspicuous (and hard to measure), chemical interactions among plants can dictate local success, diversification or extinction. They involve common use of elements that are often in short supply relative to plant demand, sharing of water supplies and the production of chemicals by one species that either impact directly or serve as signals to neighbours.

The mineral nutrients contained in plant detritus generally become fair game for any plant in the neighbourhood once they are released by decomposition of the organic matrix. Some plants, however, shortcut the process by capturing detritus behind leaf bases (e.g. some palms: Raich 1983; Putz & Holbrook 1989) or in baskets or tanks formed by leaves (e.g. bromeliads and basket-forming ferns). The ability to intercept detritus can be nutritionally significant; Reich *et al.* (2003), for example, found that bromeliads switched from atmospheric to host-tree litter sources of nitrogen as their tank diameters, and therefore their detritus-capturing capacity, increased with age.

The provision of nitrogen-rich detritus that can nourish co-occurring species is an example of a passive action by one species that benefits another. Although this phenomenon occurs across a wide range of latitudes, it is especially common in the tropics, where legumes are prominent components of many forests. Nitrogen fixation in the tropics is not restricted to those legumes (primarily in the subfamilies Papilionoideae and Mimosoideae) that have symbiotic associations with bacteria capable of reducing the diatomic nitrogen of the atmosphere. The non-legumes include plants symbiotic with actinomycetes or cyanobacteria that fix nitrogen: e.g. *Myrica* (Myricaceae) and *Casuarina* (Casuarinaceae), both symbiotic with *Frankia*; *Parasponia* (Ulmaceae; symbiotic with *Bradyrhizobium*); *Psychotria* (Rubiaceae; symbiotic with *Klebsiella*), and *Gunnera* (Gunneraceae, symbiotic with *Nostoc*).

Even though nitrogen-fixing tree species commonly resorb about half of their foliar nitrogen prior to abscission, the concentrations achieved in leaves are so high, often 4 per cent or more, that a considerable quantity reaches the forest floor. Once there, it enters the detrital food chain and some of it becomes available to co-occurring plants (e.g. DeBell *et al.* 1989; Binkley 1992). In time, this inadvertent provision of growth-stimulating nitrogen to competitors presumably can prove detrimental to the nitrogen-fixer. Nevertheless, trees that host nitrogen-fixing bacteria may exhibit their highest rates of fixation when young (e.g. Pearson & Vitousek 2001), so the benefits to competitors may be short-lived.

With relaxation of constraints caused by low nitrogen, other nutrients eventually limit growth. In stands comprising a relative-abundance gradient of a non-nitrogen fixing (*Eucalyptus saligna*) and a nitrogen-fixing tree species (*Albizia falcataria* [syn. *Falcataria moluccana* and *Paraserianthes moluccana*]), Kaye *et al.* (2000) found that although available soil nitrogen increased with increasing relative abundance of *Albizia*, available soil phosphorus declined, presumably because of uptake by trees. Whereas sequestration of nitrogen in above-ground biomass increased with increasing relative abundance of the nitrogen-fixing tree, sequestration of phosphorus in biomass peaked at a *Eucalyptus*:*Albizia* mix of 1:2, the same as carbon.

Water, like nitrogen, can be made available to a second species through the actions of a first. Hydraulic lift, whereby water taken up by deep roots flows at night from shallow roots into the surrounding soil, occurs in savannas (Scholz *et al.* 2002) and probably in other tropical ecosystems. Once in the surface soil, the water potentially becomes available to the plant that expelled it, to a competitor or to evaporative loss.

Unlike nitrogen, which can accumulate in plants and soil through biotic processes, phosphorus is limited from the start by parent material (except for the modest amounts that might enter through atmospheric deposition or flooding). As phosphorus is taken up and sequestered in vegetation, the available supply in the soil declines, until it can eventually limit plant growth. In their classic monograph on shifting cultivation, Nye and Greenland (1960) suggested that extraction of phosphorus by tree roots deep in the soil could lead to replenishment and maintenance of phosphorus in surface soils. Setting out to test this hypothesis, Kellman and Hudson (1982) severed the tap roots of pine trees in a tropical savanna, and after five years of study Kellman (1986) concluded that the tap roots had no effect on the concentrations of foliar phosphorus (or cations). Thus, the best tropical test to date of the intuitively appealing deep-pumping hypothesis has yielded null results.

Allelochemicals are an important means of interaction not only between plants and animals (see Massey *et al.*, Chapter 14, this volume), but also between plants; they constitute an exciting and very active area of chemical ecology (Dicke & Bruin 2001; Callaway 2002). Examples of chemically mediated communications between plants include wounded-plant-to-neighbours (Baldwin *et al.* 2002), root-to-root (Schenk *et al.* 1999), pollen-to-stigma (Ottaviano *et al.* 1992), and plant-smoke-to-seed (Keeley & Fotheringham 1997). Although plant communication research has not had a strong tropical component (and is unlikely to have one soon as the field becomes increasingly focused on molecular biology; e.g. Kessler & Baldwin 2002), it is inconceivable that tropical plants will not possess the full suite of signals and detectors discovered in plants of the temperate zone; they are likely, in fact, to offer some exciting surprises as well.

Support for widespread existence of allelopathy as a mechanism of plant–plant interaction has waned in the past couple of decades. Williamson (1990), however, makes a strong case for the fact that allelopathy is held to a higher standard of proof than are most interactions between plants, and he marshals evidence for both the direct production of allelochemicals by plants and the allelopathic properties of substances resulting from breakdown of non-allelopathic exudates. The species-rich genus *Eucalyptus*, widely planted throughout the tropics and subtropics, has borne the brunt of popular accusations that it is allelopathic, but the evidence is equivocal (Willis 1991). In any event, the production of allelochemicals and their exudation into the soil to impede colonization by competitors would be an expensive defence mechanism in the humid tropics, where the flow-through of soil water can easily be 75 per cent or more of rainfall; prodigious rates of chemical production would be required to keep up with losses. If allelopathy is a tropical phenomenon, it seems more logical to seek it in the dry tropics than in rainforests.

Outcomes: the three 'C's of plant interactions

The results of competition-density experiments containing two or more species can vary from no interaction between the species to complete dominance by one of them. The intermediate results, i.e. those situations in which two presumed competitors coexist, are of particular interest to community ecologists because they offer clues regarding concomitant use of resources, and therefore mechanisms that promote diversity.

Additive experiments, whereby a fully stocked stand of one species is interplanted with one or more additional plant species, are useful tools for assessing resource availability and invasibility (Snaydon 1991). The continuum of possible results from such experiments can be broken down into three broad categories: (1) competitive dominance of the stand by one species whose growth is not affected by a second species that contributes little to total productivity; (2) compensatory productivity, whereby the growth of both species is slowed by competition, but the sum of their productivity is greater than that achieved by either species in monoculture, and (3) complementarity, in which two species share resources in ways that enable at least one of them to be as productive as it would be in monoculture and resulting in total productivity greater than that achieved by either species in monoculture. In the Costa Rican experiments we have encountered all three outcomes: competition, compensation

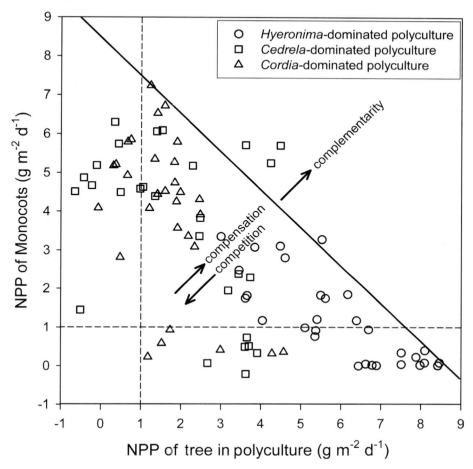

Figure 1.2 Productivity of trees and monocots when grown together in an additive design. Each data point represents one replication for one year. Ninety community-years of productivity are plotted: 3 polycultures × 3 replications × 10 years. The dashed lines differentiate low NPP values (i.e. a year of productivity at an average rate ≤ 1 g m^{-2} d^{-1}) from higher values. The diagonal connects the highest observed annual average monocot and tree NPP values (7.2 and 8.7 g m^{-2} d^{-1}, respectively).

and complementarity. Furthermore, we learned that the response depended on when observations were recorded, as relationships among species changed with age.

Ninety community-years of data (3 polycultures × 3 replications × 10 years following crown closure at age 2) are arrayed on Fig. 1.2, where the net primary productivity (NPP) of interplanted monocots (a perennial herb, *Heliconia imbricata*, and a palm, *Euterpe oleracea*) is shown in relation to NPP of the tree species on the same plot. All stands plotted were at least 2 years old and had closed canopies and root systems. The diagonal is maximum anticipated combined NPP based on the highest values of NPP observed for monocots (7.2 g m^{-2} d^{-1}) or trees in monoculture (X-axis intercept = 8.7 g m^{-2} d^{-1}). Moving away from that diagonal toward either the horizontal axis (where monocot NPP $\leq$ 0) or the vertical axis (where tree NPP $\leq$ 0), competition intensifies, and one life form or the other dominates stand productivity. Moving from the origin toward the diagonal, compensatory growth increases; competition-induced productivity decrements in one life form are more than offset by the contribution of the other life form to whole-system productivity. Those few productivity combinations that surpass the diagonal demonstrate complementarity.

If one (arbitrarily) takes an NPP of $\leq$ 1 g m^{-2} d^{-1} for either trees or monocots (demarcated by the dotted lines in Fig. 1.2) as evidence of strong competition between life forms in our systems, roughly half of the 90 data points on Fig. 1.2 fall into that category. Nevertheless, competition was manifested in stands dominated by the different tree species in very different ways. *Hyeronima*, clustered toward the lower right-hand portion of Fig. 1.2, proved to be a far more competitive tree species than the other two, and this competitive superiority was sustained throughout the study. It grows fast, has a dense canopy, is evergreen and develops a dense root system. *Cordia* and *Cedrela* each accounted for about a third of the data points that demonstrate competition, but their data points are about equally split between those close to tree NPP of 0 and those close to monocot NPP of 0: early in stand development the tree proved to be the stronger competitor, and now the palm is the stronger competitor. Thus a five-year study would have yielded conclusions quite contrary to those resulting from a ten-year study.

Most of the remaining data points (Fig. 1.2) still fall below the diagonal but illustrate combinations of NPP > 1 g m^{-2} d^{-1} for both life forms. Competition plays a role, but its effects are more than offset by the productivity of the additional species. In the case of the highly competitive *Hyeronima*, the only monocot that contributed substantially to productivity was the palm, and it developed much later there than in stands dominated by other trees. In stands of *Cedrela* and *Cordia*, early compensatory growth was provided by the herb, a role that was later usurped by the palm.

Complementarity should be a relatively rare phenomenon among immobile organisms that require the same suite of resources, and indeed only four (of 90) data points fall above the diagonal in Fig. 1.2. Three of those instances involved young (< 5 yr old) stands of *Cedrela* at a time when *Heliconia* contributed substantially to stand-level NPP but did not compete measurably with the tree. The fourth instance involved a stand dominated by *Hyeronima* that was in its tenth year. In this case, the complementary productivity was contributed by the palm.

What mechanisms can account for complementarity? There are three classes of possibilities: resource substitution, spatial segregation of resource acquisition and temporal differentiation of use (cf. Trenbath 1974).

Resource substitution

The restricted set of resources required by higher plants contrasts sharply with the dietary diversity of animals (although the dietary diversity of animals may be as much a reflection of packaging as basic chemical composition). One opportunity for using different forms of an essential resource involves nitrogen; the other involves use by understorey plants of those wavelengths of solar radiation that make it through the canopy after taller plants have skimmed the peaks off the action spectrum for photosynthesis.

The most apparent opportunity for resource substitution among terrestrial plants is the use of geochemically distinct sources of nitrogen: fixation of the diatomic form from the atmosphere by microbial symbionts, uptake of oxidized (nitrate) or reduced (ammonium) forms of mineral nitrogen, or uptake of organic nitrogen (Neff *et al.* 2003), from soil solution. Nitrogen-fixers and non-fixers use different biogeochemical pathways to acquire their nitrogen, required in prodigious quantities by both kinds of species.

Even though higher plants vary little in the quality of solar radiation that drives their photosynthesis, some plants are better adapted than others to take advantage of the shifted spectrum that reaches the forest floor. The radiation reaching the understorey of a tropical forest typically exhibits about a three-fold reduction in the red:far-red ratio (Chazdon *et al.* 1996). The fact that leaves of many understorey plants are relatively enriched in chlorophyll *b* hints that they may be able to used the red-depleted spectrum more effectively than leaves of canopy plants (Björkman 1981), although the ecological implications have not been quantified.

Spatial separation

Differential exploitation of soil depths (e.g. Jackson *et al.* 1999) for water or nutrients may be the primary mechanism that leads to resource-partitioning among plants. In recent years substantial progress has been achieved in documenting this phenomenon using stable isotopes (Dawson *et al.* 2002). Presumably all species benefit by concentrating their roots in zones of most abundant resources, making it almost impossible to determine whether differential rooting depths are inherent traits of the life forms observed or induced by competition. Growing in a semi-solid medium as they do, roots (unlike tree crowns, for example) have escaped an overriding selection for self-support. This permits an architectural flexibility that enables plants of many life forms to exploit water and nutrients from different soil depths, thereby achieving complementary resource use. Preliminary data indicate that the complementarity we observed between *Hyeronima* and *Euterpe* (Fig. 1.2) came about because the palm developed a root system capable of extracting nutrients from greater soil depth than the tree.

Nevertheless, root-system development is not entirely free of genetic constraints, even in humid tropical forests (Jeník 1978): many perennial herbs in the understorey have sparse root systems, often consisting of soft, thick roots that do not penetrate the soil very deeply; palms have cable-like roots with a very modest degree of branching and no secondary thickening; and some trees (e.g. *Cedrela odorata*, whose shoots are subject to attack by *Hypsipyla grandella*) have roots laden with starch, which enables them to recover from insect attack (Rodgers *et al.* 1995). In the seasonal tropics, variations in root systems may be especially striking because their development is influenced by requirements for storage, uptake of ephemeral surface water or water at extreme depth, and the ability to re-sprout after fires.

Unlike the trees and palms that reach the canopy, most understorey life forms studied to date, both perennial herbs and shrubs, have shallow root systems (e.g. Becker & Castillo 1990; Grainger & Becker 2001), and this may lead to some partitioning of soil resources. Nevertheless, an understorey growth response to trenching (Coomes & Grubb 1998) suggests that differential rooting alone is inadequate for partitioning of soil resources and that nutrients may be limiting in the understorey owing to the competitively superior overstorey, at least on infertile soils. Furthermore, there are interesting exceptions to the shallow-rooted understorey generalization, as in the case of *Jacquinia sempervirens*, a shrub found in neotropical dry forests that is wet-season deciduous but fully leafed in the dry season, when its long taproot exploits water deep in the soil.

Solar radiation presents resource-acquisition challenges and opportunities of a different nature. For one thing, light is a pass-through resource that is not storable unless converted to chemical energy; for another, all higher plants require the same wavelengths, so there is limited opportunity for qualitative partitioning (e.g. Lee 1997).

Optimization of light capture is closely tied to uptake and utilization of limiting mineral elements, particularly nitrogen, but also phosphorus in some situations. There is an internal tradeoff involved for an individual capable of overtopping its neighbours. The relative gains to be had from investing more nitrogen in a well-illuminated leaf must be balanced against the gains of distributing that nitrogen among many leaves including those receiving less solar radiation (Field & Mooney 1986). Some species are obligate investors in fully illuminated foliage. These include the thin-crowned, fast-growing pioneer trees (e.g. *Ochroma*, *Musanga*, *Octomeles*) as well as most nitrogen-fixing legumes. Other species sustain deep, densely foliated crowns, presumably by being effective at nitrogen acquisition or by being unable to achieve the extremely high rates of photosynthesis that make deployment of all available nitrogen to fully lighted leaves an optimal investment. Examples include *Flindersia*, *Garcinia* and *Manilkara*.

What about the space between the forest floor and tall, well-illuminated plants? Here the greater light capture that could potentially be achieved with increased stature comes at a carbon cost of investing in non-photosynthetic support tissue, an inefficient proposition in low-light environments (Chazdon 1986). It is not surprising, therefore, that the understorey of many tropical forests (particularly in the humid lowlands) is dominated not by the seedlings of plants of tall potential stature but by a guild of short-stature plants that survive in the shadows, usually by being extremely effective at photon capture. Barely at the limits of the euphotic zone, these plants often exhibit morphological or anatomical features that would be penalized if they were exposed to full sunlight and to wind, such as high specific leaf area and thin palisade parenchyma (e.g. many Rubiaceae, Melastomataceae, Piperaceae) traits often accompanied by large leaves (e.g. many Araceae, Arecaceae, Cyclanthaceae, ferns, Heliconiaceae), in contrast with their sparse root systems. Many understorey plants have non-overlapping phyllotaxy, ensuring that no leaf shades another. Still others have the thylakoid-packed leaves that greatly enhance their light-harvesting capacity, imparting an ability to sustain carbon gain in the shade, while at the same time taking advantage of ephemeral sunflecks (Chazdon 1988; Pearcy 1990). Further reducing light requirements, understory plants commonly have low rates of dark respiration, enabling them to sustain a positive carbon balance with low rates of photosynthesis (e.g. Björkman 1981; Chazdon 1986). In the extreme understorey specialization, the abaxial anthocyanin layer in leaves of many understorey species may help in the scavenging of every last photosynthetically active photon, reflecting it back up to the light-absorbing pigments (Lee 1979).

Temporal differentiation

In the tropics, as in the temperate zone, much complementarity involves differential phenology, which enables two or more species to tap the same pool of resources at different times. The resource for which temporal sharing is most obvious is solar radiation, whereby understorey life forms receive more light during seasons when overtopping plants are at least partially leafless.

Water demands at the tops of tall trees can be difficult to meet because of a combination of the length and associated resistances of the path that water must be transported internally, the greater vapour-pressure deficit and enhanced air speed with increasing height above the ground, and the force of gravity (Zimmerman 1971). Thus, as water supplies in the soil diminish and become held under increasingly greater tension, the necessary gradients cannot be sustained. The result is a top-down onset of deciduousness that, in many tropical forests, does not extend below the canopy dominants. One consequence is that during the portion of the year when the uppermost canopy trees are leafless, shorter-stature plants not only receive more solar radiation but also have exclusive access to nutrients and to any soil water that might become available through dry-season showers (Fig. 1.3).

In the fullness of time

Ecological change is an unchallenged given, yet many of us still interpret the natural world in terms of what we observe today, giving short shrift to history or to future trajectory. Our relatively short-term run (12 years) of repeated

Figure 1.3 Deciduous overstorey of *Cedrela odorata* above evergreen palm (*Euterpe oleracea*) and perennial herb (*Heliconia imbricata*) illustrates a common mechanism of resource-partitioning in the tropics. Stand age = 5 years. When the tallest components are leafless the lower-stature plants receive more solar radiation and have exclusive access to water and nutrients.

measurements on fast-growing experimental communities has provided surprises and insights, as we have witnessed predictions being overturned, trajectories redirected and answers reversed. There is no better environment for gaining time-dependent insights into the consequences of plant interactions than the humid tropical lowlands, where results come quickly. Because of the accelerated development that occurs where growth is unimpeded by harsh seasons it is feasible to observe changes within a few years that might take decades to manifest in the temperate zone.

Sustained observations of selected organisms, ecosystems and chrono-sequences will provide adequate data for understanding place-specific observations of short- to mid-duration phenomena (years to decades). Other approaches including modelling, palaeoecology, geochemistry and remote sensing will be needed to achieve understanding of ecological phenomena operating at longer time scales.

Changes in individuals

Changes in interactions accompany increases in plant age, sometimes dramatically so, and this leads to changes in the consequences of plant–plant

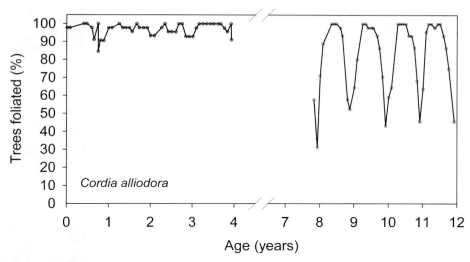

Figure 1.4 Phenological shift with age by *Cordia alliodora*. Until age 4 years, this tree is almost entirely evergreen, but once it reaches reproductive maturity, between ages 5 and 7 years (and a height of about 20 m), it becomes seasonally deciduous. A foliated tree was one carrying ≥50 per cent of its full complement of leaves.

interactions. Leaf phenology is one such example. Many tropical tree species are evergreen or nearly so when young, but become deciduous when older. For example, we found that two of four species whose phenology we observed monthly for four years, *Cordia alliodora* and *Cedrela odorata*, metamorphosed from evergreen to predictably deciduous. In the case of *Cordia*, the onset of deciduousness occurred between ages 5 and 7 years (Fig. 1.4), coincident with its age of flowering.

The stature of plants having a basal meristem and no cambium (e.g. Cyperaceae, Heliconiaceae, Musaceae, Strelitziaceae, Zingiberaceae, most Cyclanthaceae and Poaceae, and many others) is biomechanically limited, and this too can impose age-dependent constraints on their roles in the community. Although such plants can be effective competitors against plants of similar stature, they lose dominance once they are overtopped by trees and palms. For example, in two of our experimental communities containing three life forms of plants – tree (*Cordia* or *Cedrela*), palm, perennial herb – the herb (*Heliconia imbricata*) grew as tall as 6 m and contributed more than 50 per cent of NPP in the first three years following establishment. But by the fifth year, when it was overtopped by the tree and the palm, its NPP plummeted to negative values, as its losses of tissues through death and respiration exceeded its carbon gains through photosynthesis (Fig. 1.5). Thereafter its NPP rebounded, oscillating around zero as it survived, but did not thrive, in the understorey.

Development of the root system can also lead to changes in species interactions over time. Dicotyledonous trees and palms that potentially compete for soil

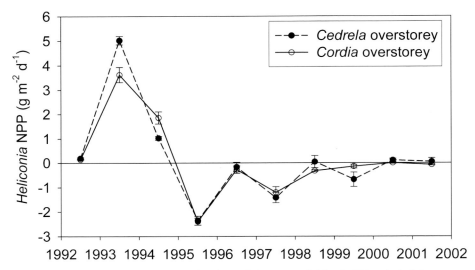

Figure 1.5 Time course of the contribution of the perennial herb, *Heliconia imbricata*, to aboveground net primary productivity (NPP) in three-life-form communities dominated by *Cordia alliodora* or *Cedrela odorata*. Early in stand development the herb contributed more than 50 per cent of total NPP, but, limited biomechanically by its basal meristem, its productivity plummeted once it was overtopped by the tree and palm, *Euterpe oleracea*.

nutrients provide an example. Many palms have geotropic stems, a behaviour that is often interpreted as an adaptation to fire or herbivores: the young palm's bud is safely below ground until a root system is well established, at which time a rapid spurt of height growth raises the bud above the height that would have made it vulnerable. In our experimental plantations we observed a similar developmental sequence by *Euterpe oleracea*, a palm of Amazonian flood plains: roots followed by stem. In this case, however, the driver is probably not fire or herbivory but delayed access by the palm to nutrients below the reach of more shallowly rooted trees.

Changes in light availability lead to physiological changes that manifest themselves in plant form, structure and functioning. Etiolation, whereby plants develop high ratios of stem height to diameter (Went 1941), is one response, and increased density of chlorophyll is another. Most plants acclimatize to some extent to changes in photon flux density in both directions, i.e. as conditions change from low light to high light (Langenheim *et al.* 1984; Demmig-Adams & Adams 1992) or from high light to low light (Björkman 1981), although low- and high-light specialists often have different capacities for accommodation. The time involved in acclimatizing typically falls in the range of weeks to months; Newell *et al.* (1993), for example, found that three species of *Miconia* adapted to high light within about four months. Like others who have studied acclimatization, they found that acclimatization required production of a new cohort of leaves, replacing those that had previously been adapted to lower light intensities. Thus, a plant capable of a particular rate of photosynthesis in one environment can often, with time, photosynthesize at a very different rate if the light environment to which it is exposed changes – a common phenomenon in tropical forests.

Changes in communities

Just as spatial patterns in soil nutrient availability can affect the spatial distribution of species with varying efficiencies of nutrient use, so too temporal patterns in soil nutrient availability may alter the outcome of species interactions by imparting different competitive abilities on co-occurring species having different nutrient-use efficiency. Species' differences in nutrient-use efficiency can arise one of two ways: they can differ in their productivity per unit of nutrient in the plant (i.e. nutrient productivity; Ågren 1983) or in the length of time that nutrients are retained in the plant. There may be evolutionary trade-offs between plant traits associated with these two components of nutrient use (Berendse & Aerts 1987), and interspecific differences in these traits can confer varying abilities to deal with nutrient limitation. Thus, species with high nutrient productivity tend to be at an advantage in fertile environments, where faster growth may be the key to rapid resource capture and there is little benefit to be derived from a conservative use of plentiful nutrients. Species with longer nutrient retention, on the other hand, tend to be at a greater advantage in resource-poor environments, where conservation of scarce nutrients (rather than rapid growth) may be the key to a species' success.

Natural distributions of the three tree species that were included in the Costa Rica study bear out this reasoning (Hiremath *et al.* 2002). *Cedrela odorata*, which has high nutrient productivity but low retention, is most often found on fertile flood plain soils. *Hyeronima alchorneoides*, with relatively high nutrient productivity and much longer nutrient retention, would be expected to occur on a wider range of soils. In fact, *Hyeronima* grows rapidly in plantations on fertile soil, but also occurs, although growing slowly, in low-resource environments (Clark & Clark 1992). *Cordia alliodora* has low nutrient productivity, but this is a consequence of high concentrations of nutrients in tissues, rather than low productivity. It also has intermediate nutrient retention. Therefore, it would be expected to grow rapidly on fertile soils but to do poorly on less-fertile soils. In fact, *Cordia* does grow vigorously on abandoned agricultural fields having fertile soils but grows only slowly on degraded soils (Butterfield 1994). In our experimental plantations the nitrogen-demanding *Cordia* was the first to manifest the effects of nutrient limitation as a consequence of competition from the co-planted monocots (Haggar & Ewel 1997).

The differential competitiveness of the three tree species in the company of the two monocots is illustrated in Fig. 1.6, where tree productivity in

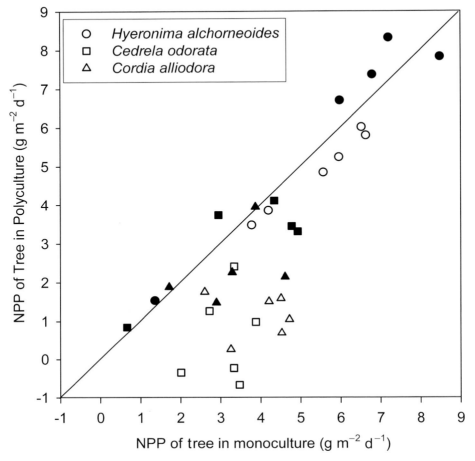

Figure 1.6 Relative above-ground net primary productivity (NPP) of stands of three tree species grown in monocultures and in three-life-form polycultures. Each point is the mean of three replications, and each of the 11 points graphed per species represents a different year-long interval from age 0 to 12 years. Filled symbols represent ages 0 to 5 years; open symbols are ages 6 to 12 years. The NPP of *Hyeronima*, arrayed along the diagonal, showed little impact of the two other life forms in the polycultures. The other two tree species showed little effect early in stand development but eventually declined in NPP in response to competition.

monoculture is compared with that of polyculture. Neither *Cordia* nor *Cedrela* ever achieved a sustained (i.e. average over a year or more) NPP greater than about 5 g m^{-2} d^{-1}, and their productivity in polyculture was consistently lower than in monoculture except during the first few years of growth (filled symbols, Fig. 1.6). *Hyeronima*, on the other hand, achieved average NPP over a year in excess of 8 g m^{-2} d^{-1}, and there was little difference between its productivity in polyculture and in monoculture.

The temporal aspects of these trends merit special note. First, there was a dramatic change in vulnerability to competition with time. Early in the experiment both *Cordia* and *Cedrela* had years when their productivity in monoculture did not differ from that in polyculture, but once the monocots increased in biomass their impacts on resource availability were reflected strongly by a decline in tree productivity. Second, the competition in these systems was clearly dominated by below-ground processes, for tree growth slowed markedly even while the trees overtopped the monocots. At this well-watered site (~ 4 m of rain per annum) competition centred on nutrients; declines in foliar nitrogen-to-phosphorus ratios (Hiremath *et al.* 2002) caused by high leaching losses of nitrate (Bigelow *et al.* 2004) indicate that the nutrient in shortest supply was nitrogen.

Legacies

Site history is easy to ignore, especially in the tropics where recovery is often fast and the impact of past occupants inconspicuous. Nevertheless, plants of the past can have an impact on soil fertility and plant-regeneration patterns observed today. Further exploration of this phenomenon is likely to prove fruitful in the tropics, just as it has in the temperate zone.

Plants (especially trees) of the past can leave a lasting imprint on the forest. This phenomenon has been studied especially well in forests dominated by large conifers, particularly in the northwestern United States, where the trunks of dead trees last many years. But *post mortem* impacts of tree death should be no less important in the tropics, even in the humid lowlands where decay rates are high. What kinds of legacies might be expected? The best-studied of these in the tropics is plant regeneration. Many tropical trees produce competition-tolerant seedlings, and these constitute the advance regeneration that captures the resources vacated by the death of the parent (e.g. Brokaw 1985). Long-dead pioneer trees such as *Trema* can affect regeneration for decades because of the large number of viable seeds they leave behind in the soil (Alexandre 1989). Gap dynamics, which gained momentum as a topic of study in the 1970s (e.g. Denslow 1987), is one of the best examples of south-to-north transfer of ecological knowledge. It is gap formation that aids niche-partitioning, particularly in the regeneration phase, among plants having different requirements for light, dispersal and (possibly) nutrient availability. Less well studied, but probably important, are the roles of dead wood in providing safe sites for regeneration of some species. In Hawaii, for example, *Metrosideros polymorpha* (the dominant tree species) regenerates predominantly on logs in alien grasslands, presumably because of lessened competition (Scowcroft 1992).

Plants sometimes change soil properties, and in doing so influence the propitiousness of a site for its own or other species. In a study of 23- to 36-year-old plantations and primary forest in the Brazilian Amazon, Smith *et al.* (1998) reported that different species modified nitrogen availability such that mineralization rates were inversely proportional to efficiency of nitrogen use. In Micronesian mangrove swamps, Gleason *et al.* (2003) found that the soil beneath *Sonneratia* was more highly reduced than the soil beneath *Rhizophora* and *Bruguiera*. This apparently occurred because the roots of *Sonneratia* conducted less oxygen into the otherwise anaerobic substrate, thereby lowering the redox potential, which in turn improves availability of phosphorus.

Global change

On longer time scales, present landscapes may be ghosts of conditions past, and the rapid global change that we are undergoing now will probably result in a future landscape very different from the one we know today. For example, a large part of now-forested West Africa may have been a much drier C_4 grassland during the mid-Pleistocene as a result of lower sea-surface temperatures off the African coast (Schefuss *et al.* 2003). So also, the reverse: the disappearance of trees from South Africa's savannas during the last glacial has been linked to low atmospheric CO_2 concentrations (in addition to fire), conditions under which the growth of fire-vulnerable trees is likely to have been slowed relative to fire-carrying grasses (Bond *et al.* 2003).

Grasses that possess the C_4 photosynthetic pathway have a distinct advantage over C_3 trees under warm, arid conditions, as their ability to concentrate CO_2 with only minimal stomatal opening enables them to tolerate extremes of drought. And where grasses and fire-vulnerable woody plants co-mingle, it takes only an ignition source to tilt community composition toward the grass (Budowski 1956). This conversion of forest to grassland, commonly triggered by agricultural activities and logging, has affected hundreds of thousands of hectares throughout the tropics. The process is reversed only when the ignition source is removed and taller-stature woody vegetation eventually suppresses the grasses through competition. In addition to low-stature monocots that fuel fires (primarily grasses and sedges, but in seasonally dry marshes including families such as Marantaceae and Typhaceae), palms often survive this conversion sequence (e.g. Anderson *et al.* 1991). Palms lack fire-vulnerable vascular cambium and their apical meristem is protected by its surrounding tissues, height above the ground or depth below the soil surface.

The grass $\rightleftarrows$ fire feedback is thought to underlie the widespread occurrence of the grassland–forest mosaics characteristic of many mountainous areas in the tropics. These mosaics are maintained by frost (Meher-Homji 1967), periodic drought (Wesche 2003), fires resulting from human activity (Ellenberg 1979; Corlett 1987), or interactions among them. Human activity is not a requisite for all such mosaics, as evidence of those in southern India goes back to at least the last glacial maximum, a period of widespread aridity (Sukumar *et al.* 1993) and a time well before any evidence of people in these areas (Misra 2001).

The increasing severity of El Niño events in the last three or four decades has resulted in increasingly more widespread fires in humid tropical forests, a result of severe drought compounded by logging, conversion to agriculture, and fragmentation (e.g. Laurance & Williamson 2001). Fires in Kalimantan in the late 1990s burned over five million hectares of wet forest, easily five times the area of forest burned in a similar El Niño period two decades before (Siegert *et al.* 2001). Although fires are by no means unprecedented in tropical wet forests (Goldammer 1992), palaeoecology and history tell us that they have been relatively infrequent phenomena, making such forests poorly adapted to fires. Even extremely low-intensity fires can result in widespread tree mortality (Cochrane 2003), very quickly leading to a landscape dominated by grasses and other fire-tolerant life forms such as palms (Anderson *et al.* 1991). With the possibility of a warmer, drier world, such forest transformations are likely to become a growing reality.

Warming-and-drying is but one potential scenario in a changing world. The reality is likely to be far more complex, with some regions becoming wetter as others become drier (e.g. Menon *et al.* 2002), and some regions becoming cooler as the globe grows warmer (Intergovernmental Panel on Climate Change (IPCC) 2001). Models based on ocean surface temperatures and circulation predict that the aerial extent, frequency and intensity of tropical storms and hurricanes could increase in some regions, leading to a greater abundance of disturbance-adapted life forms such as palms and lianas (Laurance *et al.* 2001; Pérez-Salicrup *et al.* 2001b). Palms, particularly, are well built to withstand storms, with dispensable fronds that can be shed to avoid toppling by cyclonic winds, with little damage to their ability to recover (Murphy 1916; Bannister 1970). The converse, a decline in the occurrence of tropical storms, is equally possible for some regions. This reduction in frequency of disturbance could lead to an increase in the extent of those tropical anomalies, the monodominant forests (Hart *et al.* 1989; Sheil & Burslem 2003), although there is also some evidence to suggest that monodominant forests may trace their existence to past catastrophic disturbance (e.g. Read *et al.* 1995).

Some vegetation shifts induced by climate change will undoubtedly be much more subtle than fire- and drought-induced conversions of forest to grassland, or the disturbance-wrought alterations of large landscapes. It is likely that closed-canopy forests will experience shifts in relative dominance and species composition due to changes in temperature, moisture and nutrient fluxes. Each of these abiotic factors, in turn, will influence the relative strengths of species with regard to competition, compensation and complementarity. For example, early successional species may be better able to use increased soil nutrients relative to late successional species (Huante *et al.* 1995). Similarly, rising atmospheric CO_2 may shift the balance of competition between trees and life forms such as lianas (Granados & Körner 2003). Nevertheless, because trees are long-lived, they are better buffered than most life forms against short-term fluctuations in climate.

Climate is not the only global change of ecological consequence: the human-mediated homogenization of the world's biota is an unprecedented phenomenon, one certain to have irreversible impacts on ecosystem structure and functioning. Perhaps the greatest cost will be to less-competitive species: those most likely to be displaced, some to extinction, by non-indigenous invaders. In one sense, the result of the great eco-mix in the tropics will be intriguing ecosystems containing species that are extremely effective at what they do – vines in ecosystems that have never seen climbers, shade-tolerant shrubs in once open-understorey forests, nitrogen-fixing pioneers that redirect succession, and a greater range of specialization among epiphytes, canopy dominants, and every other structural and functional niche available. Such ecosystems are likely to be as effective as the originals, and in some cases to excel, at certain functional attributes such as carbon fixation and nutrient cycling. The tragic cost of mixing, however, is the loss of unique products of evolution. With them goes much of the diversity of interspecific interactions that add great richness to tropical forests. Uniformity and functionality come at a price, one that society should not be willing to pay.

Implications for management

Interactions between plants are important not only to tropical forest diversity and functioning, but to matters of practical importance such as forest utilization, ecosystem restoration, implementation of conservation plans and the design of sustainable systems of land use. Sometimes the resource manager strives to augment certain interactions, and at other times the objective is to reduce their intensity.

Although we use value-laden terms (harmful, tolerate, etc.) below, we do not think their application is completely inappropriate in this context because it is the will of the human manager, whether a forester, a conservation biologist or a farmer, that is imposed. Nature may serve as model, but once humans become involved, natural trajectories inevitably are deflected as we tend to mould nature to our own images of what it ought to be and do.

Plant interactions sometimes lead to impacts that resource managers consider harmful. For example, loggers lament the fact that lianas tie together the crowns of multiple trees, making felling difficult and dangerous for workers, or damaging neighbouring trees, and then impeding regeneration and growth afterwards. It is not surprising, therefore, that liana-cutting has long been a standard silvicultural practice in tropical forests managed or exploited for timber (e.g. Fox 1969; Ewel & Conde 1980). Nevertheless, the costs and benefits of climber-cutting are controversial, from both economic and ecological perspectives (Pérez-Salicrup *et al.* 2001a; Schnitzer & Bongers 2002). It appears likely that, for the time being at least, this practice will be evaluated on a case-by-case basis depending on local abundance of vines, cost and effectiveness of labour, value of the residual stand and impacts on wildlife.

A common and ecologically interesting case arises when managers find it useful to substitute one life form for another. This typically happens when, in restoring forests to land dominated by grasses, establishment of a tree cover proves essential to break the cycle of burning and grassland expansion (e.g. MacDicken *et al.* 1995). Even in wet climates where fire is not a threat, tree establishment sometimes, but not always, accelerates forest recovery in grasslands. In the humid lowlands of Costa Rica, for example, surveys of replicated plantations of seven tree species and pasture plots revealed the presence of 550 plant species – a remarkably rich local flora (Powers *et al.* 1997). While some of the tree plantations had more than twice the number of species as the abandoned pasture, others were no more effective at facilitating restoration than was the grass- and fern-dominated pasture.

Plants that share a common environment sometimes tolerate one another to the detriment of neither. This can lead to enhanced species richness and to complementarity. For example, by combining plants that differ inherently in rooting depths the manager can achieve complementary resource use while enriching local diversity. Application of the rooting-depth concept is currently limited, however, because of lack of data on the rooting patterns of perennial tropical plants and the degree to which observed differences are genetic, or environmentally induced. A useful first approach might be to identify broad rooting patterns among life forms: dicotyledonous trees, palms, shrubs, vines and perennial herbs.

Combinations of plants having non-synchronous leafing phenology also offer promise for complementary resource use and risk abatement (by spreading it among several species). Where rainfall quantity and equity of distribution permit evergreen species to survive, however, the opportunity for temporal partitioning of resource use is lost (although the benefits of risk abatement still hold). In such situations, a fast-growing evergreen species almost invariably dominates a fast-growing deciduous species of similar potential stature simply because its annual period of growth is longer. We observed this in four-genus combinations in Costa Rica, where two evergreens (*Hyeronima* and two species of *Euterpe*) had better survival than either *Cedrela* (dry-season deciduous) or *Cordia* (wet-season deciduous). There were some exceptions, though, and 12 years after planting a handful of the two deciduous species still thrive in the combination.

The sharing of above-ground space provides a third opportunity to combine plants in ways that maximize resource use and augment local diversity.

As an ephemeral, pass-through resource, solar radiation lends itself to the simultaneous use of three-dimensional space by multiple species. Above-ground plant parts are subject to much greater biomechanical constraints than roots (because of the low density of air), so crown architecture is under more genetic control than is that of root systems. The resource steward can take advantage of this in two ways. First, species to be combined can be selected on the basis of stature, crown morphology and requirements for photosynthesis. This will enable them to make optimal use of available space and light, thereby enhancing diversity whether for purposes of conservation or risk reduction (or both). Second, species can be combined so that different heights and crown architectures lead to a desired bole form. For example, when slow-growing trees with dense crowns and fast-growing, light-demanding trees are growing together, the dense foliage of the shorter trees will lead to rapid limb shedding and straighter stem form of the fast-growing trees (e.g. Jennings *et al.* 2003).

In still other situations, plants can be combined such that one species aids the growth of another. The provision of nitrogen by overstorey trees (e.g. *Inga, Erythrina, Falcataria, Casuarina*) to an understorey crop such as coffee or cocoa is a common interaction that has been incorporated into land use for centuries, long before farmers understood the mechanism whereby the trees enhanced growth of the crop. In many parts of Spanish-speaking tropical America, the nitrogen-fixing tree *Gliricidia sepium* is called *madre de cacao*, or 'mother of cocoa.' The crop benefits both from the nitrogen released and the light shade cast by the tree. Nevertheless, management of the overstorey tree for timber may damage the crop when the trees are felled and extracted. As a result, shade trees in plantations are often pollarded, which allows high inputs of solar energy to the understorey until the tree re-branches, rather than tended for their value as lumber.

Although nitrogen-fixing legumes are used occasionally as nurse crops for high-value tree species, sometimes it is a high-value tree itself that fosters colonization by other species, thereby hastening within-stand diversification. In Puerto Rico, for example, Lugo (1992) found that the species richness of 50-year-old plantations of mahogany (*Swietenia macrophylla*) was comparable to that of equal-aged natural forest. Others have reported similar findings with a broad array of tropical plantation species (see papers in Parrotta & Turnbull 1997). The plantation tree co-opts resources that would have been used by life forms of smaller stature, typically grasses or vines. Once the trees achieve significant height they are less-effective inhibitors of new colonists than were the ground-covering plants that preceded them.

Sometimes it is the vine itself that is the product, not the problem. Commercially important vines are found in a number of families (e.g. rattan, Arecaceae; vanilla, Orchidaceae; black pepper, Piperaceae; wicker for basketry, Araceae). In these cases it is the supporting plant that must be chosen carefully, with particular attention paid to climbability (trellis diameter, bark traits, strength) and density of shade.

Conclusions

Whether the consequence of completely passive acts, such as the senescent palm frond crashing to the forest floor, or highly evolved mechanisms, such as a chemical signal transmitted from one species when damaged and received by an undamaged neighbour, interactions among plants are an important part of the intriguing diversity that characterizes tropical ecosystems. A common suite of essential resources leads to a range of processes when species share habitat. These can extend from competitive exclusion of one or more species to the complementary sharing of the available resources in ways that enhance total resource use and productivity.

Complementary sharing can occur qualitatively, spatially or temporally, and of these three possibilities it is the third – temporal partitioning – that is perhaps most intriguing and most often overlooked. Species' roles in a community change with time such that today's dominant might become tomorrow's victim of competition. Resource capture is a function, in part, of plant size, and as size increases with age the plant's impact on ecosystem processes is likely to increase. It is this change with time that is often overlooked, yet it is clearly important on various scales: the past, because plants formerly present leave a biogeochemical or regeneration legacy; the present, as the dominance hierarchy among cohabiting plants shifts on the scale of years to decades; and the future, as shifting climates and human-mediated dispersal change both the abiotic and the biotic ground rules over large geographic scales.

Errors in selection of plant species, choice of habitats and trajectory of community development are commonly committed in the name of conservation biology, restoration ecology and agroforestry: use of a species that might have been predicted to fail (e.g. introduction of heliophytes into the shade of old-growth forests for conservation purposes); selection of habitats that require perpetual maintenance (e.g. fire exclusion in habitats surrounded by invasive grasses in attempts at restoration); inadequate attention to the consequences of well-intended introductions (e.g. use of invasive, alien nitrogen-fixing trees in sustainable agriculture) . . . the list could go on. By learning about the ways that species interact with one another and with their local habitats, and how they fit into longer-term objectives and larger spatial scales, scientists should be able to improve the success rate of resource managers. We need to move forward from the current, and lamentably costly, situation, which consists of tackling each problem as if it were a unique case. Understanding nature's mix of species and the mechanisms involved in its maintenance is the key to our quest for generalization.

Acknowledgements

The work in Costa Rica has been generously supported by a number of organizations, most recently the US National Science Foundation, the Andrew W. Mellon Foundation and the US Forest Service. We thank the Organization for Tropical Studies, Inc., for providing the research site and associated facilities; project coordinators Fabio Chaverri, Miguel Cifuentes, Jenny Pérez and Ricardo Bedoya, plus the loyal crew of six workers and technicians who diligently assisted them with data-gathering in the field and laboratory; Thomas Cole for data management and analyses and graphics; Seth Bigelow for early suggestions; and Patrick Baker, Julie Denslow and Jack Putz for exceptionally detailed and useful comments on the manuscript.

References

Ågren, G. I. (1983) Nitrogen productivity of some conifers. *Canadian Journal of Forest Research*, **13**, 494–500.

Aide, T. M. (1987) Limbfalls: a major cause of sapling mortality for tropical forest plants. *Biotropica*, **19**, 284–285.

Alexandre, D. Y. (1989) *Dynamique de la régénération naturelle en forêt dense de Côte d'Ivoire*. Paris, France: OSTROM.

Anderson, A. B., May, P. H. & Balick, M. (1991) *The Subsidy from Nature: Palm Forests, Peasantry, and the Development of Amazon Frontier*. New York, USA: Columbia University Press.

Baldwin, I. T., Kessler, A. & Halitschke, R. (2002) Volatile signaling in plant–plant–herbivore interactions: what is real? *Current Opinions in Plant Biology*, **5**, 1–5.

Ballaré, C. L., Scopel, A. L. & Sánchez, R. A. (1990) Far-red radiation reflected from adjacent leaves: an early signal of competition in plant canopies. *Science*, **247**, 329–332.

Bannister, B. A. (1970) Ecological life cycle of *Euterpe globosa* Gaertn. *A Tropical Rain Forest* (ed. H. T. Odum). Oak Ridge, Tennessee: US Atomic Energy Commission, pp. B-18: 299–314.

Becker, P. & Castillo, A. (1990) Root architecture of shrubs and saplings in the understory of a tropical moist forest in lowland Panama. *Biotropica*, **22**, 242–249.

Berendse, F. & Aerts, R. (1987) Nitrogen-use-efficiency: a biologically meaningful definition? *Functional Ecology*, **1**, 293–296.

Bertness, M. D. & Callaway, R. (1994) Positive interactions in communities. *Trends in Ecology and Evolution*, **9**, 191–193.

Bigelow, S. W., Ewel, J. J. & Haggar, J. P. (2004). Enhancing nutrient retention in tropical tree plantations: no short cuts. *Ecological Applications*, **14**, 28–46.

Binkley, D. (1992) Mixtures of nitrogen$_2$-fixing and non-nitrogen$_2$-fixing tree species. *The Ecology of Mixed-Species Stands of Trees* (ed. M. G. R. Cannell, D. C. Malcolm & P. A. Robertson) Special Publication Number 11 of the British Ecological Society. London, UK: Blackwell Scientific Publications, pp. 99–123.

Björkman, O. (1981) Responses to different quantum fluxes. *Encyclopedia of Plant Physiology. Physiological Plant Ecology I.* (ed. O. L. Lange, P. S. Nobel, C. B. Osmond & H. Ziegler). New York, USA: Springer-Verlag, pp. 57–107.

Bond, W. J., Midgley, G. F. & Woodward, F. I. (2003) The importance of low atmospheric CO_2 and fire in promoting the spread of grasslands and savannas. *Global Change Biology*, **9**, 973–982.

Brokaw, N. V. L. (1985) Gap-phase regeneration in a tropical forest. *Ecology*, **66**, 682–687.

Budowski, G. (1956) Tropical savannas, a sequence of forest felling and repeated burnings. *Turrialba*, **6**, 23–33.

Butterfield, R. P. (1994) Forestry in Costa Rica: Status, research priorities, and the role of La Selva Biological Station. *La Selva: Ecology and Natural History of a Neotropical Rain Forest* (ed. L. A. McDade, K. S. Bawa, H. A. Hespenheide & G. S. Hartshorn). Chicago, USA: University of Chicago Press, pp. 317–328.

Calder, D. M. (1983) Mistletoes in focus: an introduction. *The Biology of Mistletoes* (ed. D. M. Calder & P. Bernhardt). San Diego, USA: Academic Press, pp. 1–18.

Calladine, A. H. & Pate, J. S. (2000) Haustorial structure and functioning of the root hemiparastic tree *Nuytsia floribunda* (Labill.) R.Br. and water relationships with its hosts. *Annals of Botany*, **85**, 723–731.

Callaway, R. M. (2002) The detection of neighbors by plants. *Trends in Ecology and Evolution*, **17**, 104–105.

Chazdon, R. L. (1986) Light variation and carbon gain in rainforest understory palms. *Journal of Ecology*, **74**, 995–1012.

(1988) Sunflecks and their importance to forest understory plants. *Advances in Ecological Research*, **18**, 1–63.

Chazdon, R. L., Pearcy, R. W., Lee, D. W. & Fetcher, N. (1996) Photosynthetic responses of tropical forest plants to contrasting light environments. *Tropical Forest Plant Ecophysiology* (ed. S. S. Mulkey, R. L. Chazdon & A. P. Smith). New York, USA: Chapman & Hall, pp. 5–55.

Clark, D. A. & Clark, D. B. (1992) Life history diversity of canopy and emergent trees in a neotropical rain forest. *Ecological Monographs*, **62**, 315–344.

Clark, D. B. & Clark, D. A. (1991) The impact of physical damage on canopy tree regeneration in tropical rain forest. *Journal of Ecology*, **79**, 447–457.

Cochrane, M. A. (2003) Fire science for rainforests. *Nature*, **921**, 913–919.

Coley, P. D. & Kursar, T. A. (1996) Causes and consequences of epiphyll colonization. *Tropical Forest Plant Ecophysiology* (ed. S. S. Mulkey, R. L. Chazdon & A. P. Smith). New York, USA: Chapman & Hall, pp. 337–362.

Coomes, D. A. & Grubb, P. J. (1998) Responses of juvenile trees to above- and belowground competition in nutrient-starved Amazonian rainforest. *Ecology*, **79**, 768–782.

Corlett, R. T. (1987) Post-fire succession on Mt. Wilhelm, Papua New Guinea. *Biotropica*, **19**, 157–169.

Dawson, T. E., Mambelli, S., Plamboeck, A. H., Templer, P. H. & Tu, K. P. (2002) Stable isotopes in plant ecology. *Annual Review of Ecology and Systematics*, **33**, 507–559.

DeBell, D. S., Whitesell, C. D. & Schubert, T. H. (1989) Using N_2-fixing *Albizia* to increase growth of *Eucalyptus* plantations in Hawaii. *Forest Science*, **35**, 64–75.

Demmig-Adams, B. & Adams, W. W. (1992) Photoprotection and other responses of plants to high light stress. *Annual Review of Plant Physiology and Plant Molecular Biology*, **43**, 599–626.

Denslow, J. S. (1987) Tropical rainforest gaps and tree species diversity. *Annual Review of Ecology and Systematics*, **18**, 431–451.

Dicke, M. & Bruin, J. (2001) Chemical information transfer between plants: back to the future. *Biochemical Systematics and Ecology*, **29**, 981–994.

Dillenburg, L. R., Teramura, A. H., Forseth, I. N. & Wigham, D. F. (1995) Photosynthetic and biomass allocation responses of *Liquidambar styraciflua* (Hammamelidaceae) to vine competition. *American Journal of Botany*, **82**, 454–461.

Drake, D. R. & Pratt, L. W. (2001) Seedling mortality in Hawaiian rain forest: the role of small-scale physical disturbance. *Biotropica*, **33**, 319–323.

Ellenberg, H. (1979) Man's influence on tropical mountain systems in South America. *Journal of Tropical Ecology*, **67**, 402–416.

Ewel, J. J. & Bigelow, S. W. (1996) Plant life forms and tropical ecosystem functioning. *Biodiversity and Ecosystem Processes in Tropical Forests* (ed. G. H. Orians, R. Dirzo & J. H. Cushman). New York, USA: Springer-Verlag, pp. 101–126.

Ewel, J. J. & Conde, L. F. (1980) *Potential Ecological Impact of Increased Intensity of Tropical Forest Utilization*. Special Publication Number 11, BIOTROP. Bogor, Indonesia: SEAMO Regional Center for Tropical Biology.

Ewel, J. J. & Hiremath, A. J. (1998) Nutrient use efficiency and the management of degraded lands. *An Anthology of Contemporary Ecological Research* (ed. B. Gopal, P. S. Pathak & K. G. Saxena). New Delhi, India: International Scientific Publications, pp. 199–215.

Ewers, F. W., Fisher, J. B. & Fichtner, K. (1991) Water flux and xylem structure in vines. *The Biology of Vines* (ed. F. E. Putz & H. A. Mooney). Cambridge, UK: Cambridge University Press, pp. 127–160.

Field, C. B. & Mooney, H. A. (1986) The photosynthesis–nitrogen relationship in wild plants. *On the Economy of Plant Form and Function* (ed. T. J. Givnish). New York, USA: Cambridge University Press, pp. 25–55.

Fox, J. E. D. (1969) Logging damage and the influence of climber cutting prior to logging in the lowland dipterocarp forest of Sabah. *Malayan Forester*, **31**, 326–347.

Gerwing, J. J. (1995) Competitive effects of three tropical tree species on two species of *Piper*. *Biotropica*, **27**, 47–56.

Gleason, S. M., Ewel, K. C. & Hue, N. V. (2003) Soil redox conditions and plant–soil relationships in a Micronesian mangrove forest. *Estuarine, Coastal and Shelf Science*, **56**, 1065–1074.

Goldammer, J. G. (1992) Ecology of natural and anthropogenic disturbance processes: an introduction. *Tropical Forests in Transition* (ed. J. G. Goldhammer). Basel, Switzerland: Birkhauser Verlag, pp. 1–16.

Goosem, S. & Lamb, D. (1986) Measurements of phyllosphere nitrogen fixation in a tropical and two sub-tropical rainforests. *Journal of Tropical Ecology*, **2**, 373–376.

Grainger, J. & Becker, P. (2001) Root architecture and root:shoot allocation of shrubs and saplings in a Bornean heath forest. *Biotropica*, **33**, 363–368.

Granados, J. & Körner, C. (2003) In deep shade, elevated CO_2 increases the vigor of tropical climbing plants. *Global Change Biology*, **8**, 1109–1117.

Haggar, J. P. & Ewel, J. J. (1994) Experiments on the ecological basis of sustainability: early findings on nitrogen, phosphorus and root systems. *Interciencia*, **19**, 347–351.

(1995) Establishment, resource acquisition, and early productivity as determined by biomass allocation patterns of three tropical tree species. *Forest Science*, **41**, 689–708.

(1997) Primary productivity and resource partitioning in model tropical ecosystems. *Ecology*, **78**, 1211–1221.

Hart, T. B., Hart, J. A. & Murphy, P. G. (1989) Monodominant and species-rich forests of the humid tropics: causes for their co-occurrence. *American Naturalist*, **133**, 613–633.

Heneghan, L., Coleman, D. C., Zou, X., Crossley, D. A. & Haines, B. L. (1999) Soil micro-arthropod contributions to decomposition dynamics: tropical–temperate comparisons of a single substrate. *Ecology*, **80**, 1873–1882.

Hiremath, A. J. (2000) Photosynthetic nutrient-use efficiency in three fast-growing topical trees with differing leaf longevities. *Tree Physiology*, **20**, 937–944.

Hiremath, A. J. & Ewel, J. J. (2001) Ecosystem nutrient use efficiency, productivity, and nutrient accrual in model tropical communities. *Ecosystems*, **4**, 669–682.

Hiremath, A. J., Ewel, J. J. & Cole, T. G. (2002) Nutrient use efficiency in three fast-growing tropical trees. *Forest Science*, **48**, 662–672.

Holbrook, N. M. & Putz, F. E. (1996) Physiology of tropical vines and hemiepiphytes: plants that climb up and plants that climb down. *Tropical Forest Plant Ecophysiology* (ed. S. S. Mulkey, R. L. Chazdon, & A. P. Smith). New York, USA: Chapman & Hall, pp. 363–394.

Holmgren, M., Scheffer, M. & Huston, M. A. (1997) The interplay of facilitation and competition in plant communities. *Ecology*, **78**, 1966–1975.

Huante, P., Rincon, E. & Chapin, F. S. (1995) Responses to phosphorus of contrasting successional tree-seedling species form the tropical deciduous forest of Mexico. *Functional Ecology*, **9**, 760–766.

Hummel, S. (2000) Understorey development in young *Cordia alliodora* plantations. *New Forests*, **19**, 159–170.

Intergovernmental Panel on Climate Change (IPCC) (2001) *Climate Change 2001: The Scientific Basis*. Contribution of Working Group I to the Third Assessment Report of the Intergovernmental Panel on Climate Change (IPCC). *Technical Summary*. Geneva, Switzerland: IPCC, pp. 26–27.

Jackson, P. C., Meinzer, F. C., Bustamante, M. *et al.* (1999) Partitioning of soil water among tree species in a Brazilian cerrado ecosystem. *Tree Physiology*, **19**, 717–724.

Jeník, J. (1978) Roots and root systems in tropical trees: morphologic and ecologic aspects. *Tropical Trees as Living Systems* (ed. P. B. Tomlinson & M. H. Zimmerman). Cambridge, UK: Cambridge University Press, pp. 323–349.

Jennings, S. M., Wilkinson, G. R. & Unwin, G. L. (2003) Response of blackwood (*Acacia melanoxylon*) regeneration to silvicultural removal of competition in regrowth eucalypt forests of north-west Tasmania, Australia. *Forest Ecology and Management*, **177**, 75–83.

Kaye, J. P., Resh, S. C., Kaye, M. W. & Chimner, R. A. (2000) Nutrient and carbon dynamics in a replacement series of *Eucalyptus* and *Albizia* trees. *Ecology*, **81**, 3267–3273.

Keeley, J. E. & Fotheringham, C. J. (1997) Trace gas emissions in smoke-induced seed germination. *Science*, **276**, 1248–1251.

Kellman, M. (1986) Long-term effects of cutting tap roots of *Pinus caribaea* growing on infertile savanna soils. *Plant and Soil*, **93**, 137–140.

Kellman, M. & Hudson, J. (1982) Nutrition of *Pinus caribaea* in its native savanna habitat. *Plant and Soil*, **64**, 381–391.

Kelty, M. J. (2000) Species interactions, stand structure, and productivity in agroforestry plantations. *The Silvicultural Basis for Agroforestry Systems* (ed. M. Ashton & F. Montagnini). Boca Raton, USA: CRC Press, pp. 183–205.

Kessler, A. & Baldwin, I. T. (2002) Plant responses to insect herbivory: the emerging molecular analysis. *Annual Review of Plant Biology*, **53**, 299–328.

Lamont, B. (1983) Mineral nutrition of mistletoes. *The Biology of Mistletoes* (ed. D. M. Calder & P. Bernhardt). San Diego, USA: Academic Press, pp. 185–201.

Langenheim, J. H., Osmond, C. B., Brooks, A. & Ferrar P. J. (1984) Photosynthetic responses to light in seedlings of selected Amazonian and Australian rainforest tree species. *Oecologia*, **63**, 215–224.

Laurance, W. F., Pérez-Salicrup, D., Delamonica, P. *et al.* (2001) Rainforest fragmentation and the structure of Amazonian liana communities. *Ecology*, **82**, 105–116.

Laurance, W. F. & Williamson, G. B. (2001) Positive feedback among forest fragmentation, drought, and climate change in the Amazon. *Conservation Biology*, **15**, 1529–1535.

Lee, D. W. (1979) Abaxial anthocyanin layer in leaves of tropical rain forest plants: enhancer of light capture in deep shade. *Biotropica*, **11**, 70–77.

(1997) Iridescent blue plants. *American Scientist*, **85**, 56–63.

Lee, D. W., Baskaran, K., Mansor, M., Mohamad, H. & Son, K. Y. (1996) Irradiance and spectral quality affect Asian tropical rain forest tree seedling development. *Ecology*, **77**, 568–580.

Lugo, A. E. (1992) Comparison of tropical tree plantations with secondary forests of similar age. *Ecological Monographs*, **62**, 1–41.

MacDicken, K., Hairiah, K., Otsamo, A., Duguma, B. & Majid, N. (1995) Shade-based control of *Imperata cylindrica*: tree fallows and cover crops. *Agroforestry Systems*, **36**, 131–149.

Mack, A. L. (1998) The potential impact of small-scale physical disturbance on seedlings in a Papuan rain forest. *Biotropica*, **30**, 547–552.

Marshall, J. D. & Ehleringer, J. R. (1990) Are xylem-tapping mistletoes partially heterotrophic? *Oecologia*, **84**, 244–248.

Meher-Homji, V. M. (1967) Phytogeography of the South Indian Hill Stations. *Bulletin of the Torrey Botanical Club*, **94**, 230–242.

Menalled, F. D., Kelty, M. J. & Ewel, J. J. (1998) Canopy development in tropical tree plantations: a comparison of species mixtures and monocultures. *Forest Ecology and Management*, **104**, 249–263.

Menon, S., Hansen, J., Nazarenko, L. & Luo, Y. (2002) Climate effects of black carbon aerosols in China and India. *Science*, **297**, 2250–2253.

Merwin, M. C., Rentmeester, S. A. & Nadkarni, N. M. (2003) The influence of host tree species on the distribution of epiphytic bromeliads in experimental monospecific plantations, La Selva, Costa Rica. *Biotropica*, **35**, 37–47.

Misra, V. N. (2001) Prehistoric human colonization of India. *Journal of Biosciences*, **26** (supplement), 491–531.

Murphy, L. S. (1916) Forests of Porto Rico, past, present, and future, and their environment. Washington, DC, USA. US Department of Agriculture Bulletin 354.

Nadkarni, N. M. (1981) Canopy roots: convergent evolution in rainforest nutrient cycles. *Science*, **214**, 1023–1024.

Nadkarni, N. M., Merwin, M. C. & Nieder, J. (2001) Forest canopies: plant diversity. *Encyclopedia of Biodiversity* (ed. S. Levin). San Diego, USA: Academic Press, pp. 27–40.

Neff, J. C., Chapin, F. S. III & Vitousek, P. M. (2003) Breaks in the cycle: dissolved organic nitrogen in terrestrial ecosystems. *Frontiers in Ecology and the Environment*, **1**, 205–211.

Newell, E. A., McDonald, E. P., Strain, B. R. & Denslow, J. S. (1993) Photosynthetic responses of *Miconia* species to canopy openings in a lowland tropical rainforest. *Oecologia*, **94**, 49–56.

Nye, P. H. & Greenland, D. J. (1960) *The soil under Shifting Cultivation.* Technical Communication No. 51. Harpenden, UK: Commonwealth Bureau of Soils.

Ottaviano, E., Mulcahy, D. L., Gorla, M. S. & Bergamini, G. (1992) *Angiosperm Pollen and Ovules: Basic and Applied Aspects.* New York, USA: Springer-Verlag.

Parrotta, J. A. & Turnbull, J. W. (1997) Catalyzing native forest regeneration on degraded tropical lands. *Forest Ecology and Management*, **99**, 1–290.

Pearcy, R. W. (1990) Sunflecks and photosynthesis in plant canopies. *Annual Review of Plant Physiology and Molecular Biology*, **41**, 421–453.

Pearson, H. L. & Vitousek, P. M. (2001) Stand dynamics, nitrogen accumulation, and symbiotic nitrogen fixation in regenerating stands of *Acacia koa*. *Ecological Applications*, **11**, 1381–1394.

Pérez-Salicrup, D. R. (2001) Effect of liana-cutting on tree regeneration in a liana forest in Amazonian Bolivia. *Ecology*, **82**, 389–396.

Pérez-Salicrup, D. R. & Barker, M. G. (2000) Effect of liana removal on water potential and growth of adult *Senna multijuga* (Caesalpinioideae) trees in a Bolivian tropical forest. *Oecologia*, **124**, 469–475.

Pérez-Salicrup, D. R., Claros, A., Guzmán, R. *et al.* (2001a) Cost and efficiency of cutting lianas in a lowland liana forest of Bolivia. *Biotropica*, **33**, 324–329.

Pérez-Salicrup, D., Sork, V. L. & Putz, F. E. (2001b) Lianas and trees in a liana forest of Amazonian Bolivia. *Biotropica*, **33**, 34–47.

Powers, J. S., Haggar, J. P. & Fisher, R. F. (1997) The effect of overstorey composition on understorey woody regeneration and species richness in 7-year-old plantations in Costa Rica. *Forest Ecology and Management*, **99**, 43–54.

Putz, F. E. (1984) The natural history of lianas on Barro Colorado Island, Panama. *Ecology*, **65**, 1713–1724.

(1991) Silvicultural effects of lianas. *The Biology of Vines* (ed. F. E. Putz & H. A. Mooney). Cambridge, UK: Cambridge University Press.

Putz, F. E. & Holbrook, N. M. (1989) Strangler fig rooting habits and nutrient relations in the llanos of Venezuela. *American Journal of Botany*, **76**, 781–788.

Raich, J. W. (1983) Understorey palms as nutrient traps: a hypothesis. *Brenesia*, **21**, 119–129.

Read, J. P., Hallam, P. & Cherrier, J.-F. (1995) The anomaly of monodominant tropical rainforests: some preliminary observations in the *Nothofagus*-dominated rainforests of New Caledonia. *Journal of Tropical Ecology*, **11**, 359–389.

Reich, A., Ewel, J. J., Nadkarni, N. M., Dawson, T. & Evans, R. D. (2003) Nitrogen isotope ratios shift with plant size in tropical bromeliads. *Oecologia*, **137**, 587–590.

Rodgers, H. L., Brakke, M. P. & Ewel, J. J. (1995) Shoot damage effects on starch reserves of *Cedrela odorata*. *Biotropica*, **27**, 71–77.

Russell, A. E., Cambardella, C. A., Ewel, J. J. & Parkin, T. B. (2004) Species, rotation-frequency, and life-form diversity effects on soil carbon in experimental tropical ecosystems. *Ecological Applications* **14**, 47–60.

Schenk, H. J., Callaway, R. M. & Mahall, B. E. (1999) Spatial root segregation: are plants territorial? *Advances in Ecological Research*, **28**, 145–180.

Schefuss, E., Schouten, S., Jansen, J. H. F. & Sinninghe Damsté, J. S. (2003) African vegetation controlled by tropical sea surface temperatures in the mid-Pleistocene period. *Nature*, **422**, 418–421.

Schmitt, J., Dudley, S. A. & Pigliucci, M. (1999) Manipulative approaches to testing adaptive plasticity: phytochrome-mediated shade-avoidance responses in plants. *American Naturalist*, **154**, S43–S54.

Schnitzer, S. A. & Bongers, F. (2002) The ecology of lianas and their role in forests. *Trends in Ecology and Evolution*, **17**, 223–230.

Scholz, F. G., Bucci, S. J., Goldstein, G., Meinzer, F. C. & Franco, A. C. (2002) Hydraulic redistribution of soil water by neotropical savanna trees. *Tree Physiology*, **22**, 603–612.

Scowcroft, P. (1992) Role of decaying logs and other organic seedbeds in natural regeneration of Hawaiian forest species on abandoned montane pasture. *General Technical Report PSW-129*. USDA Forest Service, pp. 67–73.

Sheil, D. & Burslem, D. F. R. P. (2003) Disturbing hypotheses in tropical forests. *Trends in Ecology and Evolution*, **18**, 18–26.

Siegert, F., Ruecker, G., Hinricks, A. & Hoffman, A. A. (2001) Increased damage from fires in logged forests during droughts caused by El Niño. *Nature*, **414**, 437–440.

Smith, C. K., Gholz, H. L. & Oliveira, F. (1998) Soil nitrogen dynamics and plant-induced soil changes under plantations and primary forest in lowland Amazonia, Brazil. *Plant and Soil*, **200**, 193–204.

Snaydon, R. W. (1991) Replacement or additive designs for competition studies? *Journal of Applied Ecology*, **28**, 930–946.

Sukumar, R., Ramesh, R., Pant, R. K. & Rajagopalan, G. (1993) A δ^{13}C record of late Quaternary climate change from tropical peats in southern India. *Nature*, **364**, 703–706.

Trenbath, B. R. (1974) Biomass productivity of mixtures. *Advances in Agronomy*, **26**, 129–169.

Tyree, M. T. & Ewers, F. W. (1996) Hydraulic architecture of woody tropical plants. *Tropical Forest Plant Ecophysiology* (ed. S. S. Mulkey, R. L. Chazdon & A. P. Smith). New York, USA: Chapman & Hall, pp. 217–243.

Went, F. W. (1941) Effects of light on stem and leaf growth. *American Journal of Botany*, **28**, 83–95.

Wesche, K. (2003) The importance of occasional droughts for Afro-alpine landscape ecology. *Journal of Tropical Ecology*, **19**, 197–207.

Williamson, G. B. (1990) Allelopathy, Koch's postulates, and the neck riddle. *Perspectives on Plant Competition* (ed. J. B. Grace & D. Tilman). San Diego, USA: Academic Press, pp. 143–162.

Willis, R. J. (1991) Research on allelopathy on *Eucalyptus* in India and Pakistan. *Commonwealth Forestry Review*, **70**, 279–289.

Zimmermann, M. H. (1971) Transport in the xylem. *Trees: Structure and Function* (ed. M. H. Zimmermann & C. L. Brown). New York, USA: Springer-Verlag, pp. 169–220.

CHAPTER TWO

Resource capture and use by tropical forest tree seedlings and their consequences for competition

LOURENS POORTER
Wageningen University

Introduction

All plants need the same resources (water, light and nutrients) for their survival, growth and reproduction. Paradoxically, water is a limiting resource in many tropical rain forests. Variation in community structure, composition and functioning is driven to a large extent by variation in rainfall (Medina 1999; Poorter *et al.* 2004). Species composition changes continuously and gradually along the rainfall gradient (Hall & Swaine 1981; Bongers *et al.* 1999; Bongers *et al.* 2004), and within forests species tend to sort out along slope gradients in water availability (van Rompaey 1993; Webb & Peart 2000; Harms *et al.* 2001). Within a given forest a range of species coexist that differ in their drought tolerance. These species face to a greater or lesser extent a bottleneck during the dry season, when water is in short supply. Seasonal droughts (with low soil matric potentials) occur annually in tropical moist forest and occasionally in tropical wet forest (Veenendaal *et al.* 1996a). Even in seemingly aseasonal wet forests dry periods of 15–30 days may occur every other year (Walsh & Newbery 1999).

Within a certain forest type, variation in canopy openness leads to marked gradients in irradiance, temperature and relative humidity (Brown 1993). The light available in gaps can be as high as 30% of full light, whereas in the understorey it can be as low as 1%. Light is therefore the most limiting resource in many humid tropical forests (Whitmore 1996). As with drought tolerance, within a given forest a range of species coexist that differ in their shade tolerance. Species tend to sort out along gradients in irradiance (Davies *et al.* 1998; Rose 2000), with pioneer species being more abundant in gaps, and shade-tolerant species being more abundant in the forest understorey.

Light and water availability vary over temporal and spatial scales. Tree species may adapt to different parts of these resource gradients, and are likely to differ in their ability to capture, use and conserve limiting resources. The ability of plants to capture resources affects their ability to compete successfully with other plants. Plants face a trade-off when investing limited resources (carbon,

Biotic Interactions in the Tropics: Their Role in the Maintenance of Species Diversity, ed. D. F. R. P. Burslem, M. A. Pinard and S. E. Hartley. Published by Cambridge University Press. © Cambridge University Press 2005.

nitrogen) in the acquisition of above- or belowground resources. It has, there-fore, been hypothesized that there is a trade-off between drought tolerance and shade tolerance (Smith & Huston 1989). If such a trade-off exists, then there is limited opportunity for niche differentiation. Shade-tolerant species would have a competitive advantage under moist, shaded conditions, and light-demanding species would have a competitive advantage under high irradiance conditions (moist or drought). However, if shade and drought tolerance vary independently, then there is ample opportunity for further specialization and niche differenti-ation (Coomes & Grubb 2000; Holmgren 2000; Sack & Grubb 2002).

Much of our knowledge of plant adaptations to light and water availability comes from species that differ in shade tolerance and adult stature. This might partly be because drought tolerance is more difficult to quantify (Tyree *et al.* 2003). Nevertheless, shade tolerance and adult stature may be the two major axes along which tropical tree species differentiate (Swaine & Whitmore 1988; Turner 2001). Tropical forest ecologists traditionally have paid little attention to the adult stature paradigm, but the question is whether the neglect is justified. Species that differ in adult stature face a trade-off between growth and reproduc-tion. Small-statured tree species reproduce absolutely and relatively at a smaller size (Thomas 1996a) and complete their life cycle earlier than large-statured species (Lieberman *et al.* 1985). Allocation to reproduction requires a compro-mise in height growth rate, leaving these trees behind in the race towards the canopy (Turner 2001). In this review, for convenience, I contrast shade-tolerant and pioneer species, and shrubs and canopy species. These groups represent landmarks on a continuum as shade tolerance and adult stature vary gradu-ally and continuously amongst species (Thomas 1996b; Wright *et al.* 2003), and the same life-history trade-offs are found within these groups as between these groups (Bloor & Grubb 2003; Dalling & Burslem, Chapter 3, this volume).

The aims of this chapter are threefold: (1) to describe how light and water availability vary along different spatial and temporal gradients; (2) to evaluate how species capture above- and belowground resources and how this affects their growth and survival at different parts of the resource gradient, and (3) to explore some implications for seedling competition in the field.

In this chapter I evaluate how resource acquisition affects growth, competi-tion and survival of plants in the regeneration phase. As a working definition I consider seedlings and saplings between 0.1 and 2 m height. The chapter focuses mainly on light and water. The importance of nutrients for plant functioning is acknowledged but not addressed.

Adaptations to light
Spatial and temporal gradients in irradiance
In tropical forests there is a pronounced vertical gradient in irradiance. An exponential extinction of irradiance occurs through the forest canopy, leading

a b

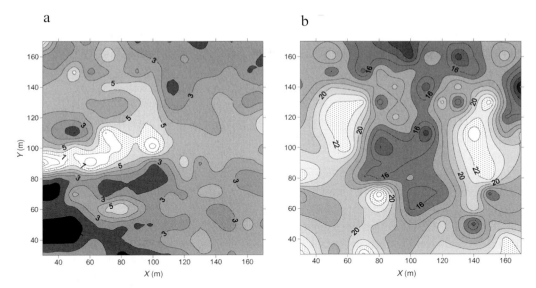

Figure 2.1 Spatial distribution map of (a) per-cent canopy openness at the forest floor as measured with hemispherical photographs, and (b) dry-season volumetric soil moisture content in the forest of El Tigre, Bolivia. Isolines connect sites with the same value for canopy openness or soil moisture content (data from Arets 1998 and Poorter & Arets 2003). Measurements were made at every intersection of a 10 × 10 m grid system.

to irradiance levels of 1%–2% of full sunlight at the forest floor (Yoda 1974; Chiariello 1986; Chazdon 1988). In absolute terms the largest changes occur in the top of the canopy, just below the crown of the canopy trees (Koop & Sterck 1994), whereas in relative terms the largest changes occur in the forest understorey. Within dense, regrowing gap vegetation such vertical gradients are even steeper than found in the understorey. A small increase in sapling height might therefore lead to a disproportional increase in light. This may put a sapling above its whole-plant light-compensation point, increase its survival and affect its competitive ability.

There is substantial horizontal variation in irradiance at the forest floor (Fig. 2.1a), because of gaps created by leaf, branch or treefalls. Understorey sites have an irradiance of 1%–2%, and large gaps an irradiance of 30% (Chazdon 1988). Temporal variation in irradiance occurs on different scales, from seconds and minutes (sunflecks) to hours (diurnal course of the Sun), months (seasonal changes in insolation) or years (gap opening and closure) (Chazdon 1988).

A carbon-balance approach

Species adaptations to irradiance should lead to a maximization of the *net* carbon gain in the environment where the plant is growing. The carbon balance of the plant is determined by the balance between carbon gain and carbon loss.

Carbon gain is determined by the irradiance levels in the microhabitat where the plant grows, and the leaf area, light capture efficiency and photosynthetic characteristics of the plant. Carbon loss is determined by respiration rate, herbivory and the turnover of plant parts. The relative growth rate (RGR) of plants summarizes the carbon balance at the whole plant level. In the following section I discuss how species acquire and conserve carbon, and how this affects their growth and survival in different light environments.

Allocation and morphology

The relative growth rate (RGR, biomass growth rate, scaled to the biomass initially present) is the product of a 'physiological component', the net assimilation rate (NAR, biomass growth per unit leaf area), and a morphological component, the leaf area ratio (LAR, leaf area per unit plant mass). The LAR can be factored into the biomass allocation to the leaves, the leaf mass fraction (LMF, leaf mass per unit plant mass) and the specific leaf area (SLA, leaf area per unit leaf mass). Veneklaas and Poorter (1998) provided a quantitative review of responses of tropical tree seedlings to the light environment. Here I build further on this review, and include the publications of the last six years. Data are presented for plant responses at low irradiance (<3% of full sunlight), typical for the understorey, and high irradiance (20–50% of full sunlight), typical for large gaps and large clearings. Data are shown for two species groups: shade-tolerant species that can establish and survive in the shade, and light-demanding species that need higher light levels for successful establishment, survival and growth. For the current analysis the light levels used (<3% and 20%–50% of full sunlight) are different from those used by Veneklaas and Poorter (1998), the group of pioneers and intermediate species have been pooled (henceforth referred to pioneers), and no corrections have been made for biomass differences amongst plants. The data presented here concern 34 studies and 550 data points. For further details see Veneklaas and Poorter (1998) and Fig. 2.2.

Plant responses to irradiance are ruled by differences in resource ratios at both ends of the light gradient. At low irradiance light is the limiting resource for plant growth. Plants enhance their light capture by investing a large part of their biomass in leaves (Fig. 2.2C). They deploy their leaf biomass efficiently, by making thin leaves with a large specific leaf area (Fig. 2.2D). As a result, they have a large leaf area per unit plant mass (Fig. 2.2E). At high irradiance, water and nutrients become the limiting factors for plant growth. High irradiance and large vapour deficits give rise to high plant-transpiration rates. To guarantee a continuous water supply, plants invest a large part of their biomass in roots (Fig. 2.2A). The increased light levels, in combination with an enhanced photosynthetic capacity, give rise to enhanced photosynthetic rates and net assimilation rates (Fig. 2.2F). Reduced light levels lead to lower growth (Fig. 2.2G) and survival rates (Augspurger 1984). Shade acclimatization allows plants to realize

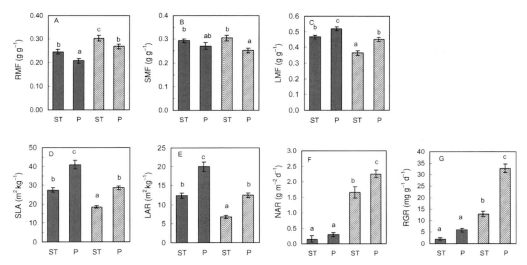

Figure 2.2 Growth and morphological responses of shade-tolerant (ST) and pioneer (P) species to low irradiance (dark bars) and high irradiance (hatched bars). (A) Root mass fraction (RMF), (B) stem mass fraction (SMF), (C) leaf mass fraction (LMF), (D) specific leaf area (SLA), (E) leaf area ratio (LAR), (F) net assimilation rate (NAR) and (G) relative growth rate (RGR). Means and standard errors are shown. Bars with a different letter are significantly different at a P-level of 0.05 (LSD test). The number of species per bar varies from 38–164, seedling biomass ranged from 1 mg to over 100 g. For references and details of methodology see Veneklaas and Poorter (1998).

substantially higher RGRs than they would have attained otherwise (Sims *et al.* 1994). Overall, such acclimatization appears to occur more through adjustment of LAR than A_{max} maximum assimilation rate A_{max} or maintenance respiration (the difference between total respiration and construction respiration) (Sims *et al.* 1994). Adjustments in LAR are mainly the results of changes in SLA, and to a minor extent the result of changes in biomass allocation (Fig. 2.2).

Shade-tolerant species have a little higher biomass fraction in roots than pioneer species (Fig. 2.2), indicating that they allocate more to the acquisition of belowground resources, or that they store more carbon in safe, belowground sites (see the section on adaptations to water stress). Pioneer species have a higher biomass fraction in leaves and a higher SLA, leading to a substantially higher LAR compared with shade-tolerant species, especially under low-light conditions (Fig. 2.2). By increasing their leaf area, pioneer species enhance their light interception and thus their RGR in a light-limited environment. Paradoxically, pioneer species are doing best what a shade-acclimatized plant is supposed to do. Why are shade-tolerant species not able to maximize their leaf area and leaf display? Apparently there is a cost associated with a high SLA and a high LAR in the forest understorey. Leaves with high SLA are more susceptible to herbivores,

pathogens and physical damage (Coley 1983). Biomass loss is especially dele-
terious in the forest understorey, as carbon gain is low and lost leaf material is
not easily replaced (Kitajima 1996). Seedling mortality in the shade is therefore
positively related to the LAR of the species (Kitajima 1994).

Pioneer and shade-tolerant species have a comparable NAR at low irradiance,
but pioneers have a higher NAR at high irradiance (Fig. 2.2F). Pioneer species
have a higher photosynthetic capacity than shade-tolerant species, but irradiance
levels in the understorey are well below the photosynthetic light saturation
point. Pioneer species therefore do not profit from their high photosynthetic
capacity in the understorey, but they do have substantially higher net assimilation rates once they grow in a region of higher irradiance.

Pioneer species maximize their growth at all costs, and outgrow the shade-
tolerant species at both high and low light levels (Fig. 2.2G). This enables them
to complete their life cycle rapidly, and to be successful in the competitive, high-
resource environments of gaps. Their high potential growth rate comes at the
cost of an increased mortality especially under suboptimal conditions, giving
rise to the growth–survival tradeoff observed for tropical tree species (Wright
et al. 2003). The high nutrient content, high SLA and low lignin content of pioneers make them more attractive and more susceptible to herbivores, and the
cheaply constructed wood makes them more susceptible to pathogens (Coley
1983; Kitajima 1996; Gilbert, Chapter 6, this volume; Marquis, Chapter 13, this
volume; DeWalt, Chapter 19, this volume). The high leaf-turnover rate of pioneers is advantageous in gaps where the gap vegetation increases rapidly in
height, but detrimental in the forest understorey, where leaf replacement can
not be sustained (King 1994; Lusk 2002). Shade-tolerant species, on the other
hand, maximize their persistence in the low-resource environment of the forest
understorey, by minimizing their carbon loss. They do so by having low respiration rates, and by investing in long-lived leaves, carbon storage and defence
(Kitajima 1996; Kobe 1997; Veneklaas & Poorter 1998).

At low light, interspecific variation in RGR is mainly determined by variation in LAR, whereas at high light it is mainly determined by variation in NAR
(Poorter 1999, cf. Fig. 2.2E–G). NAR becomes more important determinant of
growth than LAR between 15% and 20% of full light. Light levels of 15% and over
are only encountered in large gaps. If one seeks to understand interspecific differences in growth in the forest understorey and small gaps, then one has to look
for morphological components related to architecture, leaf area and leaf display.

Tree architecture

Tree architecture refers to the overall shape of the tree and the spatial positioning of its components. Tree architecture is often analysed in terms of the height
of the tree and the depth and width of the crown. Seedling architecture depends
mainly on leaf size and leaf-retention time. Only when trees grow taller do they

start to branch and build substantial crowns. The height at which species start to branch varies from 0.1 to 16 m (King 1998a). Most of the relevant architectural differentiation is therefore found in the sapling stage and beyond.

Tree architecture not only determines a tree's ability to capture light, but also the efficiency with which the tree can grow towards the canopy. Understorey species should maximize current light interception by producing wide crowns, whereas canopy species should maximize future light interception by making narrow crowns, which allow them to grow quickly towards the canopy (Kohyama & Hotta 1990; Bongers & Sterck 1998). The same argument can be made for shade-tolerant versus pioneer species. Adults of understorey species were indeed found to have larger crowns than similar-sized juveniles of overstorey species (King 1990, 1995). In a community-wide analysis of 53 Liberian rainforest tree species, it was found that stem slenderness of juvenile trees was positively correlated to the maximal height and the light demand of the species (Fig. 2.3; Poorter *et al.* 2003). By making a slender stem, tall species are rapidly able to reach their reproductive size and the canopy, at a low cost for construction and support (Sterck & Bongers 1998). A slender stem allows light-demanding species to attain or maintain a position at the top of the regrowing gap vegetation. Crown depth and crown diameter were negatively related to maximal height (Fig. 2.3), but not to the light demand of the species. Similarly, stem slenderness and crown depth were negatively correlated with maximal height for Bornean rainforest tree species (Kohyama *et al.* 2003). If light interception drives interspecific differences in tree architecture, then one would expect an especially tight relationship between crown traits and light demand. This was not the case, suggesting that the extension function of tree architecture (a mechanically cheap way to increase in height) and future light gains are more important than maximizing current light interception. The observed trade-offs facilitate the coexistence of species differing in adult stature, and allow them to partition the vertical gradient in the forest canopy. Short species reach their maximum size earlier, and invest in lateral growth and reproduction, whereas tall species grow faster and more efficiently to the canopy to reap the future benefits of high light, a long life and a greater seed production (Kohyama 1993; Turner 2001).

Leaf display

Light interception is influenced not only by the shape of the crown, but also by the geometry and distribution of the foliage. Horizontal leaves arranged in two vertical ranks (distichous) reduce self-shading and optimize light interception in a low-light environment, whereas spirally arranged leaves enhance light interception in a high-light environment (Valladares 1999). The arrangement of leaves on a stem and in relationship to each other (phyllotaxis) has a great impact on the shape of the crown as it affects the position of axillary buds and determines branching patterns (Valladares 1999). Phyllotaxis is a fixed species

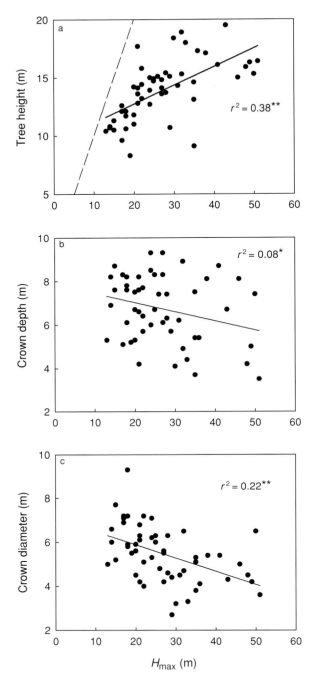

Figure 2.3 Architecture of 53 Liberian rainforest tree species differing in adult stature (H_{max}). (a) Tree height, (b) crown depth, (c) crown diameter. Architectural values are calculated for a standardized tree diameter of 15 cm. Regression lines, coefficients of determination and significance levels are shown. $^*P < 0.05$, $^{**}P < 0.01$. The broken line in the top graph indicates the isoline for which tree height equals H_{max}. Data are from Poorter *et al.* (2003).

trait, but secondary leaf orientation occurs frequently through internode-twisting, petiole-bending or pulvinus movement (Valladares 1999). Changes in leaf shape, size, and orientation and in petiole and stem length can also compensate for the negative effects of phyllotaxis (Pearcy & Yang 1998). When there are many leaves in one spiral, long petioles in older leaves and narrow leaf bases can minimize leaf overlap in certain species (Leigh 1999).

Tree species exhibit different branching patterns. Tree species with orthotropic branches bear their leaves in more or less radially symmetric arrangement on ascending axes, whereas trees with plagiotropic branches bear their leaves in planes along horizontal axes (King 1998b). Plagiotropic branches enhance light interception in a low-light environment, and such a branching pattern is commonly found among shade-tolerant species. Orthotropic branching enhances light interception in a high-light environment, and such a branching pattern is commonly found among light-demanding species (Oldeman & Van Dijk 1991; King & Maindonald 1999).

Light absorption efficiency

The light absorption efficiency (E_a) is the ratio of the photon flux density (PFD) absorbed by the plant crown relative to the PFD absorbed by a horizontal surface (Pearcy & Yang 1998). The light absorption efficiency has two components, the display efficiency and the angular efficiency. The display efficiency is the fraction of the foliage area that is not self-shaded, and varies from 1 for leaves without overlap, to nearly zero for leaves with complete overlap. Valladares *et al.* (2002) evaluated light absorption efficiencies for 24 woody and herbaceous understorey species. Most species intercepted between 60% and 75% of the available radiation, despite striking differences in life form and morphology. Most species had display efficiencies between 0.6 and 0.8, indicating that internal shading varied between 40% and 20%. Interspecific variation in display efficiency was determined most by variation in leaf angle, and to a lesser extent by specific leaf area, support biomass and internode length. The light absorption efficiencies of 11 sympatric *Psychotria* species were also rather similar, and varied from 0.5 to 0.7 (Pearcy & Valladares 2001). The shade-tolerant species did not have a higher absorption efficiency, as one might expect from species adapted to a low-light environment. Apparently there are many ways to make a living and capture light in a tropical forest environment. Compensatory adjustments at different hierarchical scales (leaf, metamer, branch and crown level) may lead to a convergence in light-capture efficiencies among tropical forest plants (Valladares *et al.* 2002).

Tracking of the light environment

For optimal resource use, plants should be able to capture and track resources in space and time (Hutchings & de Kroon 1994; Ackerly 1997). Given the small size of seedlings, they are more likely to track changes in resource availability over time than over space. Plants may track the changes in the light environment if the rate of change matches the response time of the plant organ involved. Photosynthetic induction responses occur at the scale of seconds to minutes (Rijkers *et al.* 2001), changes in the photosynthetic capacity occur from hours to days, changes in leaf morphology and leaf production take days to weeks (Ackerly 1997), and acclimatization at the whole plant level occurs at the scale of months (Popma & Bongers 1991) to years (Sterck *et al.* 1999).

Physiological responses occur more rapidly than morphological responses, and require fewer assimilates (Chazdon 1988). Sunflecks are a common phenomenon in the forest understorey, and carbon fixed during sunflecks represents a substantial part of the total daily carbon gain of understorey plants. An efficient use of sunflecks should therefore be important for shade-tolerant understorey species. Shade-tolerant species respond rapidly to these sunflecks, have faster photosynthetic induction times and lose their induction more slowly than light-demanding species (Poorter & Oberbauer 1993; Valladares *et al.* 1997).

The potential for acclimatization depends on the carbon balance of the tree. Plants therefore acclimatize much faster to an increase in irradiance than to a decrease in irradiance (Osunkoya & Ash 1991; Popma & Bongers 1991). Acclimatization at the whole-plant level occurs mainly through the formation of new leaves or roots, and the abscission of old ones. Pioneer species have higher leaf and root turnover rates than shade-tolerant species, and therefore show faster responses to changes in the light environment (Popma & Bongers 1991; Ackerly 1997). This may give them an advantage in gaps where the canopy of the gap vegetation changes rapidly and continuously over time.

Adaptations to water stress
Spatial and temporal gradients in soil moisture

In the forest soil there are pronounced vertical gradients in moisture content. During the wet season water is uniformly distributed through the soil profile, but in the dry season there is a strong predictable vertical gradient in water availability. Water availability is lowest in the topsoil, and increases exponentially with depth. Such gradients can be steep. After 4 months without rain, the soil water potential can be as low as −6 MPa in the first centimetre of the soil, then increase sharply to −1.7 MPa at 15 cm depth, whereafter it remains constant (Engelbrecht & Kursar 2003).

There is also substantial horizontal variation in soil water content (Fig. 2.1b). Soil water content might vary over centimetres because of differences in soil texture, over metres because of differences in canopy structure, and over tens of metres because of slope gradients. Soil water contents may differ between gap and understorey sites, but the direction and magnitude of the response is still not clear: soil water contents have been variously reported to be higher in gap than in understorey sites (Parker 1985; Becker *et al.* 1988; Jetten 1994; Veenendaal *et al.* 1996b; Ostertag 1998), similar (Vitousek & Denslow 1986; Howe 1990; Veenendaal *et al.* 1996b; Poorter & Hayashida-Oliver 2000; van Dam 2001), or lower (Ashton 1992; Turner *et al.* 1992).

Water availability also varies along the slope gradient. Water availability is greatest at the valley bottom and decreases towards the top. Slopes have relatively high water availability, because of water percolation from upper areas. In the dry season at Barro Colorado Island (BCI), soil water potential at 20 cm depth was about −0.02 MPa for the valley bottom, −0.06 MPa at the mid slope and −0.6 MPa at the upper plateau (Becker *et al.* 1988; cf. Daws *et al.* 2002).

Temporal variation in soil moisture occurs on different scales, from minutes (rainshower), to days (dry spells), months (dry versus wet season), and supra-annual cycles (El Niño). In a Panamanian moist forest the soil matric potential at 25 cm depth may decline from close to zero in the wet season, to −1.1 MPa in the dry season (Wright 1991), and in a Ghanaian moist forest the soil matric potential declined even to −2 MPa in the dry season (Veenendaal *et al.* 1996b).

Impact of drought on seedling survival and distribution

Seedlings with their limited root system are more susceptible to drought than taller plants. Short wet-season spells might affect the survival of tiny germinants, whereas longer dry-season spells might affect taller seedlings. In a Panamanian moist forest rainless periods of up to five days occur frequently during the wet season (Engelbrecht *et al.* 2001). Six-week-old germinants of small-seeded pioneers *Ochroma*, *Miconia* and *Cecropia* show elevated mortality rates after one week without rain (Engelbrecht *et al.* 2001). In a Bolivian moist forest, dry-season mortality decreased with seedling size for six out of seven species analysed (Poorter 1998). Similar observations have been made for Bornean heath forest species (Cao 2000). There is, therefore, a premium on fast growth in the wet season, which is ultimately paid off in the dry season. Gap plants have therefore a better dry-season water status, growth and survival than understorey plants (Veenendaal *et al.* 1996b; Poorter & Hayashida 2000) and first year's seedlings of fast-growing pioneer species may suffer less from dry-season drought than slower-growing shade-tolerant species.

Seedling mortality rates increase exponentially with a decrease in rainfall (Fig. 2.4). Prolonged droughts may therefore have strong regulatory effects on seedling dynamics and community composition. In a Panamanian moist forest Engelbrecht and Kursar (2003) did a 4.5-month droughting experiment with 5–30-cm-tall seedlings of 28 species. Sixty percent of the species showed elevated mortality rates in response to this severe drought. At the same time 40% of the species were not affected, indicating that a large proportion of this moist forest community is well adapted to drought.

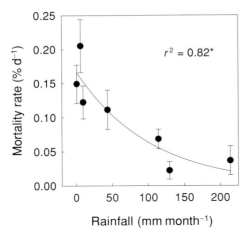

Figure 2.4 Relationship between seedling mortality rate and the rainfall during the corresponding time interval. Seedlings of eight species were transplanted to a large treefall gap in a Bolivian moist forest, and monitored for 8 months (Poorter 1998). Mortality rates are means for all species pooled. Error bars indicate one standard error of the mean. The coefficients of determination and significance level are shown.

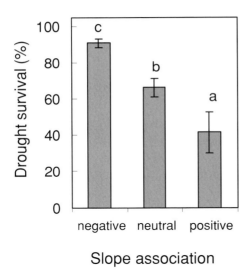

Figure 2.5 Drought survival of species that show differential preference for wet slopes in a tropical moist forest in Panama. The drought survival was determined for seedlings under controlled conditions (Engelbrecht & Kursar 2003). The slope preference was determined for saplings and taller trees. Species were assigned to different slope preference groups, based on their significant negative ($n = 6$), neutral ($n = 8$) or positive ($n = 7$) association with slopes (Harms *et al.* 2001). Means and standard errors are shown. Bars with a different letter are significantly different at a *P*-level of 0.05 (LSD test).

The drought tolerance of seedlings may determine to a large extent where a species is found in the landscape. Species with drought-intolerant seedlings are positively associated with moist slopes (Fig. 2.5), whereas species with drought-tolerant seedlings are relatively less abundant on slopes, probably because they are outcompeted under moist, shaded conditions. Similarly, in Ghana, species with drought-intolerant seedlings were largely confined to wet forests, whereas species with drought-tolerant seedlings were confined to moist forest (Veenendaal & Swaine 1998).

A water-balance approach

What features enable species to tolerate drought? Species have evolved different mechanisms (Fitter & Hay 1987). They may (1) enhance the acquisition of water, (2) increase water conservation and efficiency of water use, or (3) continue physiological plant-functioning during drought. The first two mechanisms are related to desiccation delay, whereas the latter mechanism is related to desiccation tolerance. In the following sections I discuss how species enhance water acquisition through biomass allocation and morphology, how they use water efficiently during photosynthesis, and how they continue physiological functioning through a lowered leaf water potential and resistance to xylem embolism.

Most of our evidence for drought adaptations comes from dry-forest species that differ in phenology (evergreen and deciduous species). Here I have tried to focus on recent evidence from moist and wet tropical forests. Few studies exist on the water relationships of seedlings and saplings of wetter tropical forests. Most of these studies include only a few species (five or fewer), so generalizations about plant strategies are difficult to make. Drought-related comparisons have generally focused on species differing in life form (shrubs versus trees), shade tolerance (pioneers versus shade-tolerant species) and deciduousness (deciduous versus evergreen).

Water acquisition

Biomass allocation and root morphology

Plants may acquire water by exploring deeper soil layers and/or a large soil volume. The size and shape of the root system depend on the plant size, biomass allocation to roots and root morphology. Shade-tolerant species have a significantly higher biomass fraction in roots than pioneer species (Fig. 2.2). Such differences are already established when seedlings are in the cotyledon stage (Paz 2003). Shade-tolerant species do not only invest more biomass in roots, but they also make relatively thick roots with a low root-length per unit root biomass (SRL; specific root length), indicating that these species invest carbon in storage or physical defence, to enhance survival in a low-resource environment (Paz 2003). Pioneer species, on the other hand, form thinner roots, and have a greater total root length (Huante et al. 1992; Tyree et al. 1998). This allows them to explore a larger soil volume for water and nutrients and to monopolize growth in gaps. The foraging capacity of pioneer species is also aided by their root architecture. Pioneer species tend to have a dichotomous root-branching pattern, and shade-tolerant species a herringbone branching pattern (Huante et al. 1992). The dichotomous root-branching patterns allow for more effective exploration and exploitation of zones with a high resource availability (Fitter 1985). The pioneer shade-tolerant continuum is in fact a continuum of fast versus slow growth (Veneklaas & Poorter 1998; Reich et al. 2003). RGR is therefore often used as an estimator of the light demand of a species. In a study of 15

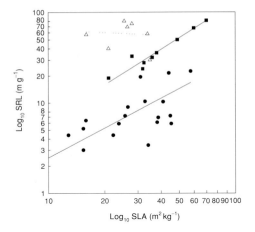

Figure 2.6 Specific root length (SRL) versus specific leaf area (SLA) for seedlings of dry tropical (triangles, Huante *et al.* 1992), moist tropical (circles, H. Paz, unpublished results) and temperate (squares, Reich *et al.* 1998) tree species. High SRL and SLA facilitate the foraging capacity for below- and aboveground resources, respectively. Significant regression lines are continuous, insignificant regression lines are broken.

Australian rainforest species, Bloor and Grubb (2003) found that there is a strong positive association between the RGR, SRL and root-branching intensity of the species.

Pioneer species have not only a higher SRL than shade-tolerant species, but also a higher SLA (Reich *et al.* 1998; Paz 2003). This indicates that they maximize the surface areas for both above- and belowground resource capture. SRL and SLA are strongly positively correlated for a wide range of woody tree species (Fig. 2.6), suggesting that aboveground and belowground foraging capacities are closely correlated (cf. Ryser 1996). Interestingly, dry-forest species have a considerably higher SRL at a given SLA. This might indicate that in an open and dry environment there is a premium on foraging for below-ground resources, which limits most plant growth.

Surprisingly few studies have analysed root morphological responses to drought. The first such study (Paz 2003) evaluates root architecture of a large number of seedlings from four different forest types. Seedlings from a seasonal forest had a larger biomass allocation to roots than seedlings from more per-humid forests. Yet their SRL and total root length per leaf area were similar to those of the perhumid forest species. Instead, they had relatively deep roots. A tentative conclusion would be that in dry environments species explore deeper, instead of larger, soil volumes in search for water.

Rooting depth
Shrubs and trees are hypothesized to differ in their rooting depth. The low light levels in the understorey do not allow for heavy investment of carbohydrates in deep and large root systems (Wright 1992). Understorey shrubs have the additional drawback that part of their carbohydrates should be allocated to reproduction, instead of new roots. To meet the nutrient demands for reproduction,

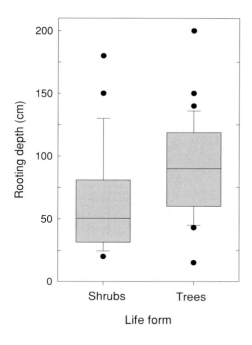

Figure 2.7 Maximal rooting depth of rain forest shrub ($n = 20$) and tree ($n = 27$) species. The boxplots indicate 10th and 90th percentile (bars), 25th and 75th percentile (lower and upper end boxes), median (central line in box), and outliers (dots). Data come from Becker and Castillo (1990); Wright *et al.* (1992); Arets (1998); Becker *et al.* (1999); Cao (2000); Grainger and Becker (2001). The two life forms differ significantly in rooting depth (*t*-test, $t = -2.1$, $P < 0.05$).

shrubs should concentrate their roots in the nutrient-rich upper stratum of the soil. It has therefore been hypothesized that understorey shrub species have shallower root systems than similar-sized saplings of tree species (Becker & Castillo 1990). This should make shrub species more susceptible to (extreme) seasonal droughts (Condit *et al.* 1995) or infertile soils and might explain why the species richness of neotropical shrubs increases with rainfall and soil fertility (Gentry & Emmons 1987; Wright 1992), and Bornean heath forests have a species-poor shrub flora (Grainger & Becker 2001). Combined data from several forest sites show that shrubs do indeed have shallower roots than saplings of canopy species (Fig. 2.7), but that they have a similar proportion of their root surface area in the upper soil layers (Grainger & Becker 2001). Floristic data from Ghana (Swaine & Becker 1999) show that the shrub species richness increased with rainfall, as predicted by Wright (1992). Yet tree species richness increased as well, so that the ratio of shrub to tree species was relatively constant over the whole rainfall gradient, or even a little enhanced under very dry conditions (Swaine & Becker 1999). A tentative conclusion would be that moist-forest shrubs are somewhat less drought-tolerant than tree saplings, but that many exceptions exist.

Jackson *et al.* (1995) made a more direct evaluation of soil water partitioning, using stable isotope techniques. By comparing the hydrogen isotope composition of xylem water with that of soil water, they inferred the depth from which plants obtain their water. Saplings of species growing in the same forest gap differed markedly in the isotopic composition of their xylem water. Evergreen species used water from deeper soil layers than deciduous species, and this allowed them to maintain their leaves during the dry season. This suggests that some degree of vertical partitioning of soil water resources takes place among co-occurring species.

The importance of deep roots is underscored by observations of Jackson *et al.* (1995) on water use by species growing in the same forest gap. Species that had access to deeper and therefore more abundant water had higher whole-plant transpiration rates and a more favourable leaf water status (less-negative leaf water potentials). Shallow-rooted species are not always at a disadvantage in terms of water acquisition. Shallow-rooted species might profit from infrequent, low rainfalls that wet only the upper parts of the soil layer (Burgess *et al.* 1998). Cao (2000) found that species with superficial roots suffered a higher mortality during drought, but recovered faster after the first rains (which only wetted the topsoil) than species with taproots. Shallow-rooted plants may profit as well from hydraulic lift by neighbouring trees, although no data are yet available for tropical moist forests.

Water-use efficiency
Wilting and leaf abscission
Water loss is most easily reduced by stomatal closure and the reduction of transpiring leaf area. The most immediate response to droughting is wilting, and thereafter the abscission of leaves. Wilting is an effective way to reduce the heat load on the leaf. By reducing the leaf angle, less light is intercepted at the leaf and crown level, decreasing transpiration rates and enhancing the water-use efficiency of the plant (Chiariello *et al.* 1987; Reekie & Wayne 1992). Midday wilting is a common phenomenon amongst pioneer trees that are regularly exposed to high radiation loads in large gaps.

Species show a large variety in wilting behaviour (Engelbrecht & Kursar 2003; Tyree *et al.* 2003). Some species pass quickly through all wilting stages whereafter they die, whereas other deciduous species quickly shed their leaves and survive. In a study with 28 moist-forest trees and shrubs, about 40% of the species showed a significant reduction in leaf area during drought (Engelbrecht & Kursar 2003). Species that are deciduous in the seedling stage may realize an enhanced dry-season survival at the expense of a reduced growth (Poorter 1998). Deciduous species are characteristically found in dry and exposed environments, such as rocky outcrops with shallow soils, in large gaps, or in the forest canopy.

Photosynthetic water-use efficiency
Photosynthetic carbon gain comes at the cost of water loss, as open stomata allow CO_2 to diffuse inwards to the sites of fixation, while water vapour

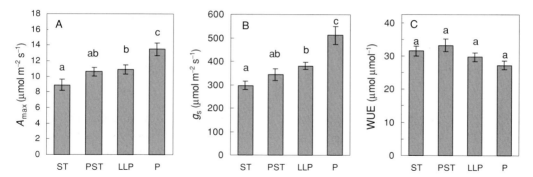

Figure 2.8 Photosynthetic characteristics of 44 Bolivian moist-forest species differing in shade tolerance. (A) Light-saturated photosynthetic rate (A_{max}), (B) stomatal conductance (g_s), (C) photosynthetic leaf water-use efficiency (WUE, calculated as A_{max}/g_s). Means and standard errors are shown. Bars with a different letter are significantly different at a P-level of 0.05 (LSD test). Species were assigned to different shade-tolerance groups: total shade-tolerant species that can grow and reproduce in the understorey (ST; $n = 11$), partial shade-tolerant species that can establish in the shade but need higher light levels for subsequent growth and reproduction (PST, $n = 10$), long-lived pioneers that need high light for establishment (LLP, $n = 18$), and short-lived pioneers that need high light for establishment (P, $n = 5$). (F. Bongers & L. Poorter, unpublished results.)

diffuses outwards into the atmosphere. Bongers and Poorter (unpublished results) evaluated the photosynthetic characteristics of 44 species differing in shade tolerance. Naturally established saplings were compared under conditions of optimal resource supply. Sun leaves of light-demanding species had both higher light-saturated photosynthetic rates (A_{max}) and higher stomatal conductances than sun leaves of more shade-tolerant species (Fig. 2.8, cf. Reich *et al.* 1995). The high carbon gain could only be sustained by an excessive water use. The ratio of assimilation over stomatal conductance is a measure of the instantaneous water-use efficiency of the leaf. It turned out that light-demanding and shade-tolerant species had similar water-use efficiencies, i.e. they fixed the same amount of carbon per unit water lost. Such measurements are made at the leaf level only, but controlled experiments with herbaceous plants indicate that the instantaneous efficiency of water use by the leaf is a good indicator of the efficiency at the whole-plant level as well (van den Boogaard & Villar 1998).

From an ecological point of view, long-term efficiency of water use is more important than instantaneous efficiency. The long-term integrated efficiency of water use can be inferred from the carbon isotope signature of the leaves. A high (less negative) $\delta^{13}C$ indicates a high intrinsic water-use efficiency. Bonal *et al.* (2000) screened sunlit leaves of a large number of large canopy trees for $\delta^{13}C$. The $\delta^{13}C$ differed markedly among species, corresponding to a threefold range in intrinsic water-use efficiency. The efficiency was negatively correlated with the midday leaf water potential of the species. Species that had the highest (least negative) midday leaf water potential had a better access to soil water resources, transpired more and used the water therefore less efficiently. Similarly, Meinzer *et al.* (1995) found a positive relationship between sapling transpiration rates and leaf water potentials. Counterintuitively, the intrinsic water-use efficiency in Bonal's study was highest for the partial shade-tolerant species, intermediate for long-lived pioneer species and lowest for complete shade-tolerant species (cf. Guehl *et al.* 1998). The group of partial shade tolerants consisted mostly of emergent species. Bonal *et al.* suggested that a high water-use efficiency is especially important for emergent species, because their emergent crowns are exposed to sunny, dry and windy conditions. Interspecific variation in intrinsic water-use efficiency was not related to the nitrogen concentration, SLA and thickness of the leaves (Guehl *et al.* 1998). This suggests that the long-term water-use efficiency is not determined by the photosynthetic capacity and length of the CO_2 diffusion pathway in the leaves. The observed patterns are determined for large canopy trees. It would be interesting to see whether the same pattern holds for seedlings and saplings that experience strikingly different environmental conditions from canopy trees, and whose shallow root systems make them more vulnerable to drought.

Drought tolerance
Water potential
Water moves down a potential gradient, from a high to a low potential. If the soil water potential declines under dry conditions, then plants should decrease their root, stem and leaf water potentials as well, to assure a continuous water flow. Most species show such a decrease in leaf water potential in response to drought (e.g. Wright *et al.* 1992; Tobin *et al.* 1999; Cao 2000). Most prominent changes in leaf water potentials are 'passive' and the result of a negative water balance. Plants can also actively lower their leaf water potential by lowering their hydrostatic pressure through changes in the cell elastic modulus, or by lowering their osmotic potential. Tyree *et al.* (2003) conducted a greenhouse experiment in which they exposed seedlings of five species to a prolonged period of drought. All species showed a decrease in relative leaf water content and leaf water potential as drought progressed. Plant cells, and especially meristems, are sensitive to a low leaf water content and a low leaf water potential. It turned out that the relative leaf water content and leaf water potential at which a species died was a good predictor of drought survival of seedlings in the field: species that were able to tolerate a very low relative leaf water content, or leaf water potential, were the ones that had the best drought survival. This led Tyree *et al.* to conclude that the drought performance of species is more related to desiccation tolerance than to desiccation delay.

Hydraulic conductance and resistance to embolism

Pioneers and shade tolerants differ perhaps most strikingly in their hydraulic architecture. Pioneers have higher root and shoot hydraulic conductance per unit leaf area than shade tolerants, enabling them to have a less-negative water potential for a given transpiration rate (Huc *et al.* 1994; Tyree *et al.* 1998). According to the Hagen–Poiseuille law, the hydraulic conductance scales to the fourth power of the vessel radius. A high hydraulic conductance is therefore aided by wide vessels. Seedling RGR of temperate tree species is positively correlated with their vessel diameters (Castro-Diez *et al.* 1998), suggesting that fast-growing pioneers realize large conductances because of their larger vessels. The increased hydraulic conductance may come at the cost of an increased chance of cavitation. Cavitation is detrimental for plant-functioning, as it impairs water supply to the leaves. Species that are less vulnerable to cavitation are therefore more drought-tolerant (Tyree & Sperry 1989; Tyree *et al.* 2003). The risk of cavitation is hypothesized to increase with the diameter of the vessels and the diameter of the pit-membrane pores that allow water and air bubbles to pass between adjacent conduits (Tyree *et al.* 1994). There is therefore a trade-off between a high conductance, abundant water use and fast growth rates on the one hand, and a risk of increased cavitation and dry-season mortality on the other hand. Pioneer species opt for the first, and risk the latter. Many light-demanding species might escape this deadlock by having a deciduous leaf habit in the dry season. This might explain why there is a close association between deciduous leaf habit and light demand (Condit *et al.* 1996).

Implications for competition

How common, or how rare, is seedling competition?

How do plant adaptations to light and water stress affect the competitive ability of seedlings in the field? Perhaps most of the competition that seedlings experience comes from large trees (Coomes & Grubb 2000; Ewel & Hiremath, Chapter 1, this volume). This competition is asymmetric, because trees have a large effect on seedlings, and seedlings have little effect on trees. Competition amongst smaller-sized individuals is likely to come from other life forms, such as palms, ferns and large-leaved monocots. These non-woody plants make up a large part of the vegetation cover in the forest understorey (Harms *et al.* 2004). Many of them are clonal, and they compete effectively with tree seedlings. The density of tree seedlings is therefore inversely related to the vegetation cover of similar-sized non-woody plants (Harms *et al.* 2004).

Competition amongst tree seedlings might be rare, as tree seedling densities in the understorey are, with the exception of conspecific seedling carpets, generally low. In neotropical rainforests, the average woody seedling density (plants between 10 and 50 cm tall) varies from 1 to 6 seedlings per m^2 (Harms *et al.* 2004). If competition amongst woody plants is important, then there should be an effect of woody plant density on seedling survival and growth. Liza Comita (unpublished results) evaluated the effect of woody plant density on one-year survival of tree seedlings (between 20 cm height and 1 cm diameter at breast height) at BCI. Of the 119 species evaluated, 72% of the species did not experience an effect of plant density, 18% were affected by conspecific plant density and 10% were affected by heterospecific plant density. The effect of heterospecific plant density was in half of the cases positive and half of the cases negative. Although competition is likely to affect growth more than survival, and although the study was relatively short, it underscores the point that competition amongst heterospecific seedlings is a relatively rare event in tropical forests (cf. Wright 2002).

Competition during gap-phase regeneration

Competition amongst tree seedlings is more likely to occur in gaps because plant densities are higher, growth rates are faster and the vegetation closes faster than in the understorey. The species that are most successful during the process of gap-phase regeneration are likely to be the ones that will dominate the patch in the future. The success of an individual in a gap depends on the interplay between resource availability and resource requirements for growth. Resource availability depends on gap size, density of the regrowing vegetation and relative size of the individual at time of gap formation, whereas species-specific traits determine resource acquisition and realized growth (*sensu* Zagt & Werger 1998).

Surprisingly few competition experiments have been carried out with tropical tree seedlings. Most of them concern trenching experiments in which seedlings are liberated from roots by neighbouring trees (summarized in Coomes & Grubb 2000). It is therefore difficult to make any generalizations. I present here results from a competition experiment with similar-sized seedlings to illustrate how species performance is affected by plant density and by species-specific traits.

Seedlings of eight tree species differing in shade tolerance were grown together in experimental seedling communities in a large tree-fall gap in the Bolivian Amazon (Poorter 1998). The gap was cleared of all natural vegetation at the onset of the experiment. Seedlings were germinated and transplanted at the beginning of the rainy season. Average initial seedling height ranged from 5 to 10 cm for the different species. The seedlings were planted in 42 experimental communities at low and high plant densities. Inter-seedling distance was 50 cm for the low-density treatment and 10 cm for the high-density treatment. Seedlings communities were harvested after one year. At final harvest the plant height, crown diameter, leaf area and length of the longest root were determined. Plants were divided into leaves, stem plus petioles, and roots, oven-dried and weighed.

Seedling communities attained an average maximal height of 1 m by the end of the experiment. The canopy of the high-density plots was closed, and therefore the irradiance decreased exponentially with depth in the stand. At the bottom of the stand, the irradiance was 5%–10% of that in full sunlight. A high plant density led to a significant decrease in plant size. Average plant biomass was reduced by 50%, and average height by 11%. Biomass was most affected by competition as it summarizes the impact of both above- and belowground competition on all plant parts. Height was less affected by competition because of its pivotal role in light capture. Since many forest ecologists measure height rather than biomass, it is likely that the strength of competitive interactions amongst forest trees is underestimated.

Allocational and morphological responses of plants in the high-density plots indicated that competition was mainly for light. When plants were compared at the same biomass, the plants in the high-density plots had a larger biomass fraction in stem and a larger specific stem length (stem length per unit stem mass), resulting in plants 25% taller than in the low-density plots. Irradiance decreased exponentially with depth in the stand, and a small increase in plant height led therefore to a disproportional increase in intercepted light. Light interception was also increased by the higher SLA and LAR of the high-density plants. Low-density plants had a higher biomass fraction in roots, suggesting that they were more limited by belowground resources.

Early successional species attained a larger biomass than late successional species in low-density and in high-density plots (Fig. 2.9a). Rapid growth during the initial phase of stand development enables early-successional species to attain a size advantage before the canopy of the stand is closed. Rapid growth in the subsequent phase secures their position in the top of the canopy. The degree of dominance that a species attained in these artificial seedling communities was positively correlated with growth, leaf turnover (Fig. 2.9b) and biomass fraction in the stem. A high leaf turnover enables plants to shed unproductive, overshaded leaves in the lower part of the stand, and replace them by new, productive leaves in the top of the stand. These results support Grime's (1977) hypothesis that the most competitive species are the ones that are able to capture and monopolize resources. The results of this competition experiment are partly corroborated by Reekie and Bazzaz (1989) who grew five tree species for 3.5 months together in a glass chamber. Species biomass at the end of the experiment was closely correlated with the mean canopy height and the SMF halfway through the experiment, indicating that competitive dominance is attained by species that compete best for light.

What is the predictive power of such experiment for gap-phase regeneration under natural conditions? What happens when plants grow for a longer time span, in smaller-sized gaps and when initial size differences occur? The biomass ranking of the species in the experiment indeed confirmed the

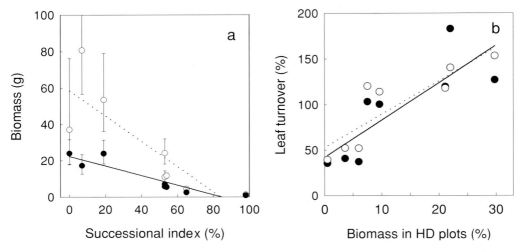

Figure 2.9 Performance of eight moist-forest species growing in experimental seedling communities at high planting density (filled symbols, continuous regression line) and low planting density (open symbols, dotted regression line). (a) Relationship between the final biomass and the successional position of the species (a successional index indicates of 0 indicates the earliest-successional species, and a successional index of 100 indicates the latest-successional species found in the area; Peña-Claros 2003). (b) Relationship between leaf turnover and the competitive ability of the species. The competitive ability is calculated as the mean proportion of community biomass occupied by a species in high-density (HD) plots (Poorter 1998).

successional order of species as observed in the field (Fig. 2.9a). However, the competition experiment lasted only one year and competitive hierarchies may change in course of time (Ewel & Hiremath, Chapter 1, this volume). Given the size-asymmetric nature of competition for light, it is likely that the established size-hierarchy will be maintained in the short term (cf. Weiner 1990). In the long term the early-successional species will eventually be replaced by later-successional species that have a longer life span or a larger size at maturity (Ewel & Hiremath, Chapter 1 this volume). It should be stressed that species traits can be important determinants of plant success in a gap environment, but that the importance of chance processes increases when gaps are smaller or initial differences in plant size are larger.

Conclusions

Pioneer species have a higher LMF, SLA, LAR and NAR than shade-tolerant species. Better light capture and higher photosynthetic rates allow the pioneer species to outgrow the shade-tolerant species at both low and high irradiance. Interspecific differences in growth in the understorey are driven by interspecific differences in light-capture, SLA and LAR. The high growth potential of the pioneer species in the understorey comes at the cost of an increased mortality, and a high herbivory and leaf turnover may exclude the pioneers from the forest understorey.

Morphologically there are many (cheap) ways in which a tree species can enhance the efficiency of light capture, and most species appear to have similar light-capture efficiencies. Architectural differences amongst species have small implications for their ability to capture light, but large implications for the efficiency with which species can grow towards the canopy. Large-statured species grow more efficiently to the canopy than small-statured species, because of their slender stems and shallow and narrow crowns.

Tropical tree species differ markedly in their capacity to acquire water, to use water efficiently and to continue plant-functioning during drought. Species differ strikingly in the depth from which they acquire their water, suggesting a high degree of vertical partitioning of soil water amongst co-occurring species. Pioneer species have higher stomatal conductances than shade-tolerant species. The higher stomatal conductances come along with higher photosynthetic rates resulting in similar photosynthetic water-use efficiencies for both groups. Species that are able to tolerate a very low leaf water content or leaf water potential are the ones that realize the best drought survival in the field.

Interspecific differences in hydraulic conductance drive to a large extent interspecific differences in water use and growth potential. Pioneer species have high hydraulic conductances because of low density wood with large vessels, and probably large pit-membrane pores. The increased hydraulic conductance may come at the cost of an increased risk of cavitation and dry-season mortality. Many light-demanding species elude this deadlock by having a deciduous leaf habit, avoiding drought and escaping dry-season mortality.

Below and aboveground foraging capacities are most determined by the lifespan and morphology (SLA, SRL) of roots and leaves. The RMF and LMF affect the foraging abilities only to a limited extent (cf. Poorter & Nagel 2000), and a carbon-based trade-off between the acquisition of above- or belowground resources may not exist. SRL and SLA are strongly positively correlated for a wide range of woody tree species, suggesting that aboveground and belowground foraging capacities are closely correlated as well (cf. Grime *et al.* 1997).

Pioneer species have a high RGR, SRL and SLA, and high root and leaf turnover, which enables them to monopolize and pre-empt resources. Pioneers will outcompete shade-tolerant species in the short term, but their fast lifestyle comes at the cost of an increased mortality and a reduced lifespan. Shade-tolerant species lag behind the pioneers in biomass and height growth during gap-phase regeneration. With their reduced growth rates they depend on long-term persistence for long-term competitive success. Low resource requirements in combination with long-lived and well-defended leaves and roots allow them to persist and wait for a brighter future in the forest canopy, when the shorter-lived pioneers eventually die off.

Acknowledgements

I thank Peter Becker, Frans Bongers, Jim Dalling and Bettina Engelbrecht for their useful comments on the manuscript, and Eric Arets, Liza Comita and Horacio Paz for kindly allowing me to use their unpublished results. This research was supported by Veni grant 863.02.007 from the Netherlands Organisation of Scientific Research (NWO).

References

Ackerly, D. (1997) Allocation, leaf display, and growth in fluctuating light environments. *Plant Resource Allocation* (ed. F. A. Bazzaz & J. Grace). New York: Academic Press, pp. 231–264.

Arets, E. J. M. M. (1998) Seedling spatial distribution patterns of 13 Bolivian tree species in relation to light availability, soil moisture content and distance to conspecific trees. Unpublished M.Sc. thesis, Utrecht University, Utrecht.

Ashton, P. M. S. (1992) Some measurements of the microclimate within a Sri Lankan tropical rain forest. *Agriculture and Forest Meteorology*, **59**, 217–235.

Augspurger, C. K. (1984) Light requirements of neotropical tree seedlings: a comparative study of growth and survival. *Journal of Ecology*, **72**, 777–795.

Becker, P. & Castillo, A. (1990) Root architecture of shrubs and saplings in the understorey of a tropical moist forest of lowland Panama. *Biotropica*, **22**, 242–249.

Becker, P., Rabenold, P. E., Idol, J. R. & Smith, A. P. (1988) Water potential gradients for gaps and slopes in a Panamanian tropical moist forest's dry season. *Journal of Tropical Ecology*, **4**, 173–184.

Becker, P., Sharbini, N. & Yahya, R. (1999) Root architecture and root:shoot allocation of shrubs and saplings in two lowland tropical forests: implications for life form composition. *Biotropica*, **31**, 93–101.

Bloor, J. M. G. & Grubb, P. J. (2003) Growth and mortality in high and low light: trends among 15 shade-tolerant tropical rain forest tree species. *Journal of Ecology*, **91**, 77–85.

Bonal, D., Sabatier, D., Montpied, P., Tremeaux, D. & Guehl, J. M. (2000) Interspecific variability of δ^{13}C among trees in rainforests of French Guiana: functional groups and canopy integration. *Oecologia*, **124**, 454–468.

Bongers, F. & Sterck, F. J. (1998) Architecture and development in rain forest trees: responses to light. In *Dynamics of tropical communities* (ed. D. M. Newbery, H. H. T. Prins & N. Brown). Oxford: Blackwell Scientific Press, pp. 125–162.

Bongers, F., Poorter, L., van Rompaey, R. S. A. R. & Parren, M. P. E. (1999) Distribution of twelve moist forest canopy tree species in Liberia and Côte d'Ivoire; response curves to a climatic gradient. *Journal of Vegetation Science*, **10**, 371–382.

Bongers, F., Poorter, L. & Hawthorne, W. D. (2004) The forests of Upper Guinea: gradients in large species composition. *Biodiversity of West African Forests. An Ecological Atlas of Woody Plant Species.* (ed. L. Poorter, F. Bongers, F. N. Kouamé & W. D. Hawthorne) Wallingford: CABI Publishing, pp. 41–52.

Brown, N. (1993) The implications of climate and gap microclimate for seedling growth conditions in a Bornean lowland rain forest. *Journal of Tropical Ecology*, **9**, 153–168.

Burgess, S. O., Adams, M. A., Turner, M. C. & Ong, C. K. (1998) The redistribution of soil water by tree root systems. *Oecologia*, **115**, 306–311.

Cao, K. F. (2000) Water relations and gas exchange of tropical saplings during a prolonged drought in a Bornean heath forest, with reference to root architecture. *Journal of Tropical Ecology*, **16**, 101–116.

Castro-Diez, P., Puyravaud, J. P., Cornelissen, J. H. C. & Villar-Salvador, P. (1998) Stem anatomy and relative growth rate in seedlings of a wide range of woody plant species and types. *Oecologia*, **11**, 57–66.

Chiariello, N. (1986) Leaf energy balance in the wet lowland tropics. *Physiological Ecology of Plants of the Wet Tropics* (ed. E. Medina, H. A. Mooney & C. Vázquez-Yánes) The Hague: Dr W. Junk Publishers, pp. 85–98.

Chiariello, N. R., Field, C. B. & Mooney, H. A. (1987) Midday wilting in a tropical pioneer tree. *Functional Ecology*, **1**, 3–11.

Chazdon, R. L. (1988) Sunflecks and their importance to forest understorey plants. *Advances in Ecological Research*, **18**, 1–63.

Coley, P. D. (1983) Herbivory and defence characteristics of tree species in a lowland tropical forest. *Ecological Monographs*, **53**, 209–233.

Condit, R., Hubbell, S. P. & Foster, R. B. (1995) Mortality rates of 205 neotropical tree and shrub species and the impact of a severe drought. *Ecological Monographs*, **65**, 419–439.

(1996) Assessing the response of plant functional types to climatic change in tropical forests. *Journal of Vegetation Science*, **7**, 405–416.

Coomes, D. A. & Grubb, P. J. (2000) Impact of root competition in forests and woodlands: a theoretical framework and review of experiments. *Ecological Monographs*, **70**, 171–207.

Davies, S. J., Palmiotto, P. A., Ashton, P. S., Lee, H. S. & LaFrankie, J. V. (1998) Comparative ecology of 11 sympatric species of *Macaranga* in Borneo: tree distribution in relation to horizontal and vertical resource heterogeneity. *Journal of Ecology*, **86**, 662–673.

Daws, M. I., Mullins, C. E., Burslem, D. F. R. P., Paton, S. R. & Dalling, J. W. (2002) Topographic position affects the water regime in a semideciduous tropical forest in Panamá. *Plant and Soil*, **238**, 79–90.

Engelbrecht, B. M. J. & Kursar, T. A. (2003) Comparative drought resistance of seedlings of 28 species of co-occurring tropical woody plants. *Oecologia*, **136**, 383–393.

Engelbrecht, B. M. J., Dalling, J. W., Pearson, T. R. H. *et al.* (2001) Short dry spells in the wet season increase mortality of tropical pioneer seedlings. *Tropical Ecosystems: Structure, Diversity, and Human Welfare. Proceedings of the International Conference on Tropical Ecosystems* (ed. K. N. Ganeshaiah, R. Uma Shaanker & K. S. Bawa) New Delhi: Oxford and IBH Publishing Co. Pvt. Ltd., pp. 665–669.

Fitter, A. H. (1985) Functional significance of root morphology and root architecture. *Ecological Interactions in Soil* (ed. A. H. Fitter, D. J. Atkinson, D. J. Read & M. B. Usher) Oxford: Blackwell, pp. 87–106.

Fitter, A. H. & Hay, R. K. M. (1987) *Environmental Physiology of Plants*. London: Academic Press.

Gentry, A. H. & Emmons, L. H. (1987) Geographical variation in fertility, phenology, and the composition of the understorey of neotropical forests. *Biotropica*, **19**, 216–227.

Grainger, J. & Becker, P. (2001) Root architecture and root:shoot allocation of shrubs and saplings in a Bruneian heath forest. *Biotropica*, **33**, 368–374.

Grime, J. P. (1977) *Plant Strategies and Vegetation Processes*. Chichester: Wiley.

Grime, J. P., Thompson, K., Hunt, R. *et al.* (1997) Integrated screening validates primary axes of specialisation. *Oikos*, **79**, 259–281.

Guehl, J. M., Domenach, A. M., Bereau, M. *et al.* (1998) Functional diversity in an Amazonian rainforest of French Guyana: a dual isotope approach (δ^{15}N and δ^{13}C). *Oecologia,* **116**, 316–330.

Hall, J. B. & Swaine, M. D. (1981) *Distribution and Ecology of Vascular Plants in a Tropical Rain Forest.* The Hague: Dr. W. Junk Publishers.

Harms, K. E., Condit, R., Hubbell, S. P. & Foster, R. B. (2001) Habitat associations of trees and shrubs in a 50-ha neotropical forest plot. *Journal of Ecology,* **89**, 947–959.

Harms, K. E., Powers, J. S. & Montgomery, R. A. (2004) Variation in small sapling density, understorey cover and resource availability in four neotropical forests. *Biotropica,* **36**, 40–51.

Holmgren, M. (2000) Combined effects of shade and drought on tulip poplar seedlings: trade-off in tolerance or facilitation? *Oikos,* **90**, 67–78.

Howe, H. F. (1990) Survival and growth of juvenile *Virola surinamensis* in Panama: effects of herbivory and canopy closure. *Journal of Tropical Ecology,* **6**, 259–280.

Huante, P., Rincon, E. & Gavito, M. (1992) Root system analysis of seedlings of seven tree species from a tropical dry forest in Mexico. *Trees,* **6**, 77–82.

Huc, R., Ferhi, A. & Guehl, J. M. (1994) Pioneer and late stage tropical rainforest tree species (French Guiana) growing under common conditions differ in leaf gas exchange regulation, carbon isotope discrimination and leaf water potential. *Oecologia,* **99**, 297–305.

Hutchings, M. J. & De Kroon, H. (1994) Foraging in plants: the role of morphological plasticity in resource acquisition. *Advances in Ecological Research,* **25**, 159–238.

Jackson, P. C., Cavelier, J., Goldstein, G., Meinzer, F. C. & Holbrook, N. M. (1995) Partitioning of water sources among plants of a lowland tropical forest. *Oecologia,* **101**, 197–203.

Jetten, V. G. (1994) *Modelling the Effects of Logging on the Water Balance of a Tropical Rain Forest. A Study in Guyana.* Tropenbos Series 6. Wageningen: The Tropenbos Foundation.

King, D. A. (1990) Allometry of saplings and understorey trees of a Panamanian forest. *Functional Ecology,* **4**, 27–32.

(1994) Influence of light level on the growth and morphlogy of saplings in a Panamanian forest. *American Journal of Botany,* **81**, 948–957.

(1995) Allometry and life history of tropical trees. *Journal of Tropical Ecology,* **12**, 25–44.

(1998a) Influence of leaf size on tree architecture: first branch height and crown dimensions in tropical rain forest trees. *Trees,* **12**, 438–445.

(1998b) Relationship between crown architecture and branch orientation in rain forest trees. *Annals of Botany,* **82**, 1–7.

King, D. A. & Maindonald, J. H. (1999) Tree architecture in relations to leaf dimensions and tree stature in temperate and tropical rain forests. *Journal of Ecology,* **87**, 1012–1024.

Kitajima, K. (1994) Relative importance of photosynthetic traits and allocation patterns as correlates of seedling shade tolerance of 13 tropical trees. *Oecologia,* **98**, 419–428.

(1996) Ecophysiology of tropical tree seedlings. *Tropical Forest Plant Ecophysiology* (ed. S. S. Mulkey, R. L. Chazdon & A. P. Smith). New York: Chapman & Hall, pp. 559–597.

Kobe, R. K. (1997) Carbohydrate allocation to storage as a basis of interspecific variation in sapling survivorship and growth. *Oikos,* **80**, 226–233.

Kohyama, T. (1993) Size-structured tree populations in gap-dynamic forest: the forest architecture hypothesis for the stable coexistence of species. *Journal of Ecology,* **81**, 131–143.

Kohyama, T. & Hotta, M. (1990) Significance of allometry in tropical saplings. *Functional Ecology*, **4**, 515–521.

Kohyama, T., Suzuki, E., Partomihardjo, T., Yamada, T. & Kubo, T. (2003) Tree species differentiation in growth, recruitment and allometry in relation to maximum height in a Bornean mixed dipterocarp forest. *Journal of Ecology*, **91**, 797–806.

Koop, H. & Sterck, F. J. (1994) Light penetration though structurally complex forest canopies: an example of a lowland tropical rainforest. *Forest Ecology and Management*, **69**, 111–122.

Leigh, E. G. (1999) *Tropical Forest Ecology. A View from Barro Colorado Island*. Oxford: Oxford University Press.

Lieberman, D., Lieberman, M., Hartshorn, G. & Peralta, R. (1985) Growth rates and age–size relationships of tropical wet forest trees in Costa Rica. *Journal of Tropical Ecology*, **1**, 97–109.

Lusk, C. H. (2002) Leaf area accumulation helps juvenile evergreen trees tolerate shade in a temperate rainforest. *Oecologia*, **132**, 188–196.

Medina, E. (1999) Tropical forests: diversity and function of dominant life-forms. *Handbook of Functional Plant Ecology* (ed. F. Pugnaire & F. Valladares). New York: Marcel Dekker, pp. 407–448.

Meinzer, F. C., Goldstein, G., Jackson, P., Holbrook, N. M., Gutiérrez, M. V. & Cavalier, J. (1995) Environmental and physiological transpiration in tropical forest gap species: the influence of boundary layer and hydraulic properties. *Oecologia*, **101**, 514–522.

Oldeman, R. A. A. & van Dijk, J. (1991) Diagnosis of the temperament of tropical rain forest trees. *Rain Forest Regeneration and Management* (ed. A. Gómez-Pompa, T. C. Whitmore & M. Hadley). Man and the Biosphere Series 6. Paris: UNESCO, pp. 21–65.

Ostertag, R. (1998) Belowground effects of canopy gaps in a tropical wet forest. *Ecology*, **79**, 1294–1304.

Osunkoya, O. O. & Ash, J. E. (1991) Acclimation to a change in light regime in seedlings of six Australian rainforest tree species. *Australian Journal of Botany*, **39**, 591–605.

Parker, G. G. (1985) The effects of disturbance on water and solute budgets of hillslope tropical rain forest in northeastern Costa Rica. Unpublished Ph.D. thesis. University of Georgia, Athens.

Paz, H. (2003) Root/shoot allocation and root architecture in seedlings: variation among forest sites, microhabitats, and ecological groups. *Biotropica*, **35**, 318–332.

Pearcy, R. W. & Valladares, F. (2001) Resource acquisition by plants: the role of crown architecture. *Physiological Plant Ecology* (ed. M. Press, J. D. Scholes & M. G. Barker). London: Blackwell Scientific Publication, pp. 45–66.

Pearcy, R. W. & Yang, W. (1998) The functional morphology of light capture and carbon gain in the Redwood forest understorey plant *Adenocaulon bicolor* Hook. *Functional Ecology*, **12**, 543–552.

Peña-Claros, M. (2003) Changes in forest structure and species composition during secondary succession in the Bolivian Amazon. *Biotropica*, **35**, 450–461.

Poorter, H. & Nagel, O. (2000) The role of biomass allocation in the growth response of plants to different levels of light, CO_2, nutrients and water: a quantitative review. *Australian Journal of Plant Physiology*, **27**, 595–607.

Poorter, L. (1998) Seedling growth of Bolivian rain forest tree species in relation to light and water availability. Unpublished Ph.D. thesis, Utrecht University.

(1999) Growth responses of fifteen rain forest tree species to a light gradient: the relative importance of morphological and physiological traits. *Functional Ecology*, **13**, 396–410.

Poorter, L. & Arets, E. J. M. M. (2003) Light environment and tree strategies in a Bolivian tropical moist forest; a test of the light-partitioning hypothesis. *Plant Ecology*, **166**, 295–306.

Poorter, L. & Hayashida-Oliver, Y. (2000) Effects of seasonal drought on gap and understorey seedlings in a Bolivian moist forest. *Journal of Tropical Ecology*, **16**, 481–498.

Poorter, L. & Oberbauer, S. F. (1993) Photosynthetic induction responses of two rainforest tree species in relation to light environment. *Oecologia*, **96**, 193–199.

Poorter, L., Bongers, F., Sterck, F. J. & Wöll, H. (2003) Architecture of 53 rain forest tree species differing in adult stature and shade tolerance. *Ecology*, **84**, 602–608.

Poorter, L., Bongers, F., Kouamé, F. N. & Hawthorne, W. D. (2004) *Biodiversity of West African Forests. An Ecological Atlas of Woody Plant Species*. Wallingford: CABI Publishing.

Popma, J. & Bongers, F. (1991) Acclimation of seedlings of three tropical rain forest species to changing light availability. *Journal of Tropical Ecology*, **7**, 85–97.

Reekie, E. G. & Bazzaz, F. A. (1989) Competition and patterns of resource use among seedlings of five tropical trees grown at ambient and elevated CO_2. *Oecologia*, **79**, 212–222.

Reekie, E. G. & Wayne, P. (1992) Leaf canopy display, stomatal conductance, and photosynthesis in seedlings of three tropical pioneer tree species subjected to drought. *Canadian Journal of Botany*, **70**, 2334–2338.

Reich, P. B., Ellsworth, D. S. & Uhl, C. (1995) Leaf carbon and nutrient assimilation and conservation in species of differing successional status in an oligotrophic Amazonian forest. *Functional Ecology*, **9**, 65–76.

Reich, P. B., Tjoelker, M. G., Walters, M. B., Vanderklein, D. W. & Buschena, C. (1998) Close association of RGR, leaf and root morphology, seed mass and shade tolerance in seedlings of nine boreal tree species grown in high and low light. *Oecologia*, **114**, 471–482.

Reich, P. B., Wright, I. J., Cavender-Bares, J. *et al.* (2003) The evolution of plant functional variation: traits, spectra, and strategies. *International Journal of Plant Sciences*, **164**, 143–164

Rijkers, T., de Vries, P. J., Pons, T. L. & Bongers, F. (2001) Photosynthetic induction in saplings of three shade-tolerant tree species: comparing understorey and gap habitats in a French Guiana rain forest. *Oecologia*, **125**, 331–340.

Rose, S. A. (2000) Seeds, seedlings and gaps – size matters. A study in the tropical rain forest of Guyana. Ph.D. thesis, Utrecht University.

Ryser, P. (1996) The importance of tissue density for growth and life span of leaves and roots: a comparison of five ecologically contrasting grasses. *Functional Ecology*, **10**, 717–723.

Sack, L. & Grubb, P. J. (2002) The combined effects of deep shade and drought on the growth and biomass allocation of shade-tolerant woody seedlings. *Oecologia*, **131**, 175–185.

Sims, D. A., Gebauer, R. L. E. & Pearcy, R. W. (1994) Scaling sun and shade photosynthetic acclimation of *Alocasia macrorrhiza* to whole-plant performance. II. Simulation of carbon balance and growth at different photon flux densities. *Plant, Cell and Environment*, **17**, 889–900.

Smith, T. & Huston, M. (1989) A theory of the spatial and temporal dynamics of plant communities. *Vegetatio*, **83**, 49–69.

Sterck, F. J. & Bongers, F. (1998) Ontogenetic changes in size, allometry, and mechanical design of tropical rain forest trees. *American Journal of Botany*, **85**, 266–272.

Sterck, F. J., Clark, D. B., Clark, D. A. & Bongers, F. (1999) Light fluctuations, crown traits, and response delays for tree saplings in a Costa Rican lowland rain forest. *Journal of Tropical Ecology*, **15**, 83–95.

Swaine, M. D. & Becker, P. (1999) Woody life-form composition and association on rainfall and soil fertility gradients in Ghana. *Plant Ecology*, **145**, 167–173.

Swaine, M. D. & Whitmore, T. C. (1988) On the definition of ecological species groups in tropical forests. *Vegetatio*, **75**, 81–86.

Thomas, S. C. (1996a) Relative size at onset of maturity in rain forest trees: a comparative analysis of 37 Malaysian species. *Oikos*, **76**, 145–154.

(1996b) Asymptotic height as a predictor of growth and allometric characteristics in Malaysian rain forest trees. *American Journal of Botany*, **83**, 556–566.

Tobin, M. F., Lopez, O. R. & Kursar, T. A. (1999) Responses of tropical understorey plants to a severe drought: tolerance and avoidance of water stress. *Biotropica*, **31**, 570–578.

Turner, I. M. (2001) *The Ecology of Trees in the Tropical Rain Forest*. Cambridge: Cambridge University Press.

Turner, I. M., Raich, J. W., Gong, W. K., Ong, J. E. & Whitmore, T. C. (1992) The dynamics of Pantei Acheh forest reserve: a synthesis of recent research. *Malayan Nature Journal*, **45**, 166–174.

Tyree, M. T. & Sperry, J. S. (1989) Vulnerability of xylem to cavitation and embolism. *Annual Review of Plant Physiology and Molecular Biology*, **40**, 19–38.

Tyree, M. T., Davis, R. D. & Cochard, D. (1994) Biophysical perspectives of xylem function: is there a tradeoff of hydraulic efficiency for vulnerability to disfunction? *IAWA Journal*, **15**, 335–360.

Tyree, M. T., Velez, V. & Dalling, J. W. (1998) Growth dynamics of root and shoot hydraulic conductance in seedlings of five neotropical tree species: scaling to show possible adaptation to different light regimes. *Oecologia*, **114**, 293–298.

Tyree, M. T., Engelbrecht, B. M. J., Vargas, G. & Kursar, T. A. (2003) Desiccation tolerance of five tropical seedlings in Panama. Relationship to a field assessment of drought performance. *Plant Physiology*, **132**, 1439–1447.

Valladares, F. (1999) Architecture, ecology, and evolution of plant crowns. *Handbook of Functional Plant Ecology* (ed. F. Pugnaire & F. Valladares). New York: Marcel Dekker, pp. 121–194.

Valladares, F., Allen, M. T. & Pearcy, R. W. (1997) Photosynthetic responses to dynamic light under field conditions in six tropical rainforest shrubs occurring along a light gradient. *Oecologia*, **111**, 505–514.

Valladares, F., Skillman, J. B. & Pearcy, R. W. (2002) Convergence in light capture efficiencies among tropical forest understorey plants with contrasting crown architectures: a case of morphological compensation. *American Journal of Botany*, **89**, 1275–1284.

van Dam, O. 2001. *Forest Filled with Gaps. Effects of Gap Size on Water and Nutrient Cycling in Tropical Rain Forest. A Study in Guyana.* Tropenbos-Guyana Series 10. Georgetown: Tropenbos-Guyana Programme.

van den Boogaard, R. & Villar, R. (1998) Variation in growth and water-use efficiency – a comparison of *Aegilops* L. species and *Triticum aestivum* L. cultivars. *Inherent Variation in Plant Growth. Physiological Mechanisms and Ecological Consequences* (ed. H. Lambers, H. Poorter & M. M. I. van Vuuren). Leiden: Backhuys Publishers, pp. 289–308.

van Rompaey, R. S. A. R. (1993) Forest gradients in West Africa. Ph.D. thesis, Wageningen Agricultural University.

Veenendaal, E. M. & Swaine, M. D. (1998) Limits to tree species distributions in lowland tropical rainforest. *Dynamics of Tropical Communities* (ed. D. M. Newbery, H. H. T. Prins & N. D. Brown). Oxford: Blackwell Scientific Ltd., pp. 163–192.

Veenendaal, E. M., Swaine, M. D., Blay, D., Yelifari, N. B. & Mullins, C. E. (1996a) Seasonal and long term soil water regime in West African tropical forest. *Journal of Vegetation Science*, **7**, 473–482.

Veenendaal, E. M., Swaine, M. D., Agyeman, V. K. *et al.* (1996b) Differences in plant and soil-water relations in and around a forest gap in West Africa during the dry season may influence seedling establishment and survival. *Journal of Ecology*, **84**, 83–90.

Veneklaas, E. J. & Poorter, L. (1998) Growth and carbon partitioning of tropical tree seedlings in contrasting light environments. *Inherent Variation in Plant Growth. Physiological Mechanisms and Ecological Consequences* (ed. H. Lambers, H. Poorter & M. M. I. van Vuuren). Leiden: Backhuys Publishers, pp. 337–361.

Vitousek, P. M. & Denslow, J. S. (1986) Nitrogen and phosphorus availability in treefall gaps of a lowland tropical rainforest. *Journal of Ecology*, **74**, 1167–1178.

Walsh, R. P. D. & Newbery, D. M. (1999) The ecoclimatology of Danum, Sabah, in the context of the world's rain forest regions, with particular reference to dry periods and their impact. *Philosophical Transactions of the Royal Society of London B*, **354**, 1869–1883.

Webb, C. O. & Peart, D. R. (2000) Habitat associations of trees and seedlings in a Bornean rain forest. *Journal of Ecology*, **88**, 464–478.

Weiner, J. (1990) Asymmetric competition in plant population. *Trends in Ecology and Evolution*, **5**, 360–364.

Whitmore, T. C. (1996) A review of some aspects of tropical rain forest seedling ecology with suggestions for further enquiry. In *The Ecology of Tropical Forest Tree Seedlings* (ed. M. D. Swaine). Man and the Biosphere Series 17. Paris: UNESCO, pp. 3–39.

Wright, S. J. (1991) Seasonal drought and the phenology of understorey shrubs in a tropical moist forest. *Ecology*, **72**, 1643–1657.

(1992) Seasonal drought, soil fertility and the species density of tropical plant communities. *Trends in Ecology and Evolution*, **7**, 260–263.

(2002) Plant diversity in tropical forests: a review of mechanisms of species coexistence. *Oecologia*, **130**, 1–14.

Wright, S. J., Machado, J. L., Mulkey, S. S. & Smith, A. P. (1992) Drought acclimation among tropical forest shrubs (Psychotria, Rubiaceae). *Oecologia*, **89**, 457–463.

Wright, S. J., Muller-Landau, H. C., Condit, R. & Hubbell, S. P. (2003) Gap-dependent recruitment, realized vital rates, and size distributions of tropical trees. *Ecology*, **84**, 3174–3185.

Yoda, K. (1974) Three-dimensional distribution of light intensity in a tropical rainforest of West Malaysia. *Japanese Journal of Ecology*, **24**, 247–254.

Zagt, R. J. & Werger, M. J. A. (1998) Community structure and demography of primary species in tropical rain forest. *Dynamics of Tropical Communities* (ed. D. M. Newbery, H. H. T. Prins & N. D. Brown). Oxford: Blackwell Scientific Ltd, pp. 193–220.

CHAPTER THREE

Role of life-history trade-offs in the equalization and differentiation of tropical tree species

JAMES W. DALLING
University of Illinois and Smithsonian Tropical Research Institute
DAVID F. R. P. BURSLEM
University of Aberdeen

Introduction

Early observations of the remarkable diversity in life history and morphology exhibited by tropical plants have strongly influenced our views on how these species coexist. Initially plants were classified according to size and life form, and subsequently into ecological species groups that recognized the importance of variation in light availability associated with the forest growth cycle (Richards 1952; Swaine & Whitmore 1988; Burslem & Swaine 2002). These ecological species groups reflect the existence of adaptive strategies that trees adopt during regeneration, and are the consequence of unavoidable trade-offs among suites of traits influencing growth, survival and fecundity.

In tropical forests, light availability has typically been identified as the primary limiting condition to growth. Adaptive strategies influencing the colonization of canopy openings and the capture or use of light are manifested as combinations of traits influencing dispersal, germination, seedling establishment, and allocation of resources to growth, storage and defence (van Steenis 1958; Budowski 1965; Whitmore 1975; Bazzaz & Pickett 1980; Coley *et al.* 1985; Kitajima 1994). This review addresses two issues: first, the extent to which individual trade-offs linking these traits can lead to the differentiation of tree species accounting for interspecific differences in the requirements for light and other resources; and second, the extent to which the action of multiple opposing trade-offs has a contrary effect of equalizing overall recruitment success so that no particular trait combination provides a recruitment advantage.

Classical views of species coexistence are founded upon niche differentiation as the mechanism preventing competitive exclusion, with limiting similarity in niche requirements providing an upper bound on local species richness (Hutchinson 1959; MacArthur & Levins 1967; Grubb 1977). Interspecific competition for limiting resources might be expected to select for adaptations that

Biotic Interactions in the Tropics: Their Role in the Maintenance of Species Diversity, ed. D. F. R. P. Burslem, M. A. Pinard and S. E. Hartley. Published by Cambridge University Press. © Cambridge University Press 2005.

maximize resource capture. However, specialized adaptations often invoke trade-offs. These may act by reducing the ability to acquire other potentially limiting resources, or may constrain the effectiveness of resource capture to particular environmental conditions. Trade-offs might therefore be expected to result in the partitioning of supply gradients of resources such as light and soil moisture (e.g. Denslow 1980; Smith & Huston 1989). Furthermore, selection affecting the acquisition of light versus moisture and nutrients may have particularly strong effects on niche differentiation as allocation to above- and below-ground resource capture may be in direct conflict (Tilman 1988).

Accumulating evidence from work on Barro Colorado Island (BCI), Panama, may be interpreted as showing that many tree species are similar in their shade-tolerance (Welden *et al.* 1991; Hubbell & Foster 1992), and that light-gradient-partitioning therefore plays a minor role in maintaining tree diversity (Hubbell *et al.* 1999). However, we caution against a premature rejection of the 'niche-differentiation' or 'niche-assembly' view for the structure of tropical tree communities based on these findings. Early acceptance of the paradigm that species coexisted through differences in light requirements precluded a more thorough examination of how species acquire and use other potentially limiting resources. Variation in supply rates of soil resources, and in the physical and biotic conditions required for seedling establishment, all contribute to variation in plant performance and may play a more important role than light availability in determining recruitment patterns, i.e. the spatial patterns of new individuals in the community.

However, in addition to the potential role played by performance-based trade-offs in aiding coexistence of species through the *differentiation* of regeneration requirements, life-history-based trade-offs may also operate with the paradoxical effect of *equalizing* the recruitment success of species. Equalizing trade-offs may arise when no recruitment advantage accrues from any particular trait combination, and may occur when fitness advantages of particular traits are balanced across life-history stages or habitat types. As a consequence, no species has an overall per-capita fitness advantage over any other (Levins 1968). Under this scenario, diversity is not maintained by niche-partitioning in the face of competitive exclusion. Instead, exclusion is avoided by competitive equivalence.

Competitive equivalence among coexisting species is an underlying assumption of neutral theories of species coexistence based on demographic stochasticity (Hubbell 1979, 2001). In addition to the equalizing effects of life-history trade-offs, Hubbell and colleagues have also argued that equivalence may arise in diverse tropical forests as a consequence of restricted seed dispersal coupled with unpredictability in the biotic composition of local neighbourhoods. Under these conditions, opportunities for specialization to particular resource conditions may be limited and thus tree species may show convergent responses to long-term average environmental conditions (Hubbell & Foster 1986; Hubbell *et al.* 1999). Determining how and when competitive equivalence occurs is therefore a critical issue in ecology.

Here we examine the role that performance-based trade-offs play in the differentiation of species resource requirements, and the role that life-history-based trade-offs play in the equalization of species recruitment success. Much of the evidence we present is based on our own studies of the regeneration ecology of neotropical pioneer species. We start by reviewing the evidence that performance-based trade-offs contribute to the partitioning of resource gradients from germination to seedling growth. We concentrate initially on the role of light availability, and then examine the more limited evidence that differences in requirements for soil-borne resources contribute to species coexistence. We then address the equalizing role of trade-offs by focusing on the early establishment stages of plant life-histories, which play a critical role in determining species abundance and distribution (Grubb 1977; Swaine *et al.* 1997). We show how trade-offs couple dispersal, germination, emergence and seedling establishment leading to similar recruitment success for species with wide variation in seed mass.

Performance trade-offs affecting germination and emergence

The seeds of most tree species in seasonally moist and aseasonal tropical forests lack dormancy and germinate within a few months of dispersal, regardless of environmental conditions (Whitmore 1975; Garwood 1983). Nonetheless, most light-demanding species, and a few small-seeded shade-tolerant species, use a variety of environmental cues to detect the presence of canopy gaps and litter-free microsites on the forest floor (Vázquez-Yanes & Orozco-Segovia 1993; Metcalfe 1996). These cues include the spectral distribution of light (Vázquez-Yanes & Smith 1982), the amplitude of diurnal temperature fluctuations (Vázquez-Yanes & Orozco-Segovia 1982) and pulses in soil nitrate concentrations (Daws *et al.* 2002a). Here we show how differences in the effectiveness of these cues under varying environmental conditions may contribute to habitat-partitioning among species.

Photoblastic species respond to shifts in the composition of light in the red and far-red spectra and are typically small-seeded (seed mass < 2 mg; Pearson *et al.* 2002). Light is an appropriate germination cue for these species because it penetrates only a few millimetres through moist soil, coinciding with the maximum burial depth from which seedlings of these species can successfully emerge (Pearson *et al.* 2002, 2003a). In addition, red light is absorbed by dead leaves that may be present on the soil surface (Vázquez-Yanes *et al.* 1990), and that provide a significant barrier to successful emergence and establishment for species with minimal seed reserves (Molofsky & Augspurger 1992; Metcalfe & Grubb 1997).

In contrast, larger-seeded light-demanding species, with the capacity to emerge successfully through several centimetres of soil, typically lack a germination response to light (Pearson *et al.* 2002). The germination cues used by some of these species remain to be determined, but many are sensitive to temperature fluctuations (Vázquez-Yanes 1974; Vázquez-Yanes & Orozco-Segovia 1982; Pearson *et al.* 2002). Elevated diurnal temperatures can be recorded at depths of >10 cm below the soil surface in well-illuminated gap microsites (Pearson *et al.* 2002). Elevated temperatures, however, only occur in gap microsites exposed to direct sun for long enough periods for soil warming to occur.

Constraints imposed by the effectiveness of these germination cues should result in predictable differences in the gap sites where light-demanding species can recruit. Large-seeded light-demanding species should germinate in large canopy gaps with open understorey allowing light penetration to soil level (Pearson *et al.* 2003b). In contrast, small-seeded species should successfully germinate across a range of gap sizes and may successfully colonize edges of gaps shaded from direct radiation. Furthermore, the capacity of photoblastic seeds to discriminate small differences in light availability from shifts in red:far-red ratios potentially allows for finer-scale niche-partitioning among species with different minimum light requirements for seedling establishment and growth (Daws *et al.* 2002a). These constraints imposed by seed size on germination behaviour appear to be congruent with size-dependent constraints on post-germination recruitment success, as dessication risk is highest for small seedlings establishing from shallow-buried seeds in high-irradiance microsites (Engelbrecht *et al.* 2001; Daws 2002).

Performance trade-offs affecting establishment and growth
Trade-offs partitioning gradients of light availability

Initial observations that tropical trees could be classified into broad regeneration guilds according to their light requirements raised hopes among temperate and tropical ecologists that fine-scale partitioning of light gradients could be responsible for much of the differentiation and coexistence of species (Ricklefs 1977; Pickett 1980; Denslow 1980, 1987). Light-gradient partitioning could be mediated by two alternate sets of trade-offs. Selection on suites of morphological and physiological traits that influence net carbon assimilation could result in narrow ranges of light conditions under which relative growth rate (RGR) is optimized. Different configurations of these traits among species could result in shifts in the ranking of species according to RGR along the light gradient (Latham 1992). If variation in RGR is sufficient to overcome the differences in initial seedling size and time of emergence among species, then recruitment patterns might vary predictably with light availability.

Alternatively, species may indirectly partition light gradients as a consequence of differing growth and mortality rates (Kitajima 1994; Kobe 1999). Competitive interactions among plants dictate that the successful colonization of high-light microsites in the forest requires rapid height growth. Once the limit of carbon-fixation efficiency is reached this can only be achieved by preferentially investing resources into new leaf area and leaf support tissue at the expense of allocation to structural support and defence (so wood density is low and leaves are thin). In contrast, under low-light conditions, where assimilation rates are low relative to the resources needed for new tissue construction, allocation to structural support and chemical defences that lower rates of damage from physical hazards, herbivores and pathogens should be favoured (Coley *et al.* 1985). These conflicting selective pressures predict that shade-adapted species lack competitive ability in high-light conditions while sun-adapted species are unable to persist for long in the shade. This trade-off is most clearly manifested as a positive relationship across species between growth rate at high irradiance and mortality rate in the shade (Hubbell & Foster 1992; Kitajima 1994).

Even though seedlings of shade-tolerant trees are clearly limited in their growth by light availability in the forest understorey, there is little evidence that differences in light response contribute to their coexistence. While some differences in growth and mortality rates have been reported for seedlings of shade-tolerant trees (Zagt 1997; Montgomery & Chazdon 2002; Bloor & Grubb 2003), the only published community-level study showed that saplings of most tropical rain forest trees had uniformly low absolute growth and mortality rates in the shade (Welden *et al.* 1991). Small differences among species in physiological responses to light availability that can be detected in short-term controlled experiments may also be unlikely to influence the outcome of interspecific competition among variable-sized individuals in the dynamic light regime of forest understorey (Barker *et al.* 1997; Rose & Poorter 2003; Clark *et al.* 2003). Therefore, it remains to be determined whether the small absolute differences in the performance of tropical forest seedlings and saplings are important when integrated over the long time intervals that these plants persist in the shade.

Differences among coexisting species in the architecture and allometry of seedlings and saplings provide an alternative perspective on the mechanism of differential species persistence in the shade over longer time scales (Kohyama 1987; King 1990; Kohyama & Hotta 1991; King 1994). Shade-tolerant species may be classified along a spectrum reflecting allocation to height growth versus lateral canopy development based on the allometric properties of individuals growing in the shade (Kohyama 1987). Species that grow quickly in height relative to their rate of canopy area development ('optimists' *sensu* Kohyama) will be at a competitive advantage in sites where irradiance increases rapidly with height and when canopy openness increases at the time of gap creation. In contrast, optimists are predicted to show a lesser ability to persist for long periods in the shade than species that invest in lateral spread of the canopy to maximize light capture and carbon gain (Kohyama's 'pessimists'). These predictions are supported by studies demonstrating positive correlations across species between maximum adult height and sapling height at a given diameter (Kohyama *et al.*

2003; Poorter *et al.* 2003). The recognition that species differ in architecture and allometry, however, does not provide direct evidence that these differences contribute to species coexistence. Further research may demonstrate that the trade-off between height growth and lateral canopy development in shade-tolerant species is another expression of the same trade-off between growth in sun versus persistence in the shade (Hubbell & Foster 1992).

A more restricted interpretation of light-gradient partitioning is that species coexist through differences in their recruitment success in gaps. The hypothesis proposes that species respond in different ways to gaps according to gap size or orientation, or according to microsites present within them (Denslow 1980, 1987; Pickett 1980; Orians 1982). The hypothesis has been tested by a number of long-term studies of seedling regeneration in natural and artificial gaps (e.g. Brokaw 1985; Uhl *et al.* 1988; Brown & Whitmore 1992), but fails to find support as a general mechanism of species coexistence among tropical forest trees (Brown & Jennings 1998; Hubbell *et al.* 1999; Brokaw & Busing 2000). Gap-partitioning, however, may be more important as a mechanism that aids coexistence among the group of light-demanding species, defined according to their requirement for high-light microsites for germination or establishment (Brokaw 1987). These species represent a small proportion of overall tree diversity (10%–20% of tree species; Dalling *et al.* 1998; Molino & Sabatier 2001), but recruit in microsites where resource availability is most variable. Total irradiance, the spectral composition of light, and the persistence of high-irradiance conditions generally covary with the size of canopy disturbances (Barton *et al.* 1989; Brown 1993; van der Meer *et al.* 1994). Thus, if species are adapted to particular irradiance regimes then they might be expected to show differential recruitment success along a gradient of canopy gap sizes.

We have looked in detail at whether differences in light requirements contribute to species coexistence of light-demanding species at the Barro Colorado Island Nature Monument (BCNM), in central Panama. A pot experiment conducted with thirteen fast-growing species failed to show much evidence for rank shifts in RGRs over the range of light conditions under which seedlings of these species are observed to recruit in the field (Dalling *et al.* 1999, 2004; Fig. 3.1). Furthermore, many of the most abundant light-demanding species had similar RGR and were rather unresponsive in growth to varying irradiance conditions. The fastest-growing species, *Cecropia*, *Cordia*, *Ochroma* and *Trema*, however, did show more variation in growth response.

In support of the growth–mortality trade-off hypothesis, our observations of the same species recruiting in natural (Dalling *et al.* 1998) and artificially created treefall gaps (Dalling & Hubbell 2002) show that growth rate is positively correlated with mortality rate, and that much of the variance in mortality could be attributed to differential susceptibility to browsing and stem-boring herbivores (Fig. 3.2). These observations suggest that a growth–mortality trade-off mediated by biotic enemies could result in gap-size partitioning among light-demanding

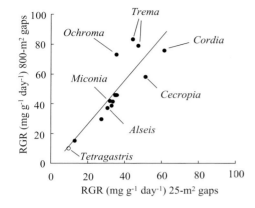

Figure 3.1 Correlation of mean relative growth rate (RGR) calculated for seedlings grown in pots in a shade house experiment simulating large (800 m²) gaps and small (25 m²) simulated gaps. Data are for thirteen light-demanding taxa (closed symbols) and one shade-tolerant species (open symbol) from Barro Colorado Island Nature Monument, Panama. For more details see Dalling *et al.* (2004).

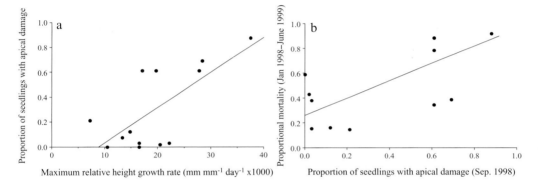

Figure 3.2 Proportional mortality of 1-year-old seedlings of pioneer species censused over 18 months in five artificially created gaps on Barro Colorado Island Panama, (a) regressed against the species-specific maximum relative height growth rate, and (b) the proportion of seedlings of each species exhibiting apical shoot damage (from stem-boring insects and mammalian browsers) and maximum relative height growth rate. Redrawn from Dalling & Hubbell (2002).

species, as has been shown by Brokaw (1987). If fast-growing species are restricted to large canopy gaps by the assimilation rate needed to replace tissues lost to herbivores, then protection of seedlings from herbivores should result in a reduction in light requirements for growth and survival. We tested this hypothesis by transplanting seedlings of three light-demanding species inside and outside mosquito-netting exclosures in artificially created small (25 m²) and large (225 m²) gaps. The three species chosen differed in growth rate and minimum gap size requirements for sapling survival: *Miconia argentea* < *Cecropia insignis* < *Trema micrantha* (Brokaw 1987). We found that while the exclosures reduced foliar herbivory and increased leaf area growth of *Trema* and *Miconia*, they had no sustained effect on seedling survivorship (Fig. 3.3, Pearson *et al.* 2003b).

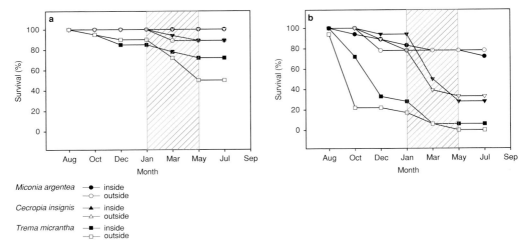

Figure 3.3 Percentage survival over time in large (a) and small (b) gaps for seedlings of *Miconia argentea*, *Cecropia insignis* and *Trema micrantha*. Filled symbols: seedlings inside enclosures; open symbols: seedlings outside enclosures. The experiment was conducted in secondary forest in the Barro Colorado Island National Monument, Panama. Hashed box indicates dry season. For more details see Pearson *et al.* (2003b).

In conclusion, light-demanding species show differences in growth and mortality rates, and in susceptibility to herbivores, but our field experiments in Panama fail to make a direct link between herbivory and variation in mortality rates in high and low light. The poor performance of the saplings of the fastest-growing species under relatively low light in the field may instead reflect the increased whole-plant light-compensation point of larger plants with greater respiratory demands associated with higher growth rates, biomass allocation to support tissue, and tissue turnover rates (Givnish 1988; King 1994). These results emphasize the need for more long-term field experiments of growth performance to complement the short-term growing-house experiments that have mostly been used so far to characterize light requirements of different species.

Trade-offs partitioning gradients of nutrient availability
Species distributions in tropical forests may respond to variation in the availability of soil-borne resources as well as light. At landscape scales, variation in underlying geology and topography may generate gradients in both soil moisture and nutrient supply (Daws *et al.* 2002b; Baker *et al.* 2003). Analyses of tree distributions along transects and in large census plots using statistical techniques that account for spatial autocorrelation in recruitment have found that habitat associations with topographic or soils variables can be quite common (Harms *et al.* 2001; Debski *et al.* 2002; Phillips *et al.* 2003; Tuomisto *et al.* 2003). At more local scales, variation in nutrient availability may arise through disturbance,

such as when sub-soils are exposed at the surface in tip-up mounds (Putz 1983) and on landslides (Dalling & Tanner 1995; Fetcher *et al.* 1996). A transient reduction in fine-root density in canopy gaps may also impose heterogeneity in nutrient concentrations in soil solution (Vitousek & Denslow 1986; Uhl *et al.* 1988), while fine-scale heterogeneity in the amount and chemistry of litterfall may influence nutrient-cycling and nutrient availability to plants that acquire nutrients directly from decomposing litter. Despite these patterns, however, the extent to which seedling growth and survival are limited by nutrients or water under field conditions remains poorly understood. Here we review evidence for resource-partitioning resulting from trade-offs in response to below-ground resource availability.

Most experimental tests of nutrient limitation of tropical tree growth have been conducted using seedlings grown in pots of amended forest soil, and are typically limited to a small number of species. One exception is the study by Huante *et al.* (1995), which examined the growth and allocation responses of thirty-four woody plant species from Mexican dry deciduous forest grown in sand containing high and low concentrations of N, P and K. This experiment failed to show any evidence for a direct trade-off among species in growth rates, as the relative growth rate (RGR) of seedlings growing in pots at low nutrient supply was positively correlated with RGR at high nutrient supply. However, species RGR in this experiment was negatively correlated with seed mass and positively correlated with specific leaf area in both nutrient treatments. This suggests that the 10-week growth period over which the experiment was run may have been too short to overcome initial differences in RGR that are constrained by seed mass and specific leaf area (Marañon & Grubb 1993; Grubb 1998).

A comparable but much longer (50-week) experiment, run by Metcalfe *et al.* (2002), gave rather different results. In this experiment, seedlings of six Australian rainforest tree species were grown in soil diluted to varying degrees with sand that contained low concentrations of N and P but relatively high concentrations of the major nutrient cations. Two of the six species that showed a significantly reduced growth rate in a medium containing only 5% soil compared with 50% soil were also two of the three species with the greatest dry mass increment in the 50% soil. In contrast, the species that was least responsive to soil dilution had the lowest growth rates in the richer soil mixes. These results provide some suggestion of cross-overs in species growth rates along a gradient of nutrient availability, but the role of trade-offs driving rank shifts in RGR in relation to nutrient supply remains unresolved, as it is for irradiance (Sack & Grubb 2001; Kitajima & Bolker 2003).

Studies of plant performance in pot experiments can perhaps be more readily interpreted when they are linked to species distributions along gradients of resource availability in the field (Lawrence 2003). Gunatilleke *et al.* (1997) compared eight species of *Shorea* section *Doona* (Dipterocarpaceae) that grow

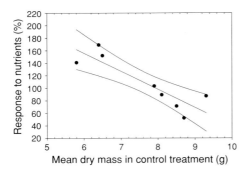

Figure 3.4 Relationship between mean dry mass (g) in an unfertilized control treatment and mean maximum dry mass yield (% of control) in response to nutrient addition for seedlings of eight species of *Shorea* section *Doona* grown in pots for 24 months at Sinharaja forest reserve, Sri Lanka. For more details see Gunatilleke *et al.* (1997).

sympatrically in lowland evergreen rain forest in Sri Lanka. They found that the four species with distributions biased towards relatively nutrient-rich soils on low slopes and valleys were more responsive to nutrient addition than the four species that occur naturally on less nutrient-rich soil. Responsiveness to nutrient addition (defined as the maximum increase in dry mass in response to nutrient addition as a percentage of the unfertilized control value) was also negatively correlated with mean dry mass in the control treatment across these eight species ($r = -0.881$; $P < 0.01$, Fig. 3.4). This 'trade-off', however, did not entirely reflect species distributions along the soil fertility gradient in the forest. Instead, a range of strategies for achieving a size advantage in response to nutrient addition were displayed among the four species of richer soil: large seed mass combined with fast relative growth rates (*S. megistophylla*), a high 'responsiveness' to nutrients (*S. trapezifolia* and *S. cordifolia*) and intermediate growth rate and responsiveness (*S. congestiflora*).

Shade-house evaluations of responses to soil resource gradients should, however, be interpreted with caution, as growth and allocation in pots may differ significantly from growth under similar mean light conditions in the field. This is largely because of variation in below-ground resource supply rates in the field, and consequent variation in patterns of seedling resource allocation (Burslem *et al.* 1994; Bloor 2003). Field manipulations of below-ground resource availability can be achieved by the removal of root competition by trenching. Although this is a robust method for testing the intensity of competition for nutrients and/or water imposed on seedlings by adult trees, it is a labour-intensive technique and most studies therefore test only a few species. Results of these studies typically support those of pot experiments, with positive growth responses for seedlings in relatively nutrient-poor oxisols or psamments (e.g. Whitmore 1966; Fox 1973; Coomes & Grubb 1998; Lewis & Tanner 2000), but not on the richer alfisols at La Selva, Costa Rica (Denslow *et al.* 1991; Ostertag 1998). Only one study, conducted by Coomes and Grubb (1998), has compared the responses of a significant number of species. They found that trenching had similar effects on the growth of saplings of thirteen species growing in a nitrogen-limited caatinga forest in Venezuela.

Trade-offs partitioning gradients of soil moisture availability

Seasonality in rainfall and in the occurrence of unpredictable droughts, coupled with topographic and edaphic variation in moisture availability, also provide scope for fine-scale resource-partitioning. Field and pot experiments have demonstrated that co-occurring species differ in survival and/or growth responses to water supply (e.g. Burslem *et al.* 1996; Poorter & Hayashida-Oliver 2000; Engelbrecht *et al.* 2000, 2002; Engelbrecht & Kursar 2003; Tyree *et al.* 2003), but only one study has compared the response of a sufficient sample of species to detect trade-offs that might partition species occurrence along gradients of soil moisture (Engelbrecht & Kursar 2003). In this study, the survival and leaf area change of twenty-eight species of woody plants growing in the understorey of a semi-deciduous forest in Panama were compared between drought and irrigated treatments. The twenty-eight species showed a continuum of response to drought that was manifested as differences in both survival and leaf area change, but growth–mortality trade-offs in treatments that varied water supply to the plants during the dry season were not observed. In some cases, species' differential response to drought predicted their local distribution along a gradient of soil water availability determined by topography on a 50-ha plot on nearby Barro Colorado Island (Becker *et al.* 1988; Harms *et al.* 2001; Daws *et al.* 2002b).

Trade-offs resulting from interactions between above- and below-ground resource acquisition

To evaluate trade-offs in resource requirements as a mechanism generating fine-scale partitioning of habitats, we require stronger experimental evidence from studies that assess interactions among potentially limiting resources. For example, when controlled independently, shade and drought have contrasting effects on the allocation of dry mass above- or below-ground. These effects might be expected to result in a growth–survival trade-off in environments that vary in both irradiance and water supply (Smith & Huston 1989). This trade-off might arise if expansion of leaf area and specific leaf area in shade occurs at the cost of reduced allocation to roots, so that more shaded plants are predicted to suffer greater mortality to drought.

Experiments conducted at BCI, Panama, in which seedlings of six pioneer species were transplanted into gaps in seasonally moist forest yielded results that are consistent with this prediction. Seedlings of the three species that maintained the highest RGR in the wet season had the lowest dry-season survival. A seasonal decline in survivorship was particularly marked for *Cecropia insignis*, the species with the highest biomass fraction allocated to leaf area when grown under similar irradiance conditions in a pot experiment (Pearson *et al.* 2003b, c; Dalling *et al.* 2004). Similar results were obtained when seedlings of two light-demanding species were transplanted to gap, edge and forest understorey sites in seasonally moist forest in Ghana (Veenendaal *et al.* 1996). Whereas seedlings survived equally well in all sites during the wet season, survival dropped to low levels in the understorey in the dry season but was unchanged in the gap. In contrast, Fisher *et al.* (1991) found no differences in survival among treatments for yearling seedlings of the shade-tolerant tree *Virola surinamensis* transplanted to gap and understorey sites on BCI, Panama, and irrigated during the dry season. However, seedlings in the understorey only showed positive growth rates when irrigated.

The strategies that plants adopt to acquire water and nutrients under limiting conditions may involve adjustments that are unrelated to plasticity in allocation of above- versus below-ground dry mass (such as changes in root form without a change in root mass). Thus it may be overly simplistic to anticipate an allocation-based growth–survival trade-off for tropical tree seedlings exposed to variable light and water regimes (Sack & Grubb 2002). Unfortunately, the diversity of mechanisms that tropical forest plants have evolved to tolerate combinations of limiting supply rates of light, nutrients and water remains poorly explored. New research in this area is important because independent effects of different resources on tree-seedling growth and survival would increase the scope for fine-scale niche-partitioning along multiple gradients of limiting soil-borne resources.

Life-history trade-offs equalizing recruitment success

The seed number/seedling survival trade-off

Trade-offs in plant performance resulting from specialized resource requirements may be less prevalent than expected because limited seed production and dispersal greatly reduce the probability that individuals encounter the microsites to which they are best adapted. Increased seed production can partially offset the effects of dispersal limitation, but constraints imposed by the finite resources available for reproduction invoke a new set of trade-offs. These life-history trade-offs arise from a negative correlation between seed mass and colonization success, which is balanced by a positive correlation between seed mass and seedling survival (described as a seed number/seedling survival trade-off by Coomes & Grubb 2003). The consequence of this trade-off is to equalize recruitment success among species with similar habitat requirements but divergent life-history characters.

The seed size/seedling survival trade-off is usually viewed as mediated by seed-size-dependent variation in colonization and competitive ability (e.g. Horn & MacArthur 1972; Tilman 1994). However, interspecific interactions among establishing seedlings still dependent upon their seed reserves are likely to be weak in forest understoreys because plant density is insufficient to permit strong direct competition for light and soil resources, and because most resources are acquired by canopy vegetation rather than by neighbouring seedlings (Wright 2002). Size-dependent seedling survival may be mediated instead by size-dependent tolerance of low overall availability of resources (Boot 1996), and to common sources of mortality affecting emerging and establishing seedlings. These include shading and physical damage from falling litter (Clark & Clark 1991), uprooting disturbance by animals (Theimer & Gehring 1999), insect seed and seedling predation (Dalling *et al.* 1997a; Harms *et al.* 1997), and shoot browsing by mammals (Harms & Dalling 1997). Whether these mortality hazards provide a sufficient recruitment advantage to large-seeded shade-tolerant species to balance their reduced reproductive output compared with small-seeded shade-tolerators remains unclear. Nonetheless, available evidence does indicate that seedling mortality scales with seed size over a wide range of seed masses (Silman 1996; Muller-Landau 2001).

Interspecific competition among recruiting seedlings might be expected to be more important in gaps than in the forest understorey. In addition to recruits from seed rain, gaps are filled by seedlings emerging from the seed bank, which can be stocked at densities of > 3000 individuals per m^2 (Garwood 1989). However, only a small fraction of seed-bank individuals emerge following gap formation. In a study tracking seedling recruitment in artificially created gaps, Dalling and Hubbell (2002) found that, on average, only 6% of seeds present in the top 3 cm of soil emerged in litter-free microsites. Natural litter cover within these gaps further reduced seedling emergence success three-fold. Moreover, small initial seedling size for most light-demanding species, coupled with high mortality in the first weeks following emergence, greatly reduces opportunities for direct competition during seedling establishment.

Detailed studies of seed dispersal and seed-bank dynamics in the 50-ha plot on BCI, and of recruitment patterns in artificially created gaps in the Dalling and Hubbell (2002) study allow us to ask, at least for light-demanding species, whether seed-size-dependent differences in seedling recruitment probabilities are sufficient alone to equalize species differences in reproductive output. In the artificial gaps study, recruitment success from the soil seed bank to first-year survival was positively correlated with seed mass and effectively balanced size-dependent variation in seed abundance in the top 0–6 cm of the soil (Fig. 3.5). As a consequence, observed first-year recruit densities were uncorrelated with seed masses ($r^2 = 0.01$). Using data on seed captures to mesh traps in the BCI 50-ha plot, Dalling *et al.* (2002) used a maximum likelihood approach to estimate fecundity parameters (seed production per unit basal area) for eleven light-demanding species. Estimated fecundity was found to scale with seed mass as a power relationship with slope of -0.92. Seed-bank densities measured for the same species at 192 sample locations in the plot also scaled with seed mass, with a somewhat shallower slope of -0.76, but slopes were not significantly different (t-test $df_{2,18} = 0.63$, $p > 0.05$; Fig. 3.6). This suggests that probabilities of seed incorporation into the soil might be largely invariant with respect to seed mass.

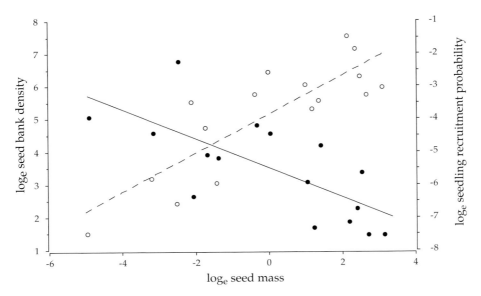

Figure 3.5 Relationship between seed mass (mg) and and soil seed bank density (seeds m^{-2}; closed symbols), and between seed mass and seedling recruitment probability (probability of successfully emerging from the seed bank and surviving one wet and dry season). Data are for sixteen light-demanding taxa recruiting into artificially created gaps in secondary forest on the Barro Colorado Nature Monument, Panama. For more details see Dalling and Hubbell (2002).

A seed size/seed persistence trade-off?

Conventional views of the relationship between seed mass and seed persistence in the soil are based on studies of weed seeds in temperate grasslands (Thompson & Grime 1979; Rees 1993; Thompson *et al.* 1993; Funes *et al.* 1999). These show that small seeds are more readily incorporated into the soil (Peart 1984), and are more likely to persist because burial provides an escape from predation by ants, beetles and rodents that forage primarily on the soil surface (Thompson 1987). Adaptive arguments have also been made that lead to the expectation of a negative relationship between size and persistence, as these traits incur a fitness cost associated with reduced reproductive output and increased generation time, respectively. To avoid incurring both these fitness costs, plants should therefore produce either small seeds that persist between unfavourable periods or large seeds that tolerate unfavourable conditions (Venable & Brown 1988).

Our observations of variation in seed persistence among light-demanding species in Panama, however, fail to support this prediction. Instead, we find the opposite pattern, whereby long-term seed persistence appears to be limited to the larger-seeded species (>3 mg seed mass). Small-seeded species (< 1 mg) that we have examined in detail typically lose viability within one year of burial (Dalling *et al.* 1997b), with mortality attributable to seed-infecting fungi (Dalling

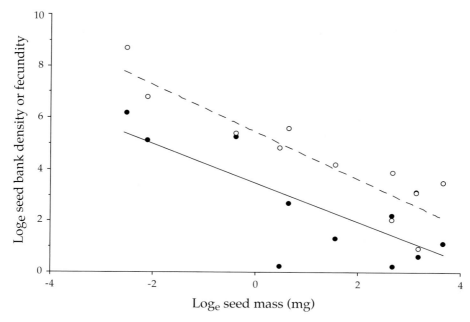

Figure 3.6 Relationship between predicted reproductive output (seeds per cm² basal area) and seed mass (open symbols, dashed regression line), and between soil seed-bank density and seed mass (closed symbols, solid regression line) for eleven light-demanding taxa. Reproductive output was calculated using maximum likelihood parameter estimation based on data obtained from seed traps arrayed in the 50-ha forest dynamics plot on BCI (Dalling *et al.* 2002). Seed-bank density was measured for the top 3 cm of soil in 192 soil samples collected in May 1993 in the BCI 50-ha plot (Dalling *et al.* 1997b).

et al. 1998; R. Gallery & J. W. Dalling, unpublished data). In contrast, larger-seeded species with thick seed coats appear to be resistant to pathogens, and may be too hard for many soil invertebrates to penetrate (cf. O'Dowd & Hay 1980). Direct ageing of the endocarps of viable seeds collected from the field on BCI using accelerator mass spectrometry of ¹⁴C (cf. Moriuchi *et al.* 2000) has shown that seeds of several of these species are capable of extraordinary persistence. Seeds of *Trema micrantha*, *Hyeronima laxiflora*, *Zanthoxylum eckmannii* and *Croton bilbergianus* buried < 3 cm below the soil surface can be >30 years old (J. W. Dalling & T. A. Brown, unpublished data).

These results, based on relatively few species, suggest therefore that in tropical moist forests, seed size and persistence are positively related for pioneer species. Under these conditions seed persistence becomes an important trait that could further offset the fecundity advantage associated with small seed mass, and for light-demanding species may be a significant component of the equalizing trade-offs that permit coexistence of species varying in seed mass. Long-term seed persistence may be particularly advantageous in tropical forests where disturbances providing recruitment opportunities for light-demanding species are infrequent and where community-level dispersal limitation is high (Murray 1998; Dalling *et al.* 2002).

Conclusions

In this chapter we have provided evidence for the view that trade-offs act both to equalize recruitment success and to differentiate habitat requirements among tropical tree species. The equalizing role is apparent early in ontogeny, and is manifested primarily as a dispersal–establishment trade-off that hinges on seed mass. At present, this trade-off is best documented for light-demanding species, but a similar relationship can also be described for shade-tolerators (Muller-Landau 2001; Daws 2002). In the absence of niche differentiation, it might be argued that the dispersal–establishment trade-off is alone sufficient to maintain tree diversity in both regeneration guilds. Evidence for declining mortality rates through ontogeny suggest that species abundance and distribution patterns are largely determined by differences in survivorship during early establishment. Equalization of recruitment success would then imply that relative abundance is determined primarily by random drift, and that community-wide dynamics are consistent with the neutral model (Hubbell 2001; Chave *et al.* 2002).

In contrast, trade-offs that result in partitioning of regeneration niches should act to stabilize the relative abundances of species in the face of random drift, according to the availability of suitable habitat (Chesson 2000). However, initial suggestions that species finely partition gradients of light availability appear to be unfounded. An unfortunate consequence of these suggestions has been a failure adequately to test for partitioning in the acquisition and use of other potentially limiting resources. In part, this may be because ecologists have assumed that life-history traits that influence the capture of secondary resources covary with shade tolerance (Bazzaz & Pickett 1980; Hubbell 2001). However, there is no a-priori reason to predict that this assumption is justified: in tropical forest environments the availability of nutrients and water varies at multiple spatial scales independently of the availability of light, and over time in response to seasonality and interannual variation in rainfall and the climatic drivers of evapotranspiration (e.g. Daws *et al.* 2002b; Baker *et al.* 2003). Thus gaps of a particular size in a wet year might not provide the same regeneration environment as gaps of the same size in a dry year. We suggest that trade-offs imposed by the conflicting demands for maximizing fitness in response to multiple limiting resources might play a much more important role in determining recruitment success than current evidence suggests. These trade-offs will only become apparent, however, once ecologists have better described spatial patterns of variation in below-ground resource supply rates as well as plant responses to them.

The future resolution of the role of equalizing versus stabilizing forces in structuring tropical tree communities is only likely to emerge from comparative studies among sites. Comparisons of relative abundances of species in tropical forests at a range of spatial scales (Pitman *et al.* 1999, 2001; Terborgh *et al.* 1996), over decadal scales during recovery from large-scale disturbance (Burslem *et al.* 2000) and in temperate forests over the past 10 000 years (Clark & McLachlan 2003) make a persuasive case for stable species assemblages in which relative-abundance distributions are conserved. Once a stronger mechanistic framework has been established to determine how seed and seedling traits influence resource capture, we will also be able to assess the extent to which trait composition varies among tree communities found on sites varying in resource supply rates. Concurrent shifts in the trait composition of communities along resource gradients may provide the strongest evidence that tropical forests are structured through niche-partitioning.

Acknowledgements

We are grateful to the collaborators, colleagues and students who have contributed to our research in Panama and elsewhere, in particular Matthew Daws, Rachel Gallery, Steve Hubbell, Eloisa Lasso, Toby Marthews, Helene Muller-Landau, Chris Mullins and Timothy Pearson. We thank Christopher Baraloto, Juliette Bloor and Lourens Poorter for many helpful suggestions on improving a draft of this manuscript. Our research in Panama was funded by the British Ecological Society, the National Science Foundation, the Natural Environment Research Council and the Leverhulme Trust.

References

Baker, T. R., Burslem, D. F. R. P. & Swaine, M. D. (2003) Associations between tree growth, soil fertility and water availability at local and regional scales in Ghanaian tropical rain forest. *Journal of Tropical Ecology*, **19**, 109–125.

Barker, M. G., Press, M. C. & Brown, N. D. (1997) Photosynthetic characteristics of dipterocarp seedlings in three tropical rain forest light environments: a basis for niche partitioning? *Oecologia*, **112**, 453–463.

Barton, A. M., Fetcher, N. & Redhead, S. (1989) The relationship between treefall gap size and light flux in a neotropical rain forest in Costa Rica. *Journal of Tropical Ecology*, **5**, 437–439.

Bazzaz, F. A. & Pickett, S. T. A. (1980) Physiological ecology of tropical succession: a comparative review. *Annual Review of Ecology and Systematics*, **11**, 287–310.

Becker, P., Rabenold, P. E., Idol, J. R. & Smith, A. P. (1988) Water potential gradients for gaps and slopes in a Panamanian tropical moist forest's dry season. *Journal of Tropical Ecology*, **4**, 173–184.

Bloor, J. M. G. (2003) Light responses of shade-tolerant tropical tree species in north-east Queensland: a comparison of forest- and shadehouse-grown seedlings. *Journal of Tropical Ecology*, **19**, 163–170.

Bloor, J. M. G. & Grubb, P. J. (2003) Growth and mortality in high and low light: trends among 15 shade tolerant rain forest tree species. *Journal of Ecology*, **91**, 77–85.

Boot, R. G. A. (1996) The significance of seedling size and growth rate of tropical rain forest tree seedlings for regeneration in canopy openings. In *The Ecology of Tropical Forest Tree Seedlings* (ed. M. D. Swaine) MAB UNESCO Series, vol. **17**. Paris: Parthenon, pp. 267–284.

Brokaw, N. V. L. (1985) Gap-phase regeneration in a tropical forest. *Ecology*, **66**, 682–687.

(1987) Gap-phase regeneration of three pioneer tree species in a tropical forest. *Journal of Ecology*, **75**, 9–19.

Brokaw, N. & Busing, R. T. (2000) Niche versus chance and tree diversity in forest gaps. *Trends in Ecology and Evolution*, **15**, 183–188.

Brown, N. D. (1993) The implications of climate and gap microclimate for seedling growth conditions in a Bornean lowland rain forest. *Journal of Tropical Ecology*, **9**, 153–168.

Brown, N. D. & Jennings, S. (1998) Gap-size niche differentiation by tropical rain forest trees: a testable hypothesis or a broken-down bandwagon. In *Dynamics of Tropical Communities* (ed. D. M. Newbery, H. H. T. Prins & N. D. Brown). Oxford: Blackwell, pp. 79–94.

Brown, N. D. & Whitmore, T. C. (1992) Do dipterocarp seedlings really partition tropical rain forest gaps? *Philosophical Transactions of the Royal Society of London B*, **335**, 369–378.

Budowski, G. (1965) Distribution of tropical American rain forest species in the light of successional processes. *Turrialba*, **15**, 40–42.

Burslem, D. F. R. P. & Swaine, M. D. (2002) Forest dynamics and regeneration. In *Foundations of Tropical Forest Biology: Classic Papers with Commentaries* (ed. R. L. Chazdon & T. C. Whitmore). Chicago: Chicago University Press, pp. 577–583.

Burslem, D. F. R. P., Turner, I. M. & Grubb, P. J. (1994) Mineral nutrient status of coastal hill dipterocarp forest and adinandra belukar in Singapore: bioassays of nutrient limitation. *Journal of Tropical Ecology*, **10**, 579–599.

Burslem, D. F. R. P., Grubb, P. J. & Turner, I. M. (1996) Responses to simulated drought and elevated nutrient supply among shade-tolerant tree seedlings of lowland tropical forest in Singapore. *Biotropica*, **28**, 636–648.

Burslem, D. F. R. P., Whitmore, T. C. & Brown, G. C. (2000) Short-term effects of cyclone impact and long-term recovery of tropical rain forest on Kolombangara, Solomon Islands. *Journal of Ecology*, **88**, 1063–1078.

Chave, J., Muller-Landau, H. C. & Levin, S. A. (2002) Comparing classical community models: theoretical consequences for patterns of diversity. *American Naturalist*, **159**, 1–23.

Chesson, P. (2000) Mechanisms of maintenance of species diversity. *Annual Review of Ecology and Systematics*, **31**, 343–366.

Clark, D. B. & Clark, D. A. (1991) The impact of physical damage on canopy tree regeneration in tropical rain forest. *Journal of Ecology*, **79**, 447–457.

Clark, J. S. & McLachlan, J. S. (2003) Stability of forest biodiversity. *Nature*, **423**, 635–637.

Clark, J. S., Mohan, J., Dietze, M. & Ibanez, I. (2003) Coexistence: how to identify trophic trade-offs. *Ecology*, **84**, 17–31.

Coley, P. D., Bryant, J. P. & Chapin, F. S. (1985) Resource availability and plant antiherbivore defense. *Science*, **230**, 895–899.

Coomes, D. A. & Grubb, P. J. (1998) Responses of juvenile trees to above- and belowground competition in nutrient-starved Amazonian rain forest. *Ecology*, **79**, 768–782.

(2003) Colonization, tolerance, competition and seed-size variation within functional groups. *Trends in Ecology and Evolution*, **18**, 283–291.

Dalling, J. W. & Hubbell, S. P. (2002) Seed size, growth rate and gap microsite conditions as determinants of recruitment success for pioneer species. *Journal of Ecology*, **90**, 557–568.

Dalling, J. W. & Tanner, E. V. J. (1995) An experimental study of regeneration on landslides in montane forest in Jamaica. *Journal of Ecology*, **83**, 55–64.

Dalling, J. W., Harms, K. E. & Aizprúa, R. (1997a) Seed damage tolerance and seedling resprout ability of *Prioria copaifera* ('El Cativo'). *Journal of Tropical Ecology*, **13**, 617–621.

Dalling, J. W., Swaine, M. D. & Garwood, N. C. (1997b) Soil seed bank community dynamics in seasonally moist lowland forest, Panama. *Journal of Tropical Ecology*, **13**, 659–680.

 (1998) Dispersal patterns and seed bank dynamics of pioneer trees in moist tropical forest. *Ecology*, **79**, 564–578.

Dalling, J. W., Lovelock, C. E. & Hubbell, S. P. (1999) Growth responses of two neotropical pioneer seedlings to simulated forest gap environments. *Journal of Tropical Ecology*, **15**, 827–839.

Dalling, J. W., Muller-Landau, H. C., Wright, S. J. & Hubbell, S. P. (2002) Role of dispersal in the recruitment limitation of neotropical pioneer species. *Journal of Ecology*, **90**, 714–727.

Dalling, J. W., Winter, K. & Hubbell, S. P. (2004) Variation in growth responses of neotropical pioneer species to simulated gaps. *Functional Ecology*, **18**, 725–736.

Daws, M. I. (2002) *Mechanisms of plant species coexistence in a semi-deciduous tropical forest in Panamá*. Unpublished Ph.D. dissertation, University of Aberdeen.

Daws, M. I., Burslem, D. F. R. P., Crabtree, L. M. *et al.* (2002a) Differences in seed germination responses may promote coexistence of four sympatric *Piper* species. *Functional Ecology*, **16**, 258–267.

Daws, M. I., Mullins, C. E., Burslem, D. F. R. P., Paton, S. & Dalling, J. W. (2002b) Topographic position affects the water regime in a semideciduous tropical forest in Panamá. *Plant and Soil*, **238**, 79–89.

Debski, I., Burslem, D. F. R. P., Palmiotto, P. A., Lafrankie, J. V., Lee, H. S. & Manokaran, N. (2002) Habitat preferences of *Aporosa* in two Malaysian forests: implications for abundance and coexistence. *Ecology*, **83**, 2005–2018.

Denslow, J. S. (1980) Gap partitioning among tropical rainforest trees. *Biotropica*, **12** (Supplement), 47–55.

 (1987) Tropical rainforest gaps and tree species diversity. *Annual Review of Ecology and Systematics*, **18**, 431–451.

Denslow, J. S., Newell, E. & Ellison, A. M. (1991) The effect of understorey palms and cyclanths on the growth and survival of *Inga* seedlings. *Biotropica*, **23**, 225–234.

Engelbrecht, B. M. J. & Kursar, T. A. (2003) Comparative drought-resistance of seedlings of 28 species of co-occurring tropical woody plants. *Oecologia*, **136**, 383–393.

Engelbrecht, B. M. J., Velez, V. & Tyree, M. T. (2000) Hydraulic conductance of two co-occuring neotropical understorey shrubs with different habitat preferences. *Annals of Forest Science*, **57**, 201–208.

Engelbrecht, B. M. J., Dalling, J. W., Pearson, T. R. H. *et al.* (2001) Short dry spells in the wet season increase mortality of tropical pioneer seedlings. *Proceedings of the Association of Tropical Biology Meeting, Bangalore, India* (ed. K. N. Ganeshaiah, R. Uma, R. Shaanker & K. S. Bawa). New Dehli: Oxford & IBH Publishing, pp. 6655–6669.

Engelbrecht, B. M. J., Wright, S. J. & de Steven, D. (2002) Survival and ecophysiology of tree seedlings during El Niño drought in a tropical moist forest in Panama. *Journal of Tropical Ecology*, **18**, 569–579.

Fetcher, N., Haines, B. L., Cordero, R. A. *et al.* (1996) Responses of tropical plants to nutrients and light on a landslide in Puerto Rico. *Journal of Ecology*, **84**, 331–341.

Fisher, B. L., Howe, H. F. & Wright, S. J. (1991) Survival and growth of *Virola surinamensis* yearlings: water augmentation in gap and understorey. *Oecologia*, **86**, 292–297.

Fox, J. E. D. (1973) Dipterocarp seedling behaviour in Sabah. *Malaysian Forester*, **36**, 205–214.

Funes, G., Basconcelo, S., Diaz, S. & Cabido, M. (1999) Seed size and shape are good predictors of seed persistence in soil in temperate mountain grasslands of Argentina. *Seed Science Research*, **9**, 341– 345.

Garwood, N. C. (1983) Seed germination in a seasonal tropical forest in Panama: a community study. *Ecological Monographs*, **53**, 159–181.

(1989) Tropical soil seed banks: a review. In *The Ecology of Soil Seed Banks* (ed. M. Leck, V. Parker & R. Simpson). San Diego: Academic Press, pp. 149–209.

Givnish, T. J. (1988) Adaptation to sun and shade, a whole-plant perspective. *Australian Journal of Plant Physiology*, **15**, 63–92.

Grubb, P. J. (1977) The maintenance of species richness in plant communities: the importance of the regeneration niche. *Biological Reviews*, **52**, 107–145.

(1998) A reassessment of the strategies of plants which cope with shortages of resources. *Perspectives in Plant Ecology, Evolution and Systematics*, **1**, 3–31.

Gunatilleke, C. V. S., Gunatilleke, I. A. U. N., Perera, G. A. D., Burslem, D. F. R. P., Ashton, P. M. S. & Ashton, P. S. (1997) Responses to nutrient addition among seedlings of eight closely related species of *Shorea* in Sri Lanka. *Journal of Ecology*, **85**, 301–311.

Harms, K. E. & Dalling, J. W. (1997) Damage and herbivory tolerance through resprouting as an advantage of large seed size in tropical trees and lianas. *Journal of Tropical Ecology*, **13**, 481–490.

Harms, K. E., Dalling, J. W. & Aizprúa, R. (1997) Cotyledonary resprouting capacity of *Gustavia superba* (Lecythidaceae). *Biotropica*, **29**, 234–237.

Harms, K. E., Condit, R., Hubbell, S. P. & Foster, R. B. (2001) Habitat associations of trees and shrubs in a 50-ha neotropical forest plot. *Journal of Ecology*, **89**, 947–959.

Horn, H. S. & MacArthur, R. H. (1972) Competition among fugitive species in a harlequin environment. *Ecology*, **53**, 749–753.

Huante, P., Rincón, R. & Acosta, I. (1995) Nutrient availability and growth rate of 34 woody species from a tropical deciduous forest in Mexico. *Functional Ecology*, **9**, 849–858.

Hubbell, S. P. (1979) Tree dispersion, abundance and diversity in a tropical dry forest. *Science*, **203**, 1299–1309.

(2001) *The Unified Neutral Theory of Biodiversity and Biogeography*. Monographs in Population Biology. Princeton: Princeton University Press.

Hubbell, S. P. & Foster, R. B. (1986) Biology, chance and history and the structure of tropical rain forest tree communities. In *Community Ecology* (ed. J. Diamond & T. J. Case). New York: Harper & Row, pp. 314–329.

(1992) Short-term dynamics of a neotropical forest – why ecological research matters to tropical conservation and management. *Oikos*, **63**, 48–61.

Hubbell, S. P., Foster, R. B., O'Brien, S. T. *et al.* (1999) Light gap disturbances, recruitment limitation, and tree diversity in a neotropical forest. *Science*, **283**, 554–557.

Hutchinson, G. E. (1959) Homage to Santa Rosalia, or why are there so many different kinds of animals? *American Naturalist*, **93**, 145–159.

King, D. A. (1990) Allometry of saplings and understorey trees of a Panamanian forest. *Functional Ecology*, **4**, 27–32.

(1994) Influence of light level on the growth and morphology of saplings in a Panamanian forest. *American Journal of Botany*, **81**, 948–957.

Kitajima, K. (1994) Relative importance of photosynthetic traits and allocation patterns as correlates of seedling shade tolerance of 13 tropical trees. *Oecologia*, **98**, 419–428.

Kitajima, K. & Bolker, B. M (2003) Testing performance rank reversals among coexisting species: crossover point irradiance analysis by Sack & Grubb (2001) and alternatives. *Functional Ecology*, **17**, 276–287.

Kobe, R. K. (1999) Light gradient partitioning among tropical tree species through differential seedling mortality and growth. *Ecology*, **80**, 187–201.

Kohyama, T. (1987) Significance of architecture and allometry of saplings. *Functional Ecology*, **1**, 399–404.

Kohyama, T. & Hotta, M. (1991) Significance of allometry in tropical saplings. *Functional Ecology*, **4**, 515–521.

Kohyama, T., Suzuki, E., Partomihardjo, T., Yamada, T. & Kubo, T. (2003) Tree species differentiation in growth, recruitment and allometry in relation to maximum height in a Bornean mixed dipterocarp forest. *Journal of Ecology*, **91**, 797–806.

Latham, R. E. (1992) Co-occurring tree species change rank in seedling performance with resources varied experimentally. *Ecology*, **73**, 2129–2144.

Lawrence, D. (2003) The response of tropical tree seedlings to nutrient supply: meta-analysis for understanding a changing tropical landscape. *Journal of Tropical Ecology*, **19**, 239–250.

Levins, R. (1968) *Evolution in Changing Environments*. Monographs in Population Biology. Princeton: Princeton University Press.

Lewis, S. L. & Tanner, E. V. J. (2000) Effects of above- and below-ground competition on growth and survival of rain forest tree seedlings. *Ecology*, **81**, 2525–2538.

MacArthur, R. H. & Levins, R. (1967) The limiting similarity, convergence, and divergence of coexisting species. *American Naturalist*, **101**, 377–385.

Marañon, T. & Grubb, P. J. (1993) Physiological basis and ecological significance of the seed size and relative growth rate relationship in Mediterranean annuals. *Functional Ecology*, **7**, 591–599.

Metcalfe, D. J. (1996) Germination of small-seeded tropical rain forest plants exposed to different spectral compositions. *Canadian Journal of Botany*, **74**, 516– 520.

Metcalfe, D. J. & Grubb, P. J. (1997) The responses to shade of seedlings of very small-seeded tree and shrub species from tropical rain forest in Singapore. *Functional Ecology*, **11**, 215–221.

Metcalfe, D. J., Grubb, P. J. & Metcalfe, S. S. (2002) Soil dilution as a surrogate for root competition: effects on growth of seedlings of Australian tropical rainforest trees. *Functional Ecology*, **16**, 223–231.

Molino, J.-F. & Sabatier, D. (2001) Tree diversity in a tropical rain forests: a validation of the intermediate disturbance hypothesis. *Science*, **294**, 1702–1704.

Molofsky, J. & Augspurger, C. K. (1992) The effect of leaf litter on early seedling establishment in a tropical forest. *Ecology*, **73**, 68–77.

Montgomery, R. A. & Chazdon, R. L. (2002) Light gradient partitioning by tropical tree seedlings in the absence of canopy gaps. *Oecologia*, **131**, 165–174.

Moriuchi, K. S., Venable, D. L., Pake, C. E. & Lange, T. (2000) Direct measurement of the seed bank age structure of a Sonoran desert annual plant. *Ecology,* **81**, 1133–1138.

Muller-Landau, H. C. (2001) Seed dispersal in a tropical forest: empirical patterns, their origins and their consequences for forest dynamics. Unpublished Ph.D. thesis, Princeton University.

Murray, K. G. (1988) Avian seed dispersal of three neotropical gap-dependent plants. *Ecological Monographs*, **58**, 271–298.

O'Dowd, D. J. & Hay, M. E. (1980) Mutualism between harvester ants and a desert ephemeral: seed escape from rodents. *Ecology*, **61**, 531–540.

Orians, G. H. (1982) The influence of tree-falls in tropical forests in tree species richness. *Tropical Ecology*, **23**, 256–279.

Ostertag, R. (1998) Belowground effects of canopy gaps in a tropical wet forest. *Ecology*, **79**, 1294–1304.

Pearson, T. R. H., Burslem, D. F. R. P., Mullins, C. E. & Dalling, J. W. (2002) Germination ecology of neotropical pioneers: interacting effects of environmental conditions and seed size. *Ecology*, **83**, 2798–2807.

(2003a) Functional significance of photoblastic germination in neotropical pioneer trees: a seed's eye view. *Functional Ecology*, **17**, 394–402.

Pearson, T. R. H., Burslem, D. F. R. P., Goeriz, R. E. & Dalling, J. W. (2003b) Interactions of gap size and herbivory on establishment, growth and survival of three species of neotropical pioneer trees. *Journal of Ecology*, **91**, 785–796.

(2003c) Regeneration niche partitioning in neotropical pioneers: effects of gap size, seasonal drought and herbivory on growth and survival. *Oecologia*, **137**, 456–465.

Peart, M. H. (1984) The effects of morphology, orientation and position of grass diaspores on seedling survival. *Journal of Ecology*, **72**, 437–453.

Phillips, O. L., Núñez Vargas, P., Lorenzo Monteagudo, A. *et al.* (2003) Habitat association among Amazonian tree species: a landscape-scale approach. *Journal of Ecology*, **91**, 757–775.

Pickett, S. T. A. (1980) Non-equilibrium coexistence of plants. *Bulletin of the Torrey Botanical Club*, **107**, 238–246.

Pitman, N. C. A., Terborgh, J. W., Silman, M. R. & Nuñez, P. (1999) Tree species distributions in an upper Amazonian forest. *Ecology*, **80**, 2651–2661.

Pitman, N. C. A., Terborgh, J. W., Silman, M. R. *et al.* (2001) Dominance and distribution of tree species in Upper Amazonian Terra Firme forests. *Ecology*, **82**, 2101–2117.

Poorter, L. & Hayashida-Oliver, Y. (2000) Effects of seasonal drought on gap and understorey seedlings in a Bolivian moist forest. *Journal of Tropical Ecology*, **16**, 481–498.

Poorter, L., Bongers, F., Sterck, F. J. & Woll, H. (2003) Architecture of 53 rain forest tree species differing in adult stature and shade tolerance. *Ecology*, **84**, 602–608.

Putz, F. E. (1983) Treefall pits and mounds, buried seeds and the importance of soil disturbance to pioneer trees on Barro Colorado Island, Panama. *Ecology*, **64**, 1069–1074.

Rees, M. (1993) Trade-offs among dispersal strategies in the British flora. *Nature,* **366**, 150–152.

Richards, P. W. (1952) *The Tropical Rain Forest.* Cambridge: Cambridge University Press.

Ricklefs, R. E. (1977) Environmental heterogeneity and plant species diversity: a hypothesis. *American Naturalist*, **111**, 376–381.

Rose, S. & Poorter, L. (2003) The importance of seed mass for early regeneration in tropical forests: a review. In *Long Term Changes in Tropical Tree Diversity: Studies from the Guyana Shield, Africa, Borneo and Melanesia* (ed. ter Steege, H.). Tropenbos Series 22. Wageningen: Tropenbos International, pp. 19–35.

Sack, L. & Grubb, P. J. (2001) Why do species of woody seedlings change rank in relative growth rate between low and high irradiance? *Functional Ecology*, **15**, 145–154.

(2002) The combined impacts of deep shade and drought on the growth and biomass allocation of shade-tolerant woody seedlings. *Oecologia*, **131**, 175–185.

Silman, M. R. (1996) Regeneration from seed in a neotropical rain forest. Unpublished Ph.D. thesis, Duke University.

Smith, T. & Huston, M. (1989) A theory of the spatial and temporal dynamics of plant communities. *Vegetatio*, **83**, 49–69.

Swaine, M. D. & Whitmore, T. C. (1988) On the definition of ecological species groups in tropical rain forests. *Vegetatio*, **75**, 81–86.

Swaine, M. D., Agyeman, V. K., Kyereh, B., Orgle, T. K., Thompson, J. & Veenendaal, E. M. (1997) *Ecology of Forest Trees in Ghana*. Overseas Development Administration Forestry Series No. 7. London: ODA.

Terborgh, J., Foster, R. B. & Nuñez, P. (1996) Tropical tree communities: a test of the non-equilibrium hypothesis. *Ecology*, **77**, 561–567.

Theimer, T. C. & Gehring, C. A. (1999) Effects of litter-disturbing bird species on tree seedling germination and survival in an Australian tropical rain forest. *Journal of Tropical Ecology*, **15**, 737–749.

Thompson, K. (1987) Seeds and seed banks. In *Frontiers of Comparative Plant Ecology* (ed. I. H. Rorison, J. P. Grime, R. Hunt, G. A. F. Hendry & D. H. Lewis). *New Phytologist*, **106** (Suppl.), 23–34.

Thompson, K. & Grime, J. P. (1979) Seasonal variation in the seed banks of herbaceous species in ten contrasting habitats. *Journal of Ecology*, **67**, 893–921.

Thompson, K., Band, S. R. & Hodgson, J. G. (1993) Seed size and shape predict persistence in soil. *Functional Ecology*, **7**, 236–241.

Tilman, D. (1988) *Plant Strategies and the Dynamics and Structure of Plant Communities*. Monographs in Population Biology. Princeton: Princeton University Press.

(1994) Competition and biodiversity in spatially structured habitats. *Ecology*, **75**, 2–16.

Tuomisto, H., Ruokolainen, K., Aguilar, M. & Sarmiento, A. (2003) Floristic patterns along a 43-km-long transect in an Amazonian rain forest. *Journal of Ecology*, **91**, 743–756.

Tyree, M. T., Engelbrecht, B. M. J., Vargas, G. & Kursar, T. A. (2003) Desiccation tolerance of five tropical seedlings in Panama. Relationship to a field assessment of drought performance. *Plant Physiology*, **132**, 1439–1447.

Uhl, C., Clark, K., Dezzeo, N. & Maquirino, P. (1988) Vegetation dynamics in Amazonian treefall gaps. *Ecology*, **69**, 751–763.

van der Meer, P. J., Bongers, F., Chatrou, L., & Riera, B. (1994) Defining canopy gaps in a tropical rain forest: effects on gap size and turnover time. *Acta Oecologica*, **15**, 701–714.

van Steenis, C. G. G. J. (1958) Rejuvenation as a factor for judging the status of vegetation types. The biological nomad theory. In *Proceedings of the Symposium on Humid Tropics Vegetation*. Paris: UNESCO.

Vázquez-Yanes, C. (1974) Studies on the germination of seeds of *Ochroma lagopus* Swartz. *Turrialba*, **24**, 176–179.

Vázquez-Yanes, C. & Orozco-Segovia, A. (1982) Seed germination of a tropical rain forest tree *Heliocarpus donnell-smithii* in response to diurnal fluctuations in temperature. *Physiologia Plantarum*, **56**, 295–298.

(1993) Patterns of seed longevity and germination in the tropical rain forest. *Annual Review of Ecology and Systematics*, **24**, 69–87.

Vázquez-Yanes, C. & Smith, H. (1982) Phytochrome control of seed germination in the tropical rain forest pioneer trees *Cecropia obtusifolia* and *Piper auritum* and its ecological significance. *New Phytologist*, **92**, 477–485.

Vázquez-Yanes, C., Orozco-Segovia, A., Rincon, E. *et al.* (1990) Light beneath the litter in a tropical forest: effect on seed germination. *Ecology*, **71**, 1952–1958.

Venable, D. L. & Brown, J. S. (1988) The selective interactions of dispersal, dormancy, and seed size as adaptations for reducing risk in variable environments. *American Naturalist*, **131**, 360–384.

Veenendaal, E. M., Swaine, M. D., Agyeman, V. K., Blay, D., Abebrese, I. K. & Mullins, C. E. (1996) Differences in plant and soil water relations in and around a forest gap in West Africa during the dry season may influence seedling establishment and survival. *Journal of Ecology*, **84**, 83–90.

Vitousek, P. M. & Denslow, J. S. (1986) Nitrogen and phosphorus availability in treefall gaps of a lowland tropical rain-forest. *Journal of Ecology*, **74**, 1167–1178.

Welden, C. W., Hewett, S. W., Hubbell, S. P. & Foster, R. B. (1991) Sapling survival, growth and recruitment: relationship to canopy height in a neotropical forest. *Ecology*, 72, 35–50.

Whitmore, T. C. (1966) The social status of *Agathis* in a rain forest in Melanesia. *Journal of Ecology*, **54**, 285–301.

(1975) *Tropical Rain Forests of the Far East.* Oxford: Oxford University Press.

Wright, S. J. (2002) Plant diversity in tropical forests: a review of mechanisms of species coexistence. *Oecologia*, **130**, 1–14.

Zagt, R. J. (1997) *Tree Demography in the Tropical Rain Forest of Guyana.* Tropenbos-Guyana Series 3. Georgetown: Tropenbos-Guyana Programme.

CHAPTER FOUR

Neighbourhood effects on sapling growth and survival in a neotropical forest and the ecological-equivalence hypothesis

MARÍA URIARTE
Institute of Ecosystem Studies, Millbrook
STEPHEN P. HUBBELL
University of Georgia and Smithsonian Tropical Research Institute
ROBERT JOHN
University of Georgia and Smithsonian Tropical Research Institute
RICHARD CONDIT
Smithsonian Tropical Research Institute
CHARLES D. CANHAM
Institute of Ecosystem Studies, Millbrook

Introduction

In 1980 S. P. Hubbell and R. B. Foster began a long-term, large-scale study of tropical forest dynamics on Barro Colorado Island (BCI), Panama. The objective of the study was to test competing hypotheses about the maintenance of high tree species richness in the BCI forest, and in tropical moist forests more generally. Hubbell and Foster established a 50-ha permanent plot on the summit plateau of BCI, within which all free-standing woody plants with a stem diameter at breast height (DBH) of a centimetre or larger were tagged, measured, mapped and identified by 1982. Subsequent complete censuses of the BCI plot have been conducted from 1985 to 2000 at 5-year intervals. In setting up the BCI plot, Hubbell and Foster (1983) reasoned that whatever diversity-maintaining mechanisms were important, they would have to operate in a spatially dependent manner in communities of sessile plants such as the BCI tree community, which meant that the trees had to be mapped. A decade earlier, Janzen (1970) and Connell (1971) had independently proposed a spatially explicit 'enemies hypothesis', now known as the Janzen–Connell hypothesis. They hypothesized that host-specific seed and seedling predators were responsible for maintaining tropical tree diversity by causing dependence on density and frequency (rare species advantage), through an interaction between seed dispersal and density-dependent seed predation.

Biotic Interactions in the Tropics: Their Role in the Maintenance of Species Diversity, ed. D. F. R. P. Burslem, M. A. Pinard and S. E. Hartley. Published by Cambridge University Press. © Cambridge University Press 2005.

In 1980, there were essentially just two principal tropical forest diversity theories to test: the enemies hypothesis and its variants, and the 'intermediate disturbance' hypothesis (Connell 1977) and its variants that invoked a role for disturbances associated with opening, growth and closure of light gaps (e.g. Ricklefs 1978; Hartshorn 1978; Orians 1982; Denslow 1987). These ideas were intellectual descendants of the 'fugitive species' concept (Hutchinson 1961) and *r-K* selection (MacArthur & Wilson 1967), which are now embodied in diversity theories that depend on a tradeoff between dispersal ability and competitive ability among species (Tilman 1994; Hurtt & Pacala 1995).

Since the plot was established, a large number of new hypotheses have been put forth to explain tree diversity in tropical forests (see Hubbell 1997, 2001; Terborgh *et al.* 2001; Wright 2002 for reviews). Many of these hypotheses are not mutually exclusive, so the challenge in discriminating among them is not qualitative, but quantitative. One of these newer hypotheses is the 'ecological equivalence' hypothesis. According to this hypothesis many if not all trophically similar species are, to at least a first approximation, demographically and competitively alike on a per-capita basis. The ecological-equivalence hypothesis arises out of symmetric neutral theory (Hubbell 2001), but has older roots (Hubbell 1979; Goldberg & Werner 1982; Schmida & Ellner 1985). This hypothesis contrasts with contemporary niche-assembly theory (Chase & Leibold 2003), which emphasizes the importance of fundamental asymmetries or differences among species. Symmetric neutral theory fits patterns of relative species abundance remarkably well at the spatial scale of the entire 50-ha BCI plot, as new analytical solutions to the theory show (Volkov *et al.* 2003), contrary to recent assertions (McGill 2003). However, there is strong empirical evidence that symmetry is broken at smaller spatial scales. Ecological dominance deviations – deviations from the expected relative abundances under symmetric neutrality – can be detected in many BCI tree species, especially at the spatial scale of hectares (100×100 m) or smaller. This suggests that species or functional group differences cause symmetry to be broken at small spatial scales and that the approximation of symmetry may apply better on larger scales, possibly because these larger scales average out environmentally driven niche differentiation.

Previously, we examined the effects of local biotic neighbourhood on the survival of focal trees and saplings at the community level (Hubbell *et al.* 2001). We found strong, pervasive, always negative, conspecific density effects, and much weaker effects of relative plant size and neighbourhood species richness. Density dependence per se does not invalidate symmetry, however, so long as all species of equivalent abundance experience the same density effects (Chave *et al.* 2002; Hubbell & Lake 2003). Recently, we have extended the neutral theory to incorporate symmetric density- and frequency-dependence (Banavar *et al.* 2003). However, in the empirical studies, we have found, in fact, that the strength of the conspecific density effects varies among functional groups of BCI species, indicating that symmetry is broken among, if not within, these life-history guilds (Hubbell *et al.* 2001). The reduction in the survival of focal plants due to conspecific neighbours was greater in canopy tree species than in shrub species, and greater in gap species than in shade-tolerant species. We evaluated the effects of neighbours at different distance intervals: 0–2.5 m, 2.5–5 m, etc., to a distance of 30 m. We found that all effects decayed to mean field (background) within 10–15 m, so the neighbourhood effects were all extremely local.

In this paper we have taken a somewhat different modelling approach to neighbourhood effects, and we analyse both survival and growth for individual species. We assume, as do most current models of such effects, that there is a direct effect of a neighbour's size, and an inverse effect of a neighbour's distance, on the focal plant's growth and survival. The primary question posed here is, to what extent is the ecological-equivalence hypothesis supported, and to what extent can it be rejected? Ecological equivalence, for the purpose of these neighbourhood analyses, means that it does not matter to what species or functional group a neighbour belongs, because all of the neighbour's significant effects on focal plant survival and growth are captured by information about the neighbour's size and distance from the focal plant. Whether the ecological-equivalence hypothesis can or cannot be rejected is likely to depend on the life-history stage, the species and the functional guild to which the species belongs. Thus, some stages, such as seedlings or small saplings, may be more sensitive to who their neighbours are, and some species and entire functional groups may also be more sensitive than others.

Suppose we assume for the moment that species are indeed ecological equivalents. How could such equivalence arise? The first way to be ecologically equivalent, or nearly so, is through common descent (Federov 1966; van Steenis 1969). This possibility is regarded as sufficiently commonplace to be a major concern for independence assumptions in hypothesis-testing in comparative evolutionary biology (Harvey & Pagel 1991). Related species are more likely to respond in the same or similar ways to the mechanisms that are ultimately responsible for diversity. Sister taxa are more likely to have similar resource requirements or share enemies that would keep them in check through Janzen–Connell effects. Tests of community assembly taking phylogenetic relationships into account do tend to show that closely related species occur together more often than expected by chance at a variety of spatial scales (Webb 2000). This finding is generally unexpected from classical niche-assembly theory, which predicts greater competition and niche separation among closely related species. A second way that equivalence may arise is through diffuse coevolution in response to a highly unpredictable and diverse neighbourhood (Connell 1980; Hubbell & Foster 1986). If species have unpredictable neighbours over their evolutionary lifespans, which will be especially true in species-rich communities, then species will tend to converge on the same life histories adapted to the long-term statistical average of the neighbourhood conditions that they all experience. If this is the case, we should see minimal, if any, variation among neighbouring species in their effects on the growth and survival of focal species. If variation among species in these effects does indeed exist, it will be among functional groups of species with distinct life histories (e.g. gap vs. shade-tolerant species).

In this chapter we address three primary questions. (1) Do all neighbour species have the same effects on focal plant growth and survival? (2) Is common descent a good predictor of the strength of the effects? (3) How does probability of coexistence as adults influence the strength of neighbourhood interactions? We note that these neighbour effects are expected to be generally negative, and they may involve direct competition, or they may be indirect competitive effects, such as density-mediated contagion of pathogens. We address these questions by comparing the ability of different models to explain variation in individual tree growth and survival of focal species. In the first model, neighbours are not identified in terms of species or functional group, but are treated as equivalent. In the second model, neighbours of the same species as the focal individual are separated from neighbours of other species. In the third model, neighbours are classified by degree of relatedness into conspecifics, neighbours belonging to the same plant family (confamilials) but not the same species, and neighbours belonging to other plant families. In the final model, neighbours are classified as conspecifics or other species, and the latter class is divided into gap species and shade-tolerant species. Uriarte *et al.* (2004a) provide a more detailed description of the model and the results for growth.

Methods

Our modelling approach follows a long tradition in forest ecology, and assumes that each species has a maximum rate of growth and survival, and that the realized growth and survival rates of the focal species are reduced additively by competitive effects, direct and indirect, from neighbours (e.g. Bella 1971; Hegyi 1974; Zeide 1993; Wimberly & Bare 1996; Wagner & Radosevich 1998; Vettenranta 1999). The general formula for growth is:

$$\text{Realized (predicted) growth} = \text{maximum growth} \cdot \exp(-C \cdot \text{NCI}^{D}) \tag{4.1}$$

The function for survival is strictly analogous, except that in this case we predict the realized probability of survival instead of realized growth rate. In the case of survival, however, the variable is not continuous, but binary, because an individual either lived (1) or died (0). We model neighbour effects using a negative exponential function. First we calculate a combined index of neighbour effects, the neighbourhood crowding index (NCI), and then we fit parameters C and D in

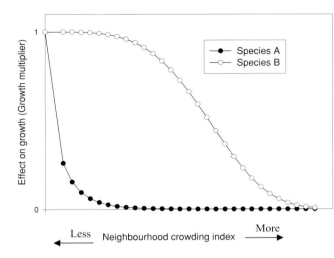

Figure 4.1 Effects of neighbourhood competition on target growth for two hypothetical species. Species A shows a sharp exponential decline in growth with a small amount of crowding. The growth of species B declines only after a minimum crowding threshold has been reached.

Eq. (4.1). In our formulation of the effects of crowding on the growth and survival of individual trees we wanted a function that could model species that respond differently to neighbours. We used Eq. (4.1) because it allows for considerable flexibility in the shape of the functional dependence of growth and survival on the NCI. For example, Fig. 4.1 illustrates very different responses by two hypothetical species, one that is very sensitive to neighbourhood competition and shows a rapid exponential decline in realized growth with increases in NCI, and another species that shows a minimum threshold response. The ratio of parameter D to parameter C is small in the first species and larger in the second species.

The neighbourhood crowding index is defined as follows:

$$\text{NCI}_{\text{focal},k} = \text{DBH}_{\text{focal},k}^{\gamma} \sum_{i=1}^{S} \sum_{j=1}^{n_i} \lambda_{ik} \frac{\text{DBH}_j^{a_k}}{\text{Distance}_j^{\beta_k}} \tag{4.2}$$

Here $\text{NCI}_{\text{focal},k}$ is the specific value of the neighbourhood crowding index for a given focal individual of species k, and $\text{DBH}_{\text{focal},k}^{\gamma}$ is the DBH of that focal individual, weighted by an exponent γ that characterizes the sensitivity of individuals of that given DBH of the focal species k to neighbourhood effects. The double sum is over S species and the n_i neighbours of each species i in the focal individual's neighbourhood of estimated maximum radius R. The parameter λ_{ik} is a pairwise competition coefficient analogous to Lotka–Volterra competition coefficients, and it estimates the per capita effects of species i on species k. Parameters α_k and β_k allow non-linear scaling of the effects of neighbour size (DBH) and distance on focal species k. The function in Eq. (4.2) leaves unspecified what the effective neighbourhood size of a focal plant is. Based on our previous analyses that showed that neighbourhood effects were undetectable beyond

about 12 m, we chose an initial neighbourhood radius of 15 m. The fitting procedure then estimates what neighbourhood radius (R) within this 15-m range best
explains patterns of growth and survival for each species.

Although Eq. (4.2) is a suitable functional form for NCI measures in temperate
forests that have relatively small numbers of tree species, it is impractical to
estimate S competition coefficients for each species in species-rich tropical tree
communities such as BCI. We therefore grouped species into two to four classes
or functional groups, and estimated the competition coefficients of these classes
with the focal species. There were seven parameters in addition to competition
coefficients to estimate for a given focal species: maximum growth (probability
of survival) rate, effective neighbourhood radius (R), α, β, γ, C and D.

We obtained the data for testing the models from the 1990 and 1995 censuses of the BCI 50-ha plot. We adopted maximum likelihood methods and
used an optimization procedure to find model parameter values that maximized the likelihood of obtaining the observed overall growth and survival data
for each focal species, given the particular model. We used simulated annealing (a global optimization procedure) to determine the most likely parameter
values (i.e. the parameter values that maximize the log likelihood), given our
observed data (Goffe *et al.* 1994). We then used Akaike's Information Criterion
(AIC_c) corrected for small sample size to identify the best model among the
set of models included in our analyses. The model with the smallest value
for AIC_c is the most parsimonious and therefore the best model among a set
of candidate models (Burnham & Anderson 2002). These maximum-likelihood
methods assume independence of focal individuals, but the growth and survival data are spatially autocorrelated on a spatial scale of about 5 m. Although
variances and confidence limits will be slightly underestimated, parameter estimates themselves, and therefore model selection, are generally unaffected by
spatial autocorrelation among the observations (Hubbell *et al.* 2001). This is an
advantage of likelihood methods over traditional parametric approaches. Further details of the likelihood and fitting methods can be found in Uriarte *et al.*
(2004b).

We evaluated the comparative fit of four models of the effects of neighbour
groups. Model 1 represented the ecological-equivalence hypothesis (species of
neighbour unimportant). In Model 2, we distinguished two classes: conspecific
neighbours and heterospecific neighbours. In Model 3, we distinguished three
classes of neighbours: conspecifics, confamilial but not conspecific neighbours,
and other more distantly related neighbours. In Model 4, we included three
classes of neighbours, but this time we distinguished conspecifics from heterospecifics divided into gap species and shade-tolerant species. Models 1, 2 and
3 were nested in that order with Model 3 being the largest, while Model 4
was distinct. We ran the four growth models on 60 focal species and survival

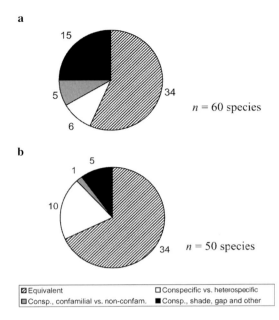

a

b

Equivalent · Consp., confamilial vs. non-confam. · Conspecific vs. heterospecific · Consp., shade, gap and other

Figure 4.2 Number of species included in the analyses supporting alternative growth (a) and survival (b) models. The equivalent-neighbours model represented the ecological-equivalence hypothesis (species of neighbour unimportant). The second model distinguished between two classes: conspecific and heterospecific neighbours. In Model 3, we distinguished three classes of neighbours: conspecifics, confamilial but not conspecific neighbours, and other more distantly related neighbours. In Model 4, we included three classes of neighbours, but this time we distinguished conspecifics from heterospecifics divided into gap species and shade-tolerant species.

models on 50 species. We analysed shade-tolerant species with > 500 individuals and gap species, which are less abundant, with > 200 individuals, and the focal individuals had a DBH of 1–4 cm. In the case of survival analyses, we analysed focal individuals of 1–2 cm DBH, the size range at which most mortality occurs. There were fewer total stems in the survival analysis, so we reduced the cutoff abundance for gap species to > 100 individuals. Among the species studied for growth, there were seven gap species and 53 shade-tolerant species. Classified by growth form, there were 22 canopy species, 21 midstorey species and 17 under-storey treelets. Among the species studied for survival, there were six gap species and 44 shade-tolerant species. By growth form, there were 18 canopy species 20 midstorey species and 12 understorey treelets.

Results

Growth

For over half of the focal species (34 out of 60), the growth model best sup-ported (with the lowest AIC_c value) was Model 1, which treated all neighbours as ecologically equivalent (Table 4.1, Fig. 4.2a). For the remaining 26 out of the 60 species in the analyses, there was striking variation in the effects of crowding (as measured by λ_s, our species-specific crowding index), depending on the iden-tity both of the focal tree and of its neighbours (Table 4.1, Fig. 4.1a). For six out of these 24 focal species, the best supported model was Model 2, which differenti-ated between conspecific and heterospecific neighbours. Conspecific neighbours

Table 4.1 *Best growth and survival models for tree species included in the analyses*

Genus	Species	Family	Light guild	Best growth model	Best survival model
Alseis	blackiana	RUBIACEAE	Shade	Conspecific/Heterospecific	Equivalence
Aspidosperma	cruenta	APOCYNACEAE	Shade	Equivalence	NA
Beilschmiedia	pendula	LAURACEAE	Shade	Conspecific/Heterospecific	Gap/Shade
Brosimum	alicastrum	MORACEAE	Shade	Equivalence	NA
Calophyllum	longifolium	CLUSIACEAE	Shade	Conspecifics/Confamilials/Others	Equivalence
Casearia	aculeata	FLACOURTIACEAE	Gap	Equivalence	NA
Cassipourea	elliptica	RHIZOPHORACEAE	Shade	Equivalence	Equivalence
Chrysophyllum	argenteum	SAPOTACEAE	Shade	Equivalence	Conspecific/Heterospecific
Cordia	bicolor	BORAGINACEAE	Gap	Gap/Shade	Conspecific/Heterospecific
Cordia	lasiocalyx	BORAGINACEAE	Shade	Equivalence	Equivalence
Coussarea	curvigemmia	RUBIACEAE	Shade	Conspecific/Heterospecific	Conspecific/Heterospecific
Croton	billbergianus	EUPHORBIACEAE	Gap	Gap/Shade	Gap/Shade
Cupania	sylvatica	SAPINDACEAE	Shade	Equivalence	Conspecifics/Confamilials/Others
Desmopsis	panamensis	ANNONACEAE	Shade	Gap/Shade	Equivalence
Drypetes	standleyi	EUPHORBIACEAE	Shade	Gap/Shade	Conspecific/Heterospecific
Eugenia	coloradensis	MYRTACEAE	Shade	Equivalence	Equivalence
Eugenia	galalonensis	MYRTACEAE	Shade	Equivalence	Equivalence
Eugenia	nesiotica	MYRTACEAE	Shade	Equivalence	Equivalence
Eugenia	oerstedeana	MYRTACEAE	Shade	Equivalence	Equivalence
Faramea	occidentalis	RUBIACEAE	Shade	Gap/Shade	Gap/Shade
Garcinia	intermedia	CLUSIACEAE	Shade	Equivalence	Equivalence
Garcinia	madruno	CLUSIACEAE	Shade	NA	Equivalence
Guarea	guidonia	MELIACEAE	Shade	Equivalence	Conspecific/Heterospecific
Guarea	'fuzzy'	MELIACEAE	Shade	Conspecifics/Confamilials/Others	Equivalent
Guatteria	dumetorum	ANNONACEAE	Shade	Equivalence	Equivalence
Heisteria	concinna	OLACACEAE	Shade	Conspecific/Heterospecific	NA
Herrania	purpurea	STERCULIACEAE	Shade	Equivalence	NA
Hirtella	triandra	CHRYSOBALANACEAE	Shade	Gap/Shade	Gap/Shade
Inga	marginata	FABACEAE:MIMOS.	Gap	Gap/Shade	Equivalence
Inga	nobilis	FABACEAE:MIMOS.	Shade	Equivalence	Conspecific/Heterospecific

Genus	species	Family	Light	Model A	Model B
Inga	*umbellifera*	FABACEAE:MIMOS.	Shade	Equivalence	Equivalence
Lacistema	*aggregatum*	FLACOURTIACEAE	Shade	Gap/Shade	Equivalence
Laetia	*thamnia*	FLACOURTIACEAE	Shade	Conspecific/Heterospecific	NA
Lonchocarpus	*latifolius*	FABACEAE:FABOID.	Shade	Gap/Shade	Equivalence
Maquira	*costaricana*	MORACEAE	Shade	Equivalence	Equivalence
Miconia	*affinis*	MELASTOMATACEAE	Gap	Equivalence	Equivalence
Miconia	*argentea*	MELASTOMATACEAE	Gap	Gap/Shade	Equivalence
Pentagonia	*macrophylla*	RUBIACEAE	Shade	Equivalence	NA
Picramnia	*latifolia*	PICRAMNIACEAE	Shade	Equivalence	Equivalence
Pouteria	*reticulata*	SAPOTACEAE	Shade	Equivalence	Equivalence
Prioria	*copaifera*	FABACEAE:CAESAL.	Shade	Gap/Shade	Conspecific/Heterospecific
Protium	*costaricense*	BURSERACEAE	Shade	Conspecifics/Confamilials/Others	Equivalence
Protium	*panamense*	BURSERACEAE	Shade	Conspecifics/Confamilials/Others	Equivalence
Protium	*tenuifolium*	BURSERACEAE	Shade	Conspecifics/Confamilials/Others	Equivalence
Pterocarpus	*rohrii*	FABACEAE:FABOID.	Shade	Gap/Shade	Equivalence
Quararibea	*asterolepis*	BOMBACACEAE	Shade	Equivalence	Equivalence
Randia	*armata*	RUBIACEAE	Shade	Equivalence	Equivalence
Simarouba	*amara*	SIMAROUBACEAE	Gap	Equivalence	Equivalence
Sloanea	*terniflora*	ELAEOCARPACEAE	Shade	Equivalence	NA
Swartzia	*simplex var. grandiflora*	FABACEAE:CAESAL.	Shade	Gap/Shade	Equivalence
Swartzia	*simplex var. ochnacea*	FABACEAE:CAESAL.	Shade	Equivalence	Equivalence
Tabernaemontana	*arborea*	APOCYNACEAE	Shade	Equivalence	Equivalence
Tachigalia	*versicolor*	FABACEAE:CAESAL.	Shade	Gap/Shade	Gap/Shade
Talisia	*nervosa*	SAPINDACEAE	Shade	Equivalence	NA
Talisia	*princeps*	SAPINDACEAE	Shade	Equivalence	Equivalence
Tetragastris	*panamensis*	BURSERACEAE	Shade	Conspecific/Heterospecific	Equivalence
Trichilia	*pallida*	MELIACEAE	Shade	Equivalence	NA
Trichilia	*tuberculata*	MELIACEAE	Shade	Gap/Shade	Conspecific/Heterospecific
Unonopsis	*pittieri*	ANNONACEAE	Shade	Equivalence	NA
Virola	*sebifera*	MYRISTICACEAE	Shade	Equivalence	Conspecific/Heterospecific
Xylopia	*macrantha*	ANNONACEAE	Shade	Equivalence	Conspecific/Heterospecific

We used data for the 1990–1995 census interval in the BCI 50-ha plot. The best model was determined using Akaike's Information Criterion corrected for small sample size. NA indicates that the species was not included in that set of analyses.

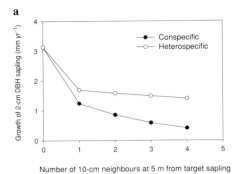
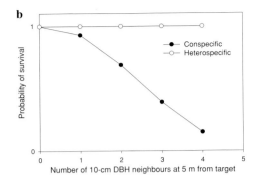

Figure 4.3 Effects of neighbours for the most parsimonious model of growth of *Alseis blackiana* (a) and survival of *Prioria copaifera* (b). Decrease in growth for *Alseis blackiana* was calculated using the following maximum-likelihood parameter values: Max. growth = 3.13, $C = 9.01$, $D = 1.01$, $\alpha = 1.30$, $\beta = 0.19$, $\gamma = -0.45$, λ (conspecific) = 1; λ (heterospecific neighbour) = 0.43.

always had much worse effects on the growth of the focal sapling than did heterospecific neighbours (Fig. 4.3a). A second group of five focal species supported Model 3, the model that distinguished among conspecific, confamilial and non-confamilial neighbours. Conspecific or confamilial effects on sapling growth of all five focal species were greater than the effect of non-confamilial neighbours. Fifteen species supported Model 4, which distinguished between conspecifics and heterspecific gap or shade-tolerant neighbours. Once again, conspecifics effects were stronger than the effects of light guild of heterospecifics. Functional group similarity was also a good predictor of the strength of neighbourhood interactions among species. The effect of gap species on the growth of shade-tolerant targets was generally weaker than the effect of other shade-tolerant neighbouring species. In contrast, the effect of gap species on the growth of gap species was always stronger than the effect of shade-tolerant neighbours (see Uriarte *et al.* (2004b) for details).

Survival

Three-quarters of focal species (34 out of 50) supported survival Model 1, the model in which all neighbours are treated as ecologically equivalent (Fig. 4.2b, Table 4.1). The remaining 16 species in the analyses showed variation in the effects of crowding, depending on the identity both of the focal tree and of its neighbours (Fig. 4.2b, Table 4.1). For 10 of these focal species, the model best supported distinguished between conspecific and heterospecific neighbours. Survival effects were consistent with the results from our growth analyses: negative effects from conspecific neighbours on survival of the focal sapling were always much stronger than heterospecific effects (Fig. 4.3b). Five species supported survival Model 4, the model that distinguished between conspecifics and two

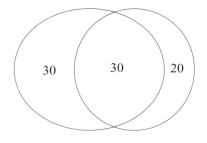

30   30   20

Growth model n = 60   Survival model n = 50

Figure 4.4 Number of species included in the analyses with similar responses to neighbour groupings for both growth and survival. Intersection shows number of species that supported the equivalent competitor model (Model 1) or a distinct neighbour model (Models 2–4) for both growth and survival.

categories of heterospecific neighbours, gap species and shade-tolerant species. In general, the negative effect of conspecifics and gap species on target survival was greater than the effect of shade-tolerant neighbours. Only one species supported Model 3, the model that distinguished between confamilial and non-confamilial heterospecific neighbours.

Given the criteria we chose to select focal species, we were able to analyse both growth and survival data for 50 species. For over 60% of this group of species (30 out of 50), data supported similar effects of broad neighbour groupings on both growth and survival (Fig. 4.4). Eleven species that showed distinct con-specific effects on growth also showed distinct effects on survival while 19 species that supported the equivalent-competitor model did not show a response to conspecific neighbours in either growth or survival. A list of the species analysed and the best-supported models for growth and survival for each is given in Table 4.1.

Discussion

We evaluated four models of neighbourhood crowding effects on the growth and survival of focal saplings (of DBH 1–4 cm) of 50–60 species in the BCI 50-ha plot over a 5-year census interval, from 1990 to 1995. The most salient result was that the ecological equivalence model (Model 1) was the model best supported by more than half of the species for growth, and three-quarters of the species for survival. The ecological-equivalence model treats all neighbours, irrespective of species, as having the same crowding effects, controlling for size and distance from the focal sapling. Of the species that supported one of the three other models, all of which included a separate class for conspecifics neighbours, the strongest effect by far (as measured by the magnitudes of the λ coefficients), was the negative effect of conspecifics neighbours. The dominating importance of conspecifics density in these analyses is congruent with similar findings by Hubbell *et al.* (1990); Hubbell *et al.* (2001); and Peters (2003). However, these previous analyses found pervasive density-dependence at the community level

(Hubbell *et al.* 2001) and could have been driven by strong density-dependence in a few dominant species (Hubbell *et al.* 1990).

Common descent also increased the strength of interaction between neighbouring species even at a higher (familial) level of taxonomic separation. Confamilial effects on sapling growth and survival for focal species that supported Model 3 were always greater than effects of neighbours that were not in the same plant family. This is consistent with the idea that related species are more likely to respond in the same or similar ways to the mechanisms that are ultimately responsible for diversity. Nevertheless, support for Model 3 was relatively weak for the group of focal species as a whole.

A few species also supported a model that distinguished between conspecifics and divided heterospecifics into gap species and shade-tolerant species. In the growth analyses, interactions among species belonging to the same light guild were always stronger than those between individuals from different light guilds. This is expected given that species are primarily surrounded by adult trees that belong to the same light guild as the focal sapling. These large neighbouring trees have the strongest effects on the growth of focal saplings. In the survival analyses, we found that heterospecific neighbours that were gap species had a stronger effect than shade-tolerant species. This is probably a correlated response in the focal saplings to being in a gap. Shade-tolerant species have much higher mortality rates in light gaps than in non-gap understorey sites (Hubbell *et al.* 1999). The correlation arises because these are the only sites where gap species occur. Taken together, these results suggest that convergent evolution along broad life-history strategies determine neighbourhood interactions for some species.

Support for the model of ecological equivalence is stronger and more pervasive than in other studies we have conducted at Luquillo Forest in Puerto Rico (Uriarte *et al.* 2004a) and in temperate forests (Canham *et al.*, in press; C. D. Canham *et al.*, unpublished data). There are several possible explanations for this difference. One is that the difference is real and due to the greater species richness and greater unpredictability of neighbourhoods around saplings of species in the BCI forest. Two individuals of the same BCI species share only about 15% of species in common among their 20 nearest neighbours (Hubbell & Foster 1986). In contrast, this percentage is much higher in species-poor temperate forests that have only 15–20 species. The strength and directionality of neighbourhood effects therefore can differ markedly for different individuals of the same species in the BCI forest. Indeed, one might expect the strength of pairwise competitive effects to decline roughly as the square of the number of species in the community.

Nevertheless, none of the models, even those that were supported best and indicated significant differences among neighbour groups, explained much of the variance in growth and survival (see Uriarte *et al.* (2004b) for details). There could be several reasons for this low explanatory power. First, the models may not be accurately capturing the true effects of neighbours in the BCI forest. Although this is always a possibility, it should be noted that these same spatially explicit neighbourhood models perform well at Luquillo and in temperate forests. Despite this low explanatory power on BCI, it is nevertheless interesting that about 60% of species included in the analyses displayed similar responses in growth and survival to the presence of conspecific neighbours. Species that showed strong conspecific effects on growth also showed strong effects on survival while species that supported the equivalent competitor model did not show a response to conspecific neighbours in either growth or survival. Similarity in the effects of different groups of neighbours on both growth and survival of focal seedlings supports the notion that the processes that determine growth and mortality in forests are tightly linked (Monserud 1976; Kobe *et al.* 1995; Kobe 1996; Wyckoff & Clark 2002). This suggests that the models are indeed capturing reality to some extent. This said, in 40% of the species there was a different best model for growth and survival.

A second possibility for low explanatory power is perhaps that the effects are stronger at a finer level of species discrimination, i.e. at the species level rather than at the functional group or taxonomic pooling levels that we tested here. The argument is that these finer levels of discrimination reflect environmentally driven niche differentiation. Although this is a possibility, we do not think it is likely given that measured heterospecific effects were much weaker than conspecific effects. The third possibility is that once again it is the high diversity and unpredictability of neighbourhoods around individual saplings of each species that obscures neighbourhood effects. These highly diverse and different neighbourhoods may elevate the 'within treatment' variance within our functional groups to the point where the 'between treatment' effects are no longer significant. If this explanation is the right one, it is nevertheless still biologically interesting. It may indicate that species are intrinsically highly variable and overlap broadly in their growth and survival responses to neighbourhood effects. Thus, even in cases where there is a small role for a deterministic 'skeleton', stochastic behaviour in neighbourhood responses remain quantitatively significant. Stochastic effects can clearly overwhelm known neighbourhood effects of light and soil nutrients on tree growth and survival (Beckage & Clark 2003). Thus, the low explanatory power of the models may be further indirect support for ecological equivalence.

The strong showing of the ecological equivalence hypothesis in this study should not be taken as demonstrating that the BCI forest is fully symmetric-neutral. There are several reasons to be cautious of over-interpretation. The most important reason is that we tested neighbourhood models for only one of the life-history stages of tropical trees, saplings of DBH 1–4 cm. We know, for example, that many BCI species differ in the strength of the density-dependent

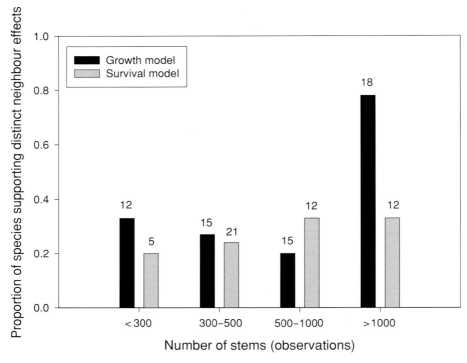

Figure 4.5 Proportion of species supporting distinct neighbour effects for growth and survival analyses. Numbers above bars indicate number of species included in the analyses for that sample size class.

mortality they experience in the seed-to-seedling transition (Harms *et al.* 2000), and seedling-to-sapling recruitment (Condit *et al.* 1994a; Wills *et al.* 1997). In general, density-dependent effects appear to weaken or disappear as seedlings are recruited into the larger size classes (Connell *et al.* 1984; Harms *et al.* 2000; but see Condit *et al.* 1994b; Peters 2003). Studies elsewhere have demonstrated that density-dependent effects can range from allelopathic to facilitative through a plant's life history (Goldberg *et al.* 2001). This kind of analysis has not been completed for tropical forests. Also, species differ in their mean growth and survival rates, which will affect their residency times in the forest and the likelihood that they will occupy new sites when these become available (Chesson & Warner 1981).

Finally, detection of ecological equivalence may be a function of the sample size of stems available for different species. If this is the case, the ability to detect interactions should increase with sample size. Figure 4.5 shows the percentage of species supporting the equivalent competitor growth and survival models as a function of sample size class. Although a few species with large sample sizes support the ecological-equivalence growth model (e.g. *Garcinia intermedia* and

Swartzia simplex var. *ochnaceae)*, the ability to detect distinct neighbour effects on focal-sapling growth increases dramatically for sample sizes greater than 1000 stems. Thus, there are likely to be non-symmetrical 'hidden' interactions that cannot be detected with this dataset. Curiously, the probability of detecting distinct neighbour effects on focal-sapling survival was fairly constant regardless of sample size. Previous analyses of density dependence in this forest (Hubbell *et al.* 1990) and elsewhere (Uriarte *et al.* 2004a) found that being next to a conspecific was more likely to affect growth than survival. Self-thinning and sheer 'crowding' may be the most important drivers of sapling survival regardless of the identity of neighbours, particularly for common species (Hubbell *et al.* 2001). In contrast, the effects of neighbours on focal-sapling growth may be more complex (e.g. differentiation in soil resource requirements or neighbour effects on depletion of soil resources). Incorporating these differences into models of tropical forest dynamics should provide insights into the importance of these effects for the maintenance of diversity.

References

Banavar, J. R., Hubbell, S. P. & Maritan, A. (2003) Frequency dependence and the statistical mechanics of relative species abundance. *Science* **424**: 1034–1037.

Beckage, B. & Clark, J. S. (2003) Seedling survival and growth of three forest tree species: the role of spatial heterogeneity. *Ecology* **84**: 1849–1861.

Bella, I. E. (1971) A new competition model for individual trees. *Forest Science* **17**: 364–372.

Burnham, K. P. & Anderson, D. R. (2002) *Model Selection and Multimodel Inference: A Practical Information-theoretic Approach.* New York: Springer.

Canham, C. D., LePage, P. T. & Coates, K. D. In press. A neighbourhood analysis of canopy tree competition: effects of shading versus crowding. *Canadian Journal of Forest Research* **34**: 778–787.

Chase, J. & Leibold, M. (2003) *Ecological Niches: Linking Classical and Contemporary Approaches.* Chicago: University of Chicago Press.

Chave, J., Muller-Landau, H. C. & Levin, S. A. (2002) Comparing classical community models: theoretical consequences for patterns of diversity. *American Naturalist* **159**: 1–23.

Chesson, P. L. & Warner, R. R. (1981) Environmental variability promotes coexistence in lottery competitive systems. *American Naturalist* **117**: 923–943.

Condit, R., Hubbell, S. P. & Foster, R. B. (1994a) Recruitment near conspecific adults and the maintenance of tree and shrub diversity in a neotropical forest. *American Naturalist* **140**: 261–286.

(1994b) Density dependence in two understorey tree species in a neotropical forests. *Ecology* **5**: 671–680.

Connell, J. H. (1971) On the role of natural enemies in preventing competitive exclusion in some marine animals and in rain forest trees. In P. J. den Boer and G. R. Gradwell, eds., *Dynamics of Populations.* Wageningen: Centre for Agricultural Publishing and Documentation, pp. 298–312.

(1978) Diversity in tropical rain forests and coral reefs. *Science* **199**: 1302–1310.

(1980) Diversity and the coevolution of competitors: the ghost of competition past. *Oikos* **35**: 131–138.

Connell, J. H., Tracey, J. G & Webb, L. W. (1984) Compensatory recruitment, growth and mortality as factors maintaining rain forest tree diversity. *Ecological Monographs* **54**: 141–164.

Denslow, J. S. (1987) Tropical rainforest gaps and tree species diversity. *Annual Review of Ecology and Systematics* **18**: 431–451.

Federov, A. A. (1966) The structure of the tropical rain forest and speciation in the humid tropics. *Journal of Ecology* **54**: 1–11.

Goffe, W. L., Ferrier, G. D. & Rogers, J. (1994) Global optimization of statistical functions with simulated annealing. *Journal of Econometrics* **60**: 65–99.

Goldberg, D. E. & Werner, P. A. (1982) Equivalence of competitors in plant communities: a null hypothesis and a field experimental approach. *American Journal of Botany* **70**: 1098–1104.

Goldberg, D. E., Turkington, R., Olsvig-Whittaker, L. & Dyer, A. R. (2001) Density dependence in an annual plant community: variation among life history stages. *Ecological Monographs* **71**: 423–446.

Harms, K. E., Wright, J. S., Calderón, O., Hernández, A. & Herre, E. A. (2000) Pervasive density-dependent recruitment enhances seedling diversity in tropical forests. *Nature* **404**: 493–495.

Hartshorn, G. S. (1978) Treefalls and tropical forest dynamics. In P. B. Tomlinson and M. H. Zimmerman, eds., *Tropical Trees as Living Systems*. Cambridge: Cambridge University Press, pp. 617–628.

Harvey, P. H. & Pagel, M. D. (1991) *The Comparative Method in Evolutionary Biology*. Oxford: Oxford University Press.

Hegyi, F. (1974) A simulation model for managing jack-pine stands. In J. Fries, ed., *Growth Models for Tree and Stand Simulation*. Stockholm: Royal College of Forestry, pp. 74–90.

Hubbell, S. P. (1979) Tree dispersion, abundance and diversity in a tropical dry forest. *Science* **203**: 1299–1309.

(1997) A unified theory of biogeography and relative species abundance and its application to tropical rain forests and coral reefs. *Coral Reefs* **16**: s9–s21.

(2001) *The Unified Theory of Biodiversity and Biogeography*. Princeton Monographs in Population Biology. Princeton, NJ: Princeton University Press.

Hubbell, S. P. & Foster, R. B. (1983) Diversity of canopy trees in a neotropical forest and implications for the conservation of tropical trees. In S. J., Sutton, T. C. Whitmore and A. C. Chadwick, eds., *Tropical Rain Forest: Ecology and Management*. Oxford: Blackwell, pp. 25–41,

(1986) Biology, chance and history and the structure of tropical rain forest tree communities. In J. M. Diamond and T. J. Case, eds., *Community Ecology*. New York: Harper and Row, pp. 314–329.

Hubbell, S. P. & Lake, J. (2003) The neutral theory of biogeography and biodiversity: and beyond. In T. Blackburn and K. Gaston, eds., *Macroecology: Concepts and Consequences*. Oxford: Blackwell, pp. 45–63.

Hubbell, S. P., Condit, R. & Foster, R. B. (1990) Presence and absence of density dependence in a neotropical community. *Philosophical Transactions of the Royal Society of London B* **330**: 269–281.

Hubbell, S. P., Foster, R. B., O'Brien, S. *et al.* (1999) Light-gap disturbances, recruitment limitation, and tree diversity in a tropical forest. *Science* **283**: 554–557.

Hubbell, S. P., Ahumada, J. A., Condit, R. & Foster, R. B. (2001) Local neighbourhood effects on long-term survival of individual trees in a neotropical forest. *Ecological Research* **16**: 859–875.

Hurtt, G. C. & Pacala, S. W. (1995) The consequences of recruitment limitation: reconciling chance, history, and competitive differences between plants. *Journal of Theoretical Biology* **176**: 1–12.

Hutchinson, G. E. (1961) The paradox of the plankton. *American Naturalist* **95**: 137–145.

Janzen, D. (1970) Herbivores and the number of tree species in tropical forests. *American Naturalist* **104**: 501–528.

Kobe, R. K. (1996) Intraspecific variation in sapling mortality and growth predicts genographic variation on forest composition. *Ecological Monographs* **66**: 181–201.

Kobe, R. K, Pacala, S. W, Silander, J. A. & Canham, C. D. (1995) Juvenile tree survivorship as a component of shade tolerance. *Ecological Applications* **5**: 517–532.

MacArthur, R. H. & Wilson, E. O. (1967) *The Theory of Island Biogeography*. Monographs in Population Biology. Princeton, NJ: Princeton University Press.

McGill, B. J. (2003) A test of the unified neutral theory of biodiversity. *Nature* **422**: 881–885.

Monserud, R. A. (1976) Simulation of forest tree mortality. *Forest Science* **22**: 438–444.

Orians, G. H. (1982) The influence of tree-falls in tropical forests in tree species richness. *Tropical Ecology* **23**: 255–279.

Peters, H. (2003) Neighbour-regulated mortality: the influence of positive and negative density dependence on tree populations in species-rich forests. *Ecology Letters* **6**: 757–765.

Ricklefs, R. E. (1977) Environmental heterogeneity and plant species diversity: a hypothesis. *American Naturalist* **111**: 376–381.

Schmida, A. & Ellner, S. (1985) Coexistence of plant species with similar niches. *Vegetatio* **58**: 29–55.

Terborgh, J., Pitman, N. C. A., Silman, M. R, Schlichter, H. & Núñez, P. (2001) Maintenance of diversity in tropical forests. In D. J. Levey, W. R. Silva and M. Galetti, eds., *Seed Dispersal and Frugivory: Ecology, Evolution and Conservation*. Oxfordshire: CAB International Press.

Tilman, D. (1994) Competition and biodiversity in spatially structured habitats. *Ecology* **75**: 2–16.

Uriarte, M., Canham, C. D., Thompson, J. & Zimmerman, J. K. (2004a). A maximum-likelihood, spatially-explicit analysis of tree growth and survival in a hurricane-driven tropical forest. *Ecological Monographs* **71**: 591–614.

Uriarte, M., Condit, R., Canham, C. D. & Hubbell, S. P. H. (2004b). A spatially-explicit model of sapling growth in a tropical forest. Does the identity of neighbours matter? *Journal of Ecology* **92**: 348–360.

van Steenis, C. G. G. J. (1969) Plant speciation in Malesia, with special reference to the theory of non-adaptive salutatory evolution. *Biological Journal of the Linnean Society* **1**: 97–133.

Vettenranta, J. (1999) Distance-dependent models for predicting the development of mixed coniferous stands in Finland. *Silva Fennica* **33**: 51–72.

Volkov, I., Banavar, J. R., Hubbell, S. P. & Maritan, A. (2003) Neutral theory and relative species abundance in ecology. *Nature* **424**: 1035–1037.

Wagner, R. G. & Radosevich, S. R. (1998) Neighbourhood approach for quantifying interspecific competition in coastal Oregon forests. *Ecological Applications* **8**: 779–794.

Webb, C. O. (2000) Exploring the phylogenetic structure of ecological communities: an example for rain forest trees. *American Naturalist* **156**: 145–155.

Wills, C., Condit, R., Foster, R. B. & Hubbell, S. P. (1997) Strong density- and diversity-related effects help maintain tree species diversity in a neotropical forest. *Proceedings of the National Academy of Sciences* **94**: 1252–1257.

Wimberly, M. C. & Bare, B. B. (1996) Distance-dependent and distance-independent models of Douglas fir and western hemlock basal area growth following silvicultural treatment. *Forest Ecology and Management* **89**: 1–11.

Wright, S. J. (2002) Plant diversity in tropical forests: a review of mechanisms of species coexistence. *Oecologia* **130**: 1–14.

Wyckoff, P. H. & Clark, J. S. (2002) The relationship between growth and mortality for seven co-occurring tree species in the southern Appalachian Mountains. *Journal of Ecology* **90**: 604–615.

Zeide, B. (1993) Analysis of growth equations. *Forest Science* **39**: 594–616.

CHAPTER FIVE

Ecological drift in niche-structured communities: neutral pattern does not imply neutral process

DREW W. PURVES AND STEPHEN W. PACALA

Princeton University

Introduction

The neutral vs. structure debate

We can define a neutral community as one in which all species, and so all individuals, are equivalent, in the sense that they are interchangeable at all times and under all conditions. In contrast, we can define a structured community as one in which species are not equivalent, and species-specific differences affect the population dynamics, and therefore the behaviour, of the community.

This distinction is an important one, because in a neutral community the biodiversity, as measured by species richness and abundance patterns, has nothing to do with the biogeochemical functioning of the community (e.g. carbon fixation and nutrient-cycling). In fact, in a truly neutral community one could eliminate all but one species without affecting the biogeochemical functioning of the community at all.

In contrast, much of the species-specific variation in biological traits observed in reality (see below) has direct relevance for the functioning of the community. For example, the short-term carbon uptake of a forest depends on the growth rates of the individual trees, and the long-term carbon storage depends on adult life-span and wood density, and there is wide species-specific variation in these traits. In niche-structured communities, the biodiversity and functioning are intimately linked, and some combination of at least some species is required to maintain the functioning of the community. In the most highly structured community possible there is no equivalence between any of the species, which is the so-called 'one species one niche' idea so prevalent in the history of ecology: in such a community, removing just one species has a significant impact on the dynamics and functioning of the community.

Which of these two pictures of communities – neutral or structured – is nearer to the truth? Is it 'one species one niche' or 'all species one niche'? This is the neutral vs. structure debate, and it continues apace because there is

Biotic Interactions in the Tropics: Their Role in the Maintenance of Species Diversity, ed. D. F. R. P. Burslem, M. A. Pinard and S. E. Hartley. Published by Cambridge University Press. © Cambridge University Press 2005.

good evidence for both sides of the argument. For some important recent contributions to this argument see Bell (2001); Hubbell (2001); Chave *et al.* (2002); Condit *et al.* (2002); Duivenvoorden *et al.* (2002); Clark and McLachlan (2003); Fargione *et al.* (2003); McGill (2003); Phillips *et al.* (2003); Tuomitso *et al.* (2003) and Volkov *et al.* (2003).

Evidence for structure

There is an enormous amount of detailed evidence that species differ significantly in many different biological traits, and that these differences affect the dynamics and functioning of communities. First, at a global scale the principal features of vegetation can be predicted easily from climate and other external forcings (Walter 1973, 2002; Archibold 1995), and the examples of these 'biomes' on different continents often share few or no species in common: this sort of convergence cannot possibly be explained by ecological drift (and indeed, no-one has suggested that it is). At the global scale, then, species are not close to being neutral, and this prompts the question: at what spatial scale do communities become collections of equivalent species? Most field ecologists will say that no such scale exists, and that there are always important differences between some of the species in a given community.

Focusing on the example of moist tropical forests (which compared with many other biomes have been poorly studied) there is good evidence for species-specific variation along at least four very different trade-off axes. First and strongest of these is the trade-off between growth rate in high light and survival in low light (e.g. Fig. 5.7a, b), which is known to drive succession in many temperate forests (Bazzaz 1996; Pacala *et al.* 1996). The successional status of species in tropical forests is correlated with this trade-off (Poorter & Arets 2003), as seen in the well-known neotropical example of the fast-growing, shade-intolerant pioneer genus *Cecropia*, contrasted with the shade-tolerant, long-lived late-successionals that dominate old-growth stands (Foster & Brokaw 1982). Second, different species segregate along a number of different environmental gradients including soil drainage, topography and nutrient levels (e.g. Harms *et al.* 2001; Pyke *et al.* 2001; Pelissier *et al.* 2002; Phillips *et al.* 2003). Third, species adopt different positions along the competition–colonization trade-off, whereby inferior competitors can coexist with superior competitors because they are better colonizers (Levins & Culver 1971), and which is most simply seen in the inevitable trade-off of seed size against seed number (e.g. Dalling & Hubbell 2002; see Coombes and Grubb 2003 for a review). Fourth, different species are resistant to different pests, putting the seedlings and saplings of locally rare species at an advantage, the so-called Janzen–Connell effect (Janzen 1970; Connell 1971; for tropical forests see Wright 2002 and Peters 2003).

None of these niche mechanisms in isolation is able to explain the coexistence of communities with such high diversity as is seen in tropical forests;

their relative importance is likely to depend on scale and location; and their interactions are poorly understood. Nonetheless, the literature makes a strong case that there is at least some structure according to each of these mechanisms, and several others, operating in any tropical forest community (Wright 2002).

Evidence for neutrality

Despite all this species-specific complexity there are some aggregate properties of communities that appear to follow fairly general patterns. Specifically, the distribution of population sizes within a community often follows either a log-series distribution (Fisher *et al.* 1943; Hubbell 2001) or a log-normal distribution (Preston 1948; Sugihara 1980; Hubbell 2001). This observation alone is enough to suggest that the distribution of population sizes does not depend on every detail of the natural history of the different species, but instead depends on some more general processes.

In part to explain this observation, and in part to explain the extremely high diversity seen in some communities (especially tropical forests), a small number of researchers began to consider models where all the species were equivalent, to give a null expectation against which real patterns could be compared. Surprisingly, they found that these neutral models could explain both the distribution of population sizes, and many other widespread patterns cherished by community ecologists. As Bell (2001) and Hubbell (2001) point out, what is surprising is how little work has been attempted within the neutral paradigm, especially considering how important it has been in population genetics over the same time period.

The first of these was the observation by May (1975) that the log-normal distribution was the expected outcome from a collection of separate populations each undergoing multiplicative random walks (double one year, halve the next, $\times 1.37$ the next year, etc.), analogous to the normal distribution being the expectation from a number of random additive processes. Unfortunately, the total size of a collection of separate multiplicative random walks also grows without bound through time, which in the long term is obviously not realistic, and is completely unacceptable in communities where the total size is close to constant through time (e.g. closed-canopy forests, and many other space-limited communities of sessile organisms).

Around the same time, Caswell (1976) introduced a simple neutral model that is (crucially) based on the fundamental ecological processes of births, deaths and immigration: the model is neutral because the per-capita probabilities attached to these events are the same for individuals of any species. This idea was to become the framework for what Bell (2001) has termed 'neutral community models' or NCM, the best known of which is Hubbell's unified theory of biodiversity and biogeography (Hubbell 2001, and see below). These models are almost identical in form to neutral models developed in population genetics, and in fact some mathematical results apply to both sets of models. However, the NCM for which Caswell (1976) gave results (Model I) could not provide a good fit to observed data, and it predicted that the community would grow without bound through time (as did May's simple idea, see above). Caswell's (1976) Models II and III included the critical constraint of a constant community size, but results for these models were not given (see Hubbell 2001, p. 49).

Much more recently, Hubbell introduced a very simple neutral model which was (crucially) based on the fundamental processes of births, deaths, migrations and speciation (Hubbell 2001). The model is almost identical to that of Caswell (1976), and includes the all-important addition of a constant community size (as did Caswell's Models II and III). Amazingly, Hubbell was able to show that this model produced the log-series, and the log-normal, as special cases of a more varied family of curves (the so-called zero-sum multinomial or ZSM). By tuning only one or two biological parameters, Hubbell was able to get a near-perfect fit to the pattern of abundances in many communities, and to the tree communities in tropical forests in particular (Hubbell 2001). Furthermore, Hubbell was able to show that this theory could reproduce some of the most commonly observed features of species-area curves, including the so-called triphasic shape of these curves when plotted over a wide range of scales (see Hubbell 2001, p. 161). In addition Bell (2001) has shown that a very similar NCM can reproduce a surprising number of other aggregate patterns seen in community ecology, including the distribution of range sizes, the range–abundance relationship and turnover in community structure in space.

Resolving the conflict

How do we begin to resolve this apparent conflict? We believe that much of the neutral vs. structure debate has been mistaken, because it has assumed that the fit between neutral models and observed patterns implies that real communities are neutral (although see Bell 2001 for an explicit consideration of this assumption). In contrast, we demonstrate here that the introduction of several different forms of extremely strong structure into the most important neutral model in the literature (Hubbell's zero-sum ecological drift: Hubbell 2001) can leave its key prediction (the distribution of species abundances) completely unchanged. This tells us that, whilst a real pattern that matches a neutral model is consistent with a neutral community, it is also consistent with extremely strong structure according to one or more known mechanisms. Therefore the functional equivalence of species cannot be inferred from aggregate patterns, but has to be demonstrated directly; and since these direct tests have provided abundant evidence of non-equivalence, we conclude that real communities are structured, not neutral.

However, we also show that this match between a structured community and a neutral community depends on the dimensionality of niche structure being low compared with the number of species in the community, or more simply, that within-guild diversity is sufficiently great, and this in turn implies some functional equivalence in the community. Our final conclusions, then, are as follows. (1) Purely neutral patterns of species abundance are robust, in important cases perfectly so, to the introduction of deterministic factors that produce deterministic structure. Both the factors and the deterministic patterns they produce can be arbitrarily strong. This greatly strengthens the case that apparently random distributions of abundances are caused by neutral ecological drift: round one to the neutralists. (2) Apparently neutral distributions of abundance do provide evidence of the presence of random ecological drift among at least a subset of the species present (e.g. members of the same guild). (3) On the other hand, distributions of abundance that mirror predictions of neutral theory do not provide any evidence whatsoever for an absence of deterministic mechanisms and deterministic structure. The only way to understand these processes is to study them directly.

Ecological drift under niche structure

In this section we present an analytical demonstration that adding strong niche structure into Hubbell's (2001) neutral community model can leave the pattern of species abundances completely unchanged. We then introduce extensions of Hubbell's (2001) model, with three different forms of niche structure (successional niches, lottery niches and habitat specialization). We demonstrate that in diverse communities the pattern of species abundances predicted by these models is the same as that predicted by the purely neutral model, no matter how strong the niche structure is. But in communities where the species diversity is low in comparison with the dimensionality of niche structure (i.e. within-guild diversity is low), and where the niche structure is sufficiently strong, the pattern of species abundances given by the niche-structured models differs from that given by the neutral model. In all cases the predictions of the three different niche-structured models are the same, so even where the community diversity is low enough, and niche structure strong enough, to give non-neutral patterns of species abundances, the species abundances cannot be used to infer which form of niche structure is operating. The generality of these results is discussed in the following section.

Hubbell's neutral model

Hubbell's theory has at its core the extremely simple concept of zero-sum ecological drift (Hubbell 2001). This is best understood from a simple cartoon (Fig. 5.1). Assume (1) a community of J sites, each occupied by one individual; (2) that all sites have the same probability per unit time of being disturbed, whereby the individual occupying the site is removed (creating a gap); (3) that after this removal, the gap immediately becomes occupied by a new individual;

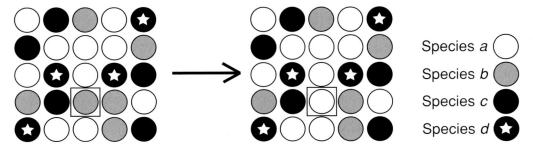

Figure 5.1 A cartoon to illustrate Hubbell's zero-sum ecological drift model. In a community of fixed size (in this case 25), a randomly chosen site is disturbed (the square indicates the disturbed site). The individual in this site is then removed, and replaced by an individual from a species within the community, or (with low probability) it is replaced from outside the community. For replacement from within, the probability that the gap is captured by species *i* is proportional to the abundance of species *i* in the community.

and (4) that the probability that this new individual is from species *i* is simply the fraction of the overall community occupied by species *i*. The process is termed zero-sum because an increase in the abundance of one species (by capturing a gap) is accompanied by an equal decrease in the abundance of another species (from a local death), so the sum of the changes in the different species' abundance is always zero. It is assumption (4) that makes Hubbell's theory neutral rather than structured: the probability of a species coming to occupy a gap is simply proportional to that species' overall abundance, and so is irrespective of species identity, implying that all species are equal.

Figure 5.1 illustrates the process: a population of 25 individuals is initially made up of up 10, 6, 5 and 4 individuals of species *a*, *b*, *c* and *d* respectively. An iteration of the model consists of removing one individual (which in this example happened to be of species *b*), then replacing this individual with species *a*, *b*, *c* or *d* with probabilities in proportion to their abundances in the population. Since in this example the lost individual was of species *b*, the probability of replacement by species *a*, *b*, *c* and *d* was respectively 10/24, 5/24, 5/24 and 4/24. A random number is drawn to decide which species comes to occupy the gap (in this case species *a*), and the iteration is complete.

Left to its own devices, the drift process described by assumptions (1)–(4) leads inevitably to a one-species community, because once a species disappears from the community it cannot re-appear (although this can take a surprisingly long time even in relatively small communities: Hubbell & Foster 1986). However, Hubbell's model contains an extra assumption: (5) that occasionally a gap is captured from outside the community, rather than from within. For local dynamics, this is taken to mean that the site is captured by an immigrant individual taken from a larger community called the metacommunity. For the metacommunity itself, 'outside' means that the site becomes occupied by an individual of a new species that has arisen via speciation.

With the addition of assumption (5), rather than approaching monodomi-nance, the population reaches an equilibrium where the loss of species through local extinction is matched by the arrival of new species from outside the community (via immigration or speciation). Moreover, once this equilibrium is reached, the distribution of population sizes takes on a characteristic shape (the family of curves that Hubbell calls the zero-sum multinomial or ZSM). In the high-diversity limit (a large metacommunity with no dispersal limitation) this shape tends to the log-series, which appears as a straight line on a chart of log abundance vs. rank abundance (Hubbell 2001: for example see Fig. 5.3 bottom right).

Niche structure can look like drift: a simple proof

How do the predictions of zero-sum ecological drift change when niche struc-ture is included? In the next section, we introduce zero-sum models including different forms and strengths of niche structure, and use simulations to exam-ine how the niche structure affects the pattern of species abundances. But before we leap into the simulations, in this section we give a simple analytical proof that it is possible for extremely strong niche structure to give exactly the same rank abundance curve as neutral ecological drift. This proof is not necessary to understand the modelling results presented in the rest of this chapter, so the reader may wish to skip to the simulations at this stage.

Hubbell (2001) showed that in the limit of a large community size (which gives a large diversity), the distribution of relative abundances from zero-sum ecological drift takes the form of a log-series (Fisher *et al.* 1943: e.g. see Fig. 5.3 bottom right), which appears as a straight line on a plot of log abundance vs. rank abundance. The formula for the log-series is:

$$I_n = (\alpha \, x^n)/n \tag{5.1}$$

where I_n is the number of species with an abundance of n individuals, α is a constant that determines the slope of the rank abundance curve and x is a constant in the range 0–1 (Hubbell 2001, p. 32). The total number of individuals in the sample N (found by the sum of Eq. 5.1 over all n) is $\alpha x/(1 - x)$, and therefore $x = 1/[1 + (\alpha/N)]$, so Eq. 5.1 can be written as:

$$I_n = \left(\frac{\alpha}{n}\right) \left(\frac{1}{1 + (\alpha/N)}\right)^n \tag{5.2}$$

Note that for a given community size N, the distribution is controlled completely by the value of α.

Hubbell was able to show that, in the high-diversity limit, α was identical to his fundamental biodiversity number $\theta = 2 J_m v$, where J_m is the number of individuals in the metacommunity, and v is the speciation rate (Hubbell 2001, p. 165; the same result was known from neutral models in population genetics: Watterson 1974). Therefore we can re-write Eq. (5.2) for Hubbell's model, as follows. First, by equating J_m with N, and equating α with θ, we note that the α/N term becomes $\theta/J_m = 2v$. Second, substituting θ for α in eq. (5.2) gives:

$$I_n = \left(\frac{\theta}{n}\right) \cdot \left(\frac{1}{1+2v}\right)^n = \left(\frac{2J_m v}{n}\right) \cdot \left(\frac{1}{1+2v}\right)^n \tag{5.3}$$

Note that in this case, for a given community size J_m, the distribution depends only on the speciation rate v.

Adding niche structure

To understand what the pattern of relative abundance of species may look like in a niche-structured analogue of Hubbell's drift model, consider a very simple, but extreme, form of niche structure. Assume that the metacommunity is divided equally into two halves with very different physical characteristics (say swamp and dryland); that there are two guilds, each containing species specialized to one habitat type only; and that this specialization is so strong that a swamp specialist can never survive in the dryland, and vice versa. In this case, we really have two completely separate *sub-communities* (the swamp and the dryland), within which the populations of the different species are free to drift up and down exactly as before (because within a guild all species are equivalent to each other). Considering one of these sub-communities in isolation (e.g. the swamp), we have a sub-community that is half the size of the metacommunity, and so we can write:

$$I'_n(2) = \left(\frac{2(J_m/2)v}{n}\right)\left(\frac{1}{1+2v}\right)^n = \left(\frac{J_m v}{n}\right)\left(\frac{1}{1+2v}\right)^n \tag{5.4}$$

where $I'_n(2)$ is the number of species in one sub-community with an abundance of n individuals, given that the metacommunity is divided into two such sub-communities. But we note that since there are two independent drift processes, each described by Eq. (5.4), the pattern of species abundances for the metacommunity as a whole is given by simply summing two independent processes, to give the surprising result that:

$$I''_n(2) = 2I'_n(2) = \left(\frac{2J_m v}{n}\right)\left(\frac{1}{1+2v}\right)^n = I_n \tag{5.5}$$

where $I''_n(2)$ is the number of species in the overall metacommunity with an abundance of n individuals, given that the metacommunity is divided into two separate sub-communities. Equation (5.5) proves that as long as the distribution of abundances in each sub-community is a log-series, dividing the metacommunity into completely separate sub-communities does not affect the distribution of abundances in the metacommunity as a whole. Furthermore this result is not special to our example of dividing into two sub-communities, but generalizes easily to a number of sub-communities u:

$$I_n''(u) = u I_n'(u) = u \left(\frac{2(J_m/u)v}{n} \right) \left(\frac{1}{1 + 2v} \right)^n$$
$$= \left(\frac{2J_m v}{n} \right) \left(\frac{1}{1 + 2v} \right)^n = I_n \tag{5.6}$$

that is, if we take a metacommunity and divide into an arbitrary number of perfectly separate sub-communities, and let these sub-communities reach their equilibrium distribution of relative abundances, then add up the results, we get exactly the same answer as if all of these sub-communities operated as one large metacommunity undergoing nothing but ecological drift.

This result has an immediate and important implication: a fit between observed relative abundances, and relative abundances from Hubbell's neutral model, does not imply ecological drift, because it is also consistent with extremely strong niche structure. For this reason, the rank abundance curve cannot be used to infer the strength, or absence, of niche structure.

This result is in no way specific to the particular form of niche structure that we chose for illustration (habitat specialization). It applies equally to communities structured by cyclic replacement processes such as succession, and communities structured by lottery replacements, because in both cases the result is a division of the metacommunity into different guilds, strong regulation of the number of individuals in each of these guilds and ecological drift of species abundances within guilds (see the simulation results below).

However, it is important to realize that under some circumstances, niche structure can give a pattern of relative abundances that looks nothing at all like drift. This is simple to see by considering an extreme case when there are so many sub-communities (say 100) that each is small enough to contain only one species (i.e. each sub-community is so small that it quickly drifts to monodominance). Since species from one sub-community cannot stray into another sub-community (because they are specialized for one habitat type only), the metacommunity consists of 100 species all of equal abundance, which is fantastically unlikely to arise from ecological drift alone.

The reason that Eq. (5.6) does not apply in this case is because it presupposes that the distribution within each sub-community is close to a log-series, which is true only when the diversity of each sub-community is sufficiently great. That is, Eq. (5.6) is based on an approximation that breaks down when the within-guild diversity is low. This allows us to predict the conditions under which niche structure will leave a signature on the rank abundance curve, namely, when within-guild diversity is low. Conversely, for this reason we suspect that in many cases observed in nature, Eq. (5.6) will hold, and niche structure will not affect the rank abundance curve. This is because in high-diversity communities, the number of guilds will generally be very low in comparison with the number of species, and so the within-guild diversity will be high.

Also, our proof only applies exactly when the niche structure is so strong that there is no 'leakage' between guilds. Under weaker niche structure, the replacement dynamics will be less predictable, and there will be leakage from one guild to another. This leakage will increase the importance of drift relative to niche structure and make the species abundances more closely resemble the produce of a neutral process. Thus the distribution of species abundances could look neutral even in cases where the within-guild diversity is low, as long as the niche structure is sufficiently weak.

The simulation results below allow a closer assessment of the conditions under which the rank abundance curve from a niche-structured community differs from the expectation from ecological drift.

Niche-structured models

We now introduce a family of models that allows us to add three different types of niche structure into Hubbell's zero-sum ecological drift model – successional niches, lottery dynamics and habitat specialization. The simplest (and at least for forests the most reasonable) way to introduce structure into Hubbell's model is to alter assumption (4) (above) to make the probability of gap capture by species i depend in some way on the identity of i, and leave all the other assumptions exactly as before. We introduce a single parameter ω to tune the strength of niche-structuring: $\omega = 1$ gives no niche structure and therefore pure ecological drift; $\omega = 0$ gives deterministic niche structure, where the guild of the species capturing the gap is perfectly predictable; $0 < \omega < 1$ gives a model with random replacements, but with the probability of replacements weighted according to species identity. The models we present are all zero-sum models, but they are neutral only in the special case of $\omega = 1$.

Cyclic-succession model

Our first model is *cyclic succession*. We imagine four successional niches (any number could be used – we use four as the upper limit of what is usually observed for this form of niche-structuring in real communities), with four corresponding guilds of species, each specialized to one of the niches. Each guild contains one or more species. When an individual from a species in guild 1 dies, it is more likely to be replaced by an individual from a species in guild 2 than by an individual from a species in guilds 1, 3 or 4; when an individual from a species in guild 2 dies, it is more likely to be replaced by an individual from a species in guild 3; same for 3 to 4; and finally guild 4 is more likely to be replaced by guild 1 (Fig. 5.2).

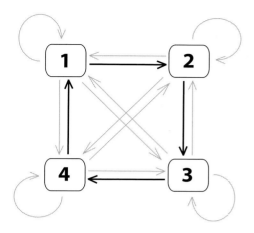

Figure 5.2 Cyclic-succession model. Each guild 1–4 contains several species. Each arrow is a transition (replacement of an individual from one guild by an individual from another guild). All transitions are possible, including self-transitions from guild x to x, but the transitions in black are more likely than those in grey. Our parameter ω tunes the relative strength of the black and grey arrows: in the limit when they have the same strength, the model is Hubbell's neutral drift model; at the other extreme, when the grey arrows have strength zero, all transitions follow the black arrows and the model has deterministic replacement of one guild by another.

The model is implemented as follows. When an individual from species i dies (i.e. is removed from a site), we set the probability of replacement by species k, $P(k, i)$, to be:

$$P(k, i) = \frac{H^{cs}\left(g_k^{cs}, g_i^{cs}\right) J_k}{\sum_n H^{cs}\left(g_n^{cs}, g_i^{cs}\right) J_n} \tag{5.7}$$

where the index cs denotes the cyclic-succession model, to differentiate it from other niche-structured models introduced below; J_k is the abundance (number of individuals) of species k in the population; g_i^{cs} is the successional guild of species i (1, 2, 3 or 4); and $H^{cs}(g_k^{cs}, g_i^{cs})$ is a weighting for transitions from successional guild g_i^{cs} to successional guild g_k^{cs} (the different values of $H^{cs}(g_k^{cs}, g_i^{cs})$ can be visualized as the strength of the different arrows in Fig. 5.2). We set the value of $H^{cs}(g_k^{cs}, g_i^{cs})$ to 1.0 for transitions following the direction of the cyclic succession (the dark arrows in Fig. 5.2), and to ω^{cs} for all other transitions. The value of ω^{cs} then tunes the model between deterministic replacements ($\omega^{cs} = 0$, i.e. transitions other than the cyclic succession have probability 0) and pure drift ($\omega^{cs} = 1$, probability of replacement by species i proportional to J_i, same as Hubbell's drift model). Finally, as in Hubbell's model, there is a small probability of replacement from outside the community by speciation (parameter v). For simplicity, we restrict the modelling to the case of a single, large, well-mixed metacommunity of 20 000 individuals. The two other main forms of the model – a small island population with immigration from a large mainland population; and a metacommunity divided into separate local communities connected to each other by limited near-neighbour dispersal – are more complex, and there is much scope for further work here.

One iteration of the metacommunity model is as follows: first, pick one individual at random and remove it from the population; second, draw a random number to decide if this individual will be replaced from within the community (probability $1 - v$), or from outside (probability v); if the former, draw a second random number and select the species of the new individual according to the probabilities given by Eq. (5.7); if the latter, make a new species, give it a guild g_i^{cs} of 1, 2, 3 or 4 at random (all four values equally likely), and give the new species i a population size of 1 individual. We run the model for 4×10^8 iterations, which equals 20 000 average lifetimes. During the final 5000 lifetimes, we calculate a rank abundance curve every 100 lifetimes, and present the average of these curves.

Results

Figure 5.3 compares the rank abundance curves from the cyclic succession model with different values of ω^{cs} (from 1.0 = pure drift, to 0.0 = perfect deterministic cyclic succession) for metacommunities of different diversity (given by varying the speciation probability v). The first result is that, at low diversity, the curves from the niche-structured model ($\omega^{cs} < 1$) are very different from pure drift ($\omega^{cs} = 1$). For low-diversity communities, zero-sum ecological drift always predicts dominance by one species, with a steep drop-off in abundance with rank (Fig. 5.3 top left, bold black line). In contrast the niche-structured model always gives a set of four dominant species, one from each guild, with roughly equal abundance, followed by a steep drop-off in abundance for the remaining species (interestingly, the tail of the rank abundance curve from the niche-structured model is parallel to the tail of the curve from the drift process, similar to a pattern from forests in Manu, Peru, seen by Hubbell (Hubbell 2001, p. 338)).

It is easy to understand why cyclic succession cannot allow dominance by one species. Imagine that species i, say in guild 2, was the dominant species (see Fig. 5.2). Most deaths would then be from guild 2, and so any species in guild 3 would be at a strong advantage. Guild 3 species would then become more common (as guild 2 species became rarer), in turn favouring guild 4 species. The community can only equilibrate with all four guilds represented equally, and in a community with only a few species, this has to mean equal abundance of four dominant species.

In contrast, in a high-diversity community, the niche-structuring has negligible impact on the rank abundance curve (Fig. 5.3 bottom right). This is a surprising result with an immediate and important implication for the interpretation of fits between observations and Hubbell's ecological-drift model, namely that the rank abundance curves cannot be used to assess the strength of niche structure. The reason that the niche-structuring can have so little impact on the rank abundance curve in high-diversity communities is as follows. Although the guild of the individual that dies predicts the guild of the replacement individual (in the case of $\omega = 0$, this prediction is perfect every time), if there is a large number of species within each guild, there is still plenty of opportunity for the abundances of the different individual species to drift up and down.

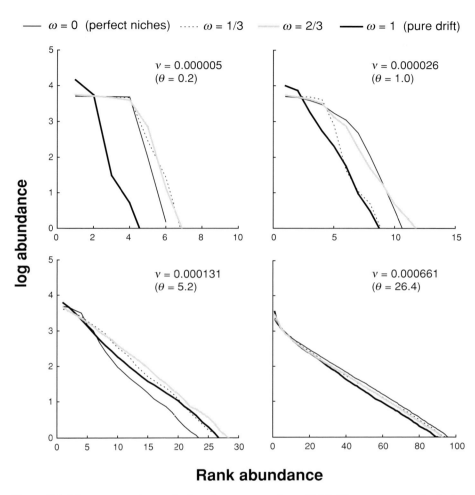

Figure 5.3 Metacommunity rank abundance curves from Hubbell's zero-sum ecological drift (bold black lines) compared with those from models with different strengths of niche-structuring according to the cyclic-succession model (lower ω gives stronger niche structure: Fig. 5.2). Results are shown from metacommunities with different diversity, given by different speciation rates (v). For comparison with Hubbell (2001), the fundamental biodiversity number θ is given for each case. At high diversity (bottom right) even perfect niche structure (non-bold black lines) does not affect the rank abundance curve.

Consequently, although the number of individuals within each guild is tightly regulated (see above), the number of individuals in different species within each guild is not regulated at all.

A simple way to understand this is to consider the model with deterministic replacement ($\omega = 0$), where once a site has been disturbed four times the replacement will be from the same guild as the original individual (guild $1 \to 2 \to 3 \to 4 \to 1$, same as $1 \to 1$). However, *which* species from the original guild comes to occupy the gap is random, with probabilities in proportion to the species abundances within the guild. This means that, within a guild, the abundances of the different species drift around exactly as in Hubbell's pure-drift model. The only difference at the community level is that there are four such drift processes operating simultaneously, which are then summed to produce the rank abundance curve for the whole community. In fact, this was the logic used above to show analytically that in the high-diversity limit, the pattern of species abundances from the deterministic niche-structured model and pure drift will be exactly the same.

Lottery model

Our second model also has four guilds, but in this case there is no predictable sequence of replacement of one guild by another. Rather, we assume that the species in the four different guilds are specialized to one of four different environmental conditions which appear at random within a gap immediately after disturbance (for example, gaps appearing at different times of year, or in years with different environmental conditions). After each individual dies, it leaves behind an environmental condition $E = 1, 2, 3$ or 4, that makes the replacement individual more likely to come from guild 1, 2, 3 or 4 respectively (i.e. if the dead individual leaves behind environmental condition 1, then the replacement individual is more likely to come from guild 1, etc.). We set the probability that the next gap is captured by species k, $P(k)$, to be:

$$P(k) = \frac{H^{lt}\left(g_k^{lt}, E\right) J_k}{\sum_n H^{lt}\left(g_n^{lt}, E\right) J_n} \tag{5.8}$$

where the index lt denotes the lottery model; E is a randomly chosen environmental condition taking the value 1, 2, 3 or 4 with equal probability; and $H^{lt}(G_k^{lt}, E)$ is the weighting given to a species from guild G_k^{lt} encountering environmental condition E (other notation as in Eq. 5.7). Note that, unlike the cyclic-succession model, $P(k)$ does not depend on the species of the individual that just died (i). We set the value of H^{lt} at 1.0 where $E = G_k^{lt}$ (e.g. environmental condition $= 3$ and guild $= 3$), and to ω^{lt} otherwise. Therefore as before, $\omega^{lt} = 1$ gives pure drift, and $\omega^{lt} = 0$ gives a deterministic process whereby the guild of the replacement species is perfectly predictable from the environmental condition E.

Since the environmental condition E is chosen completely at random, it is not obvious at first sight that this model differs from drift at all, but a simple example shows that the lottery model has non-neutral dynamics if $\omega < 1$. Assume that a metacommunity has a large number of individuals of species a, J_a, and one individual of species b. Assume also that the guilds of species a and b are 1 and 4 respectively. The probability that the next gap will be captured by a particular individual of species a, $P(a)$, is:

$$P(a) = P(E = 1) \cdot P(a|E = 1) + P(E = 2 \text{ or } 3) \cdot P(a|E = 2 \text{ or } 3)$$
$$+ P(E = 4) \cdot P(a|E = 4)$$
$$= \left(\frac{1}{4}\right) \cdot \left(\frac{1}{J_a + \omega^{lt}}\right) + \left(\frac{1}{2}\right) \cdot \left(\frac{\omega^{lt}}{\omega^{lt}J_a + \omega^{lt}}\right)$$
$$+ \left(\frac{1}{4}\right) \cdot \left(\frac{\omega^{lt}}{\omega^{lt}J_a + 1}\right) \approx \frac{1}{J_a} \tag{5.9}$$

As expected, since almost all individuals are from species a, the chance that any particular individual of species a will capture the next gap is approximately equal to the reciprocal of the community size (the probability that the next gap is captured by species a is the per-capita probability $P(a)$ multiplied by the population size of a, J_a). In contrast, for species b:

$$P(b) = P(E = 1) \cdot P(b|E = 1) + P(E = 2 \text{ or } 3) \cdot P(b|E = 2 \text{ or } 3)$$
$$+ P(E = 4) \cdot P(b|E = 4)$$
$$= \left(\frac{1}{4}\right) \cdot \left(\frac{\omega^{lt}}{J_a + \omega^{lt}}\right) + \left(\frac{1}{2}\right) \cdot \left(\frac{\omega^{lt}}{\omega^{lt}J_a + \omega^{lt}}\right)$$
$$+ \left(\frac{1}{4}\right) \cdot \left(\frac{1}{\omega^{lt}J_a + 1}\right) \approx \frac{1}{2J_a} \cdot (\omega^{lt} + 1/\omega^{lt}) \tag{5.10}$$

and since $(\omega^{lt} + 1/\omega^{lt}) > 2$ if $\omega^{lt} < 1$, the per-capita probability of gap capture for species b, $P(b)$, is greater than that for species a. More generally, in the lottery model any individual in a guild that is currently underrepresented in the community is at an advantage over individuals in guilds that are overrepresented, and this acts to regulate the proportion of individuals in different guilds. This can be thought of as a rare-guild advantage. Importantly, however, there is no rare-*species* advantage, because a rare species in an overrepresented guild has its fitness reduced by exactly the same amount as a common species in the same guild.

Results

As for the cyclic-succession model, in a low-diversity metacommunity the rank abundance curves from the lottery model are very different from pure drift, with the lottery model always giving a set of four dominant species of roughly equal abundance, in contrast to the dominance by one species given by pure drift (Fig. 5.4). Also as before, in high-diversity communities the niche structure has negligible impact on the rank abundance curves (Fig. 5.4). Once again, with sufficient diversity, the rank abundance curve cannot be used to assess the strength of niche-structuring in these model communities.

Furthermore, where there are significant differences between the curves from the pure drift and lottery models, the rank abundance curves from the lottery

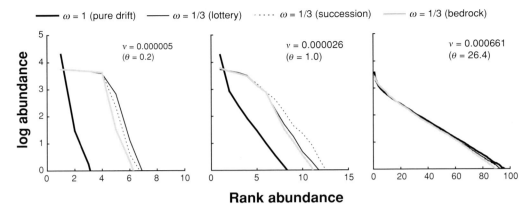

Figure 5.4 Metacommunity rank abundance curves from Hubbell's zero-sum ecological drift (bold black line) compared to those from models with strong niche structure according to cyclic succession, a lottery or habitat specialization (bedrock model). With high diversity all the models give the same curve as pure drift (right). Where the models deviate from drift (left two panels), the three models give the same curve.

model are very close to those given by the cyclic-succession model (Fig. 5.4: the two curves are also very close to the curve from the bedrock model introduced below). This can be understood in principle from the explanations of the model dynamics (see above): both models have a strong regulating force leading to an equal number of individuals in four guilds, where within a guild the different species are free to drift in abundance. However, the two models could hardly be more different in their biological assumptions. In the cyclic-succession model, the guild of the replacement is predictable from the guild of the individual that just died, such as might be expected when a pioneer tree dies, releasing the mid-successional seedlings underneath (by definition pioneer seedlings cannot tolerate the shade cast by pioneers), or when a late-successional tree dies, leaving a very open gap that can be colonized by a pioneer (the equivalent to the $4 \rightarrow 1$ transition in the cyclic-transition model). In contrast, in the lottery model the guild of the replacement individual is completely unpredictable from the identity of the individual that died, as might be expected where species are specialized to gaps appearing at different times of year, or in years with different environmental conditions. What is surprising is that the rank abundance curves cannot be used to assess which form of niche structure is present in these model communities.

Habitat specialization: bedrock model

We now consider a third form of niche structure, namely guilds specialized to different habitat types (e.g. high vs. low altitude, dry areas vs. swamps, different bedrock types): for convenience we call this the bedrock model. Habitat specialization can be incorporated into the zero-sum model by making the probability of replacement by species from different guilds depend on the identity of site q:

$$P(k, q) = \frac{H^{\mathrm{br}}(g_k^{\mathrm{br}}, b_q)J_k}{\sum\limits_{n} H^{\mathrm{br}}(g_n^{\mathrm{br}}, b_q)J_n} \tag{5.11}$$

where the index br denotes the bedrock model; b_q is the bedrock type of site q (1, 2, 3 or 4); and $H^{\mathrm{br}}(g_k^{\mathrm{br}}, b_q)$ is weighting given to a species from guild g_k^{br} encountering a gap with bedrock type b_q. Note that $P(k, q)$ is a function of the site q, but it does not depend on the species of the individual that just died (i). We set H^{br} to 1.0 where $b_q = g_k^{\mathrm{br}}$, and to ω^{br} otherwise. Then as before, $\omega^{\mathrm{br}} = 1$ gives pure drift and $\omega^{\mathrm{br}} = 0$ gives deterministic niche-structuring, where the guild of the replacement individual is perfectly predictable from the bedrock type at site q.

Results
The results from the bedrock model are the same as the previous models (Fig. 5.3): (a) in a low-diversity metacommunity the niche structure appears in the rank abundance curves as four dominant species of roughly equal abundance; (b) in a high-diversity metacommunity the niche structure does not affect the rank abundance curve; (c) where the niche structure is strong enough (and the metacommunity low enough in diversity) to give a rank abundance curve that does not look like drift, the bedrock model gives the same rank abundance curve as both the cyclic-succession model and the lottery model.

Generality
In this section we discuss first the robustness of the analytical proof and modelling results given above to the assumption that species can be divided into discrete guilds. We introduce models with continuous variation in life-history traits, and show that these also produce patterns of relative abundances that match the predictions of neutral models. We conclude that this result should extend to any competition model with the fundamental property that species' vital rates are continuous functions of the model parameters – a realistic property shared by most of the niche-based competition models in the literature. We then discuss two important models that often do not show this property – the competition–colonization trade-off and the Janzen–Connell effect – and conclude that in the former case our results hold anyway, and in the latter case they may or may not hold.

Continuous niches
The model and formal proof given above assumed that the species can be divided into discrete guilds. However, in real communities, many forms of niche differentiation appear as more or less continuous variation in biological traits

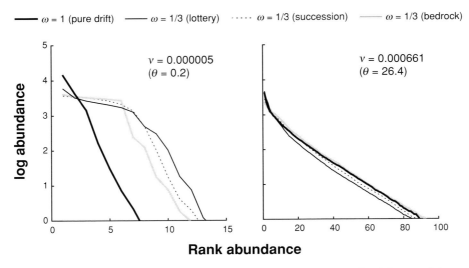

Rank abundance

Figure 5.5 Rank abundance curves from continuous zero-sum models with one of three forms of niche structure (see below), compared with neutral ecological drift.

(e.g. see Fig. 5.7). Do the conclusions of the modelling above depend critically on the assumption of discrete guilds? To assess this possibility we introduce continuous analogues of the three discrete models above, and give simulation results from these models which show that: (1) the introduction of continuous life-history variation does not change the fundamental result that with sufficiently high diversity the distribution of abundances looks like the outcome of a pure-drift model (Fig. 5.5); (2) in low-diversity communities, where there is a departure between structured and neutral models, structured models with different forms of niche structure give very similar predictions (Fig. 5.5). The model descriptions are given below, but conclusions (1) and (2) can be expected to be insensitive to the details of how the continuous niches are implemented.

Continuous cyclic succession model

Let each species i have a randomly assigned successional position Z_i^{cs} in the range $0 - Z_{max}^{cs}$, where Z_{max}^{cs} is the number of niches in the corresponding discrete cyclic-succession model (in our case 4: we chose a uniform distribution for Z_i^{cs}). When an individual of species i is removed from a site, it leaves behind an environmental condition $S^{cs} = Z_i^{cs} + 1 + \epsilon$, where ϵ is a random variable (we set ϵ to a uniform distribution with range $\pm 2/3$). If $S^{cs} > Z_{max}^{cs}$ then we set $S^{cs} = 0$. We then set the weight for species k encountering environment S^{cs}, $\tilde{H}^{cs}(Z_k^{cs})$, to be a Gaussian function of the difference between the niche position and the environmental condition, $|Z_k^{cs} - S^{cs}|$,

$$\tilde{H}^{cs}\left(Z_k^{cs},\, S^{cs}\right) = 1 - [1 - \omega^{cs}] \cdot \left[1 - \exp\left(|Z_k^{cs} - S^{cs}|/\sigma\right)^2\right] \tag{5.12}$$

(the tilde on $\widetilde{H}^{cs}$ is to differentiate this weight from the discrete models discussed elsewhere in the text). Equation (5.12) sets the weight $\widetilde{H}^{cs}$ equal to 1 when $Z_k^{cs} = S^{cs}$, and equal to ω^{cs} when $|Z_k^{cs} - S^{cs}|$ is very large; the parameter σ sets how quickly the weight decreases with increasing $|Z_k^{cs} - S^{cs}|$: we set σ to $1/3$, giving a value of $\widetilde{H}^{cs}$ close to ω^{cs} when $|Z_k^{cs} - S^{cs}| = 1$. Finally, the probability that the gap is captured by species k is calculated as before using Eq. (5.7), but substituting $\widetilde{H}^{cs}(Z_i^{cs}, S^{cs})$ for $H^{cs}(g_k^{cs}, g_i^{cs})$.

Continuous lottery model

Let each species i have a randomly assigned lottery preference Z_i^{lt} in the range 0 to Z_{max}^{lt}, where Z_{max}^{lt} is the number of niches in the corresponding discrete lottery model (in this case 4: we chose a uniform distribution for Z_i^{lt}). When the next site is vacated, we draw a random variable S^{lt}, also in the range 0 to Z_{max}^{lt}, and also uniformly distributed. We then set the weight for species k encountering environment S^{lt}, $\widetilde{H}^{lt}(Z_k^{lt}, S^{lt})$, with Eq. (5.12), but substituting $|Z_k^{lt} - S^{lt}|$ for $|Z_k^{cs} - S^{cs}|$. Finally, the probability that the gap is captured by species k is calculated as before using Eq. (), but substituting $\widetilde{H}^{lt}(Z_k^{lt}, S^{lt})$ for $H^{lt}(g_k^{lt}, E)$.

Continuous bedrock model

Let each species i have a randomly assigned habitat preference Z_i^{br} in the range 0 to Z_{max}^{br}, where Z_{max}^{br} is the number of bedrock types in the corresponding discrete bedrock model (in our case 4: we chose a uniform distribution for Z_i^{br}). In addition let each site q have a habitat score S_q^{br}, also in the range 0 to Z_{max}^{br}, also uniformly distributed. When site q is vacated, we set the weight for species k capturing gap q, $\widetilde{H}^{br}(Z_k^{br}, S_q^{br})$, from Eq. (5.12), but substituting $|Z_k^{br} - S^{br}|$ for $|Z_k^{cs} - S^{cs}|$. Finally, the probability that the gap is captured by species k is calculated as before using Eq. (5.11), but substituting $\widetilde{H}^{br}(Z_k^{br}, S_q^{br})$ for $H^{br}(g_k^{br}, b_q)$.

Continuity in continuous models

The key difference between models with discrete guilds and models with continuous variation in life history is that in the latter case no two species are exactly alike. This means that continuous models have no functional equivalence in the community, in the sense that no two species are exactly interchangeable. Therefore these models have the potential to show deterministic, structured dynamics between each and every pair in the community, which would seem to imply that ecological drift cannot operate. This conclusion is not correct, however, for two reasons: first, because in any model with a constant or maximum community size, higher diversity gives lower population sizes, which increases the importance of drift compared with selection; second, because vital rates in continuous models are continuous functions of the model parameters, as explained below.

A function is called continuous if you can trace it without lifting your pencil from the paper, or more precisely, a function is said to be continuous at a point x_0 if, for all x sufficiently close to x_0, the value of the function at x differs from the value at x_0 by an arbitrarily small amount. Why is continuity important in niche-structured competition models? Simply, these models assign a competitive difference to two species that depends on the difference between the traits of those two species, so the difference in competitive ability of the two species is a continuous function, such that the competitive difference tends to zero as the trait difference tends to zero. And because at high diversity any species k will have a set of species with traits that are very close to its own, its competitive advantage over those species will be very close to zero, and the abundance of k will drift relative to the species within this set. In effect, each species becomes surrounded by a guild of very similar species, within which the abundances drift, in the same way as the discrete models above exhibited drift within guilds. This is true even though there are no discrete guilds in the continuous models, and no two species are exactly interchangeable.

Also, because each species drifts relative to its neighbours (i.e. similar species), it also drifts relative to its neighbours' neighbours, and so on, so that each species in a high-diversity community is free to drift relative to any other species. However, even though any pair of species drifts in abundance relative to any other, the distribution of individuals in 'trait space', that is the proportion of individuals with traits of different types, is strongly regulated. In our simulations if one selects a region of trait space that is reasonably large, the total number of individuals with traits that fall within that region is very predictable, and strongly regulated: but whether this is achieved via one common species with traits in that region, or by lots of rare species, is determined by the balance between random walks in abundance, local extinction, and speciation and immigration, i.e. ecological drift.

We therefore suspect that niche-structured models will generally predict drift-like distributions of abundance in high-diversity communities simply because species' vital rates are continuous functions of the parameters in these models. Examples range from the early species-packing literature (e.g. the work of MacArthur, May and Levins) to more mechanistic models like JABOWA-FORET (Shugart 1984; Botkin 1992); ALLOCATE (Tilman 1988) and SORTIE (Pacala *et al.* 1996).

Two important processes that, as usually implemented, do not have vital rates as continuous functions of parameters are discussed below: these are the competition–colonization trade-off (CC), and the Janzen–Connell (JC) effect, both of which are thought to be important in tropical forest communities (Dalling & Hubbell 2002; Wright 2002; Peters 2003). Because these processes are usually implemented in models that do not exhibit continuity, they have the potential to produce rank abundance curves that are not drift-like. Indeed Chave *et al.*

(2002) begin with a neutral community model much like Hubbell's (2001), and add a form of CC trade-off, with or without JC mortality; importantly, the site-capture process in all of their models awards the site to the superior competitor at the site with certainty, no matter how small the difference in competitive ability, i.e. their models do not exhibit continuity (see appendices to Chave *et al.* 2002). In this case both the CC and JC processes caused significant changes to the rank abundance curves. However, we argue below that, implemented differently, both the CC and JC processes can exhibit continuity, and can therefore exhibit ecological drift and give rank abundance curves that look like the outcome of a purely neutral model.

Competition–colonization trade-off

Models of the competition–colonization (CC) trade-off predict that inferior competitors can coexist with superior competitors, if the inferior competitors are better at getting to newly opened gaps (Levins & Culver 1971; Tilman 1994). CC models typically assume a perfect competitive hierarchy: contests for local sites are *always* won by the species with the highest competitive rank of those present at the site. Because of this property, species' vital rates are not continuous functions of the parameters. Any change in a parameter, no matter how small, that causes a reversal in the competitive ranking of two species will produce a large change in the vital rates of the species concerned.

However, it turns out that even without the continuity property, the CC model exhibits the same combination of regulation along the niche axis, and ecological drift of the abundances of individual species, that leads to the key results discussed for the models above (Hurtt & Pacala 1995; Kinzig *et al.* 1999). As its name implies, the competition–colonization model assumes that colonizing ability is limited. The probability that a species k will arrive at a site in time interval Δt is simply proportional to the abundance of species k, J_k: $F_k J_k \Delta t$ where F_k is a species-specific measure of colonizing ability. Because the total number of individuals in the community is fixed (just as in Hubbell's model and our niche models) population sizes decrease as diversity increases, so the probability of arrival for each species decreases. Thus at high diversity, species very rarely compete with near neighbours in the competitive hierarchy and this means that, overall, species abundances are still free to drift with respect to one another, as in continuous models (Hurtt & Pacala 1995; Kinzig *et al.* 1999).

It should be noted also that the CC trade-off can readily be implemented in a form that exhibits continuity, most simply by adding some stochasticity into the gap-capture process, in which case it can support the coexistence of a much smaller set of species (see Adler & Mosquera 2000). We predict that neutral community models with the addition of this more realistic form of the CC effect will exhibit ecological drift, and produce rank abundance curves that look like neutral drift.

Janzen–connell effect

The Janzen–Connell (JC) effect, as it is usually explained, is simply that species that are locally common will experience higher loads of species-specific pests, leading to a rare-species advantage (Janzen 1970; Connell 1971). The niche space in this case governs resistance to different pests, and whether or not this model exhibits the continuity property depends on assumptions about cross-resistance to pests. At one extreme, what if pest resistance is simply about the tannin concentration in leaves? Then we can arrange all species on an axis of tannin concentration, with resistant (high-tannin) species at one end and vulnerable (low-tannin) species at the other. This case has a one-dimensional trait space. At high diversity there would be many species near to each other in trait space (similar tannin concentration), and those species would have similar resistance and therefore similar competitive abilities. A model of this case would exhibit the continuity property, and so would produce drift-like distributions of abundances at high diversity. At the other extreme, what if each species has its own species-specific pest and there is no cross-resistance? As each new species invades a community it creates its own niche axis, so the trait space is potentially infinite in dimensionality. A model of this case would not exhibit the continuity property simply because all species-specific differences in pest resistance are assumed to be equal, and large. (This is the form of the JC effect implemented in Chave *et al.* 2002, referred to there as density dependence, and this helps to explain why those models did not show drift-like abundances.) In between these two extremes would be a trait space of intermediate diversity, for example with functional groups for pest resistance.

However, even in the case of species-specific pests it is still possible for ecological drift to operate in JC models (Fig. 5.6). Species coexist in JC models because of a rare-species advantage: as a species becomes rare it loses its natural enemies and gains a competitive advantage. As diversity increases, and the abundance of each species becomes small, species could come to have very similar competitive abilities, if the rare-species advantage saturates with rarity (curve A in Fig. 5.6). As long as the community is diverse enough, ecological drift will keep all species within the range of abundances where the competitive abilities are close to equal, and the community will reach a drift/extinction/speciation balance. If instead the rare-species advantage increases sufficiently sharply with rarity (curve B in Fig. 5.6) there will always be large differences in competitive ability at reasonable levels of diversity, and the abundances will not drift.

It is too early to tell how important the JC effect generally is (Givnish 1999; Wright 2002; Peters 2003), although there are good reasons to suspect that its strength has been underestimated (Wright 2002); and there is no information on how the rare-species advantage depends on rarity. Furthermore the degree of cross-resistance to the pests that drive the JC effect, that is the dimensionality of

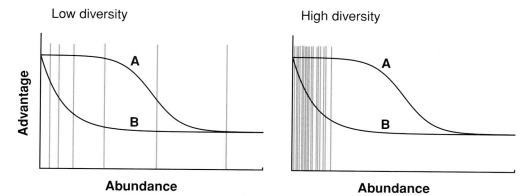

Figure 5.6 When the Janzen–Connell (JC) effect operates, community diversity and the form of the rarity advantage interact to determine whether the species abundances (grey lines) drift. In a low-diversity community (left), the rare-species advantage will lead to differences in fitness whatever the form of the rarity advantage (curve A or B). In a high-diversity community (right) all species will be quite rare, so if the rare-species advantage saturates with rarity (curve A) the JC effect will not lead to important differences in fitness and the abundances can drift, as long as the abundances stay within the region where the rare species advantage is similar (left plateau of curve A).

the JC trait space, has apparently not been considered to be important: empirical and theoretical treatments alike have assumed the JC effect is species-specific, corresponding to the potentially infinite-dimensional case discussed above. For these reasons we cannot say whether the JC effect as it operates in reality is likely to give patterns that look like neutral drift, or not. But the observed match between patterns of species abundances in tropical forests and neutral models (Hubbell 2001) implies that either the JC effect is weak, or it operates such that it does not affect the distribution of abundances.

Implications
Coexisting species are not equivalent

Hubbell introduced a very simple neutral model and, surprisingly, found that it could give a near-perfect fit to the pattern of species abundances in various communities. This provided strong evidence that ecological drift was an important feature of the population dynamics of these communities. However, as we have shown above, it is possible for all sorts of forms and strengths of niche structure to leave no signature on the rank abundance curve, and so the fact that a neutral model can reproduce these curves does not imply that species are all equivalent. We therefore disagree with Hubbell when he concludes that his results imply that species are mostly equivalent to each other:

One of the great surprises to me in developing this theory has been just how well, in fact, it does work. It works astonishingly well in spite of making what might appear to be a false and crippling assumption, namely, that all individuals are identical. As I will argue below, however, this assumption is far closer to the truth than might at first be appreciated.   Hubbell 2001, p. 320

Hubbell's main argument for why species must be mostly equivalent is that different coexisting species must all have the same fitness, or most of the species would quickly become extinct. But what of the wide variation in physiology and life history observed within communities of plants and animals, much of which could be expected a priori to have immediate impact on fitness? Hubbell argues that the different species represent different combinations of these various biological traits that all confer the same overall fitness (Hubbell 2001, p. 322). For example, the gain in fitness from being a faster-growing tree is cancelled by an increased mortality rate, or the gain in nitrogen from being a legume is balanced by the costs of supporting nitrogen-fixing bacteria, in both cases leaving no net change in fitness. Of course, there are combinations of traits that are possible but sub-optimal (e.g. a slow-growing tree with a high mortality rate, or a plant supporting bacteria that do not produce nitrogen), but species exhibiting these trait combinations would quickly be lost, leaving behind a set of different, but equivalent, optimal solutions (the species in the community). This final set of coexisting optimal species are said to occupy a 'life-history manifold'. Figure 5.7 gives two examples of the shade-tolerance manifold for forests: in both cases the different species occupy a continuum from fast-growing, shade-intolerant species (bottom right) to very slow-growing, shade-tolerant species (top left).

The second part of Hubbell's argument is that, once the sub-optimal species outside the life-history manifold have been lost from a community (which should happen quickly), and the species that remain are therefore equivalent in fitness, the only process remaining to operate is ecological drift. This argument is mistaken, however, because it assumes that the fitness of a species is fixed by its biological traits alone (this is the metaphor of the fitness landscape, which has a 'height', equal to fitness, for each 'location', meaning combination of traits). In truth, fitness is not fixed but depends crucially on the community composition, and because that composition changes through time, so does fitness (the fitness landscape is therefore not fixed but depends on how individuals are distributed across trait-space at any one time – more like a water-bed than a mountain range).

For illustration, consider a forest stand dominated by light-demanding, early-successional species (bottom right in Fig. 5.7a, b). In this stand the more shade-tolerant, mid-successional species (middle in Fig. 5.7a, b) have a higher fitness than the average tree, because they can survive in the shade of the pioneers, and so capture the majority of new gaps: these species will therefore increase in

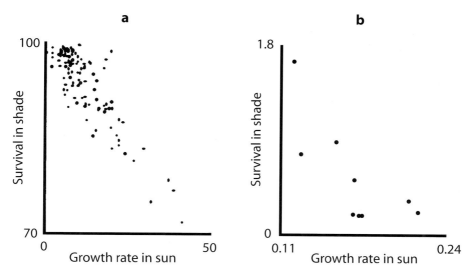

Figure 5.7 Life-history manifolds for trees defined by the trade-off between growth rate in high light, and survival in shade: (a) lowland tropical rainforest (Barro Colorado Island, from Hubbell 2001); (b) temperate deciduous forest (northeastern USA, from Kobe *et al.* 1995). The units in (a) and (b) have different interpretations so the exact position and shape of the two manifolds cannot be compared.

abundance. Conversely, consider a stand dominated by mid- and late-successional species. These species sometimes leave large gaps, with light levels on which the mid- and late-successional species are unable to capitalize: therefore in this stand the early-successional species have a higher fitness than the other species. Thus the relative fitness of the different successional strategies changes through time as the community composition changes, and as in our simple cyclic-succession model (see above), the community tends toward some characteristic mix of successional strategies. It is only when this equilibrium distribution is reached that the fitness of the different species is equalized.

To see the fundamental difference in predictions between this view of fitness (in which the composition of the community affects or dominates fitness) and the view Hubbell takes in his book (in which fitnesses are determined by the life-history trade-offs of a species, and so are fixed; see Hubbell 2001 p. 325), consider the simplest and best-known ecological model of competition: the two-species Lotka–Volterra competition equations. In these equations, if one species is made superior to the other in the sense that it has a higher density at equilibrium, nevertheless once that equilibrium is reached, the fitnesses of the two species are exactly the same: in fact they are 1.0, meaning no expected long-term change in density. This is a fundamental behaviour of all population-dynamics models with a stable equilibrium of any kind: by definition, at this equilibrium all species have the same fitness, but also by definition, any perturbation from the equilibrium changes the fitnesses such that the dynamics head back toward the equilibrium. Thus in this model, and (we believe) in real communities, the observation that the long-term fitness of different species is equal does not imply that the populations are free to drift. They could just as well be very tightly regulated.

More practically, according to Hubbell's argument, since equal fitness means that the only process left to operate is ecological drift (Hubbell 2001, p. 327), the community could drift to any composition. For example a tropical forest stand could drift to become all pioneers, or all late-successional species, just by chance. And although probabilistically this may be unlikely based on drift alone, nevertheless, according to this view, it is only drift that determines the mix of successional strategies: there should be no 'typical' mix of types under drift, except a 'typical' random sample from the metacommunity; and there should be no predictable change in the mix of types found in areas of contrasting disturbance, productivity, soil nutrients, topography or any other external influence. In contrast, under our view of fitness, differences in these external influences will translate into predictable differences in community composition; communities will show predictable responses to perturbation; and communities will show a directed return to their previous state after the perturbation is removed. Importantly, however, the formal proof and modelling results given above show that even in the face of all this structure, the distribution of species abundances may look like the outcome of a purely neutral process.

Does drift matter?

It is important to realize that in our non-neutral models, as in Hubbell's original model, the process that leads to the log-series-like distribution of abundances is the same, namely a balance between zero-sum random walks, local extinction and speciation. Furthermore for the pure-drift model and the structured models to give the same pattern of abundances, there has to be sufficient within-guild diversity in the community, implying a high degree of functional equivalence (although a pair of species picked at random from the community will tend to exhibit very different traits, and so have a substantial and predictable difference in competitive ability in any given situation). Thus our results support Hubbell's assertion that ecological drift, due to functional equivalence, appears to be the dominant process in determining species abundance patterns in high-diversity forests. Furthermore, the assumptions of Hubbell's zero-sum model seem to apply so well to forest communities (with the exception of the equivalence between species, which as we have shown does not affect the rank-abundance patterns anyway), that for forest communities Hubbell's ZSM distribution should be considered superior to other neutral models that make assumptions that are demonstrably untrue for forests, even where the ZSM might be slightly inferior in its fit to observations (McGill 2003; although see Volkov *et al.* 2003).

The application of Hubbell's theory to animal communities, where the key assumption of a constant number of individuals is not so obviously valid, is more contentious. However, even where this assumption is not valid, many animal communities depend on plants as resources, so we might expect the distribution of animal abundances to follow the plants anyway (consider an extreme case where every insect species is a specialist for one species of tree). This idea is very similar to the 'broken stick' model of MacArthur (1960), and the 'niche pre-emption' model of Motomura (1932; see also Whittaker 1965), in which a fixed resource is divided among species according to a simple set of rules (Hubbell 2001, p. 37–40). The niche-pre-emption model produces a straight line on a plot of log abundance against rank abundance, as does the log-series, and since in both cases the total number of individuals has to equal the community size, both models have to give the same relationship between the slope and intercept, so they add up to the same thing in the end; and the broken-stick model produces an approximately straight line over some range.

These models differ from the neutral models discussed elsewhere in this chapter because the distribution of species abundances in them is determined perfectly by the distribution of niche sizes. For example, imagine taking the bedrock model, with many bedrock types, one species per type, and with a log-series (or log-normal) distribution of areas for the different bedrock types. Then even in a purely deterministic model, the species abundances would themselves follow the log-series (or log-normal), and as such look exactly like neutral, stochastic drift.

In effect, this is how the broken-stick and niche-pre-emption models work, but these models have always begged the question: why should the resources be divided up in this way? Indeed, for the factors influencing plants, there is no reason to think that the resources should be divided according to any simple rules, but Hubbell's theory, by showing that zero-sum ecological drift can lead to a log-series or log-normal distribution of species abundances in the plant community, suggests that models of the broken-stick type may be relevant to animal populations that depend on plants as resources. This transfer between trophic levels provides yet another reason for suspecting that the neutral distributions highlighted by Hubbell (2001) and others (e.g. Bell 2001) should be common in nature.

Does niche structure matter?

Since pure drift, and strong (even perfect) niche structure, can both end up giving the same patterns of species abundance, and since drift-like patterns are so common, one could be tempted to dismiss the neutral vs. structured debate as irrelevant. If patterns of species abundances can be understood without recourse to non-neutral structuring mechanisms like niche differentiation, then doesn't parsimony imply that much of community ecology doesn't need niches? But it is important to understand that, just as niche structure does not affect the process of ecological drift, ecological drift has no effect on the function and regulation of communities. This is because the aspects of the community that are dominated by ecological drift (the pattern of species abundances, and also the number of species at the drift/speciation balance) are by definition functionally neutral, and therefore all the functionally important aspects of the community are regulated by processes other than drift.

For example, in the bedrock model as presented above, the fraction of individuals in each of the four guilds is regulated at 1/4. If one changed the fraction of the sites in each bedrock type (remember that 'bedrock' is just the name – the model corresponds to any form of habitat specialization including climatic conditions), the fraction of individuals in the different guilds would change accordingly, until the fraction in the guilds matched the fraction in each bedrock type. Examples of this include the obvious changes in vegetation that have resulted from human-induced changes in hydrology and nutrient status; coherent vegetation responses to historical changes in climate (Wick 2000; Tinner & Lotter 2001; Williams *et al.* 2002); and predicted future changes in vegetation from climate change (e.g. Cramer & Leemans 1993; Kirilenko *et al.* 2000). These predictable community-level responses to perturbation have everything to do with the niche structure and regulation of the relative abundance of guilds, and nothing at all to do with ecological drift. However, just as for niche structure itself, changes in the relative abundance of guilds may leave no signature in the pattern of species abundances (as visualized for example in a rank abundance curve).

Similar arguments apply to secondary succession in forests. After stand-scale disturbance, forest communities become dominated by pioneers, followed by mid- then late-successional species, until the community reaches a composition typical of old-growth forest (this mix of species still contains a variety of successional strategies: Bazzaz 1996; Pacala *et al.* 1996). The same pattern is repeated in mesic forests all over the world, and of very different diversity. As expected, the mix of successional types is responsive to the disturbance regime, so any changes in disturbance cause predictable changes in the relative abundances of species with different successional strategies. For example Slik and Eichorn (2003) show that stands of tropical rainforest in Borneo with a higher fire frequency have a higher proportion of pioneer species. Once again this predictability has nothing to do with ecological drift, but nevertheless the overall pattern of species abundances (i.e. the rank abundance curve) may be unaffected by these changes, and take exactly the same shape as expected from pure drift.

Therefore the reason that the niche vs. structure is relevant is that the functioning of communities with stronger niche structure (1) is more strongly regulated, (2) will show more pronounced, and more predictable, responses to perturbation, and (3) will be better able to recover after the perturbation is removed. But in communities that are sufficiently diverse, the pattern of species abundances does not respond to niche structure, and so this pattern, and the rank abundance curve in particular, tells us nothing about the extent or the nature of the regulating forces in a community. This is the main point of the work presented here.

Therefore, we would argue that any attempts to understand the relationship between community composition, biodiversity and ecosystem-functioning, and responses of ecosystems to climate change and other anthropogenic disturbances, will gain little from neutral theories, and would be better to begin with a theory based on some concept of functional groups (for good reason, this is how global ecosystem models are formulated); or where the life-history variation is continuous, with a model of the distribution of individuals along the dominant life-history manifolds. In principle, such theories are capable of predicting the functional response of a community to perturbations, without making any predictions about species-level phenomena such as diversity, species–area curves and patterns of relative abundances.

Along these lines we believe there is a general and important message to come from neutral theory, and Hubbell's theory in particular. This message is that in community ecology the species need not be the basic unit of study, and in fact a concentration on species biology, and species-specific differences and interactions, can distract from the basic biological principles underlying an ecosystem, and blind us to the fact that much of the variation apparent to us as investigators may not matter very much, either for the population dynamics underlying the community, or from the larger perspective of the function and regulation of the community and ecosystem. With sufficient diversity, and with low-dimensional niches (the only kind that can be strongly regulated), the distributions of species abundances are determined by ecological drift, independent of structure; and function is determined by niche structure, independent of drift.

Dispersal

All of the model results we present here are from a single, large well-mixed community with global dispersal, but dispersal in plant communities operates over restricted distances (Harper 1977; Willson 1993; Levine & Murrell 2003). Neutral community models that include local dispersal can make predictions about the spatial distributions of individual species and patterns of community turnover in space, and these have been shown to match observed patterns in some cases (Bell 2001; Hubbell 2001). Local dispersal also affects the pattern of relative species' abundances, whether implemented as a collection of local communities linked by migration events (Hubbell 2001), or as local dispersal within a community of contiguous patches (Bell 2001; Chave et al. 2002). Neutral community models with local dispersal are much more difficult to analyse, and take much longer to simulate, than those with global dispersal, and much further work is needed before we understand the interaction between ecological drift, dispersal limitation and niche-structuring processes (Chave et al. 2002). This is especially so because at the spatial scales at which the predictions of neutral community models can be tested there is always significant variation in climate and other abiotic factors (Condit *et al.* 2002; Duivenvoorden *et al.* 2002; Tuomisto *et al.* 2003). Therefore whilst the patterns of relative abundance of species may tell us little or nothing about the strength or nature of niche structure in real communities (as we argue here), these same patterns may contain information about the nature and importance of dispersal, but this remains to be demonstrated convincingly (Levine & Murrell 2003).

Acknowledgements

We thank Professors Jerôme Chave and Helene Muller-Landau for reviewing an earlier version of this chapter. This work was supported by the Andrew Mellon Foundation (DWP).

References

Adler, F. R. & Mosquera, J. (2000) Is space necessary? Interference competition and limits to biodiversity. *Ecology*, **81**, 3226–3232.

Archibold, O. W. (1995) *Ecology of World Vegetation*. New York: Chapman & Hall.

Bazzaz, F. A. (1996) *Plants in Changing Environments*. New York: Cambridge University Press.

Bell, G. (2001) Neutral macroecology. *Science*, **293**, 2413–2418.

Botkin, D. B. (1992) *Forest Dynamics: An Ecological Model*. Oxford: Oxford University Press.

Caswell, H. (1976) Community structure: a neutral model analysis. *Ecological Monographs*, **46**, 327–354.

Chave, J., Muller-Landau, H. C. & Levin, S. A. (2002) Comparing classical community models: theoretical consequences for patterns of diversity. *American Naturalist*, **159**, 1–23.

Clark, J. S. & McLachlan, J. S. (2003) Stability of forest biodiversity. *Nature*, **423**, 635–638.

Condit, R., Pitman, N., Leigh, E. G. Jr *et al.* (2002) Beta diversity in tropical forest trees. *Science*, **295**, 666–669.

Connell, J. H. (1971) On the role of natural enemies in preventing competitive exclusion in some marine animals and in rain forest trees. In B. J. den Boer & G. R. Gradwell, eds., *Dynamics of Populations*. Wageningen: Centre for Agricultural Publishing and Documentation, pp. 298–310.

Coombes, D. A. & Grubb, P. J. (2003) Colonization, tolerance, competition and seed-size variation within functional groups. *Trends in Ecology and Evolution*, **18**, 283–291.

Cramer, W. P. & Leemans, R. (1993) Assessing impacts of climate change on vegetation using climate classification systems. In A. M. Solomon & H. H. Shugart, eds., *Vegetation Dynamics and Global Change*. IIASA, New York: Chapman & Hall, pp. 190–217.

Dalling, J. W. & Hubbell, S. P. (2002) Seed size, growth rate and gap microsite conditions as determinants of recruitment success for pioneer species. *Journal of Ecology*, **90**, 557–568.

Duivenvoorden, J. F., Svenning, J.-C. & Wright, S. J. (2002) Beta diversity in tropical forests. *Science*, **295**, 636–637.

Fargione, J., Brown, C. S. & Tilman, D. (2003) Community assembly and invasion: an experimental test of neutral vs. niche processes. *Proceedings of the National Academy of Sciences*, **100**, 8916–8920.

Fisher, R. A., Corbet, A. S. & Williams, C. B. (1943) The relation between the number of species and the number of individuals in a random sample of an animal population. *Journal of Animal Ecology*, **12**, 42–58.

Foster, R. B. & Brokaw, N. V. L. (1982) Structure and history of the vegetation of Barro Colorado Island. In E. G. Leigh, A. S. Rand & D. M. Windsor, eds., *The Ecology of a Tropical Forest*. Smithsonian Institution, pp. 67–81.

Givnish, T. J. (1999) On the causes of gradients in tropical tree diversity. *Journal of Ecology*, **87**, 193–210.

Harms, K. E., Condit, R., Hubbell, S. P. & Foster, R. B. (2001) Habitat associations of trees and shrubs in a 50-ha neotropical forest plot. *Journal of Ecology*, **89**, 947–959.

Harper, J. L. (1977) *Population Biology of Plants*. London: Academic Press.

Hubbell, S. P. (2001) *The Unified Neutral Theory of Biodiversity and Biogeography*. Princeton: Princeton University Press.

Hubbell, S. P. & Foster, R. B. (1986) Biology, chance and history, and the structure of tropical tree communities. In J. M. Diamond & T. J. Case, eds., *Community Ecology*. New York: Harper & Row, pp. 314–324.

Hurtt, G. C. & Pacala, S. W. (1995) The consequences of recruitment limitation: reconciling chance, history, and competitive differences between plants. *Journal of Theoretical Biology*, **176**, 1–12.

Janzen, D. H. (1970) Herbivores and the number of tree species in tropical forests. *American Naturalist*, **104**, 501–508.

Kinzig, A. P., Levin, S. A., Dushoff, J. & Pacala, S. W. (1999) Limiting similarity, species packing, and system stability for hierarchical competition models. *American Naturalist*, **153**, 371–383.

Kirilenko, A. P., Belotelov, N. V., Bogatyrev, B. G. (2000) Global model of vegetation migration: incorporation of climatic variability. *Ecological Modelling*, **132**, 125–133.

Kobe, R. K., Pacala, S. W., Silander, J. A. & Canham, C. D. (1995) Juvenile tree survivorship as a component of shade tolerance. *Ecological Applications*, **5**, 517–532.

Levine, J. M. & Murrell, D. J. (2003) The community-level consequences of seed dispersal patterns. *Annual Reviews of Ecology and Systematics*, **34**, 549–574.

Levins, R. & Culver, D. (1971) Regional coexistence of species and competition between rare species. *Proceedings of the National Academy of Sciences*, **68**, 1246–1248.

MacArthur, R. H. (1960) On the relative abundance of species. *American Naturalist*, **94**, 25–36.

May, R. M. (1975) Patterns of species abundance and diversity. In M. L. Cody & J. M. Diamond, eds., *Ecology and Evolution of Communities*. Wageningen: Centre for Agricultural Publishing and Documentation, pp. 298–310.

McGill, B. J. (2003) A test of the unified neutral theory of biodiversity. *Nature*, **422**, 881–885.

Motomura, I. (1932) A statistical treatment of associations. *Zoological Magazine Tokyo*, **44**, 379–383.

Pacala, S. W. Canham, C. D., Saponara, J., Silander, J. A., Kobe, R. K. & Ribbens, E. (1996) Forest models defined by field measurements: estimation, error analysis and dynamics. *Ecological Monographs*, **66**, 1–43.

Pelissier, R., Dray, S. & Sabatier, D. (2002) Within-plot relationships between tree species occurrences and hydrological soil constraints: an example in French Guiana investigated through canonical correlation analysis. *Plant Ecology*, **162**, 143–156.

Peters, H. A. (2003) Neighbour-regulated mortality: the influence of positive and negative density dependence on tree populations in species-rich tropical forests. *Ecology Letters*, **6**, 757–765.

Phillips, O. L., Vargas, P. N., Monteeagudo, A. L. *et al.* (2003) Habitat association among Amazonian tree species: a landscape-scale approach. *Journal of Ecology*, **91**, 757–775.

Poorter, L. & Arets, E. J. M. M. (2003) Light environment and tree strategies in a Bolivian moist forest: an evaluation of the light partitioning hypothesis. *Plant Ecology*, **166**, 295–306.

Preston, F. W. (1948) The commonness, and rarity, of species. *Ecology*, **29**, 254–283.

Pyke, C. R., Condit, R., Aguilar, S. & Lao, S. (2001) Floristic composition across a climatic gradient in a neotropical lowland forest. *Journal of Vegetation Science*, **12**, 553–566.

Shugart, H. H. (1984) *A Theory of Forest Dynamics.* New York: Springer-Verlag.

Slik, J. W. F. & Eichhorn, K. A. O. (2003) Fire survival of lowland tropical rain forest trees in relation to stem diameter and topographic position. *Oecologia*, **137**, 446–455.

Sugihara, G. (1980) Minimal community structure: an explanation of species abundance patterns. *American Naturalist*, **117**, 770–787.

Tilman, D. (1988) *Plant Strategies and the Dynamics and Structure of Plant Communities.* Princeton: Princeton University Press.

(1994) Competition and biodiversity in spatially structured habitats. *Ecology,* **75**, 2–16.

Tinner, W. & Lotter, A. F. (2001) Central European vegetation response to abrupt climate change at 8.2 ka. *Geology*, **29**, 551–554.

Tuomisto, H., Ruokolainen, K. & Yli-Halla, M. (2003) Dispersal, environment, and floristic variation of western Amazonian forests. *Science*, **299**, 241–244.

Volkov, I., Banavar, J. R., Hubbell, S. P. & Maritan, A. (2003) Neutral theory and relative species abundance in ecology. *Nature*, **242**, 1035–1037.

Walter, H. (1973) *Vegetation of the Earth in Relation to Climate and Eco-physiological Conditions* (trans. J. Wieser). New York: Springer-Verlag.

(2002) *Walter's Vegetation of the Earth: The Ecological Systems of the Geobiosphere* (trans. G. Lawlor & D. Lawlor). Berlin: Springer.

Watterson, G. A. (1974) The sampling theory of selectively neutral alleles. *Advances in Applied Probability*, **6**, 463–488.

Whittaker, R. H. (1965) Dominance and diversity in land plant communities. *Science* **147**, 250–260.

Wick, L. (2000) Vegetational response to climatic changes recorded in Swiss late glacial lake sediments. *Palaeogeography Palaeoclimatology Palaeoecology*, **159**, 231–250.

Williams, J. W., Post, D. M., Cwynar, L. C., Lotter, A. F. & Levesque, A. J. (2002) Rapid and widespread vegetation responses to past climate change in the North Atlantic region. *Geology*, **30**, 971–974.

Willson, M. F. (1993) Dispersal mode, seed shadows, and colonization patterns. *Vegetatio*, **107/108**, 216–280.

Wright, S. J. (2002) Plant diversity in tropical forests: a review of mechanisms of species coexistence. *Oecologia*, **130**, 1–14.

PART II

Plant–microbe interactions

CHAPTER SIX

Dimensions of plant disease in tropical forests

GREGORY S. GILBERT

University of California, Santa Cruz, and Smithsonian Tropical Research Institute, Balboa

Introduction

Pest pressure is the inevitable, ubiquitous factor in evolution which makes for an apparently pointless multiplicity of species in all areas in which it has time to operate.

(Gillett 1962)

At a symposium 44 years ago, J. B. Gillett proposed the Theory of Pest Pressure, whereby plant pathogens and pests were responsible for the genesis and maintenance of high plant diversity in tropical forests and other high-diversity systems. In his conclusion to the paper produced from that talk he hoped that this 'new theory may be useful in stimulating discussion and research' on the roles of pests and pathogens as a force in plant diversity. Apparently his theory has been useful, as much has happened in the last four decades to explore and expand on his idea. My aim is to review key research on the effects of plant pathogens in tropical forests since Gillett's seminal paper with emphasis on the special case of his Theory of Pest Pressure known as the Janzen–Connell Hypothesis. I will also suggest critical areas that need to be explored as we continue to discuss and research the role of pathogens in the maintenance of species diversity in tropical forests.

'*Parcere subiectis et debellare superbos*'

Gillett (1962) paraphrased Virgil's famous formula for the greatness of Rome, 'Spare the lowly and conquer the haughty' (*Aeneid* VI: 853), when he first introduced the idea that if plant pests have a greater impact on more common species than on rare species, the rare species should then increase in relative frequency, providing a mechanism for the maintenance of diversity in species-rich communities. Gillett's proposal is essentially that plant pests prevent competitive exclusion through density-dependent disease or herbivory. The competitive exclusion principle predicts that in a heterogeneous environment, one species will exclude all others (Gause 1934). Species can coexist by partitioning resources

Biotic Interactions in the Tropics: Their Role in the Maintenance of Species Diversity, ed. D. F. R. P. Burslem, M. A. Pinard and S. E. Hartley. Published by Cambridge University Press. © Cambridge University Press 2005.

(Tilman 1982), but since plants generally have similar requirements for the same few limiting resources (e.g. light, water, N, P, K), Gillett wondered 'can it be seriously suggested that a rather uniform area of Amazonian rain forest provides, in 3.5 hectares of land, anything like 179 separate ecological niches for trees?' (Gillett 1962). Density-dependent mortality caused by host-specific pests and pathogens got the lead role in Gillett's 'Theory of Pest Pressure' for the maintenance of diversity in plant communities. A decade later D. H. Janzen and J. H. Connell (Janzen 1970; Connell 1971) further explored the spatial component of pest impacts on seedling shadows to popularize some of the ideas in Theory of Pest Pressure in what is now known as the Janzen–Connell Hypothesis.

The Janzen–Connell Hypothesis – a special case

It will be seen that this invasiveness of taxa, due to pest pressure, applies as well to the spread of species from one plant association to another . . . When we see that by moving from one formation to the other it escapes to a great extent from its own pests, while entering into competition with plants which are still burdened down by theirs, the phenomenon is easy to understand.

(Gillett 1962)

The Janzen–Connell Hypothesis (a special case of the Theory of Pest Pressure) begins by recognizing that the fruits of most tree species in tropical forests do not, on average, disperse very far, leading to a great density of conspecific seedlings close to the mother tree. Within the seedling shadow, the high density of that plant species may exclude the growth of other species. Pathogens or other pests that are either transmitted from the mother tree or that develop in a density-dependent way on the locally dominant species can make scarce water, light or nutrients available to rare, but pest-resistant plant species. Similarly, escaping pest pressure may provide a competitive advantage to those rare seeds that disperse out of their maternal neighbourhood and into the neighbourhood of a heterospecific mother tree where the dominant species of seedlings will be burdened down by its own pathogens. In this way, pathogens or other pests provide an advantage to the 'lowly' rare (and resistant) species over the 'haughty' (but susceptible) dominants, helping to maintain local tree-species diversity.

Density-dependent mortality caused by host-specific pathogens and pests is at the heart of the Theory of Pest Pressure. There should be strong selection pressure for seeds to disperse out of their maternal neighbourhood (where they are disadvantaged by local pathogens) into the 'disease neighbourhood' of a heterospecific mother tree. The Janzen–Connell Hypothesis leads to several clear predictions of the Theory: (1) disease incidence should be greater at higher host density and/or among juveniles close to the mother tree, (2) through non-random mortality, the spatial distribution of the host species will become less clumped through time, (3) there should be greater recruitment of non-susceptible species within the mother's disease neighbourhood, and (4) over time, plant diversity should be greater than expected compared with random survival of seeds.

From a collection of studies on Barro Colorado Island in Panama, we have significant empirical evidence to support the first of these predictions, that disease development should be greatest at high host density and/or among juveniles close to the maternal tree. Damping-off diseases are density- and/or distance-dependent for seedlings of a range of host species (Augspurger 1983a, b; Augspurger & Kelly 1984; Augspurger 1984, 1990; Davidson 2000). Similarly, canker diseases have density- or distance-dependent rates of infection on saplings of susceptible species (Gilbert *et al.* 1994; Gilbert & de Steven 1996). As a general principle for both agronomic and wild systems, nearly all fungal diseases except heteroecious rusts show density-dependent development (see reviews in Burdon & Chilvers 1982; Gilbert 2002). As such, density-dependent disease development, at least for fungal and oomycete-caused diseases, is probably the rule in both tropical and temperate systems, but there are exceptions. Davidson (2000) found that for two pathogens of *Anacardium* seedlings, *Phytophthora heveae* showed density-dependent effects on the seedling populations whereas the related *Pythium* spp. did not. Pathogens have as much life-history variation as any other group of organisms in a tropical forests; not all pathogens will cause density-dependent mortality on all hosts, although most will. Those that do not will be unlikely to contribute to Janzen–Connell effects.

Similarly, several studies have demonstrated that diseases can mediate shifts in the spatial pattern of host trees so that they become less clumped over time, as predicted by Janzen (1970) and Connell (1971). Once again, damping-off and canker diseases provide the classic examples. The median distance from seedlings of *Platypodium elegans* to the mother tree shifted from 15 m to 30 m, following non-random mortality caused by damping-off pathogens (Augspurger 1983a). Even among decades-old juveniles, phytophthora canker of *Ocotea whitei* caused a significant shift in distribution of survivors away from conspecific adults (Gilbert *et al.* 1994).

To date, however, no studies in tropical forests have provided direct evidence for the last two predictions – that disease-related density-dependent mortality of one plant contributes to the greater survival of heterospecific, non-susceptible hosts, leading to greater local plant diversity than expected. The most complete demonstration of a disease fulfilling the diversity-maintaining Janzen–Connell predictions is that of damping-off caused by *Pythium* in a temperate forest in the mid-western United States (Packer & Clay 2000). There, *Pythium* causes density- and distance-dependent mortality of wild cherry (*Prunus serotina*) seedlings around maternal trees, causing a reduction in clumping over time, and allowing a greater survival of rarer but less-susceptible tree species in the area.

These case studies clearly show the potential for pathogens to reduce the local dominance of tree species and permit greater establishment of rare species (see also review of impacts of herbivores in Clark & Clark 1984). Collectively, they demonstrate that pathogens can (1) cause density- or distance-dependent mortality of the locally dominant tree species, (2) leading to a shift in density over time away from the maternal tree, while (3) allowing greater survival of less common, resistant species, and (4) ultimately enhancing local tree diversity, fulfilling the specific predictions of the Janzen–Connell version of the Theory of Pest Pressure. What is not yet clear, however, is whether such impacts are common enough to have a significant impact on diversity of tropical forests.

For plant diseases to be useful in explaining the maintenance of tree diversity in tropical forests through the mechanisms proposed in the Janzen–Connell Hypothesis and the Theory of Pest Pressure, several additional requirements must be met. For disease to be a driving force, (1) pathogens must be the 'inevitable, ubiquitous factors' suggested by Gillett (1962), (2) their impacts must be significant, but not devastating, (3) there must be adequate host specificity among the pathogens to create different selective pressures on different host species, (4) dominant plant species must be affected more strongly, and (5) their effects must be variable in space or time. I will discuss each of these in turn.

'The inevitable, ubiquitous factor'

For pathogens to be important drivers in tropical tree diversity, plant disease must be common and widespread in tropical forests. Clearly, if pathogens caused disease only rarely and on few plant species, they might be important components of the life history of individual species, but would be unlikely to be important factors in maintaining overall tree-species diversity. Gillett's (1962) claim of the ubiquity of plant diseases came at a time when there were very few studies on plant pathogens in tropical forests beyond a handful of economically important diseases. In developing his theory he consciously followed Darwin's lead and relied on case studies and agricultural analogies to propose general principles. However, in assessing the relevance of the Theory of Pest Pressure to real tropical forest diversity we must recognize that real plant pathogens come in many forms. The most familiar are foliar pathogens that cause necrosis, chlorosis or deformation of leaves. Pathogens can also kill seeds and seedlings, plug the vascular system, cause cankers on stems and trunks, decay wood in roots and trunks, and attack flowers and developing fruits – often with potentially significant impacts on the host populations in natural ecosystems (see recent review in Gilbert 2002). The life histories, diversities and potential impacts of each of these groups of plant pathogens vary widely, and with potentially large consequences for their roles in maintenance of tropical tree-species diversity. If we are to understand the role of pathogens, it is crucial that they not be lumped into a single black box. We would not blithely consider harpy eagles and jaguars to be ecological equivalents in a tropical forest simply because they are both predators, and similarly, we stand to learn much more about the impacts of pathogens by considering the kinds of interactions particular pathogens have with their hosts.

Studies of individual, charismatic plant diseases can provide useful case studies of the potential effects of diseases, but to assess Gillett's claim of ubiquity of diseases we need to examine comparative studies across a range of plant species within a forest. Selecting study species because they have obvious diseases leads to a potentially misleading form of 'publication bias'. Much more useful are comparative studies across host species chosen either arbitrarily or for their ecological significance. Fortunately, there are now a few such quantitative, comparative studies for some kinds of plant pathogens in tropical forests, which allow us to begin to assess whether plant pathogens really are Gillett's inevitable and ubiquitous selective agents. I will review such studies grouped by broad life-history strategies of the pathogens.

Foliar diseases

Foliar diseases, which reduce the photosynthetic area available to plants, are the most widely studied diseases in tropical forests. In the most detailed comparative survey to date, García-Guzmán and Dirzo (2001) surveyed understorey plant species in the Los Tuxtlas tropical rain forest in Mexico and found that 69% of 57 understorey plant species, and 45% of all examined leaves, were damaged by fungal pathogens. Similarly, all 10 species examined on Barro Colorado Island (BCI) in Panama (Barone 1998), three species examined in Brazilian rain forest (Benitez-Malvido et al. 1999), all five species examined in a seasonal dry forest in Panama (Gilbert 1995), and all three species present in a Panamanian mangrove forest (Gilbert et al. 2001) suffered foliar diseases. However, disease incidence (proportion of leaves with at least some disease) varied widely across species (10% to 32%; Benitez-Malvido et al. 1999). From these studies, which included both high- and low-diversity tropical forests and both moist and dry sites, it is clear that most tropical tree species suffer from foliar diseases but that the incidence of disease varies widely across species.

Seed and seedling diseases

Diseases of seeds and seedlings have a major impact on individual plants, usually leading to a quick death. Fungal attack caused 47% and 39% annual mortality of seeds in the soil seed bank of Miconia argentea (Melastomataceae) and Cecropia insignis (Moraceae), two pioneer tree species in Panama (Dalling et al. 1998), and, based on studies in temperate systems, I would expect pathogenic fungi to have widespread effects on other tropical species with seed banks. Seedlings are probably the most vulnerable stage in the life of a tropical tree, and a number of damping-off pathogens (often the fungus-like oomycetes Phytophthora and *Pythium*) cause dramatic losses of seedlings. In pioneering studies of the impacts of pathogens on the survival of tropical forest plants, Augspurger and coworkers, working on BCI, found that damping-off affected 80% of all tested species, and was the leading cause of death for seedlings of six of nine focal species (Augspurger 1983b; Augspurger & Kelly 1984; Augspurger 1984; Kitajima & Augspurger 1989).

Cankers, wilts and diebacks

Wilt and canker diseases affect the growth and survival of juvenile and adult woody plants, but have been little studied in tropical forests. The responsible pathogens use a variety of mechanisms to disrupt the plant's vascular system. On BCI, a canker associated with *Phytophthora* sp. affects 9 of 10 species of Lauraceae, including 73% of the individuals of *Ocotea whitei* (Gilbert *et al.* 1994, and G. S. Gilbert, unpublished data). Other case studies indicate that canker, wilt and dieback diseases can have significant effects on particular host species (botryosphaeria canker on *Tetragastris panamensis* (Burseraceae), Gilbert and de Steven (1996); fusarium wilt on the Hawaiian endemic *Acacia koa* (Fabaceae), Anderson *et al.* (2001)), but there have been no comparative studies across a diverse range of host species or in different forests to assess how widespread and common these pathogens are in tropical forests.

Wood-decay fungi

Wood-decay fungi, particularly the basidiomycete polypore fungi, play a variety of roles in tropical forests. Some can be aggressive pathogens, destroying root systems or cambial tissue and killing trees directly, whereas others consume only dead wood, hollowing out the centres of living trees and making them more susceptible to damage from wind and rain. Still others may infect living trees, but really only colonize and decay extensively when the tree is dead or dying from other factors. Finally, some polypore fungi may play several of these roles sequentially. Recent studies provide some assessment of the ubiquity of wood-decay fungi not only on fallen logs but on standing trees in tropical forests, although the studies do not consistently differentiate among different ecological roles. For instance, in a survey of 10 focal tree species on BCI, the percentage of live trees with polypore fruiting bodies ranges from zero to 33% (with a mean of 7%), while 56% (ranging from 38% to 93%) of all dead trees had polypores (Gilbert *et al.* 2002). Comparable systematic study of macroscopic ascomycete wood-decay fungi has been rare in tropical systems, but for three tree species at the same site, ascomycete fruiting bodies were about two-thirds as common as polypore fungi (Ferrer & Gilbert 2003). In a Caribbean mangrove forest, 3% to 19% of all live and dead trunks of the three dominant tree species had fruiting bodies of polypore fungi (Gilbert & Sousa 2002). Most, if not all, tropical trees have associated wood-decay polypore fungi, and for some tree species, a high proportion of live hosts may be attacked.

Flower and fruit diseases

Although any pathogens that affect the growth or survival of their host plants can reduce the host's reproductive output, a number of pathogens attack flowers and developing fruits directly. Very little work and no comparative studies have been done in tropical forests on such diseases. In one example, the rust *Aecidium farameae* attacks developing ovaries of the understorey treelet *Faramea occidentalis* (Rubiaceae) on BCI, but over the last decade has only appeared in outbreaks every several years (Travers *et al.* 1998). Of course, fungal colonization of seed may also be an important mechanism for vertical transmission of endophytes or pathogens (Bayman *et al.* 1998), so a clear picture of the frequency and ecological role of fruit- and seed-infecting fungi may be complex.

Missing diseases

García-Guzmán and Dirzo (2001) note specifically that they found no rusts, smuts or powdery mildews in their survey of plant diseases at Los Tuxtlas, and this corresponds well with my experiences elsewhere. Rusts have been known to cause problems in tropical tree plantations (Lee 1999), and the introduction of coffee rust (*Hemileia vastatrix*) has had devastating effects in tropical America (Wellman & Echandi 1981). The importance of rusts in many wild temperate systems (Bella & Navratil 1988; Burdon & Jarosz 1991; Davelos *et al.* 1996) leads us to wonder about their importance in the tropics. Nevertheless, there are few examples of rusts in tropical forests (Arthur & Cummings 1933; Gardner 1994; Chen *et al.* 1996). Smuts as well can have large impacts on natural plant populations in temperate systems (Alexander & Antonovics 1988; García-Guzmán *et al.* 1996; Carlsson-Graner 1997). However, although smuts have been recorded on tropical trees in the Sterculiaceae, Araliaceae, Tiliaceae and Piperaceae, only 11 (of 1450) smut species are known from woody plants and their overall importance to tropical trees is probably minimal (Vánky 2002). Powdery mildews cause serious problems in tropical agriculture including on fruit trees (Schoeman *et al.* 1995), but I am unaware of any studies of their incidence or importance in tropical forests. Other 'missing' plant pathogens such as viruses and phytoplasmas are of great importance in tropical agriculture, but remain essentially undocumented in tropical forests owing to the inherent technical difficulties in their detection and study. Whether the lack of records of rusts, smuts and powdery mildews reflects a similar blind spot among those working in tropical forests, the paucity of researchers in this area or a biological reality is not yet clear. However, all these missing pathogens are obligate biotrophs; that is, they complete their life cycles only on living hosts, for the most part cannot grow apart from a living plant, and in many cases show strong host specificity. It is possible that conditions in diverse tropical forests are not supportive of such specialized pathogens, a possibility I will return to later.

Endophytic and epifoliar fungi

Two groups of plant-infecting fungi are often excluded from discussion of diseases, but they share so many life-history traits with other pathogens, and the potential for interaction with other pathogens is so great, that they must be included here. Leaf endophytes are fungi that invade living plant tissue and cause unapparent and asymptomatic infections within healthy plant tissues (Wilson 1995). Endophytes include mutualists and commensals, but many endophytes are latent pathogens that may cause disease at a later time depending on environmental conditions or plant stress. For instance, *Pestalotiopsis* (= *Pestalotia*) *subcuticularis* on *Hymenaea courbaril* (Fabaceae) stays as a benign endophyte unless leaves are wounded by cutting or scraping (Fail & Langenheim 1990). Recent studies clearly demonstrate the ubiquity of endophytes in tropical forests; generally 90%–100% of all leaves are infected by endophytic fungi (Lodge *et al.* 1996; Bayman *et al.* 1998; Arnold *et al.* 2000; Bethancourt 2000; Arnold *et al.* 2001; Gilbert *et al.* 2001). On a finer scale, the area of individual leaves that is infected varies widely, from 98% of all 2-mm^2 leaf fragments from plants in nine families on BCI (Arnold *et al.* 2001) to only 25% of leaf discs of the Amazonian palm *Euterpe oleraceae* (Rodrigues 1994). Variation in infection rates may be due to different anti-fungal defences across plant species (Gilbert *et al.* 2001) or to genotypic or environmental variation among individuals within a tree species (Bethancourt 2000). In fact, Langenheim and Stubblebine (1983) proposed that differences in chemical defences between adults trees and nearby offspring may reduce pressure on juvenile trees from pests of the adults, and offset the distance effects of the Janzen–Connell Hypothesis. This proposal has yet to be tested for tropical plant diseases, however.

On the surface of many leaves are epifoliar (or epiphyllous) fungi (e.g. sooty moulds) that may act entirely as superficial saprotrophs, consuming honeydew, leaf exudates and detritus. Some species receive nutritional support from the plant by penetrating the leaf cuticle and forming haustoria within host-plant cells. Epifoliar fungi do not cause apparent disease symptoms, but may intercept light (but see Anthony *et al.* 2002), or interact synergistically or antagonistically with pathogens to affect disease development (Leben 1965). In a recent comparative study at sites in moist tropical forests in Cape Tribulation in Queensland, Australia and on the Caribbean slope of Panama, 36% (of 182) and 65% (of 81), respectively, of the surveyed understorey plant species were colonized by epifoliar fungi (D. Reynolds and G. S. Gilbert, unpublished data). Similarly, epifoliar fungi were found in 51% and 81% of 120 understorey sampling locations at the Australian and Panamanian sites, respectively. Their commonness, their potential for direct effects, and the possibility for augmenting or inhibiting the effects of other foliar pathogens lend epifoliar fungi potential importance in tropical forests beyond what is usually recognized.

Asymptomatic fungal infection of leaves (as well as of epiphyte roots (Richardson & Currah 1995; Bayman *et al.* 1997), and bark (Suryanarayanan & Rajagopal 2000)) is clearly ubiquitous in tropical forests, but their importance to the plants is almost entirely unknown. In particular, understanding how many of these fungi are latent pathogens, how they affect diseases caused by other pathogens, or how many may provide other direct benefits to host plants may be key to understanding many aspects of plant diseases.

The importance of plant diseases in tropical forests has only entered the mainstream of thought in plant ecology in the last two decades, and few plant pathologists have ventured into the forest with Petri plate or PCR machine in hand. There is of course a rich history, too vast to review here, of collecting and describing fungi from tropical forests (see e.g. Stevens 1927; Weston 1933), but the collecting has traditionally been haphazard and qualitative, and because the focus was on the fungi, often lacked adequate information about host plants. Despite our patchwork understanding of plant pathogens in tropical forests, a growing body of literature clearly indicates that a plant without disease is a biological anomaly, but that the types of pathogens and that the appearance of disease may vary greatly among hosts and across sites. Pathogens are ubiquitous, but particular plant–pathogen interactions are not. Overall, however, plant pathogens clearly conform to Gillett's vision of an 'inevitable, ubiquitous factor'. The question remains whether they have the impact on plants necessary for them to really affect species diversity in tropical forests.

Significant but not devastating impacts

It is inevitable that each tropical tree will suffer at least some disease over the course of its life, and being diseased is probably the normal state for most trees. However, what is not clear is how large an impact disease has on host survival, growth and reproduction in tropical forests, nor how the impact on individual plants translates into effects on populations or species. For disease to be an important driver of tropical diversity, pathogens must be capable of causing diseases severe enough to limit (but not extinguish) populations of susceptible plant species either directly or by placing them at a competitive disadvantage to other plant species. However, although a highly virulent pathogen with a catholic appetite may have a dramatic impact on forest ecosystems (e.g. *Phytophthora cinnamomi* in temperate Western Australian forests (Weste & Marks 1987)), it is unlikely that such a pathogen will help maintain local plant-species diversity. It is the subtle, not the sledgehammer, that is most likely to be important in tropical diversity.

Foliar diseases
Although foliar diseases affect the majority of species studied and may be responsible for between 2% and 75% of all identifiable leaf damage (mean 34%) (Barone
1998), the proportion of leaf tissue actually damaged is often low. García-Guzmán
and Dirzo (2001) found that although 45% of leaves suffered damage from fungal pathogens, damage was generally less than 6% of the leaf area. Similarly,
Benitez-Malvido *et al.* (1999) found that average damage from fungal pathogens
on three host plants in Amazonian forest never exceeded 1.5% of the leaf area.
The authors concluded that 'fungal infection is so rare that it is unlikely to
affect seedling performance'. However, even very small reductions in leaf area
of understorey seedlings may have large impacts on seedling survival (Clark &
Clark 1985), and foliar pathogens can cause a large reduction in host growth
(Esquivel & Carranza 1996). Clearly, even low levels of foliar disease have the
potential for large impacts on plants, but the realized impacts will vary greatly
among hosts, pathogen species and sites.

Seed and seedling diseases
By definition, damping-off diseases have a large impact (death) on individual
seedlings. On BCI, Augspurger (1984) found that the proportion of seedlings that
died from damping-off exceeded 70% for two species of Bombacaceae, but that
mortality across eight focal species varied greatly, averaging 35% in the shade.
Similarly, Infante (1999) applied fungicide drenches to control damping-off in
tree seedlings in the rainforest interior. Five of nine species showed significantly
greater survival with fungicide treatment, with an overall 31% increase in mean
seedling survival time. However, damping-off was negligible in high-light environments such as light-gaps (Augspurger 1984) or the edges of forest fragments
(Infante 1999). Susceptible host species in conducive environments may suffer
dramatic mortality from damping-off pathogens, but like foliar diseases, the
impacts will vary among host species and sites.

Cankers, wilts and diebacks
Canker, wilt and dieback pathogens may have large impacts on the growth and
survival of both juvenile and mature tropical trees, but there are only a limited
number of studies with direct measurements of impact. *Fusarium* dieback of *Acacia koa* in Hawaii (a wilt disease) quickly reduces stomatal conductance and foliar
growth, and usually leads to death of the tree (Anderson *et al.* 2001). Inoculation
of juveniles of *Tetragastris panamensis* with the canker pathogen *Botryosphaeria
dothidea* led to a three-fold increase in mortality, and among survivors, limited
growth to 59% of controls (Gilbert & De Steven 1996). Studying the impacts of
canker, wilt and dieback pathogens on host growth and mortality is much more
difficult than for foliar or seedling diseases because there is often a long latent
period between infection and symptom development, and because large, woody

trees have inherently slow rates of growth. Long-term and broadly based studies of canker diseases in tropical forests are needed to make more than an educated guess as to their potential impact on species diversity.

Wood-decay fungi

Wood-decay fungi, particularly those that cause root- and butt-rots of large trees, have been shown to play key roles in the population dynamics and community structure of temperate forests (Hansen & Goheen 2000) and to reduce the growth rate of infected trees (Alexander *et al.* 1981; Bloomberg & Morrison 1989). For tropical forests, although we know that polypore and other wood-decay fungi are common in standing trees (Meza 1992; Gilbert *et al.* 2002; Gilbert & Sousa 2002; Ferrer & Gilbert 2003), we have few data on their importance in either killing trees directly, or making trees susceptible to breaking or falling from excessive wind and water. Although in temperate forests such disease is often an important precursor to trees falling or breaking (e.g. Worrall & Harrington 1988), only 13% of snapped trees on BCI had heart rot; incidence of root rot was not determined for uprooted trees (Putz *et al.* 1983). Recent studies of the diversity and incidence of wood-decay fungi have highlighted their potential importance in tropical forests, but there is much need for work on their ecological effects.

Flower and fruit diseases

A number of plant venereal diseases have been shown to have large impacts in temperate systems (see review in Gilbert 2002), and we might expect similar effects in tropical forests. The potential for large effects of flower and fruit diseases can be illustrated by the ovary-attacking rust *Aecidium farameae*, which reduced fruit set by 75% for infected individuals of the understorey treelet *Faramea occidentalis* (Rubiaceae) on BCI (Travers *et al.* 1998). Similarly, 85% of fertilized ovules of *Anacardium excelsum* (Anacardiaceae) were killed by the fungus *Cladosporium* (Sánchez Garduño *et al.* 1995). Based on these examples, and the high rate of tropical agricultural fruit loss to pathogens (Opoku *et al.* 2002), it would be surprising if a significant proportion of fruits and developing seeds of tropical forest trees were not lost to diseases. However, not enough work has been done on flower and fruit diseases in tropical forests to make an informed guess about their overall importance.

We have good evidence that plant pathogens – and plant–microbe interactions in general – are extremely common in tropical forests. It is also clear that some can have significant impacts on individual plants. However, the effects of pathogens on plants will only be important in the dynamics of tree populations or in maintaining species diversity in the forest if disease reduces the contribution of that population to the next generation. Compensatory responses may prevent even strong effects of disease on individual fitness from affecting numerical population responses (see reviews in Alexander and Holt 1998; Gilbert 2002).

Integrating effects of disease into plant population dynamics across multiple generations is the next key step in understanding the real impact of diseases on individual plant populations, and will ultimately be central to understanding the importance of disease in tree diversity.

Host specificity

The diversity of fungi (including pathogens) in tropical forests is staggering, and efforts to estimate the true number of fungal species have involved much deliberation over the scale of host specificity (Hawksworth 2001). May (1991) suggested that the most successful plant-infecting fungi in species-rich tropical forests should be non-specialists, because specialists would have difficulty with transmission among individuals of rare host species. However, despite its importance, our estimates of fungal specificity in tropical forests are currently little more than guesses. As for most tropical taxa, fungal diversity takes the form of many rare species and just a few common species (Bills & Polishook 1994; Polishook *et al.* 1996; Huhndorf & Lodge 1997; Arnold *et al.* 2000; Bethancourt 2000; Lindblad 2000; Arnold *et al.* 2001; Gilbert *et al.* 2001; Gilbert *et al.* 2002). The rarity of most species makes determining the scale of host specificity extremely difficult. Most existing literature is only minimally helpful in determining host ranges, since phytopathological and mycological investigation has focused strongly on plants of economic value (Cannon & Hawksworth 1995; Clay 1995), and traditional fungal collecting in forests has been haphazard and non-quantitative. Recently, however, a few systematic, comparative studies (aided by developments in molecular characterization of pathogens) are beginning to provide useful data on the scale of host specificity among plant-infecting fungi in the tropics.

Arnold and coworkers (Arnold *et al.* 2000; Arnold *et al.* 2001) found evidence for host specialization by endophytes from the leaves of two co-occurring but unrelated understorey shrubs, *Heisteria concinna* (Olacaceae) and *Ouratea lucens* (Ochnaceae). Of the 140 morphospecies found more than once, almost two-thirds were found on one host or the other, but not both. Although suggestive, the power of this study is limited because a small number of hosts were studied, the species-accumulation curves had not yet saturated, and there was difficulty in differentiating fungal species in largely undescribed groups. Focusing instead on a single fungal genus, Laesse and Lodge found that more than half of 13 species of *Xylaria* isolated from leaves and fruits in Puerto Rico were restricted to a single host-plant genus or family (Laesse & Lodge 1994; Lodge 1997). These studies suggest the potential for ecologically significant specialization among leaf-inhabiting fungi, but much more extensive sampling is needed together with additional use of molecular classification to distinguish among morphologically similar species and to recognize the unity of phenotypically plastic ones.

Recent studies of tropical wood-decay fungi, particularly polypores, also provide a glimpse into the scale of host specificity for another group of fungi. Polypore fungi are taxonomically among the best known of tropical fungi, are largely confined to woody substrates, and produce persistent, showy fruiting bodies convenient for broad systematic surveys. In dry tropical forest in Costa Rica, Lindblad (2000) found that only 3 of 32 common species of polypore fungi showed host specificity, and that there was a strong correlation between the number of hosts a fungus was found on and the commonness of the fungus. In a similar study in seasonal moist tropical forest on BCI, Gilbert *et al.* (2002) found no statistical evidence for host specialization among the 17 species found multiple times. For the five most common fungi, each was found on multiple host species from multiple families (Fig. 6.1a). Although the most common fungi are clearly non-specialists, the assemblages of polypore and ascomycete wood-decay fungi on three host species were readily distinguishable and persistent through time, suggesting that there is sufficient host specialization for host composition to influence the types of fungi present in the forest (Ferrer & Gilbert 2003).

In trying to understand host specificity in species-rich tropical forests, it may be most useful to consider what happens when tree diversity is naturally low. In a Caribbean mangrove forest about 25 km from BCI, only three tree species make up the entire community. In sharp contrast to the high diversity and low dominance of polypore species found on BCI (Gilbert *et al.* 2002; Ferrer & Gilbert 2003), 88% of all polypore collections (nine species total) were comprised of just three fungal species (Gilbert & Sousa 2002). In fact, of the five polypore species found multiple times, all showed high host specificity (Fig. 6.1b). The combination of these polypore studies provides support for May's (1991) prediction of low host specificity in species-rich forests. It may be that in a low-diversity forest where the density of individual host species is high, reliable transmission among hosts allows for the success of host-specialized fungi. In contrast, in high-diversity forests with a low density of individual host species, host non-specialists will be more successful. Similar tests with other groups of fungi should be a high priority for study.

The mangrove polypore study also highlights the importance of scale in the discussion of host specificity. Many herbivorous insects have generalized diets over the entire range of the species, but have much more restricted diets in a local context (Fox & Morrow 1981). Similar behaviour might be expected from plant pathogens. In fact, all the fungi described from the mangroves are described in the literature as attacking non-mangrove hosts (Gilbert & Sousa 2002). Within the local context of a three-species mangrove forest, the fungi are specialists, but globally they are not. It is important to determine the appropriate geographic scale when investigating host specialization.

Although the studies of endophytes and polypores provide a glimpse into host specificity of tropical plant-infecting fungi, only a portion of the included species

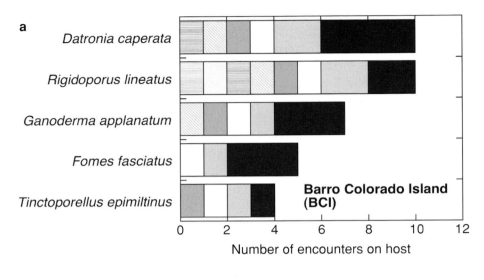

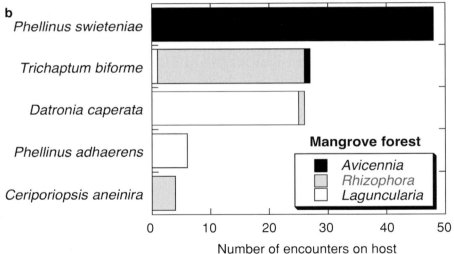

Figure 6.1 Host specificity of the five most common polypore fungi in (a) the seasonal moist tropical forest of Barro Colorado Island and (b) the Caribbean mangrove forest at Punta Galeta, Republic of Panama. For BCI, different shadings within a bar indicate a different host species, but shades are not consistent across fungi. Polypore fungi in the species-rich forest on BCI show no host specificity, whereas in the low-diversity mangrove forest host specificity is high. Data are from Gilbert *et al.* (2002), and Gilbert and Sousa (2002).

are plant pathogens. Systematic studies of host range for pathogens in tropical forests are rare, but one interesting study of a pathogen with a broad host range comes from the study of an unusual luminescent mushroom. The agaric *Mycena citricolor* causes American leaf spot disease of introduced coffee (*Coffea* spp., Rubiaceae), a devastating disease through much of tropical America. It also has an exceptionally large host range. Sequiera (1958) found it causing leaf spots on 150 host plants from 45 families in Costa Rica, including many forest tree species. Although it can cause disease on many hosts, *M. citricolor* may reproduce readily on only a small number of hosts (Sequiera 1958). Lodge and Cantrell (1995) noted that *M. citricolor* became rare in the Luquillo Mountains of Puerto Rico when coffee production was abandoned, but that outbreaks of *M. citricolor* followed the sharp increase in density of native Rubiaceae after hurricane damage to the forest. It may be important to consider source–sink dynamics when pathogens can infect and cause disease on a range of hosts, but are able to reproduce readily on only some subset of those hosts. In a recent temperate example, the spread of sudden oak death (caused by *Phytophthora ramorum*) through California forests has had devastating impacts on oaks and tanoak (*Quercus* spp. and *Lithocarpus densiflorus*, Fagaceae), but much of the source of inoculum appears to come from infections on the sympatric *Umbellularia californica* (Lauraceae), a host that suffers only minimal damage but supports luxuriant reproduction of the pathogen (Rizzo & Garbelotto 2003). Such asymmetrical disease interactions may lead to apparent competition (Alexander & Holt 1998) and must be considered when looking at the importance of host range in the role of diseases in maintaining plant diversity. Our poor understanding of the extent of host specificity, and the potential for linking host population dynamics through shared pathogens, is probably the most important gap in our ability to assess the importance of pathogens in tropical tree diversity.

Effects on dominant plant species

Host specificity is not sufficient for virulent pathogens to prevent competitive exclusion, as envisaged by Gillett. The pathogens must also have greater effects on the locally dominant plant species. There are currently no clear studies from tropical forests of the relative impacts of either individual pathogens across a range of susceptible species or the overall pathogen load of a suite of species across a range of abundance. The ubiquity of density-dependent disease development discussed above suggests that whatever tree species is locally abundant at a particular time is likely to suffer significantly more disease pressure than more rare species. Since density effects are nearly always examined within a single host species, there is little empirical support for this expectation. However, in a broad test across 10 tree species on BCI, the more common tree species were more often colonized by wood-decay fungi (Gilbert *et al.* 2002), suggesting the strong possibility that density-dependent disease development may be observable

even across host species with very different phylogenetic, physiological and ecological backgrounds.

Variability in space or time

Plant diseases meet the requirements of inevitability and potential impact, but
for diseases to be important factors in explaining the 'pointless multiplicity
of species' in tropical forests, there must be variation in how pathogens affect
different tree species. Host specificity may be the most fundamental dimension
of variation, but two other dimensions may be equally important ecologically:
variation across space due to environmental factors or dispersal limitation, and
change over time.

Environmental heterogeneity

It will be noted, however, that pest pressure is not uniform in all areas; it will be less
where there is a cold winter, or a long dry season, than in constantly warm and humid
areas where the pests can live and breed all the year round.   (Gillett 1962)

Gillett considered that warm temperatures, abundant moisture and aseasonality
would favour populations of pests and pathogens, increasing their impacts in
areas such as aseasonal moist tropical forests. This could in part explain latitudinal gradients in plant diversity, as well as diversity differences along moisture
gradients within tropical regions. Givnish (1999) expanded on this idea, and proposed that increased rainfall, increased soil fertility and decreased seasonality
would not only favour pests and pathogens, but should decrease plant investment in anti-herbivore (or anti-pathogen) defences. This should in turn lead to
greater pathogen/pest-driven density-dependent mortality, supporting a greater
diversity of tree species in moist tropical communities than elsewhere. He suggests that plant competition should play a greater role in determining plant
diversity at high latitudes, and that pest pressure should play a relatively more
important role at low latitudes. There is growing evidence for the importance
in environmental heterogeneity in causing differential disease impacts at scales
from very local to geographical.

  The predictions of Gillett and Givnish depend on differential disease development along environmental gradients, with generally greater disease pressure
under wetter and more constant climatic conditions. There is a large literature from agricultural systems and temperate-zone natural ecosystems showing
greater disease pressure with increased moisture (e.g. Bradley *et al.* 2003 and references therein). In tropical forests, damping-off disease is reduced under low relative humidity associated with light gaps and forest fragment edges (Augspurger
& Kelly 1984; Augspurger 1984; Infante 1999). Rodrigues (1994) found greater
rates of endophyte infection in understorey saplings than in canopy adults of

the palm *Euterpe oleraceae*, as well as differences in the types of fungi present. She speculates that the differences may be driven by different microclimates experienced by understorey saplings and taller palms. Environmental variation can affect not only the abundance or activity of particular pathogens, but the kinds of fungi present. Different kinds of diseases developed on leaves in the subcanopy shade than in the sun for *Anacardium excelsum* (Gilbert 1995). Finally, Lindblad (2001) found little overlap in polypore species among dry, moist and wet tropical forests in Costa Rica. Across different scales, environmental variation is associated with different kinds and intensity of disease development.

Latitudinal gradient

For the Theory of Pest Pressure to be an effective explanation for differences in plant diversity along the largest environmental gradient of all (from the Equator to the poles), disease pressure must vary along a latitudinal gradient. As supportive evidence, a large number of studies point to the pervasiveness and importance of density-dependent mortality in tropical forests (e.g. Webb & Peart 1999; Harms *et al.* 2000). However, Hille Ris Lambers *et al.* (2002) compared a number of published studies of density-dependent mortality in temperate and tropical forests, and concluded that the proportion of species affected does not change systematically along a latitudinal gradient. They suggest that for density-dependent mortality to explain high diversity of tropical forests, its strength must be greater in the tropics. This idea has yet to be tested.

A formal comparison of the diversity and severity of diseases in forests along a latitudinal gradient or along an environmental gradient within the tropics would be the ideal test of Gillett's and Givnish's predictions, but such a study is complicated by changes in the plant species along those same gradients. Wellman (1968) provides a surrogate measure of disease pressure along a latitudinal gradient. From a variety of sources and field observations, he catalogued the number of diseases on certain important crop species that were grown under similar conditions both in the temperate United States and in tropical Latin America. For 11 of 12 crops there were significantly more diseases at tropical latitudes than temperate (Fig. 6.2).

Change through time

Every time that pest pressure helps a species to spread out of the region in which it evolved and to whose conditions it is adapted, a new evolutionary process is started to adapt the species to the conditions of its new habitat.   (Gillett 1962)

Some temperate grasses (van der Putten & Peters 1997), herbs (Bever 1994; Bever *et al.* 1997) and trees (Hansen & Goheen 2000) have been shown to build up soil pathogens over time that reduce their own competitiveness and allow for less

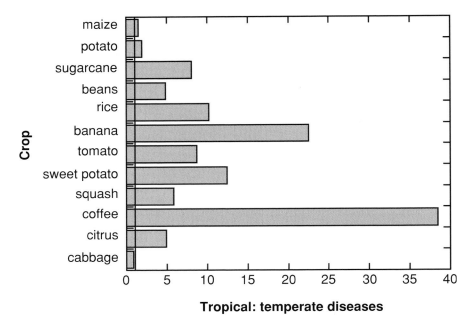

Figure 6.2 The ratio of number of diseases on important crops grown in both tropical and temperate American regions. All crops except cabbage had more diseases in the tropics. The vertical line indicates 1:1 ratio. Data are from Wellman (1968).

susceptible species to replace them. This temporal buildup, followed by a gradual decrease in pathogen pressure, is the basis for crop rotation in agricultural systems. In tropical forests, dispersal of the host to pathogen-free space gives temporal respite from disease pressures, and may help maintain plant-species diversity in a dynamic mosaic driven by pathogen feedback. Such feedback may be driven primarily by numerical increase in host-specific pathogens. It can also involve rapid evolution of increased virulence of a pathogen on a particular host. This is particularly likely when the life cycle of the pathogen is much shorter than the life cycle of the host, which would be the case for many pathogens of trees. Selection for greater virulence can occur very quickly; virulence of individual isolates of flax rust (*Melampsora lini*) increased significantly through a single growing season of a wild population of *Linum marginale* (Burdon & Jarosz 1991).

Gillett also recognized that pest pressure could affect plant diversity at levels other than species diversity: 'The effect of pest pressure in producing genetic diversity applies at the intraspecific as well as at the specific, generic, or family level' (Gillett 1962). When pathogens have more-limited dispersal than their hosts, local adaptation to particular genotypes may become important. Local adaptation of a pathogen that completes numerous generations on a single mother tree could accentuate the impact of the pathogen on offspring of that tree, while providing a possible rare-genotype advantage to seedlings from distant mothers of the same species. However, in a reciprocal transplant test of this idea for three species in Panama, Davidson (2000) found no evidence to support a rare-genotype advantage for resistance to seedling diseases. She concluded that for tropical tree species, long-range gene movement and heterogeneity in the rainforest environment are likely to be more important to maintenance of genetic diversity within host species.

Gillett (1962) also proposed that as pathogens force plant species out of disease-ridden areas, the plants are forced to adapt to different environmental conditions and to competition with a different set of neighbouring plant species. In this way, pathogen pressure could drive plant evolution even beyond selection for resistant species and genotypes. Givnish (1999) also suggests that a random walk over evolutionary time through changing pressures from pests and pathogens may work better than Hubbell's ecological-drift model (Hubbell 1997, 2001) to account for non-random dominance by certain plant families, or for the predictable shifts in species composition across environmental gradients. Most work to date has considered the role of pathogens in maintenance of diversity, but consideration is due to the role of pathogens in the genesis of species diversity in tropical forests as well.

Concluding remarks

Gillett's 1962 address is too often overlooked in discussions of diseases as driving forces in the maintenance of diversity in tropical forests. It established the framework for later theoretical developments by Janzen (1970), Connell (1971), and Givnish (1999), and outlined the needed empirical work provided by a diverse and growing number of researchers. Since 1962 we have accumulated a significant but still thin body of literature that shows that in tropical forests, plant pathogens could be Gillett's 'inevitable, ubiquitous factors', that they have significant impacts on host performance, and that their effects are variable in both space and time. Some data indicate significant host specificity in tropical forests, but there is still great need for improved understanding of the scale of host specificity and the degree to which dominant plant species must be affected more strongly. Careful, community-based, comparative work is key to advancing our appreciation for the various dimensions of influence of plant pathogens in tropical forests.

Acknowledgements

Thanks to I. M. Parker, B. Ayala-Orozco, and three anonymous reviewers for critical comments that improved this paper.

References

Alexander, H. M. & J. Antonovics. 1988. Disease spread and population dynamics of anther-smut infection of *Silene alba* caused by the fungus *Ustilago violacea*. *Journal of Ecology* **76**:91–104.

Alexander, H. M. & R. D. Holt. 1998. The interaction between plant competition and disease. *Perspectives in Plant Ecology Evolution and Systematics* **1**:206–220.

Alexander, S. A., J. M. Skelly & R. S. Webb. 1981. Effects of *Heterobasidion annosum* on radial growth in southern pine beetle-infested loblolly pine. *Phytopathology* **71**:479–481.

Anderson, R. C., D. E. Gardner, C. C. Daehler & F. C. Meinzer. 2001. Dieback of *Acacia koa* in Hawaii: ecological and pathological characteristics of affected stands. *Forest Ecology and Management* **5576**:1–14.

Anthony, P. A., J. A. M. Holtum & B. R. Jackes. 2002. Shade acclimation of rainforest leaves to colonization by lichens. *Functional Ecology* **16**:808–816.

Arnold, A. E., Z. Maynard & G. S. Gilbert. 2001. Fungal endophytes in dicotyledonous neotropical trees: patterns of abundance and diversity. *Mycological Research* **105**:1502–1507.

Arnold, A. E., Z. Maynard, G. S. Gilbert, P. D. Coley & T. A. Kursar. 2000. Are tropical fungal endophytes hyperdiverse? *Ecology Letters* **3**:267–274.

Arthur, J. C. & G. B. Cummings. 1933. New species of Uredinales. *Annales Mycologica* **31**:41.

Augspurger, C. K. 1983a. Offspring recruitment around tropical trees – changes in cohort distance with time. *Oikos* **40**:189–196.

1983b. Seed dispersal of the tropical tree, *Platypodium elegans*, and the escape of its seedlings from fungal pathogens. *Journal of Ecology* **71**:759–771.

1984. Seedling survival of tropical tree species – interactions of dispersal distance, light-gaps, and pathogens. *Ecology* **65**:1705–1712.

1990. Spatial patterns of damping-off disease during seedling recruitment in tropical forests. In J. J. Burdon & S. R. Leather, eds., *Pests, Pathogens, and Plant Communities*. Oxford: Blackwell Scientific Publications, pp. 131–144.

Augspurger, C. K. & C. K. Kelly. 1984. Pathogen mortality of tropical tree seedlings: experimental studies of the effects of dispersal distance, seedling density, and light conditions. *Oecologia* **61**:211–217.

Barone, J. A. 1998. Host-specificity of folivorous insects in a moist tropical forest. *Journal of Animal Ecology* **67**:400–409.

Bayman, P., P. Angulo-Sandoval, Z. Báez-Ortiz & D. J. Lodge. 1998. Distribution and dispersal of *Xylaria* endophytes in two tree species in Puerto Rico. *Mycological Research* **102**:944–948.

Bayman, P., L. L. Lebron, R. L. Tremblay & D. J. Lodge. 1997. Variation in endophytic fungi from roots and leaves of *Lepanthes* (Orchidaceae). *New Phytologist* **135**:143–149.

Bella, I. E. & S. Navratil. 1988. Western gall rust dynamics and impact in young lodgepole pine stands in west-central Alberta. *Canadian Journal of Forest Research* **18**:1437–1442.

Benitez-Malvido, J., G. García-Guzman & I. D. Kossmann-Ferraz. 1999. Leaf-fungal incidence and herbivory on tree seedlings in tropical rainforest fragments: an experimental study. *Biological Conservation* **91**:143–150.

Bethancourt, E. 2000. Un método de muestreo para el estudio de la diversidad de hongos endófitos asociado a especies arbóreas en un bosque tropical. Unpublished Master's thesis, Universidad Tecnológica de Panamá.

Bever, J. D. 1994. Feedback between plants and their soil communities in an old field community. *Ecology* **75**:1965–1977.

Bever, J. D., K. M. Westover & J. Antonovics. 1997. Incorporating the soil community into plant population dynamics: the utility of the feedback approach. *Journal of Ecology* 85:561–573.

Bills, G. F. & J. D. Polishook. 1994. Abundance and diversity of microfungi in leaf litter of a lowland rain forest in Costa Rica. *Mycologia* 86:187–198.

Bloomberg, W. J. & D. J. Morrison. 1989. Relationship of growth reduction in Douglas-fir to infection by Armillaria root disease in southeastern British Columbia [Canada]. *Phytopathology* 79:482–487.

Bradley, D. J., G. S. Gilbert & I. M. Parker. 2003. Susceptibility of clover species to fungal infection: the interaction of leaf surface traits and environment. *American Journal of Botany* 90:857–864.

Burdon, J. J. & G. A. Chilvers. 1982. Host density as a factor in plant disease ecology. *Annual Review of Phytopathology* 20: 143–166.

Burdon, J. J. & A. M. Jarosz. 1991. Host-pathogen interactions in natural populations of *Linum marginale* and *Melampsora lini*. 1. Patterns of resistance and racial variation in a large host population. *Evolution* 45:205–217.

Cannon, P. F. & D. L. Hawksworth. 1995. The diversity of fungi associated with vascular plants: the known, the unknown and the need to bridge the knowledge gap. *Advances in Plant Pathology* 11:277–302.

Carlsson-Graner, U. 1997. Anther-smut disease in *Silene dioica*: variation in susceptibility among genotypes and populations, and patterns of disease within populations. *Evolution* 51:1416–1426.

Chen, W.-Q., D. E. Gardner & D. T. Webb. 1996. Biology and life cycle of *Atelocauda koae*, an unusual demicyclic rust. *Mycoscience* 37:91–98.

Clark, D. A. & D. B. Clark. 1984. Spacing dynamics of a tropical rain forest tree: evaluation of the Janzen–Connell model. *American Naturalist* 124:769–788.

Clark, D. B. & D. A. Clark. 1985. Seedling dynamics of a tropical tree: impacts of herbivory and meristem damage. *Ecology* 66:1884–1892.

Clay, K. 1995. Correlates of pathogen species richness in the grass family. *Canadian Journal of Botany* 73:S42–S49.

Connell, J. H. 1971. On the role of natural enemies in preventing competitive exclusion in some marine animals and in rain forest trees. In P. J. Boer & G. R. Graadwell, eds., *Dynamics of Numbers in Populations (Proceedings of the Advanced Study Institute, Osterbeek 1970)*. Wageningen: Centre for Agricultural Publication and Documentation, pp. 298–312.

Dalling, J. W., M. D. Swaine & N. C. Garwood. 1998. Dispersal patterns and seed bank dynamics of pioneer trees in moist tropical forest. *Ecology* 79:564–578.

Davelos, A. L., H. M. Alexander & N. A. Slade. 1996. Ecological genetic interactions between a clonal host plant (*Spartina pectinata*) and associated rust fungi (*Puccinia seymouriana* and *Puccinia sparganioides*). *Oecologia* 105:205–213.

Davidson, J. M. 2000. Pathogen-mediated maintenance of diversity and fitness consequences of near versus far pollination for tropical trees. Unpublished Ph.D. dissertation, University of California, Davis.

Esquivel, R. E. & J. Carranza. 1996. Pathogenicity of *Phylloporia chrysita* (Aphyllophorales: Hymenochaetaceae) on *Erythrochiton gymnanthus* (Rutaceae). *Revista de Biologia Tropical* 44:137–145.

Fail, G. L. & J. H. Langenheim. 1990. Infection processes of *Pestalotia subcuticularis* on leaves of *Hymenaea courbaril*. *Phytopathology* 80:1259–1265.

Ferrer, A. & G. S. Gilbert. 2003. Effect of tree host species on fungal community composition in a tropical rain forest in

Panama. *Diversity and Distributions* **9**:455–468.

Fox, L. R. & P. A. Morrow. 1981. Specialization: species property or local phenomenon? *Science* **211**:887–893.

García-Guzmán, G., J. J. Burdon, J. E. Ash & R. B. Cunningham. 1996. Regional and local patterns in the spatial distribution of the flower-infecting smut fungus *Sporisorium amphilophis* in natural populations of its host *Bothriochloa macra*. *New Phytologist* **132**:459–469.

García-Guzmán, G. & R. Dirzo. 2001. Patterns of leaf-pathogen infection in the understorey of a Mexican rain forest: incidence, spatiotemporal variation, and mechanisms of infection. *American Journal of Botany* **88**:634–645.

Gardner, D. E. 1994. The native rust fungi of Hawaii. *Canadian Journal of Botany* **72**:976–989.

Gause, G. F. W. 1934. *The Struggle for Existence*. Baltimore: Williams and Wilkins.

Gilbert, G. S. 1995. Rain forest plant diseases: the canopy-understorey connection. *Selbyana* **16**:75–77.

2002. Evolutionary ecology of plant diseases in natural ecosystems. *Annual Review of Phytopathology* **40**:13–43.

Gilbert, G. S. & D. de Steven. 1996. A canker disease of seedlings and saplings of *Tetragastris panamensis* (Burseraceae) caused by *Botryosphaeria dothidea* in a lowland tropical forest. *Plant Disease* **80**:684–687.

Gilbert, G. S., A. Ferrer & J. Carranza. 2002. Polypore fungal diversity and host density in a moist tropical forest. *Biodiversity and Conservation* **11**:947–957.

Gilbert, G. S., S. P. Hubbell & R. B. Foster. 1994. Density and distance-to-adult effects of a canker disease of trees in a moist tropical forest. *Oecologia* **98**:100–108.

Gilbert, G. S., M. Mejía-Chang & E. Rojas. 2001. Fungal diversity and plant disease in mangrove forests: salt excretion as a possible defense mechanism. *Oecologia* **132**:278–285.

Gilbert, G. S. & W. P. Sousa. 2002. Host specialization among wood-decay polypore fungi in a Caribbean mangrove forest. *Biotropica* **34**:396–404.

Gillett, J. B. 1962. Pest pressure, an underestimated factor in evolution. *Systematics Association Publication Number* **4**:37–46.

Givnish, T. J. 1999. On the causes of gradients in tropical tree diversity. *Journal of Ecology* **87**:193–210.

Hansen, E. M. & E. M. Goheen. 2000. *Phellinus weirii* and other native root pathogens as determinants of forest structure and process in western North America. *Annual Review of Phytopathology* **38**: 515–539.

Harms, K. E., S. J. Wright, O. Calderon, A. Hernandez & E. A. Herre. 2000. Pervasive density-dependent recruitment enhances seedling diversity in a tropical forest. *Nature* **404**:493–495.

Hawksworth, D. L. 2001. The magnitude of fungal diversity: the 1.5 million species estimate revisited. *Mycological Research* **105**:1422–1432.

Hille Ris Lambers, J., J. S. Clark & B. Beckage. 2002. Density-dependent mortality and the latitudinal gradient in species diversity. *Nature* **417**:732–735.

Hubbell, S. P. 1997. A unified theory of biogeography and relative species abundance and its application to tropical rain forests and coral reefs. *Coral Reefs* **16**: S9–S21.

2001. *The Unified Neutral Theory of Biodiversity and Biogeography*. Princeton, NJ: Princeton University Press.

Huhndorf, S. M. & D. J. Lodge. 1997. Host specificity among wood-inhabiting pyrenomycetes (Fungi, Ascomycetes) in a wet tropical forest in Puerto Rico. *Tropical Ecology* **38**:307–315.

Infante, L. A. 1999. Complex interactions: exploring the role of soilborne plant pathogens in tropical seedling communities. Unpublished M.Sc. thesis, University of California, Berkeley.

Janzen, D. H. 1970. Herbivores and the number of tree species in tropical forests. *The American Naturalist* **104**:501–527.

Kitajima, K. & C. K. Augspurger. 1989. Seed and seedling ecology of a monocarpic tropical tree, *Tachigalia versicolor*. *Ecology* **70**:1102–1114.

Laesse, T. & D. J. Lodge. 1994. Three host-specific *Xylaria* species. *Mycologia* **86**:436–446.

Langenheim, J. H. & W. H. Stubblebine. 1983. Variation in leaf resin composition between parent tree and progeny in *Hymenaea*: implications for herbivory in the humid tropics. *Biochemistry Systematics and Ecology* **11**:97–106.

Leben, C. 1965. Epiphytic microorganisms in relation to plant disease. *Annual Review of Phytopathology* **3**:209–230.

Lee, S. S. 1999. Forest health in plantation forests in south-east Asia. *Australasian Plant Pathology* **28**:283–291.

Lindblad, I. 2000. Host specificity of some wood-inhabiting fungi in a tropical forest. *Mycologia* **92**:399–405.

    2001. Diversity of poroid and some corticoid wood-inhabiting fungi along the rainfall gradient in tropical forests, Costa Rica. *Journal of Tropical Ecology* **17**:353–369.

Lodge, D. J. 1997. Factors related to diversity of decomposer fungi in tropical forests. *Biodiversity and Conservation* **6**:681–688.

Lodge, D. J. & S. Cantrell. 1995. Fungal communities in wet tropical forests: variation in time and space. *Canadian Journal of Botany* **73**:S1391–S1398.

Lodge, D. J., P. J. Fisher & B. C. Sutton. 1996. Endophytic fungi of *Manilkara bidentata* leaves in Puerto Rico. *Mycologia* **88**:733–738.

May, R. M. 1991. A fondness for fungi. *Nature* **352**:475–476.

Meza, M. R. 1992. La pudrición del duramen de los Arboles en pie en Venezuela. El caso de la Mora de la Guayana. Unpublished Master's thesis, Universidad de Los Andes.

Opoku, I. Y., A. Y. Akrofi & A. A. Appiah. 2002. Shade trees are alternative hosts of the cocoa pathogen *Phytophthora megakarya*. *Crop Protection* **21**:629–634.

Packer, A. & K. Clay. 2000. Soil pathogens and spatial patterns of seedling mortality in a temperate tree. *Nature* **404**:278–281.

Polishook, J. D., G. F. Bills & D. J. Lodge. 1996. Microfungi from decaying leaves of two rain forest trees in Puerto Rico. *Journal of Industrial Microbiology* **17**:284–294.

Putz, F. E., P. D. Coley, K. Lu, A. Montalvo & A. Aiello. 1983. Uprooting and snapping of trees: structural determinants and ecological consequences. *Canadian Journal of Forest Research* **13**:1011–1020.

Richardson, K. A. & R. S. Currah. 1995. The fungal community associated with the roots of some rainforest epiphytes of Costa Rica. *Selbyana* **16**:49–73.

Rizzo, D. M. & M. Garbelotto. 2003. Sudden oak death: endangering California and Oregon forest ecosystems. *Frontiers in Ecology and the Environment* **1**:197–204.

Rodrigues, K. F. 1994. The foliar fungal endophytes of the Amazonian palm *Euterpe oleracea*. *Mycologia* **86**:376–385.

Sánchez Garduño, C., S. J. Wright & C. Potvin. 1995. Seed predation. In S. J. Wright & M. Colley, eds., *Accessing the Canopy*. Panama: United Nations Environment Programme/ Smithsonian Tropical Research Institute, pp. 18–19.

Schoeman, M. H., B. Q. Manicom & M. J. Wingfield. 1995. Epidemiology of powdery mildew on mango blossoms. *Plant Disease* **79**:524–528.

Sequiera, L. 1958. The host range of *Mycena citricolor* (Berk. & Curt.) Sacc. *Turrialba* 8:136–147.

Stevens, F. L. 1927. Fungi from Costa Rica and Panama. *Illinois Biological Monographs* 11:1–255.

Suryanarayanan, T. S. & K. Rajagopal. 2000. Fungal endophytes (Phellophytes) of some tropical forest trees. *Indian Forester* 126:165–170.

Tilman, D. 1982. *Resource Competition and Community Structure*. Princeton, NJ: Princeton University Press.

Travers, S. E., G. S. Gilbert & E. F. Perry. 1998. The effect of rust infection on reproduction in a tropical tree (*Faramea occidentalis*). *Biotropica* 30: 438–443.

van der Putten, W. H. & B. A. M. Peters. 1997. How soil-borne pathogens may affect plant competition. *Ecology* 78: 1785–1795.

Vánky, K. 2002. *Illustrated Genera of Smut Fungi*, 2nd edn. St Paul, MN: APS Press.

Webb, C. O. & D. R. Peart. 1999. Seedling density dependence promotes coexistence of Bornean rain forest trees. *Ecology* 80:2006–2017.

Wellman, F. L. 1968. More diseases on crops in the tropics than in the temperate zone. *Ceiba* 14:17–28.

Wellman, F. L. & E. Echandi. 1981. The coffee rust situation in Latin America in 1980. *Phytopathology* 71:968–971.

Weste, G. & G. C. Marks. 1987. The biology of *Phytophthora cinnamomi* in Australasian forests. *Annual Review of Phytopathology* 25:207–229.

Weston, W. H. 1933. The fungi of Barro Colorado. *Science Monthly* 36.

Wilson, D. 1995. Endophyte – the evolution of a term, and clarification of its use and definition. *Oikos* 73.

Worrall, J. J. & T. C. Harrington. 1988. Etiology of canopy gaps in spruce-fir forests at Crawford Notch, New Hampshire [USA]. *Canadian Journal of Forest Research* 18:1463–1469.

CHAPTER SEVEN

Mycorrhizas and ecosystem processes in tropical rain forest: implications for diversity

I. J. ALEXANDER
University of Aberdeen

S. S. LEE
Forest Research Institute of Malaysia

Introduction

The roots of almost all species of tropical rainforest trees contain mycorrhizal fungi (Alexander 1989a). Our aim here is to demonstrate that not only are these fungi central to ecosystem processes such as carbon- and nutrient-cycling, but that they also have the potential to influence biotic interactions between species, and so help to shape the structure and composition of forest communities. As such, they should be of interest to all ecologists, not just those who are primarily concerned with nutrient dynamics.

Mycorrhizas have continued to be the subject of intensive research in the 15 years since we last reviewed their role in tropical rain forest (Alexander 1989a). The processes by which mycorrhizal fungi access mineral nutrients in natural substrates are more fully understood, and important functional differences between types of mycorrhiza have been recognized (Read & Perez-Moreno 2003). There have been major advances in our understanding of the role of mycorrhizal fungi in forest carbon cycles. In boreal forest, for example, 20%–30% of current assimilate is consumed by mycorrhizal fungi (Söderström 2002), over 50% of CO_2 released from soils is accounted for by the respiration of tree roots and their associated mycorrhizal fungi (Högberg *et al.* 2001) and 30% of the soil microbial biomass is the extraradical hyphae of mycorrhizal fungi (Högberg & Högberg 2002). There is also growing evidence that mycorrhizal associations are multifunctional, and that benefit to the host may not accrue solely or entirely through enhanced capture of mineral nutrients (Newsham *et al.* 1995a). We also know that the diversity of mycorrhizal fungi and the extent of non-random association with host species is much greater than was once thought (Vandenkoornhuyse *et al.* 2003), and that not all mycorrhizal fungi, even within a particular

Biotic Interactions in the Tropics: Their Role in the Maintenance of Species Diversity, ed. D. F. R. P. Burslem, M. A. Pinard and S. E. Hartley. Published by Cambridge University Press. © Cambridge University Press 2005.

mycorrhizal type, are functionally similar (e.g. Jakobsen *et al.* 1992; Agerer 2001). In addition, because mycorrhizal fungi can have differential effects on the performance of individual higher plant species, they have the potential directly to influence the composition and structure of higher plant communities (e.g. van der Heijden *et al.* 1998a, b; Hartnett & Wilson 2002). While the research cited above has not been done in tropical rain forest or with rainforest species, the general principles are likely to hold true, and serve to emphasize the point that understanding the role of mycorrhizas in rain forests is highly relevant to understanding rainforest ecology in general.

Most tropical tree taxa form arbuscular mycorrhizas (Alexander 1989a) with fungi in the order Glomales, division Glomeromycota (Schüssler *et al.* 2001). An ecologically important minority of taxa are colonized by ectomycorrhizal (ECM) fungi, most of which are basidiomycetes or ascomycetes. This minority of tree taxa includes, *inter alia*, the Dipterocarpaceae (Lee 1998) and Fagaceae (Corner 1972), many legumes in Caesalpinioideae (Alexander 1989b; Bakarr & Janos 1996; Henkel *et al.* 2002) and Myrtaceae in the subfamily Leptospermoideae (Moyersoen *et al.* 2001). A consequence of this taxon-based occurrence, as Janos (1996) has pointed out, is that ectomycorrhizas are more common in palaeo-than in neotropical rain forest. Some ECM-forming hosts have been found also to harbour arbuscular mycorrhizal (AM) fungi (Lee 1998; Moyersoen & Fitter 1999; Tawaraya *et al.* 2003).

In temperate and boreal vegetation the relative abundance of AM and ECM host species in different places is fairly predictable, and is thought to reflect ecological gradients of nitrogen and phosphorus limitation of plant growth (Read & Perez-Moreno 2003) and the rate of ecosystem carbon turnover (Cornelissen *et al.* 2001). In temperate vegetation, ectomycorrhizas are typical of plant communities where carbon turnover is slow and nitrogen is the limiting element to plant growth, while arbuscular mycorrhizas predominate in communities where carbon turnover is more rapid, and phosphorus limits growth. This generalization has not been properly evaluated with respect to tropical rain forest.

With these ideas in mind about mycorrhizal function and distribution, derived primarily from temperate studies, our objectives here are:

- To review what is known about the benefits of mycorrhizal infection to tropical rainforest trees, and to speculate on the functional basis of the symbiosis in tropical conditions.
- To examine the diversity of mycorrhizal fungi in tropical rain forest, to consider the significance of that diversity for ecosystem processes, and to identify the threat posed by forest clearance.
- To consider whether the distribution of ECM host species in tropical forest reflects the same ecological gradients thought to determine their distribution in temperate and boreal regions.

- To speculate about how feedbacks between mycorrhizal fungi and their hosts might affect competition between tree species and so impact on forest composition and dynamics.

Throughout this chapter we attempt to identify those areas where research is most urgently needed. We use the term tropical rain forest in the broad sense to include evergreen and semi-deciduous, moist to humid forests.

The benefits of mycorrhizal infection

The conventional model of mycorrhizal function is that host plants benefit from increased uptake of soil nutrients, particularly inorganic phosphorus. Over the past 20 years this simple model has been increasingly challenged. Firstly, attention has turned away from the uptake of mineral ions from soil solution to the ability of mycorrhizal fungi to access phosphorus and nitrogen in a wide range of natural and semi-natural substrates. Secondly, there has been growing interest in the possible non-nutritional benefits of mycorrhizal infection, namely resistance to toxins, improved water relations, altered interactions with herbivores, and resistance to fungal pathogens (Newsham *et al.* 1995a). Coupled with these considerations about the nature of the benefits of mycorrhizal infection are questions about whether all plants benefit to the same extent, and whether all fungi provide the same sorts of benefit. In this section we summarize how these ideas have advanced since our last review (Alexander 1989a), and consider the extent to which they are relevant to tropical rain forest, drawing where possible on work with tropical forest taxa.

Nutrient-uptake mechanisms

Phosphorus

The mechanisms of inorganic phosphorus (P_i) capture by mycorrhizal fungi are well known (Bolan 1991; Marschner & Dell 1994; Smith & Read 1997), and clearly relevant to the acid oxisols and ultisols that support much tropical rain forest, and in which concentrations of soluble inorganic phosphorus in soil solution are exceedingly low. However, the importance of surface organic horizons as a source of mineralizable P, and the existence of a sizeable pool of labile, predominantly organic, P in the mineral soil of tropical rain forests has been increasingly recognized (Newbery *et al.* 1997; Johnson *et al.* 2003). There is no doubt that the transformations and fate of P derived from organic matter are more important to P cycles in tropical forest than the release of P_i in weathering processes (Tiessen *et al.* 1994; Johnson *et al.* 2003). However, there is likely to be intense competition between plant roots, mycorrhizal fungi and free-living microbes (Cleveland *et al.* 2002) for labile phosphorus. Moreover, all these biotic sinks will be in competition with the chemical and physical processes that remove P from the labile pool and render it unavailable to organisms in the short term. In these circumstances the ability of mycorrhizal fungi to access labile organic P is likely to be of critical importance.

There are very few studies of the ability of AM fungi, the most widespread symbionts in tropical forest, to access this type of phosphorus. In sterilized soil Tarafdar and Marschner (1994) demonstrated that the contribution of AM external mycelium to P uptake by wheat was 50% higher when the P was supplied as phytate as opposed to inorganic P, and that soil phosphatase activity was strongly correlated with the length of the external mycelium. In a similar experiment with clover, Feng et al. (2003) found that uptake of P via AM hyphae from organic sources was comparable to that from orthophosphate. Koide and Kabir (2000) have shown that the hyphae of an AM fungus can hydrolyse organic P under axenic conditions. However, Joner et al. (2000) argue that, in unsterile soil, the contribution of phosphatases from AM hyphae to total soil phosphatase cannot be quantitatively significant, and that the primary effect of AM hyphae is to contribute to efficient uptake of P mineralized by other soil microorganisms. Using ^{32}P-labelled plant residues as a P source, Joner and Jakobsen (1995) demonstrated that P uptake from the residues was indeed greater into AM than non-mycorrhizal plants. However, there were no consistent differences in P specific activity in the plant tissue between mycorrhizal and non-mycorrhizal treatments, supporting the contention that AM hyphae did not mineralize P over and above what would be mineralized in their absence. It remains unclear, therefore, whether or not AM fungi could be actively involved in the mineralization of labile organic P in tropical forest, or whether they merely make their hosts competitive in the acquisition of inorganic P-cycling in the labile pool. What are urgently needed are studies of organic P utilization in soils using AM fungal isolates typical of the surface organic layers of tropical soils, rather than the few AM isolates from temperate agriculture soils currently favoured for experimentation.

In contrast, the use of phytates as a P source by temperate ECM fungi is well documented (e.g. Antibus et al. 1992) although the significance of phytates as a P source for ectomycorrhizas in soil has also been questioned (Colpaert et al. 1997). More persuasively, both field (Griffiths et al. 1991) and microcosm (Bending & Read 1995; Perez-Moreno & Read 2000) studies show a reduction in the P content of litter and other natural substrates colonized by ECM fungi, with the extent of P depletion different between fungal isolates. This sort of experimental work has not been done with tropical isolates; however, it seems likely that similar potential to access organic P is present in tropical ECM fungi. Indeed, Newbery et al. (1997) have argued that the ability of ECM fungi to access labile organic phosphorus gives ECM legumes in the Atlantic rain forest of Cameroon a considerable advantage over AM trees, and leads to their local dominance and the establishment of a distinctive P cycle based on deep P-rich soil surface organic layers.

Nitrogen

In most lowland tropical rain forest N is more available for uptake by trees than P, and certainly more available than in most temperate and boreal forest (Vitousek & Sanford 1986; Martinelli *et al.* 1999). The exceptions to this generalization are tropical heath or white sand forests where there are deep surface organic horizons, and where tree growth appears to be N-limited (Vitousek & Sanford 1986; Cuevas & Medina 1988). In N-rich forests (the majority of tropical rain forests) it could be argued that N-capture mechanisms by mycorrhizal fungi are not important. However, mycorrhizally mediated N capture or competition for N between species could still be significant where N supply fluctuates in time and/or space (Hodge 2003), and where leaching losses of N are potentially high.

Recent evidence challenges the long-held view that AM fungi do not contribute to plant N uptake. The hyphae of AM fungi have been shown to contribute substantially to the uptake of mineral N (NH_4^+ and NO_3^-) by tomatoes (Mader *et al.* 2000) in model systems, and AM grass can access simple forms of organic N added to field soils (Persson *et al.* 2003). AM mycorrhizal hyphae of tropical forest trees may therefore be able to compete effectively with nitrifiers for NH_4^+ or soluble organic N, and leaching and denitrification losses will thereby be reduced. Furthermore, Hodge *et al.* (2001) describe an experiment in which AM hyphae were shown to accelerate decomposition of plant litter and increase N capture by *Plantago lanceolata*. In subsequent work, AM mycorrhizal colonization increased the ability of competing *Lolium perenne* and *Plantago lanceolata* to capture N from organic material (Hodge 2003). It is not clear to what extent the AM fungus in Hodge's experiments was itself responsible for decomposition, rather than priming the activities of other soil organisms. Nevertheless, these results indicate that when AM hyphae are intimately associated with decomposing organic matter, as they are in tropical rain forest, they are likely to be taking up N and transferring it to their host plants.

While the role of AM mycorrhizas in the uptake of organic N remains conjectural, no such doubts surround the role of some ECM fungi in temperate and boreal forest in accessing N from protein–phenolic complexes (Bending & Read 1996) or from a variety of natural substrates (Read & Perez-Moreno 2003). It is clear that many ECM fungi possess the ability to break down the structure of plant litter and absorb N (and P, see above) from plant polymers. Read has argued persuasively that this explains the dominance of ECM trees in the forests of the boreo-temperate zone where the availability of mineral N limits plant growth (Read 1991; Read & Perez-Moreno 2003). Unfortunately, there are no comparable studies of the saprotrophic abilities of ECM fungi from tropical rain forest, and we have only indirect indications of the involvement of tropical rainforest ectomycorrhizas in the N nutrition of their hosts. Högberg and Alexander (1995) compared the $\delta^{15}N$ signature in leaves of co-existing AM and ECM trees in Cameroon, arguing that a difference in the signature would reflect differential use of N sources in the soil by the two mycorrhizal types. In the event they found no differences, and concluded that there was no evidence for different N usage. However, in light of the doubts that have subsequently been cast on the use of $\delta^{15}N$ as an indication of N source (Emmerton *et al.* 2001a, b), this may no longer be a safe assumption. Brearley *et al.* (2003) grew ECM seedlings of three dipterocarp species with and without litter additions. The litter had a lower $\delta^{15}N$ signature than the soil. The seedlings grew better with litter, and there was a negative relationship between the $\delta^{15}N$ signature of the seedlings and the extent of ECM infection. This could be interpreted as evidence that ectomycorrhizas increase uptake of litter-derived N.

Do all species respond the same way to mycorrhizal infection?

Differences in seedling response to mycorrhizal infection could be an important way in which mycorrhizal infection influences interactions among plants, and therefore the structure and composition of plant communities. Since the pioneering work of Janos (1980a) it has been recognized that the seedlings of tropical AM trees differ in their responsiveness to mycorrhizal infection, and in the extent to which uninfected individuals can survive and grow in unamended forest soil. Although they do not address rainforest species, the large datasets of Siqueira *et al.* (1998) and Zangaro *et al.* (2003) show a clear trend in the response of woody species in different successional groups from southern Brazil. The AM response was much greater in the pioneer and early successional groups than in late successional or climax species. Overall, response to infection was negatively correlated with seed weight and root:shoot ratio, and positively with relative growth rate. In these pot experiments fast-growing species, with few seed reserves and low investment in root systems, were able better to respond to AM infection than slower-growing species. Interestingly Zangaro *et al.* (2003) found that response in the pot trials was strongly and directly related not only to the extent of colonization of the root systems in the experiment, but also to colonization of the species in the field. Unfortunately there are no comparable datasets for rainforest species, but Gehring (2003) provides an interesting, smaller-scale interspecific comparison. She compared the response to AM infection of four late successional rainforest canopy species from Queensland at 3% and 10% PAR. One responded in both light regimes, one only in high light and two in neither. Clearly more studies are required before a consistent picture emerges, but Gehring's experiment suggests that the relationships between mycorrhizal response, growth rates and light intensity are likely to be complex in AM tree species. It is also important to consider (Janos 1980a, 1996) that unresponsive, often slow-growing species may be dependent on mycorrhizal infection to grow at all.

Because of the phylogenetic constraints (ectomycorrhizas are restricted to relatively few tree taxa) one could hypothesize that interspecific variation in species response to ECM infection might be less than between the multitude of unrelated AM species. There are no published data that allow a test of this hypothesis, but indirect evidence indicates that it may not, in fact, be the case and that similar variability in response will be found. Seedlings of dipterocarps, for example, display strong interspecific variation in response to light and nutrients (Gunatilleke *et al.* 1997; Bungard *et al.* 2002), and it is likely that these traits will be partial correlates to mycorrhizal response. Similarly, Green and Newbery (2001) showed that seedlings of three co-occurring and closely related ECM legumes from Cameroon differed markedly in survival and growth at low PAR.

Is there a nutritional basis for the growth response?

A few authors have attempted to use woody species from tropical rain forest to test, directly or indirectly, whether the growth response to mycorrhizal infection is indeed related to increased nutrient uptake. These experiments need to be interpreted with care not only because, inevitably, they have been done in pots with seedlings over relatively short time spans, but also because the substrate (usually mineral soil low in organic matter), the inoculum (various forms of field inoculum, or known model species) and the growth conditions (particularly light intensity) may all have influenced the outcome. In these experiments, effects of mycorrhizal infection on nutrient uptake have been inferred from foliar analysis and/or by including a nutrient treatment, usually superphosphate addition, in the experimental design.

In the datasets of Siqueira *et al.* (1998) and Zangaro *et al.* (2003) those species that responded to inoculation (mostly pioneer and early successional species, see above) showed increased uptake of P, and in some cases also Ca and K. Bereau *et al.* (2000) found that the increased growth of *Dicorynia guianensis* in reponse to AM inoculation was associated with increased P uptake. Siqueira and Saggin-Junior (2001) found a range of responses in the woody species they studied. Some responded in the expected manner: the response to AM infection was reduced when the concentration of inorganic P in the soil solution increased. Others, however, responded to increased P only when mycorrhizal, indicating a high degree of dependency on infection for P uptake. In contrast, Alexander *et al.* (1992) found that *Paraseriathus falcataria* and *Parkia speciosa* responded better to AM inoculation than to P addition.

In the case of ECM species, growth stimulation of *Hopea odorata* and *H. helferi* following ECM infection by a *Pisolithus* isolate was associated with improved P uptake (Yazid *et al.* 1994). In a separate experiment with the same two species, Lee and Alexander (1994) found that field inoculum also increased P uptake, but in this case the response to ECM infection was greater than the reponse to added P. Ba *et al.* (1999, 2002) could not find consistent trends in foliar chemistry following ECM inoculation of the moist savanna species, *Afzelia africana* and *A. quanzensis*, and concluded that non-nutritional effects were present.

Moyersoen *et al.* (1998) compared the response of two contrasting species from Cameroon rain forest to P addition. *Oubanguia alata*, a widespread understorey/canopy species, forms arbuscular mycorrhizas, whereas *Tetraberlinia moreliana*, an emergent, is ECM. Phosphorus uptake by *O. alata* was only related to AM colonization at low P availability, but growth was positively related to AM colonization at both low and high P. i.e. growth promotion was not simply a function of P uptake. In contrast, P uptake of *T. moreliana* (the ECM species) was related to mycorrhizal colonization at both low and high P availability – but there was no growth response. Alexander *et al.* (1992) also tried to compare the effect of arbuscular mycorrhizas and ectomycorrhizas, but in their case by using a species *Intsia palembanica*, which is capable of forming both mycorrhizal types. They applied AM or ECM inoculum at a range of P availabilities, and found that both inocula improved the growth of the seedlings and increased P uptake over a range of P availabilities – but once again the response to mycorrhizal infection was greater than the response to P. In their experimental conditions, AM infection stimulated growth and P uptake by *Intsia* more than ECM infection.

There are relatively few direct tests of nutritional benefit of mycorrhizas to tropical tree seedlings (particularly rainforest species). However, there are many experiments where nutrients have been applied to tree seedlings growing in tropical forest soil in pots to discover which elements, if any, are limiting to growth. Because there are no non-mycorrhizal seedlings in these experiments and the extent of mycorrhizal infection is usually not assessed, the nutritional role of mycorrhizas can only be inferred indirectly from the results. However, if a nutrient is shown consistently to limit seedling growth in tropical forest soil then it could be argued that there will have been selection on the mycorrhizal mutualism to alleviate that limitation. It is important to remember when reviewing these experiments that other constraints, such as low irradiance, may impose overriding limitation on growth, and that as discussed above there are likely to be marked differences between species in the intrinsic capacity to respond to increased nutrients (Huanté *et al.* 1995). Lawrence (2003) provides a useful meta-analysis of the response to complete (usually NPK) nutrient addition of 91 predominantly AM species, from 15 different studies in which seedlings were grown under relatively high irradiance. She found that 73% of the 'light-demanders' responded positively in either biomass or relative growth rate to added NPK, whereas only 60% of the 'shade-tolerant' species did so. A number of pot trials give some indication that where species do not respond to P they may respond to K, Mg or Ca (Denslow *et al.* 1987; Burslem *et al.* 1995).

Taken together the direct tests of mycorrhizal benefit in nutrient uptake and the inferences that can be drawn from pot trials of nutrient addition suggest that the seedlings of many tropical trees growing in forest soil are capable of responding to nutrient additions, particularly phosphorus, when irradiance is not limiting, and that in these circumstances mycorrhizal infection increases phosphorus uptake. However, a substantial number, particularly the slower-growing shade-tolerant species, do not show a growth response, although mycorrhizas may increase uptake of the added nutrient. In some situations, for example where P supply is adequate, species may be limited by specific cations, and while there is no direct evidence for mycorrhizal involvement in the supply of these nutrients, it must be a possibility. Differences between species appear to be very important.

A number of authors (e.g. Whitbeck 2001) have commented that enhanced nutrient uptake by an unresponsive shade-tolerant or light-limited seedling as a result of mycorrhizal infection is of advantage because that nutrient store can be used to fuel seedling response rapidly when light does become available. Indeed, we go further and suggest that the long-term fitness (i.e. the ability to leave off-spring) of seedlings in shade may be improved by harbouring mycorrhizal fungi that provide no current benefit to nutrient uptake, if possessing the mycorrhiza improves the ability to respond to the increased nutrient demand which results from increased growth when canopy openings occur.

Pot experiments which test the response of individual seedlings to mycorrhizal colonization and/or nutrient addition do not take into account how response might differ where root competition exists. Coomes and Grubb (2000) have argued that in rain forest, the effects of root competition will be less in shade than in greater irradiance, and greater in those species with a capacity to respond to increased irradiance. In shade, root competition is only to be expected where available nutrients are very low. Various authors have attempted to crystallize the possible role of mycorrhizal fungi in root competition depending on the mycorrhizal dependence of the competing species, the degree of infection and the availability of nutrients (Janos 1980b; Allen & Allen 1990; Hart *et al.* 2003). We agree that the level of nutrient availability at which the effects of root competition are felt will depend on how much the species relies on mycorrhizal infection for efficient nutrient capture. Similarly whether the mycorrhizal infection improves the success of a species in competitive nutrient capture will depend not only on how much that species relies on mycorrhizas for efficient nutrient capture, but also the extent to which the competition is capable of depleting the nutrient supply. These ideas could be tested experimentally. Perhaps the most critical question is 'Can a species afford to be non-mycorrhizal if it encounters other species whose mycorrhizal symbionts make them more competitive under some conditions of nutrient supply, even if these occur only rarely?' If the answer is 'no', we may be closer to explaining why mycorrhizas are ubiquitous in tropical rain forest.

Do mycorrhizas provide nutritional benefits to mature trees? Certainly for canopy species there is less chance that the ability to use absorbed nutrient is constrained by low irradiance. Of course, there is no direct evidence to draw on, because the experimental difficulties of manipulating mycorrhizal infection on large trees are overwhelming, but if the growth of mature trees is, say, P limited, then one could argue that the mycorrhizal condition is a response to that selection pressure. Newbery *et al.* (2002) applied P to a grove of ECM *Microberlinia bisulcata* in Cameroon and followed P in litterfall, as well as seedling estab-lishment and the growth of small and medium-sized trees over 5 years. They found no effect on establishment and growth of seedlings of either AM or ECM species, and no effects on tree growth (girth increment), although the added P was taken up because P concentration in litter and in seedling foliage increased. Mirmanto *et al.* (1999) applied N and P factorially to lowland dipterocarp (ECM) forest in Central Kalimantan. They found that litterfall mass and P concentra-tions increased after fertilization, indicating that the added P was taken up and had probably increased leaf production. However, overall tree growth (girth increment) over a 5-year period was not affected, although there was an indi-cation that the relatively faster-growing 'red meranti' (*Shorea* spp.) dipterocarps might have responded. Both the trials described above took place in lowland forest with a high percentage of ECM trees. Experiments in predominantly AM forest have been done largely on tropical mountains and have been summarized by Tanner *et al.* (1998) who concluded that nutrient limitation of tree growth in montane forest is widespread, but largely due to N rather than P. So far then, fertilizer experiments have not supported the idea that P generally limits tropi-cal forest. Nevertheless, P limitation has been widely accepted to be the case for those forests growing on oxisols and ultisols, both intuitively because of soil P status, and on the basis of litter and foliage analysis (Vitousek 1984; Vitousek & Sanford 1986). On the other hand, that P may not be limiting was also the conclusion of Johnson *et al.* (2003), because their meta-analysis showed that the pool of labile P in tropical forest soils is several times larger than the annual forest P requirement and P demand on the soil. We suggest that mycorrhizal infection improves access to this pool, as argued above, and effectively prevents P limitation of tree growth in closed forest on what are ostensibly P-poor soils.

Growth and replacement of tissue is not the only sink for absorbed nutrient in mature trees. Reproduction has a nutrient 'cost' and for mast-fruiting species in particular (a high proportion of tropical ECM trees are mast fruiters) the nutri-ent demand of reproduction may at times be very high. Newbery *et al.* (1997) found that P concentration in litterfall declined in the year following a mast-fruiting of ECM legumes and suggested that one of the roles of ectomycorrhizas might be to build up P reserves in the plant to support mast-fruiting. Koide and Dickie (2002) have summarized the experimental evidence that AM colonization of plants increases reproduction (via both male and female functions) and off-spring survival in model populations. This is an aspect of the mycorrhizal role in determining fitness that demands greater attention.

Resistance to toxins

There is a large literature on the role of mycorrhizal fungi in protecting their hosts from the toxic effects of metals (Jentschke & Godbold 2000; Meharg &

Cairney 2000; Perotto & Martino 2001). The most relevant aspect of this, in the context of relatively undisturbed rain forest, is the possibility that mycorrhizas confer enhanced resistance to aluminium toxicity (Cumming & Weinstein 1990; Rufyikiri et al. 2000; Cumming & Ning 2003). The high concentration of phyto-toxic monomeric aluminium in soil solution is one of the major constraints to crop growth on acid tropical soils cleared of forest. It follows that the highly productive forest plants must have evolved a high level of tolerance to free aluminium, possibly involving their mycorrhizal symbionts. In addition, decom-position of organic matter under permanent forest cover gives rise to soluble humic molecules and low-molecular weight organic acids that complex with monomeric aluminium and reduce its phytotoxicity (Haynes & Mokolobate 2001). This has the added advantage that phenolic and low-molecular-weight organic acids can also be adsorbed to aluminium and iron oxide surfaces and so reduce the capacity of the soil to fix labile phosphorus (see above). The interactions between tropical forest plants, mycorrhizal fungi, organic matter, aluminium and phosphorus seem to us to be a largely unexplored area of research of fun-damental importance to understanding how tropical rain forest functions.

Resistance to fungal pathogens

Fungal pathogens have been shown to be important contributory agents in the mortality of seedlings in tropical rain forest (Augspurger & Kelly 1984; Gilbert 1994) and presumably also affect, to an unknown extent, the vigour of adult trees. The interactions between soil-borne fungal pathogens and myc-orrhizal fungi have received increasing attention in recent years and there is compelling evidence that mycorrhizal infection reduces the incidence, or detri-mental effects, of root diseases (Borowicz 2001; Graham 2001). The great major-ity of published work deals with interactions between AM infection and root pathogens, and a number of mechanisms have been proposed to account for the observed reduction in disease. These range from simple increase in host resistance as a result of the improved nutrition associated with AM infection in nutrient-poor soils (Smith 1988), through increased antagonism to pathogens in the myco-rhizosphere (Thomas et al. 1994), to the up-regulation of pathogenesis-related defence responses (Morandi 1996; Blee & Anderson 2000) or competition for resource within the root cortex between the mycorrhizal fungus and the pathogen (Larsen & Bodker 2001). There has been much less research into the interactions between ECM fungi and root pathogens, although clearly the same sorts of mechanisms are possible with, in addition, the potential for direct antag-onism between ECM fungi and pathogens and the creation of a physical barrier to penetration by the formation of the ECM fungal sheath (Hwang et al. 1995; Branzanti et al. 1999).

Clearly this is an important subject, but surprisingly little is known of the effect of mycorrhizal infection on disease incidence and impact in natural ecosystems. All the work referred to above has been done *in vitro*, or in green-houses, or in managed systems using monocultures of economically important crops or trees. The notable exception is the work of Newsham *et al.* (1995b). They were able to show that inoculation with a *Glomus* sp. protected plants of the grass *Vulpia ambigua* from the effects of the root pathogen *Fusarium oxysporum* when the inoculated plants were transplanted into natural populations in the field. The AM fungus had no effects on *Vulpia* growth in the absence of the pathogen and did not increase P uptake. Newsham *et al.* (1995a) suggested that AM fungi offer a continuum of benefit to their hosts, and that species that are efficient in nutrient capture in the absence of mycorrhizas, largely those with highly branched root systems, may well derive most benefit from infection via increased pathogen resistance.

It seems to us that this is an area of mycorrhizal research that could offer new insights into tropical forest dynamics. The Janzen–Connell hypothesis proposes that the inoculum potential of host-specific diseases will be higher close to parent trees (Janzen 1970; Connell 1971), and there may also be a build-up of more generalist pathogens in the soil around parent trees as generations of seedlings have effectively 'cultured' the microbial community (Bever 1994). There is indeed evidence that seedling mortality is higher close to adult con-specifics (Augspurger 1990; Gilbert 1994). On the other hand, parent trees are a source of mycorrhizal inoculum for seedlings. If the mycorrhizal inoculum from parents offers some advantages to seedlings (assuming seedlings pick up these fungi, see below), these would run counter to distance-related pathogen effects, particularly if the mycorrhizal species present in the parent-derived inoculum had been selected for its anti-pathogen role. Hood (2002) has addressed this problem. She compared the growth and survival of seedlings of *Milicia regia* potted in soil collected from beneath adult female *Milicia* trees in Ghanaian rain forest and from points at least 100 m away. She grew the seedlings at 2% and 20% photosynthetically active radiation (PAR). Mortality (largely caused by oomycetes) was higher in soil collected under adults, but at 20% PAR the sur-vivors in soil collected beneath adults grew better than those in soil collected remote from adults. These seedlings also had a higher percentage root length colonized by AM fungi, and growth was correlated with AM fungal coloniza-tion. These striking results introduce a new level of complexity into consid-eration of how spatially variable biotic factors interact to influence seedling recruitment.

Mycorrhizas and water relations

This is a complex and somewhat controversial area of research. The greatest body of work has examined AM plants, but many of the general principles are likely to apply equally to ECM systems. Because mycorrhizal infection alters plant nutrient status and growth, many of the observed effects of AM infection on plant–water relations can be explained by changes in P acquisition or in modified plant size or development. However, there is evidence that AM infection can alter water relations and drought physiology in ways unrelated to P nutrition, although these alterations may be small and transient (Augé 2001). They include effects on stomatal conductance, rates of transpiration, hydraulic conductivity and leaf hydration, and the postulated mechanisms include hormonal action, stimulation of gas exchange, improved root–soil contact or direct contributions of soil hyphae to water uptake. From an ecological point of view, whether effects on water relations are direct or indirect is less important than the possibility that AM infection can improve plant fitness through improved water relations when a nutritional benefit is absent as well as in addition to, or as a consequence of, improving nutrition. The role of drought-resistance in niche-partitioning and population dynamics in tropical forest seedlings (Delissio & Primack 2003; Engelbrecht & Kursar 2003; Pearson *et al.* 2003) is increasingly of interest. In this context, the most relevant mycorrhizal studies are those that demonstrate improved drought resistance in AM plants through the maintenance of higher internal water potentials in drying soils (see Augé 2001). Different AM fungi can be expected to have different effects, possibly related to the extent of internal colonization of the roots and the amount of external mycelium (Marulanda *et al.* 2003).

As indicated above, these ideas also apply in principle to ECM systems, although the body of relevant work is much smaller. However, in one respect the potential of ECM systems to alter water relations is much greater because many ECM fungi (including those of tropical rain forests) produce highly structured hyphal strands that can transport water over many centimetres (Brownlee *et al.* 1983). The importance of these strands in maintaining water transport from moist substrate to plants growing in dry soil has been elegantly demonstrated (Read & Boyd 1986). There must also be a strong possibility that the hydrophobic (Unestam & Sun 1995) nature of the fungal sheath of many ectomycorrhizas serves to reduce fine root-death during drought events in the surface layers of the soil. There is some suggestion that drought stress selects for more drought-resistant ECM fungi (Shi *et al.* 2002).

Perhaps not surprisingly, in view of the complex and demanding nature of the experimental procedures, little has been published that explicitly addresses these questions with tropical species in a tropical context. However, Kyllo *et al.* (2003) have investigated the interactions between AM infection and light in water uptake by understorey shrubs in the genera *Piper* and *Psychotria* from Barro Colorado Island. They found that AM infection altered root conductance in species from both genera. For shade-tolerant species growing at low PAR mycorrhizal infection reduced the total and proportional production of fine roots, but improved their water uptake efficiency by 60–70%. This might be critical for the survival of shallow-rooted species in the dry season.

Interactions with herbivores

Above- and below-ground herbivores can affect growth and survival of tropical tree seedlings, particularly when herbivory occurs in concert with another environmental stress such as drought or fungal attack (Bebber *et al.* 2002). The effects of herbivory on the fitness of adult trees are more difficult to discern, but judging by the allocation of resource to anti-herbivore defences, they are not insignificant (Coley & Barone 1996). There has been some interest in the extent to which there are positive or negative interactions between herbivores and mycorrhizal fungi, and how these might impact on plant communities (Gehring & Whitham 1994). Negative effects of herbivory on mycorrhizal colonization, for example because of reduced photosynthesis, are often noted and these may reinforce the reduction in competitive ability or stress toleration of the attacked plant. On the other hand, because mycorrhizal infection alters the chemistry of plant leaves and roots, mycorrhizas could in theory stimulate herbivory, or conversely improve resistance to or recovery from it (Jones & Last 1991; Goverde *et al.* 2000). A number of studies with AM plants have indeed shown reduced insect performance on AM plants (e.g. Rabin & Pacovsky 1985; Gange & West 1994; Gange 2001). This is clearly a complex area, because not all types of insect react the same way, and it also seems to be important which AM fungi are present. There is little experimental work examining the impact of ectomycorrhizas on herbivores (Gehring *et al.* 1997). Many years ago Janos (1975) reported that seedlings of two AM species from Costa Rica were better able to recover from herbivore attack when mycorrhizal, but we are not aware of any subsequent research into herbivore–mycorrhiza interactions in tropical forest.

Do all fungi provide the same benefits?

Mycorrhizal fungi do not all do the same things. This may seem self-evident, but until relatively recently there has been a tendency in the literature to assume that all AM fungi in particular were functionally very similar. For both ECM and AM fungi we now know that this is emphatically not the case. One of the consequences of this realization is that great caution must be exercised in extrapolating the results of experimental work using the 'lab rats' of the mycorrhizal world (e.g. *Glomus mosseae, Glomus intraradices, Paxillus involutus, Pisolithus tinctorius sensu lato*) to the generality of mycorrhizal fungi. The very fact that these species are easily manipulated and persist in culture collections over long periods could indicate that they are functionally not representative of those species that are more difficult to isolate or maintain in the laboratory, but are abundant in natural vegetation.

How do we know that not all mycorrhizal fungi are functionally the same? The simplest evidence is that of growth and morphology, assuming that form and function are related. For instance, different AM fungal isolates colonize the root system of a single host plant in different ways (Sanders *et al.* 1977) and produce different amounts and types of extraradical mycelium (Jakobsen *et al.* 1992). Different ECM fungal species produce markedly different sheaths on host roots (Agerer 1986) and different amounts and forms of extraradical mycelium and hyphal strands. This means that ECM fungal species fall into clear functional groups with respect to the way that they explore and interface with the substrate (Agerer 2001). Then there are one-on-one experiments where different mycorrhizal isolates can be shown differentially to affect growth, nutrient uptake, disease resistance, water relations and herbivore interactions (e.g. Thomson *et al.* 1994; Gange 2001; Helgason *et al.* 2002; Marulanda *et al.* 2003) of a given host. In the case of ECM fungi, many of which can be cultured in the laboratory, functional groups can be recognized on the basis of, for example, utilization of organic nitrogen (Abuzinadah & Read 1986), although there is clearly considerable intraspecific physiological variation within ECM fungal species (Cairney 1999).

In recent years AM research in particular has moved on from characterizing somewhat arbitrary differences between mycorrhizal isolates with respect to how they affect individual plant species in the laboratory, to dissecting how naturally occurring assemblages of AM fungal species interact with co-occurring hosts (van der Heijden *et al.* 1998a; Helgason *et al.* 2002). Such studies demonstrate that individual AM fungal species display host preference, that different AM species have significantly different effects on plant growth, and that these effects are not the same on each host species. Because individual plant species also vary in the extent to which they depend on AM fungi in a given situation, it follows that AM fungi have the potential to influence the outcome of plant competition and thereby determine plant community structure and composition (van der Heijden *et al.* 1998b; Hartnett & Wilson 2002; O'Connor *et al.* 2002).

Studies like these on tropical rainforest plants have not been published, but they are urgently required, because the outcomes described above, if applicable outwith herbaceous plant communities, have clear implications for the structure, composition and dynamics of tropical forest. So far there are a number of studies, many of which are in the 'grey' literature, which confirm that, as might be expected, different AM or ECM fungal species, or combinations of species, affect the growth of tropical tree seedlings in different ways. The most thought-provoking study is that of Kiers *et al.* (2000). The authors tested the infectivity and growth-promoting potential of AM inoculum derived from the chopped roots of adults of a range of tree species from Barro Colorado Island. The inoculum was tested against seedlings of six species representing a range of life forms found in the forest. Summarizing three separate experiments, they found that AM root colonization varied depending on the source of the inoculum and the identity of the host seedling, and that different inocula produced different growth responses depending on host species. Intriguingly, although there was a suggestion that the inocula showing the highest colonization came from the roots of conspecifics, the greatest growth responses were not observed with conspecific inocula. If this outcome were to be confirmed, it could drive the type of negative feedback on host abundance mediated through AM mycorrhiza described by Bever (2002a, b) and contribute to the diversity of forest where AM trees dominate.

Conclusions

To summarize, it is clear that mycorrhizal fungi possess mechanisms that allow them to improve the uptake of nutrients by tropical rainforest trees, and that this potentially extends to more than the uptake of inorganic P from soil solution. It is also clear that different tree species, by virtue of different intrinsic relative growth rates, seed reserves and responses to PAR, have different capacities to respond to mycorrhizal infection, and different dependence on infection for survival in the field in the seedling state. The precise role of mycorrhizas in the nutrition of adult trees, in below-ground competition and in relieving nutrient limitation of forest growth is unclear. However, it does seem likely that mycorrhizas can provide non-nutritional benefits to trees instead of, or in addition to, any nutritional benefit. Finally there is ample evidence that there is great functional diversity among species of mycorrhizal fungi, and that this diversity of function, coupled with host preference and diversity of effects on fitness across different host species, could influence the diversity and distribution of hosts. It is appropriate therefore to ask, just how diverse are communities of mycorrhizal fungi in tropical forest and are the threats to below-ground diversity similar to those above ground? These questions are addressed in the next section.

The diversity of mycorrhizal fungi in tropical rain forest

Plant diversity in tropical rain forest is greater than in any other vegetation type. Considering only the tree flora, there are as many species (1175) in a 0.52-km^2 plot in Borneo as there are in all of northern hemisphere temperate forests (Wright 2002). This is much greater host diversity than mycorrhizal researchers are accustomed to dealing with. Does tropical mycorrhizal fungal diversity reflect this enormous above-ground diversity?

Views of mycorrhizal fungal diversity in general have been greatly influenced in recent years by the application of molecular methods to description of mycorrhizal fungal communities. AM fungi in environmental samples could previously be identified morphologically only in the spore stage, and spores do not accurately reflect the presence or relative abundance of fungi in the vegetative state. By extracting AM fungal DNA directly from roots and using partial sequences from the small-subunit rRNA gene it is possible to recognize sequence 'types' corresponding to clades of closely related sequences. Although these sequence types do not correspond perfectly to traditional species based on spore morphology, they form the basis of a growing body of work on AM fungal diversity (Clapp *et al.* 2002; Redeker 2002). One consequence of these developments is a challenge to the received wisdom that AM fungal communities are species-poor relative to plant communities (Bever *et al.* 2001). In temperate grassland and forest ecosystems these approaches have shown the co-occurrence of several different AM fungal types within a plant root system (Helgason *et al.* 1999, 2002) and the non-random association of different AM types with co-occurring host species (Vandenkoornhuyse *et al.* 2003). This approach has recently been applied to the AM fungal communities in the roots of seedlings of two co-occurring species from the forest on Barro Colorado Island (Husband *et al.* 2002a, b). There was significant spatial heterogeneity in the AM fungal community as well as clear non-random associations of AM types with the two hosts. In addition, AM types that were dominant in newly germinated seedlings were replaced by other types in those seedlings which survived to the following year, and seedlings of different ages sampled at the same time supported significantly different AM communities. This temporal, spatial and host-associated variation in AM fungal distribution, taken with the work of Kiers *et al.* (2000) on seedling growth in the same forest (see above), reinforces the suggestion that AM fungi could influence seedling recruitment processes. An overview of the mycorrhizal research on Barro Colorado Island is presented elsewhere in this volume (Herre *et al.*, Chapter 8, this volume).

Husband looked at the roots of 48 plants and found 30 AM fungal types, of which 17 had not been previously recorded (Husband *et al.* 2002a). She compared her data with the diversity of AM types recorded in other studies using the same methodology (Table 7.1) and found that both the 'species' richness and the diversity were higher in the tropical site. There are a number of caveats with this comparison, not least the confounding effect of varying sampling effort. Although the number of roots and species sampled are comparable, Husband pointed out that she had screened around 30% fewer fragments per root than in the temperate grassland study, but still found 25% more types. Censuses of many more sites will be needed before any secure general conclusions can be drawn, but from this initial study it looks as if tropical forest AM communities are highly diverse and that there are many new species to be discovered.

The realization that diversity, spatial variability and host preference in AM fungal communities of tropical rain forest is much greater than previously appreciated should instill a sense of urgency to determine the effects of forest clearance on this genetic resource. It has been known for some time that logging can reduce AM inoculum potential in tropical forest soils (Alexander *et al.* 1992) and that this might affect the progress of forest regrowth (Janos 1980b). If, as previously assumed (Smith & Read 1997), the total number of AM fungal species is low and host specificity or preference does not occur, one

Table 7.1 *The number of AM fungal types, and the diversity of the AM fungal community, in a number of ecosystems, based on phylogenetic analysis of partial SSU rRNA sequences*

Ecosystem	No. of roots sampled	No. of host species	No. of AM fungal types	Diversity[a] (H')	Reference
Tropical forest, Panama	48	2	30	2.33	Husband et al. (2002a)
Temperate grassland, UK	47	2	24	1.71	Vandenkoornhuyse et al. (2002)
Temperate woodland, UK	49	5	11	1.44	Helgason et al. (1998)
Arable fields, UK	79	4	8	1.16	Daniell et al. (2001)

Taken from Husband et al. (2002a).
[a] Shannon–Weiner diversity index.

might expect the full compliment of AM fungal species on a site to re-establish fairly quickly. However, under the new scenario the loss of specialized host- or environment-selected AM fungal genotypes might have more serious consequences for forest recovery or re-establishment. In temperate forest, loss of AM fungal species does occur on conversion to agriculture (Helgason et al. 1998). It could be that the same process is going on, but to a much greater extent, in tropical forest. This topic is discussed in detail by Janos (1996).

The application of molecular techniques, in this case the use of the internally transcribed spacer (ITS) region of nuclear rDNA, has similarly improved our understanding of the below-ground ecology of ECM fungi (Horton & Bruns 2001). Even before the advent of molecular techniques there was evidence from morphological descriptions of mycorrhizas (Taylor & Alexander 1990) that sporocarp surveys did not accurately reflect the community of ECM fungi below ground. That has now been confirmed (Gardes & Bruns 1996), and the reason seems to be that many ECM fungi produce sporocarps infrequently or cryptically. Below-ground surveys in temperate and boreal forest using molecular techniques show that members of the Thelephoraceae, Corticiaceae and Sebacinaceae, in particular, may be important components of the community of ECM fungi on host roots (Horton & Bruns 2001), but are absent or under-represented in sporocarp records. Thelephoraceae have been described from tropical forest (Corner 1968; Natarajan & Chandrashekara 1978; Roberts & Spooner 2000) and there are already indications that these fungi can be abundant on root systems of dipterocarps in Southeast Asia (Ingleby et al. 2000; Supaart et al. 2003; Lee Su See, unpublished observations). More detailed studies applying molecular techniques to below-ground communities of ECM fungi are urgently required.

In the absence of such studies, the diversity of ECM fungi in tropical rain forest, and comparisons with temperate and boreal forest, can only be inferred from sporocarp surveys and investigations based on the morphotyping (Agerer 1986) of ectomycorrhizas. Lee *et al.* (2003) recorded ECM basidiomata during nine 3-day visits over a 7-year period (1991–97) to about 20 ha of lowland rain forest at Pasoh in Malaysia containing many host species in the Dipterocarpaceae and Fagaceae. They found 296 species of potentially ECM fungi, of which 66% were new to science. This is 55% more species than recorded in a survey over a similar time scale (1981–87) in oak/beech/hornbeam plots at Lainzer Tiergarten near Vienna in Austria (Straatsma & Krisai-Greilhuber 2003). In this case the area surveyed was smaller (7 ha) but the sampling effort (130 days) greater, and 189 potentially ECM species were recorded. The Pasoh number is greater even than the 265 potentially ECM species recorded on weekly visits May–November over 21 years in oak/beech/conifer plots in Switzerland (Straatsma *et al.* 2001). All these surveys were based primarily on conspicuous epigeous basidiomata, so the real diversity of ECM fungi was undoubtedly greater. Nevertheless, assuming that the discrepancy is roughly the same in all cases, there is a clear inference that ECM fungal diversity could be considerably higher in lowland dipterocarp forest than in temperate European forest.

Three recent studies have used morphotyping to characterize communities of ectomycorrhizas on dipterocarp seedlings in Southeast Asia. Lee and Alexander (1996), Ingleby *et al.* (1998) and Lee *et al.* (1996) recognized 23, 26 and 61 morphotypes respectively. It is difficult to make definite comparisons with the number of morphotypes on seedlings in temperate and boreal forest because of the variation in sampling design (number of sites, treatments, etc.) and sampling effort (number of seedlings, time scale). However, representative studies using a similar type of approach have recorded fewer morphotypes (Dahlberg & Stenström 1991; Jonsson *et al.* 1999; Bradbury 1998), so ECM diversity on dipterocarp seedlings is at least as great as that on temperate and boreal species, and possibly greater. There is no doubt that our understanding of the ecology of ECM fungi in temperate and boreal forest has increased significantly as databases of mycorrhizal descriptions have been established (Agerer 1987–2002; Agerer *et al.* 1996–2002; Goodman *et al.* 1996–2002) and molecular and morphotyping approaches have been combined to analyse ECM communities and their response to experimental manipulation (Horton & Bruns 2001). Similar initiatives are required in the tropics.

The study of Lee *et al.* (1996) mentioned above is interesting in that it examines the effect of logging on ECM diversity, and also tests for the existence of host specificity. Further analysis of the data from that study are presented here. In brief, at three sites in Danum valley, Sabah, Malaysia, adjacent areas of logged and undisturbed lowland dipterocarp forest were identified. At each site, six 16-m^2 plots were established, three in logged and three in undisturbed

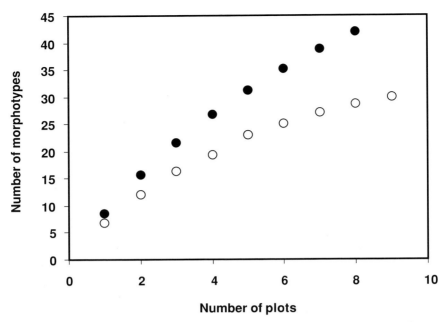

Figure 7.1 The number of ectomycorrhizal morphotypes on 6-month-old seedlings ($n = 13$) of *Hopea nervosa* in undisturbed (solid symbols) or logged (open symbols) forest in Danum Valley, Sabah, in relation to the number of plots sampled.

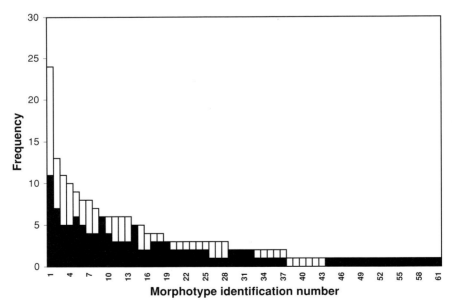

Figure 7.2 The frequency of ectomycorrhizal morphotypes in plots containing 6-month-old seedlings of *Hopea nervosa* (black) or *Shorea leprosula* (white bars). Morphotypes are ranked in order of combined frequency in all plots. $F_{\mathrm{max}\ Hopea} = 17$, $F_{\mathrm{max}\ Shorea} = 16$, $F_{\mathrm{max\ combined}} = 33$.

forest. Fifteen non-mycorrhizal seedlings of *Hopea nervosa* and *Shorea leprosula* were planted into each plot, and six months later the ECM morphotypes on their roots were recorded. Survival of the light-demanding species, *S. leprosula*, was poor (33%) in the undisturbed forest (effectively reducing sampling effort) so the comparison of ECM fungal richness is most robust when the comparison is restricted to morphotypes found on *H. nervosa*, which survived equally well (85%) in both situations. ECM infection (60%) was the same in both undisturbed and logged forest, but there were 40% more morphotypes (42) on the seedlings in the undisturbed forest than in the logged areas (30). Projection of the collector's curves (Fig. 7.1) indicates that the real difference would be even greater (second-order jack-knife estimates (Palmer 1991) are 79 and 43 respectively). As is normally the case, most of the morphotypes found in the study were uncommon (Fig. 7.2) and twice as many of these uncommon types were found only in the undisturbed forest (Lee *et al.* 1996): it is the rare fungi that are most likely to be lost following logging. The communities of ECM fungi were different at the three sites, and clearly different between logged and undisturbed forest, but there were no differences between the two host species (Fig. 7.3). The common fungi were equally frequent on both hosts (Fig. 7.2), but because these analyses are based on the presence or absence of a morphotype on each host, the possibility of host preference, i.e. differing relative abundances of colonization, cannot be discounted. These data, therefore, do not support the notion of widespread host specificity in dipterocarp ECM fungi, at least on seedlings, but they do show that logging may lead to a loss in ECM fungal diversity.

In conclusion then, the diversity of mycorrhizal fungi in tropical rain forest is likely to be high, and a major effort is needed to catalogue this diversity, to relate it to geographic origin, host species, environment and management, and to define the functional diversity it represents. Although logged areas still retain mycorrhizal inoculum, it is possible that forest destruction has already resulted in the loss of an unknown number of mycorrhizal fungal taxa. From the point of view of forest rehabilitation, it is important that we discover whether the fungi of the undisturbed forest play a special role in ecosystem processes, or whether more widespread generalist fungi are equally effective.

The distribution of ECM host species in tropical forest

ECM trees are in the minority in tropical rain forest where, over the whole biome, the overwhelming majority of taxa and individuals form arbuscular mycorrhizas. Can the distribution of ECM trees in tropical rain forest be explained by some functional trait of ectomycorrhizas that sets them apart from arbuscular mycorrhizas, and that confers selective advantage in certain ecological situations? We have already addressed this question, to some extent, by reviewing current ideas about nutrient-acquisition mechanisms of the two mycorrhizal types. It is worth re-emphasizing that these ideas are based on work in

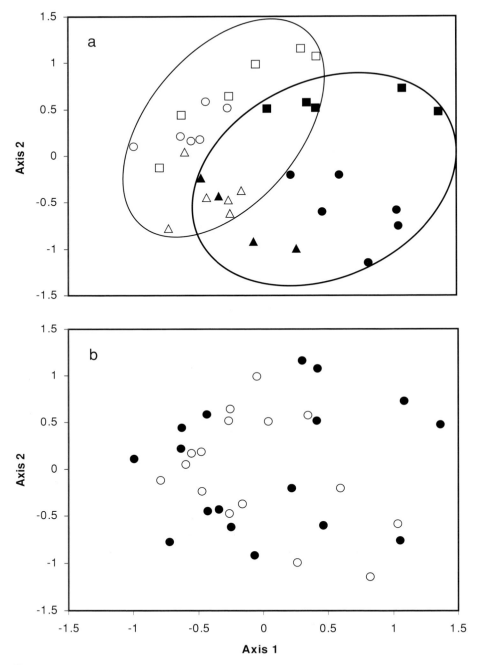

Figure 7.3 (a) Non-metric multidimensional scaling (NMS) ordination of presence/absence data for 61 ectomycorrhizal morphotypes in 33 plots of 6-month-old seedlings of *Hopea nervosa* and *Shorea leprosula* at three sites (■, ●, ▲) in undisturbed (solid symbols) and logged (open symbols) forest in Danum Valley, Sabah. (b) The same ordination but indicating plots of *Hopea nervosa* (solid circles) or *Shorea leprosula* (open circles).

temperate and boreal ecosystems using relatively few fungi, and that the functional diversity of tropical ECM fungi is unexplored.

Read (1991) proposed that, on a global scale, AM plants were characteristic of ecosystems where the limiting nutrient to plant growth was phosphorus, and that AM fungi were efficient in the capture of inorganic phosphorus from mineral soils. In such ecosystems litter decomposition was relatively rapid and nitrogen freely available in mineral form. ECM plants on the other hand were characteristic of soils of lower pH where litter decomposition, for reasons of climate or litter quality, was slower, causing nitrogen to be the limiting factor to plant growth. Where this tendency for slow litter breakdown was most pronounced, ECM fungi capable of accessing nitrogen directly from complex organic sources were important. These ideas sit fairly comfortably with the distribution of vegetation along latitudinal and altitudinal gradients in the northern hemisphere outwith the tropical rainforest biome. Read and Perez-Moreno (2003) extended these arguments, drawing particular attention to the abilities of ECM fungi to mobilize N and P from organic residues. On a more local scale, Cornelissen *et al.* (2001) compared seedling relative growth rate, foliar N and P concentration and leaf-litter decomposability among species of known mycorrhizal status in the United Kingdom. Within a subset of 32 woody species, they found that ECM and AM hosts did not differ in seedling RGR or foliar N and P, but ECM species had significantly lower mean litter decomposition rates. They concluded that there were strong links between plant mycorrhizal status and ecosystem carbon dynamics, with arbuscular mycorrhizas being associated with faster, and ectomycorrhizas with slower, carbon turnover. On an even finer scale of resolution ectomycorrhizas have long been known to be abundant in surface organic horizons of temperate and boreal forest (Frank 1894; Meyer 1967; Harvey *et al.* 1978) and when a host forms both ectomycorrhizas and arbuscular mycorrhizas (e.g. *Populus, Eucalyptus*), ECM colonization is greater in the surface organic layers and AM colonization in the mineral soil (Reddell & Malajczuk 1984; Neville *et al.* 2002).

Three fundamental questions arise. Are ECM species more common than AM species in tropical rain forest where nitrogen rather than phosphorus is thought to limit plant growth? Do ECM species in tropical rain forest produce litter with a slower decomposition rate than that of AM species? Where ECM and AM hosts occur together do the roots of the ECM species preferentially exploit surface organic horizons and those of AM species the mineral soil?

It is important to remember that ECM status amongst tropical trees has a very strong phylogenetic basis. All Dipterocarpaceae so far examined are ECM (Lee 1998), regardless of the ecological conditions in which they occur. In Leguminoseae, ECM species are particularly prevalent in the tribe Amherstieae of the Caesalpinioideae (Alexander 1989b). In Myrtaceae, ectomycorrhizas are restricted to the subfamily Leptospermoideae (Moyersoen *et al.* 2001). The Fagaceae are always ECM. We must therefore be careful when making ECM/AM comparisons, or in attributing distribution and ecology to ECM status. Ectomycorrhizas are likely to be linked to other related or unrelated traits within taxa, which could be equally or more constraining to host ecology and distribution.

Are ECM species more common where nitrogen limits growth?

Where do ECM trees occur in tropical rain forest? Probably the most cited examples are the legumes that give rise to monodominant stands in parts of Africa and South America, e.g. *Gilbertiodendron dewevrei* (Gerard 1960) and *Dicymbe corymbosa* (Henkel 2003). Much has been written about the factors leading to the establishment and persistence of these forests and the possible role of ectomycorrhizas (Alexander 1989b; Connell & Lowman 1989; Hart *et al.* 1989; Torti & Coley 1999; Torti *et al.* 2001; Henkel 2003). Their occurrence is not determined by gross edaphic features, as they are found on a range of soil types (Hart *et al.* 1989; Henkel 2003). However, there is evidence for a strong interaction between the dominant ECM species and the site, because litter layers are deeper and litter decomposition slower in the ECM-dominated forest than in surrounding AM-dominated mixed forest (Torti *et al.* 2001; Henkel 2003). In the case of *Gilbertiodendron*, this was linked to lower mineral N availability (Torti *et al.* 2001). These results are reminiscent of the reports by Singer (Singer & Araujo 1979; Singer & Araujo-Aguiar 1986) of the accumulation of deep surface organic layers in campinarana forest dominated by ECM trees. It is important to note that litter accumulation and slow decomposition in tropical lowland forest are not restricted to ECM monodominant stands, but occur in other situations (notably heath forests), including monodominant stands of AM trees (Moyersoen 1993).

Some features of these monodominant stands are shared by the groves of ECM legumes that have been intensively studied in Korup National Park, Cameroon, by Newbery and co-workers (e.g. Newbery *et al.* 1988, 1997, 2002a, b). Although there are no dramatic differences in litter accumulation and decomposition between the groves and the surrounding mixed AM forest, there are more subtle changes. The intensively rooted layer at the top of the mineral soil is deeper and richer in organic carbon in the groves, indicating that decomposition processes are indeed different between the two forest types, and the pools of P, including labile organic P, in this layer are 50% higher than in the AM forest. The ECM trees apparently give rise to conditions where the P capital of the site is concentrated in, and efficiently recycled through, the organic-matter-enriched surface layer of the soil. There is no evidence that nitrogen is limiting at this site (Högberg & Alexander 1995).

ECM trees are also widespread, locally abundant, or attain 'family' dominance in the lowland dipterocarp forests of Southeast Asia through the representation of Dipterocarpaceae, Fagaceae and some genera of Myrtaceae. Dipterocarpaceae itself is a large and diverse family, with great ecological amplitude, and ECM

Table 7.2 *The abundance of ectomycorrhizal trees and ectomycorrhizal roots in two heath forests and two neighbouring mixed dipterocarp forests in Brunei*

Forest type	Location	% ECM Basal area	% ECM Root density	% ECM Infection (tips)	% AM Infection (RLC)
Heath forest	Badas	10	8	90	54
	Sawat	37	42		
Mixed dipterocarp	Andalau	33	29	90	27
	Aarhus	32	46		

Data from Moyersoen *et al.* (2001).

RLC = root length colonized

Root and mycorrhizas were counted in the organic horizons and top 10 cm of mineral soil.

dipterocarps are found in a wide range of forest types, growing under a wide range of nutrient regimes. While there is some evidence that dipterocarp abundance, or the distribution of certain species, could for instance be related to soil cations (Baillie *et al.* 1987; Sharif & Miller 1990) we are not aware of any evidence, or suggestion in the literature, that these ECM trees of Southeast Asia are particularly associated with nitrogen-limiting conditions. The same can be said of those other ECM hosts, in diverse genera, that occur throughout various tropical rain forest formations (Janos 1983; Alexander 1989a, b).

Where do nitrogen-limiting conditions occur in the rainforest biome? The two obvious situations are tropical montane forest (Tanner *et al.* 1998) and tropical heath forest (Vitousek & Sanford 1986). Tanner *et al.* (1998) considered that 'with regard to N, tropical montane forests appear to function more like many temperate and boreal forests than like most lowland tropical forests' and drew attention to the build up of soil surface organic matter, low pH and low N concentrations in foliage and litter of montane forest. If Read's hypothesis (see above) applies equally to tropical forest, then we would expect to find greater abundances of ECM trees in montane than in lowland forest. We can find no evidence that this is the case, and some of the most-studied tropical montane forests (Tanner 1977; Tanner *et al.* 1992; Vitousek *et al.* 1995) probably contain no ECM trees. Moyersoen *et al.* (2001) tested the hypothesis that ECM trees would be more abundant in heath forest (kerangas) than in mixed dipterocarp forest in Brunei, and found that it was not the case (Table 7.2). This confirmed findings from heath forest in Venezuela (Moyersoen 1993), and Guyana (Bereau *et al.* 1997), where most of the dominant tree species were AM. However, as discussed above, Henkel (2003) has described monodominant ECM forests with structural similarities to heath forest on sandy soils in Guyana. It seems likely that a number of factors interact to determine the floristics of these extremely nutrient-deficient forests, but on balance ECM trees are no more likely to dominate in heath forests than AM trees, and we conclude that Read's hypothesis does not apply in general to tropical heath forest.

Does the litter of ECM species decompose more slowly than that of AM species?

There are surprisingly few published data that might answer this question. Chuyong *et al.* (2002) showed that the litter of four ECM legumes decayed significantly more slowly than that of three (unrelated) AM species from Korup, although the former had higher tissue N and P concentrations to begin with. This supports the view of Cornelissen *et al.* (2001) that ectomycorrhizal species are linked to lower ecosystem carbon turnover, although the phylogenetic constraint on making ECM/AM comparisons is particularly relevant here. Torti *et al.* (2001) found decomposition to be slower in ECM forest than the surrounding mixed AM forest, but this seemed to be a function of site rather than litter quality because there were no differences between the rates of decay of *Gilbertiodendron* litter and mixed litter from AM trees. More studies are required.

Do ectomycorrhizas exploit surface organic horizons, and arbuscular mycorrhizas the mineral soil?

Where ECM tree species occur in tropical rain forest, they co-exist with AM species. This is true even in 'monodominant' ECM forest (Torti & Coley 1999; Henkel 2003). If ectomycorrhizas function primarily to access nutrients from decomposing organic matter, and arbuscular mycorrhizas primarily to capture inorganic phosphorus from mineral soil, we would expect to see some vertical separation in the soil profile when they occur together. Accepting that total fine-root density declines with depth, the relative abundance of ectomycorrhizas, or ECM colonization of roots, should be higher in surface layers and the relative abundance of arbuscular mycorrhizas, or AM colonization, higher in the mineral soil.

That ectomycorrhizas are abundant in surface organic layers of tropical soils is not in doubt (Singh 1966; Newbery *et al.* 1988; Fassi & Moser 1991). Neither is the occurrence of arbuscular mycorrhizas in these layers, where AM fractional colonization may be higher than in the mineral soil (St John & Machado 1978; Rose & Paranka 1987). However, comparative studies of ECM and AM distribution are rare. Moyersoen *et al.* (1998) looked at the vertical distribution of ectomycorrhizas and arbuscular mycorrhizas in Korup National Park, Cameroon. They did not measure the relative abundance of ECM and AM fine roots in different horizons, but they did show that fractional colonization of roots by either type of mycorrhizas was unaffected by horizon. In a later study Moyersoen *et al.* (2001) found that there were no differences in the relative abundances of ECM and AM fine roots in organic layers and mineral soil in either heath forest or mixed dipterocarp forest in Brunei. AM fractional colonization (of non-ECM roots) was significantly higher in the more organic, lower-pH, more nutrient-stressed heath forest. These studies do not support the idea that there is simple vertical niche differentiation where both mycorrhizal types occur. However, both studies did demonstrate that where a soil volume was occupied by the ectomycorrhizas it was less likely also to be occupied by arbuscular mycorrhizas, raising the possibility of spatial competition between the two mycorrhizal types. More studies are required and, critically, we need to find out whether the AM fungi that colonize roots in the organic layers of tropical soils conform to the temperate stereotype (i.e. accessing only inorganic nutrients) or whether they have enzymatic capabilities comparable with the ECM fungi with which they co-exist.

Feedbacks between mycorrhizal infection and interspecific competition

Bever (2003) has provided a general conceptual framework which can be used to analyse the feedbacks between mycorrhizal fungi and their hosts. The presence of a plant species encourages the development of a mycorrhizal fungal community composed of fungal species that do well on that host. As the mycorrhizal fungal community changes so it may alter, positively or negatively, the performance of both the host and the competing species. The net effect of the mycorrhizal community on plant community dynamics then depends on the relative magnitude of the positive and negative effects on competing hosts. One of the factors most likely to affect the relative magnitude of effects on different hosts (and the effect of different hosts on the mycorrhizal community) is the non-random association of fungi with hosts.

Taking first the case of arbuscular mycorrhizas, different fungi have differential effects on competing hosts and so affect the outcome of competition (van der Heijden *et al.* 1998a). Different hosts 'culture' different mycorrhizal fungal communities (Vandenkoornhuyse *et al.* 2003), and the feedback between those fungi and their hosts can be positive (Helgason *et al.* 2002) or negative (Bever 2002b). In theory, therefore, AM fungi can reduce or promote co-existence between competing species. The little relevant information there is from tropical rain forest has been discussed above. There is non-random association of fungi with hosts (Husband *et al.* 2002a) and the fungi that associate with adults may not best promote the growth of conspecific seedlings (Kiers *et al.* 2000). Much more experimental work is required in this area.

In the case of ectomycorrhizas, there is also evidence for differential effects of fungi on hosts. However, although certain ECM fungi in the temperate zone are restricted to particular taxa, most can associate in the field with multiple hosts (Horton & Bruns 1998), and, so far, this also seems to be the case in the tropics. The relative abundance of ECM fungi on different hosts may be different, but we are not aware of reports of negative feedback between ECM fungi and hosts or marked differences in the ECM community of seedlings and adult trees in close proximity. On the basis of work in temperate and boreal forest we would therefore expect the feedbacks between ECM fungal communities and their hosts to be positive, and to reinforce existing community composition. This effect could be amplified if, as suggested above, ECM trees promote decomposition dynamics in which ectomycorrhizas are the best competitors. There have been two attempts to test this hypothesis in tropical rain forest, both in Cameroon, and both giving equivocal results. Newbery et al. (2000) found that seedlings of only one of three ECM legume species survived better where there were more ECM trees, although they did acknowledge that ECM inoculum was present throughout the experimental area. Onguene and Kuyper (2002) showed that contact with the roots of adult ECM hosts improved ECM colonization and survival of *Paraberlinia* seedlings. This was not unexpected, and confirmed earlier reports from dipterocarp forest of the importance of networks for seedling colonization (Alexander et al. 1992). More controversially, they thought that survival might be poorer under a conspecific (i.e. negative feedback), but as the adults of different species were not replicated, and were at different sites, other factors cannot be excluded.

How might feedbacks affect competition between ECM and AM trees? Clearly co-existence is possible, but equally it is apparent that in the majority of instances where ECM trees occur in tropical rain forest they do so either in 'monodominant' stands, or stands where they attain a disproportionately high percentage of basal area, or they attain 'family' dominance over very large areas (Alexander 1989a). Without claiming ECM fungi as primary causal agents, it does appear that they function in concert with one or more of a suite of other characters (large seeds, shade tolerance, mast-fruiting, litter quality) to promote the success of their hosts. One aspect of this could be strong negative feedback between ECM fungi and AM hosts.

In the discussion above we have avoided the vexed question of carbon transfer from adults to seedlings in the shaded understorey via mycorrhizal mycelium. If such transfer occurs it would be an extreme case of positive feedback, reinforcing existing community composition, or at least favouring species capable of joining and benefiting from the network. The importance of carbon transfer to seedlings is a controversial area (Read 1997; Robinson & Fitter 1999; Newbery et al. 2000) and more-convincing experimental evidence that significant amounts of carbon can move between the above-ground parts of autotrophic plants is required. On the other hand, myco-heterotrophy has evolved in a number of herbaceous taxa (Bidartondo & Bruns 2001; Bidartondo et al. 2002), involving both ECM and AM fungi as conduits of carbon from illuminated autotrophic trees to heterotrophic, often shaded, partners. Is it possible that the intermediate condition, of partial dependence, could be widespread? If so, tropical rain forest, where seedlings of both ECM and AM hosts persist for long periods in deep shade, would be a good place to look.

Acknowledgements

We would like to thank David Janos for his comments on an earlier version of this review.

References

Abuzinadah, R. A. & Read, D. J. (1986) The role of proteins in the nitrogen nutrition of ectomycorrhizal plants. 1. Utilization of peptides and proteins by ectomycorrhizal fungi. *New Phytologist*, **103**, 481–493.

Agerer, R. (1986) Studies on ectomycorrhizae. 2. Introducing remarks on characterization and identification. *Mycotaxon*, **26**, 473–492.

(2001) Exploration types of ectomycorrhizae – a proposal to classify ectomycorrhizal mycelial systems according to their patterns of differentiation and putative ecological importance. *Mycorrhiza*, **11**, 107–114.

ed. (1987–2002) *Colour Atlas of Ectomycorrhizae*. Schwäbisch Gmünd: Einhorn-Verlag.

Agerer, R., Danielson, R. M., Egli, S., Ingleby, K., Luoma, D. & Treu, R., eds. (1996–2002) *Descriptions of Ectomycorrhizae*, Vol. 1–6. Schwäbisch Gmünd: Einhorn-Verlag.

Alexander, I. J. (1989a) Mycorrhizas in tropical forests. In *Mineral Nutrients in Tropical Forest and Savanna Ecosystems* (ed. J. Proctor). Oxford: Blackwell Scientific Publications, pp. 169–188.

(1989b) Systematics and ecology of ectomycorrhizal legumes. *Advances in Legume Biology, Monographs in Systematic Botany*, **29**, 607–624.

Alexander, I. J., Ahmad, N. & Lee, S. S. (1992) The role of mycorrhizae in the regeneration of some Malaysian forest trees. *Philosophical Transactions of the Royal Society B*, **335**, 379–388.

Allen, M. F. & Allen, E. B. (1990) The mediation of competition by mycorrhizae in successional and patchy environments. *Perspectives on Plant Competition* (ed. J. B.

Grace & D. Tilman). New York: Academic Press, pp. 367–389.

Antibus, R. K., Sinsabaugh, R. L. & Linkins, A. E. (1992) Phosphatase-activities and phosphorus uptake from inositol phosphate by ectomycorrhizal fungi. *Canadian Journal of Botany*, **70**, 794–801.

Augé, R. M. (2001) Water relations, drought and vesicular–arbuscular mycorrhizal symbiosis. *Mycorrhiza*, **11**, 3–42.

Augspurger, C. K. (1990) Spatial patterns of damping-off disease during seedling recruitment in tropical forests. *Pests, Pathogens and Plant Communities* (ed. J. Burdon & S. Leather). Oxford: Blackwell Scientific Publications, pp. 3–14.

Augspurger, C. K. & Kelly, C. K. (1984) Pathogen mortality of tropical tree seedlings – experimental studies of the effects of dispersal distance, seedling density, and light conditions. *Oecologia*, **61**, 211–217.

Ba, A. M., Sanon, K. B., Duponnois, R. & Dexheimer, J. (1999) Growth response of *Afzelia africana* Sm. seedlings to ectomycorrhizal inoculation in a nutrient-deficient soil. *Mycorrhiza*, **9**, 91–95.

Ba, A. M., Sanon, K. B. & Duponnois, R. (2002) Influence of ectomycorrhizal inoculation on *Afzelia quanzensis* Welw. seedlings in a nutrient-deficient soil. *Forest Ecology and Management*, **161**, 215–219.

Baillie, I. C., Ashton, P. S., Court, M. N., Anderson, J. A. R., Fitzpatrick, E. A. & Tinsley, J. (1987) Site characteristics and the distribution of tree species in mixed dipterocarp forest on tertiary sediments in central Sarawak, Malaysia. *Journal of Tropical Ecology*, **3**, 201–220.

Bakarr, M. I. & Janos, D. P. (1996) Mycorrhizal associations of tropical legume trees in Sierra Leone, West Africa. *Forest Ecology and Management*, **89**, 89–92.

Bebber, D., Brown, N. & Speight, M. (2002) Drought and root herbivory in understorey *Parashorea* Kurz (Dipterocarpaceae) seedlings in Borneo. *Journal of Tropical Ecology*, **18**, 795–804.

Bending, G. D. & Read, D. J. (1995) The structure and function of the vegetative mycelium of ectomycorrhizal plants. 5. Foraging behaviour and translocation of nutrients from exploited litter. *New Phytologist*, **130**, 401–409.

(1996) Nitrogen mobilization from protein–polyphenol complex by ericoid and ectomycorrhizal fungi. *Soil Biology & Biochemistry*, **28**, 1603–1612.

Bereau, M., Gazel, M. & Garbaye, J. (1997) Mycorhizal symbiosis in trees of the tropical rainforest of French Guiana. *Canadian Journal of Botany*, **75**, 711–716.

Bereau, M., Barigah, T. S., Louisanna, E. & Garbaye, J. (2000) Effects of endomycorrhizal development and light regimes on the growth of *Dicorynia guianensis* Amshoff seedlings. *Annals of Forest Science*, **57**, 725–733.

Bever, J. D. (1994) Feedback between plants and their soil communities in an old field community. *Ecology*, **75**, 1965–1977.

(2002a) Host-specificity of AM fungal population growth rates can generate feedback on plant growth. *Plant and Soil*, **244**, 281–290.

(2002b) Negative feedback within a mutualism: host-specific growth of mycorrhizal fungi reduces plant benefit. *Proceedings of the Royal Society of London B*, **269**, 2595–2601.

(2003) Soil community feedback and the coexistence of competitors: conceptual frameworks and empirical tests. *New Phytologist*, **157**, 465–473.

Bever, J. D., Schultz, P. A., Pringle, A. & Morton, J. B. (2001) Arbuscular mycorrhizal fungi: more diverse than meets the eye, and the ecological tale of why. *Bioscience*, **51**, 923–931.

Bidartondo, M. I. & Bruns, T. D. (2001) Extreme specificity in epiparasitic Monotropoideae (Ericaceae): widespread phylogenetic and geographical structure. *Molecular Ecology* **10**, 2285–2295.

Bidartondo, M. I., Redecker, D., Hijri, I. *et al.* (2002) Epiparasitic plants specialized on arbuscular mycorrhizal fungi. *Nature*, **419**, 389–392.

Blee, K. A. & Anderson, A. J. (2000) Defense responses in plants to arbuscular mycorrhizal fungi. *Current Advances in Mycorrhizal Research* (ed. G. K. Podila & D. D. Douds Jr). St Paul, Minnesota: American Phytopathological Society, pp. 27–44.

Bolan, N. S. (1991) A critical review on the role of mycorrhizal fungi in the uptake of phosphorus by plants. *Plant and Soil*, **134**, 189–207.

Borowicz, V. A. (2001) Do arbuscular mycorrhizal fungi alter plant–pathogen relations? *Ecology*, **82**, 3057–3068.

Bradbury, S. M. (1998) Ectomycorrhizas of lodgepole pine (*Pinus contorta*) seedlings originating from seed in Southwestern Alberta cut blocks. *Canadian Journal of Botany*, **76**, 213–217.

Branzanti, M. B., Rocca, E. & Pisi, A. (1999) Effect of ectomycorrhizal fungi on chestnut ink disease. *Mycorrhiza*, **9**, 103–109.

Brearley, F. Q., Press, M. C. & Scholes, J. D. (2003) Nutrients obtained from leaf litter can improve the growth of dipterocarp seedlings. *New Phytologist*, **160**, 101–110.

Brownlee, C., Duddridge, J. A., Malibari, A. & Read, D. J. (1983) The structure and function of mycelial systems of ectomycorrhizal roots with special reference to their role in assimilate and water transport. *Plant and Soil*, **71**, 433–443.

Bungard, R. A., Zipperlen, S. A., Press, M. C. & Scholes, J. D. (2002) The influence of nutrients on growth and photosynthesis of seedlings of two rainforest dipterocarp species. *Functional Plant Biology*, **29**, 505–515.

Burslem, D. F. R. P., Grubb, P. J. & Turner, I. M. (1995) Responses to nutrient addition among shade-tolerant tree seedlings of lowland tropical rain-forest in Singapore. *Journal of Ecology*, **83**, 113–122.

Cairney, J. W. G. (1999) Intraspecific physiological variation: implications for understanding functional diversity in ectomycorrhizal fungi. *Mycorrhiza*, **9**, 125–135.

Chuyong, G. B., Newbery, D. M. & Songwe, N. C. (2002) Litter breakdown and mineralization in a Central African rain forest dominated by ectomycorrhizal trees. *Biogeochemistry*, **61**, 73–94.

Clapp, J. P., Rodriguez, A. & Dodd, J. C. (2002) Glomales rRNA gene diversity – all that glistens is not necessarily Glomalean? *Mycorrhiza*, **12**, 269–270.

Cleveland, C. C., Townsend, A. R. & Schmidt, S. K. (2002) Phosphorus limitation of microbial processes in moist tropical forests: evidence from short-term laboratory incubations and field studies. *Ecosystems*, **5**, 680–691.

Coley, P. D. & Barone, J. A. (1996) Herbivory and plant defences in tropical forests. *Annual Review of Ecology and Systematics*, **27**, 305–335.

Colpaert, J. V., Vanlaere, A., Vantichelen, K. K. & Vanassche, J. A. (1997) The use of inositol hexaphosphate as a phosphorus source by mycorrhizal and non-mycorrhizal Scots pine (*Pinus sylvestris*). *Functional Ecology*, **11**, 407–415.

Connell, J. H. (1971) On the role of natural enemies in preventing competitive exclusion in some marine animals and in rain forest trees. *Dynamics of Numbers in Populations* (ed. P. J. den Boer & G. R. Gradwell). Wageningen: Centre for Agricultural Publishing and Documentation, pp. 298–312.

Connell, J. H. & Lowman, M. D. (1989) Low-diversity tropical rain forests – some possible mechanisms for their existence. *American Naturalist*, **134**, 88–119.

Coomes, D. A. & Grubb, P. J. (2000) Impacts of root competition in forests and woodlands: a theoretical framework and review of experiments. *Ecological Monographs*, **70**, 171–207.

Cornelissen, J. H. C., Aerts, R., Cerabolini, B., Werger, M. J. A. & van der Heijden, M. G. A. (2001) Carbon cycling traits of plant species are linked with mycorrhizal strategy. *Oecologia*, **129**, 611–619.

Corner, E. J. H. (1968) A monograph of Thelephoraceae. *Nova Hedwigia*, **27**, 1–110. (1972) *Boletus in Malaysia*. Singapore: Government Printing Office.

Cuevas, E. & Medina, E. (1988) Nutrient dynamics within Amazonian forests. 2. Fine-root growth, nutrient availability and leaf litter decomposition. *Oecologia*, **76**, 222–235.

Cumming, J. R. & Ning, J. (2003) Arbuscular mycorrhizal fungi enhance aluminium resistance of broomsedge (*Andropogon virginicus* L.). *Journal of Experimental Botany*, **54**, 1447–1459.

Cumming, J. R. & Weinstein, L. H. (1990) Aluminum-mycorrhizal interactions in the physiology of pitch pine seedlings. *Plant and Soil*, **125**, 7–18.

Dahlberg, A. & Stenström, E. (1991) Dynamic changes in nursery and indigenous mycorrhiza of *Pinus sylvestris* seedlings planted out in forest and clearcuts. *Plant and Soil*, **136**, 73–86.

Daniell, T. J., Husband, R., Fitter, A. H. & Young, J. P. W. (2001) Molecular diversity of arbuscular mycorrhizal fungi colonising arable crops. *FEMS Microbiology Ecology* **36**, 203–209.

Delissio, L. J. & Primack, R. B. (2003) The impact of drought on the population dynamics of canopy-tree seedlings in an aseasonal Malaysian rain forest. *Journal of Tropical Ecology*, **19**, 489–500.

Denslow, J. S., Vitousek, P. M. & Schultz, J. C. (1987) Bioassays of nutrient limitation in a tropical rain-forest soil. *Oecologia*, **74**, 370–376.

Emmerton, K. S., Callaghan, T. V., Jones, H. E., Leake, J. R., Michelsen, A. & Read, D. J. (2001a) Assimilation and isotopic fractionation of nitrogen by mycorrhizal fungi. *New Phytologist*, **151**, 503–511.

(2001b) Assimilation and isotopic fractionation of nitrogen by mycorrhizal and nonmycorrhizal subarctic plants. *New Phytologist*, **151**, 513–524.

Engelbrecht, B. M. J. & Kursar, T. A. (2003) Comparative drought-resistance of seedlings of 28 species of co-occurring tropical woody plants. *Oecologia*, **136**, 383–393.

Fassi, B. & Moser, M. (1991) Micorrize nelle foreste naturali nell'Africa tropicale e nei neotropici. *Funghi, Plante e Suolo*. Turin: Centro di Studio sulla Micologia del Torreno, pp. 157–202.

Feng, G., Song, Y. C., Li, X. L. & Christie, P. (2003) Contribution of arbuscular mycorrhizal fungi to utilization of organic sources of phosphorus by red clover in a calcareous soil. *Applied Soil Ecology*, **22**, 139–148.

Frank, A. B. (1894) Die Bedeutung der Mycorrhizapilze für de gemeine Kiefer. *Forstwissenschaftliche Centralblat*, **16**, 1852–1890.

Gange, A. C. (2001) Species-specific responses of a root- and shoot-feeding insect to arbuscular mycorrhizal colonization of its host plant. *New Phytologist*, **150**, 611–618.

Gange, A. C. & West, H. M. (1994) Interactions between arbuscular mycorrhizal fungi and foliar-feeding insects in *Plantago lanceolata* L. *New Phytologist*, **128**, 79–87.

Gardes, M. & Bruns, T. D. (1996) Community structure of ectomycorrhizal fungi in a *Pinus muricata* forest: above- and below-ground views. *Canadian Journal of Botany*, **74**, 1572–1583.

Gehring, C. A. (2003) Growth responses to arbuscular mycorrhizae by rain forest seedlings vary with light intensity and tree species. *Plant Ecology*, **167**, 127–139.

Gehring, C. A. & Whitham, T. G. (1994) Interactions between aboveground herbivores and the mycorrhizal mutualists of plants. *Trends in Ecology & Evolution*, **9**, 251–255.

Gehring, C. A., Cobb, N. S. & Whitman, T. G. (1997) Three-way interactions among ectomycorrhizal mutualists, scale insects, and resistant and susceptible Pinyon pines. *American Naturalist*, **149**, 824–841.

Gerard, P. (1960) Etude écologique de la fôret dense à *Gilbertiodendron dewevrei* dans la région de l'Uelé. *Publication de l'Institut National pour l'Etude Agronomique du Congo Belge. Série Scientifique*, **87**, 1–159.

Gilbert, G. S. (1994) Density- and distance-to-adult effects of a canker disease of trees in a moist tropical forest. *Oecologia*, **98**, 100–108.

Goodman, D. M., Durall, D. M., Trofymow, J. A. & Berch, S. M., eds. (1996–2002) *A Manual of Concise Descriptions of North American Ectomycorrhizae*. Sidney, BC: Mycologue Publications.

Goverde, M., van der Heijden, M. G. A., Wiemken, A., Sanders, I. R. & Erhardt, A. (2000) Arbuscular mycorrhizal fungi influence life history traits of a lepidopteran herbivore. *Oecologia*, **125**, 362–369.

Graham, J. H. (2001) What do root pathogens see in mycorrhizas? *New Phytologist*, **149**, 357–359.

Green, J. J. & Newbery, D. M. (2001) Light and seed size affect establishment of grove-forming ectomycorrhizal rain forest tree species. *New Phytologist*, **151**, 271–289.

Griffiths, R. P., Ingham, E. R., Caldwell, B. A., Castellano, M. A. & Cromack, K. (1991) Microbial characteristics of ectomycorrhizal mat communities in Oregon and California. *Biology and Fertility of Soils*, **11**, 196–202.

Gunatilleke, C. V. S., Gunatilleke, I. A. U. N., Perera, G. A. D., Burslem, D. F. R. P., Ashton, P. M. S. & Ashton, P. S. (1997) Responses to nutrient addition among seedlings of eight closely related species of *Shorea* in Sri Lanka. *Journal of Ecology*, **85**, 301–311.

Hart, M. M., Reader, R. J. & Klironomos, J. N. (2003) Plant coexistence mediated by arbuscular mycorrhizal fungi. *Trends in Ecology & Evolution*, **18**, 418–423.

Hart, T. B., Hart, J. A. & Murphy, P. G. (1989) Monodominant and species-rich forests of the humid tropics – causes for their co-occurrence. *American Naturalist*, **133**, 613–633.

Hartnett, D. C. & Wilson, G. W. T. (2002) The role of mycorrhizas in plant community structure and dynamics: lessons from grasslands. *Plant and Soil*, **244**, 319–331.

Harvey, A. E., Larsen, M. F. & Jurgensen, M. F. (1978) Distribution of ectomycorrhizae in a mature Douglas-fir/larch forest soil in Western Montana. *Forest Science*, **22**, 393–398.

Haynes, R. J. & Mokolobate, M. S. (2001) Amelioration of Al toxicity and P deficiency in acid soils by additions of organic residues: a critical review of the phenomenon and the mechanisms involved. *Nutrient Cycling in Agroecosystems*, **59**, 47–63.

Helgason, T., Daniell, T. J., Husband, R., Fitter, A. H. & Young, J. P. W. (1998) Ploughing up the wood-wide web? *Nature*, **394**, 431.

Helgason, T., Fitter, A. H. & Young, J. P. W. (1999) Molecular diversity of arbuscular mycorrhizal fungi colonising *Hyacinthoides non-scripta* (Bluebell) in a seminatural woodland. *Molecular Ecology*, **8**, 659–666.

Helgason, T., Merryweather, J. W., Denison, J., Wilson, P., Young, J. P. W. & Fitter, A. H.

(2002) Selectivity and functional diversity in arbuscular mycorrhizas of co-occurring fungi and plants from a temperate deciduous woodland. *Journal of Ecology*, **90**, 371–384.

Henkel, T. W. (2003) Monodominance in the ectomycorrhizal *Dicymbe corymbosa* (Caesalpiniaceae) from Guyana. *Journal of Tropical Ecology*, **19**, 417–437.

Henkel, T. W., Terborgh, J. & Vilgalys, R. J. (2002) Ectomycorrhizal fungi and their leguminous hosts in the Pakaraima mountains of Guyana. *Mycological Research*, **106**, 515–531.

Hodge, A. (2003) N capture by *Plantago lanceolata* and *Brassica napus* from organic material: the influence of spatial dispersion, plant competition and an arbuscular mycorrhizal fungus. *Journal of Experimental Botany*, **54**, 2331–2342.

Hodge, A., Campbell, C. D. & Fitter, A. H. (2001) An arbuscular mycorrhizal fungus accelerates decomposition and acquires nitrogen directly from organic material. *Nature*, **413**, 297–299.

Högberg, P. & Alexander, I. J. (1995) Roles of root symbioses in African woodland and forest – evidence from N^{15} abundance and foliar analysis. *Journal of Ecology*, **83**, 217–224.

Högberg, M. N. & Hogberg, P. (2002) Extramatrical ectomycorrhizal mycelium contributes one-third of microbial biomass and produces, together with associated roots, half the dissolved organic carbon in a forest soil. *New Phytologist*, **154**, 791–795.

Högberg, P., Nordgren, A., Buchmann, N. *et al.* (2001) Large-scale forest girdling shows that current photosynthesis drives soil respiration. *Nature*, **411**, 789–792.

Hood, L. A. (2002) Effects of pathogenic and mycorrhizal fungi on regeneration of two tropical tree species. Unpublished Ph.D. thesis, University of Aberdeen.

Horton, T. R. & Bruns, T. D. (1998) Multiple-host fungi are the most frequent and abundant ectomycorrhizal types in a mixed stand of Douglas fir (*Pseudotsuga menziesii*) and bishop pine (*Pinus muricata*). *New Phytologist*, **139**, 331–339.

(2001) The molecular revolution in ectomycorrhizal ecology: peeking into the black-box. *Molecular Ecology*, **10**, 1855–1871.

Huanté, P., Rincon, E. & Chapin, F. S. (1995) Responses to phosphorus of contrasting successional tree-seedling species from the tropical deciduous forest of Mexico. *Functional Ecology*, **9**, 760–766.

Husband, R., Herre, E. A., Turner, S. L., Gallery, R. & Young, J. P. W. (2002a) Molecular diversity of arbuscular mycorrhizal fungi and patterns of host association over time and space in a tropical forest. *Molecular Ecology*, **11**, 2669–2678.

Husband, R., Herre, E. A. & Young, J. P. W. (2002b) Temporal variation in the arbuscular mycorrhizal communities colonising seedlings in a tropical forest. *FEMS Microbiology Ecology*, **42**, 131–136.

Hwang, S. F., Chakravarty, P. & Chang, K. F. (1995) The effect of two ectomycorrhizal fungi, *Paxillus involutus* and *Suillus tomentosus*, and of *Bacillus subtilis*, on *Fusarium* damping-off in Jack pine seedlings. *Phytoprotection*, **76**, 57–66.

Ingleby, K., Munro, R. C., Noor, M., Mason, P. A. & Clearwater, M. J. (1998) Ectomycorrhizal populations and growth of *Shorea parvifolia* (Dipterocarpaceae) seedlings regenerating under three different forest canopies following logging. *Forest Ecology & Management*, **111**, 171–179.

Ingleby, K., Thuy, L. T. T., Phong, N. T. & Mason, P. A. (2000) Ectomycorrhizal inoculum potential of soils from forest restoration sites in South Vietnam. *Journal of Tropical Forest Science*, **12**, 418–422.

Jakobsen, I., Abbott, L. K. & Robson, A. D. (1992) External hyphae of vesicular-arbuscular mycorrhizal fungi associated with *Trifolium subterraneum* L.1. Spread of hyphae and phosphorus inflow into roots. *New Phytologist*, **120**, 371–380.

Janos, D. P. (1975) Effects of vesicular-arbuscular mycorrhizae on lowland tropical rainforest trees. *Endomycorrhizas* (ed. F. E. Sanders, B. Mosse & P. B. Tinker) London: Academic Press, pp. 437–446.

(1980a) Vesicular-arbuscular mycorrhizae affect lowland tropical rainforest plant growth. *Ecology*, **61**, 151–162.

(1980b) Mycorrhizae influence tropical succession. *Biotropica*, **12** (supplement), 56–64.

(1983) Tropical mycorrhizas, nutrient cycles and plant growth. *Tropical Rain Forest: Ecology and Management* (ed. S. L. Sutton, T. C. Whitmore & A. C. Chadwick) Oxford: Blackwell Scientific Publications, pp. 327–345.

(1996) Mycorrhizas, succession, and the rehabilitation of deforested lands in the humid tropics. *Fungi and Environmental Change* (ed. J. C. Frankland, N. Magan & G. M. Gadd) Cambridge: Cambridge University Press, pp. 129–162.

Janzen, D. H. (1970) Herbivores and the number of trees in tropical forests. *The American Naturalist*, **104**, 501–528.

Jentschke, G. & Godbold, D. L. (2000) Metal toxicity and ectomycorrhizas. *Physiologia Plantarum*, **109**, 107–116.

Johnson, A. H., Frizano, J. & Vann, D. R. (2003) Biogeochemical implications of labile phosphorus in forest soils determined by the Hedley fractionation procedure. *Oecologia*, **135**, 487–499.

Joner, E. J. & Jakobsen, I. (1995) Uptake of P^{32} from labelled organic matter by mycorrhizal and nonmycorrhizal subterranean clover (*Trifolium subterraneum* L). *Plant and Soil*, **172**, 221–227.

Joner, E. J., van Aarle, I. M. & Vosatka, M. (2000) Phosphatase activity of extra-radical arbuscular mycorrhizal hyphae: a review. *Plant and Soil*, **226**, 199–210.

Jones, C. G. & Last, F. T. (1991) Ectomycorrhizae and trees: implications for above-ground herbivory. *Microbial Mediation of Plant-Herbivore Interactions* (ed. P. Barbosa, V. A. Krischik & C. G. Jones) Chichester: Wiley, pp. 65–103.

Jonsson, L., Dahlberg, A., Nilsson, M. C., Karen, O. & Zackrisson, O. (1999) Continuity of ectomycorrhizal fungi in self-regenerating boreal *Pinus sylvestris* forests studied by comparing mycobiont diversity on seedlings and mature trees. *New Phytologist,* **142**, 151–162.

Kiers, E. T., Lovelock, C. E., Krueger, E. L. & Herre, E. A. (2000) Differential effects of tropical arbuscular mycorrhizal fungal inocula on root colonization and tree seedling growth: implications for tropical forest diversity. *Ecology Letters,* **3**, 106–113.

Koide, R. T. & Dickie, I. A. (2002) Effects of mycorrhizal fungi on plant populations. *Plant and Soil,* **244**, 307–317.

Koide, R. T. & Kabir, Z. (2000) Extraradical hyphae of the mycorrhizal fungus *Glomus intraradices* can hydrolyse organic phosphate. *New Phytologist,* **148**, 511–517.

Kyllo, D. A., Velez, V. & Tyree, M. T. (2003) Combined effects of arbuscular mycorrhizas and light on water uptake of the neotropical understoroy shrubs, *Piper* and *Psychotria. New Phytologist,* **160**, 443–454.

Larsen, J. & Bodker, L. (2001) Interactions between pea root-inhabiting fungi examined using signature fatty acids. *New Phytologist,* **149**, 487–493.

Lawrence, D. (2003) The response of tropical tree seedlings to nutrient supply: meta-analysis for understanding a changing tropical landscape. *Journal of Tropical Ecology,* **19**, 239–250.

Lee, S. S. (1998) Root symbiosis and nutrition. *A Review of Dipterocarps: Taxonomy, Ecology and Silviculture* (ed. S. Appanah & J. M. Turnbull) Bogor: CIFOR, pp. 99–114.

Lee, S. S. & Alexander, I. J. (1994) The response of seedlings of two dipterocarp species to nutrient additions and ectomycorrhizal infection. *Plant & Soil* **163**, 299–306.

(1996) The dynamics of ectomycorrhizal infection of *Shorea leprosula* (Miq) seedlings in Malaysian rain forests. *New Phytologist* **132**, 297–305.

Lee, S. S., Alexander, I. J., Moura-Costa, P. & Yap, S. W. (1996) Mycorrhizal infection of dipterocarp seedlings in logged and undisturbed forests. *Proceedings of Fifth Round-Table Conference on Dipterocarps* (ed. S. Appanah & K. C. Khoo). Kuala Lumpur: Forest Research Institute of Malaysia, pp. 157–164.

Lee, S. S., Watling, R. & Turnbull, E. (2003) Diversity of putative ectomycorrhizal fungi in Pasoh Forest Reserve. *Pasoh: Ecology of a Lowland Rain Forest in Southeast Asia* (ed. T. Okuda, N. Manokaran, Y. Matsumoto, K. Niiyama, S. C. Thomas & P. S. Ashton) Tokyo: Springer, pp. 149–159.

Mader, P., Vierheilig, H., Streitwolf-Engel, R. et al. (2000) Transport of N^{15} from a soil compartment separated by a polytetra-fluoroethylene membrane to plant roots via the hyphae of arbuscular mycorrhizal fungi. *New Phytologist* **146**, 155–161.

Marschner, H. & Dell, B. (1994) Nutrient uptake in mycorrhizal symbiosis. *Plant and Soil,* **159**, 89–102.

Martinelli, L. A., Piccolo, M. C., Townsend, A. R. et al. (1999) Nitrogen stable isotopic composition of leaves and soil: tropical versus temperate forests. *Biogeochemistry,* **46**, 45–65.

Marulanda, A., Azcon, R. & Ruiz-Lozano, J. M. (2003) Contribution of six arbuscular mycorrhizal fungal isolates to water uptake by *Lactuca sativa* plants under drought stress. *Physiologia Plantarum,* **119**, 526–533.

Meharg, A. A. & Cairney, J. W. G. (2000) Co-evolution of mycorrhizal symbionts and their hosts to metal-contaminated environments. *Advances in Ecological Research*, **30**, 69–112.

Meyer, F. H. (1967) Feinwurzelnverteilung bei Waldbäume in Abhängigkeit vom Substrat. *Forstarchiv*, **38**, 286–290.

Mirmanto, E., Proctor, J., Green, J., Nagy, L. & Suriantata (1999) Effects of nitrogen and phosphorus fertilization in a lowland evergreen rainforest. *Philosophical Transactions of the Royal Society of London B*, **354**, 1825–1829.

Morandi, D. (1996) Occurrence of phytoalexins and phenolic compounds in endomycorrhizal interactions, and their potential role in biological control. *Plant & Soil* **185**, 241–251.

Moyersoen, B. (1993) Ectomicorrizas y micorrizas vesiculo-arbusculares en Caatinga Amazonica del Sur de Venezuela. *Scientia Guianea* **3**, 1–82.

Moyersoen, B. & Fitter, A. H. (1999) Presence of arbuscular mycorrhizas in typically ectomycorrhizal host species from Cameroon and New Zealand. *Mycorrhiza*, **8**, 247–253.

Moyersoen, B., Alexander, I. J. & Fitter, A. H. (1998) Phosphorus nutrition of ectomycorrhizal and arbuscular mycorrhizal tree seedlings from a lowland tropical rain forest in Korup National Park, Cameroon. *Journal of Tropical Ecology*, **14**, 47–61.

Moyersoen, B., Fitter, A. H. & Alexander, I. J. (1998) Spatial distribution of ectomycorrhizas and arbuscular mycorrhizas in Korup National Park Rain Forest, Cameroon, in relation to edaphic parameters. *New Phytologist*, **139**, 311–320.

Moyersoen, B., Becker, P. & Alexander, I. J. (2001) Are ectomycorrhizas more abundant than arbuscular mycorrhizas in tropical heath forests? *New Phytologist*, **150**, 591–599.

Natarajan, K. & Chandrashekara, K. V. (1978) A new species of *Tomentella* from south India. *Mycologia* **70**, 1294–1297.

Neville, J., Tessier, J. L., Morrison, I., Scarratt, J., Canning, B. & Klironomos, J. N. (2002) Soil depth distribution of ecto- and arbuscular mycorrhizal fungi associated with *Populus tremuloides* within a 3-year-old boreal forest clear-cut. *Applied Soil Ecology*, **19**, 209–216.

Newbery, D. M., Alexander, I. J., Thomas, D. W. & Gartlan, J. S. (1988) Ectomycorrhizal rain-forest legumes and soil phosphorus in Korup National Park, Cameroon. *New Phytologist*, **109**, 433–450.

Newbery, D. M., Alexander, I. J. & Rother, J. A. (1997) Phosphorus dynamics in a lowland African rain forest: the influence of ectomycorrhizal trees. *Ecological Monographs*, **67**, 367–409.

(2000) Does proximity to conspecific adults influence the establishment of ectomycorrhizal trees in rain forest? *New Phytologist*, **147**, 401–409.

Newbery, D. M., Chuyong, G. B., Green, J. J., Songwe, N. C., Tchuenteu, F. & Zimmermann, L. (2002a) Does low phosphorus supply limit seedling establishment and tree growth in groves of ectomycorrhizal trees in a Central African rainforest? *New Phytologist*, **156**, 297–311.

Newbery, D. M., Songwe, N. C. & Chuyong, G. B. (2002b) Phenology and dynamics of an African rainforest at Korup, Cameroon. *Dynamics of Ecological Communities* (ed. D. M. Newbery, H. H. T. Prins & N. D. Brown). Oxford: Blackwell Scientific Publications, pp. 267–308.

Newsham, K. K., Fitter, A. H. & Watkinson, A. R. (1995a) Multi-functionality and biodiversity in arbuscular mycorrhizas. *Trends in Ecology & Evolution*, **10**, 407–411.

(1995b) Arbuscular mycorrhiza protect an annual grass from root pathogenic fungi in the field. *Journal of Ecology*, **83**, 991–1000.

O'Connor, P. J., Smith, S. E. & Smith, E. A. (2002) Arbuscular mycorrhizas influence plant diversity and community structure in a semiarid herbland. *New Phytologist*, **154**, 209–218.

Onguene, N. A. & Kuyper, T. W. (2002) Importance of the ectomycorrhizal network for seedling survival and ectomycorrhiza formation in rain forests of South Cameroon. *Mycorrhiza*, **12**, 13–17.

Palmer, M. W. (1991) Estimating species richness: the second-order jackknife reconsidered. *Ecology*, **72**, 1512–1513.

Pearson, T. R. H., Burslem, D. F. R. P, Goeriz, R. E. & Dalling, J. W. (2003) Regeneration niche partitioning in neotropical pioneers: effects of gap size, seasonal drought and herbivory on growth and survival. *Oecologia*, **137**, 456–465.

Perez-Moreno, J. & Read, D. J. (2000) Mobilization and transfer of nutrients from litter to tree seedlings via the vegetative mycelium of ectomycorrhizal plants. *New Phytologist*, **145**, 301–309.

Perotto, S. & Martino, E. (2001) Molecular and cellular mechanisms of heavy metal tolerance in mycorrhizal fungi: what perspectives for bioremediation? *Minerva Biotecnologica*, **13**, 55–63.

Persson, J., Hogberg, P., Ekblad, A., Hogberg, M. N., Nordgren, A. & Nasholm, T. (2003) Nitrogen acquisition from inorganic and organic sources by boreal forest plants in the field. *Oecologia*, **137**, 252–257.

Rabin, L. B. & Pacovsky, R. S. (1985) Reduced larva growth of two Lepidoptera (Noctuidae) on excised leaves of soybean infected with a mycorrhizal fungus. *Journal of Economic Entomology*, **78**, 1358–1363.

Read, D. J. (1991) Mycorrhizas in ecosystems. *Experientia*, **47**, 376–391.

(1997) The ties that bind. *Nature*, **388**, 517–518.

Read, D. J. & Boyd, R. (1986) Water relations of mycorrhizal fungi and their host plants. *Water, Fungi and Plants* (ed. P. G. Ayres & L. Boddy). Cambridge: Cambridge University Press, pp. 287–303.

Read, D. J. & Perez-Moreno, J. (2003) Mycorrhizas and nutrient cycling in ecosystems: a journey towards relevance? *New Phytologist*, **157**, 475–492.

Reddell, P. & Malajczuk, N. (1984) Formation of mycorrhizae by jarrah (*Eucalyptus marginata* Donn ex Smith) in litter and soil. *Australian Journal of Botany*, **32**, 511–520.

Redeker, D. (2002) Molecular identification and phylogeny of arbuscular mycorrhizal fungi. *Plant & Soil*, **244**, 67–73.

Roberts, P. J. & Spooner, B. M. (2000) Cantherelloid, clavarioid and thelephoroid fungi from Brunei Darussalam. *Kew Bulletin*, **55**, 843–851.

Robinson, D. & Fitter, A. (1999) The magnitude and control of carbon transfer between plants linked by a common mycorrhizal network. *Journal of Experimental Botany*, **50**, 9–13.

Rose, S. L. & Paranka, J. E. (1987) Root and VAM distribution in tropical agricultural and forest soils. *Mycorrhizae in the Next Decade* (ed. D. M. Sylvia, L. L. Hung & J. H. Graham) Gainesville: University of Florida, p. 165.

Rufyikiri, G., Declerck, S., Dufey, J. E. & Delvaux, B. (2000) Arbuscular mycorrhizal fungi might alleviate aluminium toxicity in banana plants. *New Phytologist*, **148**, 343–352.

Sanders, F. E., Tinker, P. B., Black, R. L. B. & Palmerley, S. M. (1977) The development of endomycorrhizal root systems: I. Spread of infection and growth-promoting effects with four species of vesicular-arbuscular endophyte. *New Phytologist*, **78**, 257–268.

Schüssler, A., Gehrig, H., Schwarzott, D. & Walker, C. (2001) Analysis of partial Glomales SSU rRNA gene sequences: implications for primer design and phylogeny. *Mycological Research*, **105**, 5–15.

Sharif, A. H. M. & Miller, H. G. (1990) *Shorea leprosula* as an indicator species for site fertility evaluation in dipterocarp forests of Peninsular Malaysia. *Journal of Tropical Forest Science*, **3**, 101–110.

Shi, L. B., Guttenberger, M., Kottke, I. & Hampp, R. (2002) The effect of drought on mycorrhizas of beech (*Fagus sylvatica* L.): changes in community structure, and the content of carbohydrates and nitrogen storage bodies of the fungi. *Mycorrhiza*, **12**, 303–311.

Singer, R. & Araujo, I. (1979) Litter decomposition and ectomycorrhiza in Amazonian forests 1. A comparison of litter decomposing and ectomycorrhizal basidiomycetes in latosol-terra-firme rain forest and white podsol campinarana. *Acta Amazonica*, **9**, 25–41.

Singer, R. & Araujo-Aguiar, I. (1986) Litter decomposition and ectomycorrhizal basidiomycetes in an Igapó forest. *Plant Systematics and Evolution*, **153**, 107–117.

Singh, K. G. (1966) Ectotrophic mycorrhiza in equatorial rain forests. *Malayan Forester*, **29**, 13–18.

Siqueira, J. O. & Saggin-Junior, O. J. (2001) Dependency on arbuscular mycorrhizal fungi and responsiveness of some Brazilian native woody species. *Mycorrhiza*, **11**, 245–255.

Siqueira, J. O., Carneiro, M. A. C., Curi, N., Rosado, S. C. S. & Davide, A. C. (1998) Mycorrhizal colonization and mycotrophic growth of native woody species as related to successional groups in southeastern Brazil. *Forest Ecology and Management*, **107**, 241–252.

Smith, G. S. (1988) The role of phosphorus nutrition in interactions of vesicular-arbuscular mycorrhizal fungi with soil-borne nematodes and fungi. *Phytopathology*, **78**, 371–374.

Smith, S. E. & Read, D. J. (1997) *Mycorrhizal Symbiosis*, 2nd edn. San Diego: Academic Press.

Söderström, B. (2002) Challenges for mycorrhizal research into the new millennium. *Plant and Soil*, **244**, 1–7.

St John, T. V. & Machado, A. D. (1978) Efeitos da profundidade e do sistema de manejo de um solo de terra firme em infestaceos por micorrizes. *Acta Amazonica*, **8**, 139–141.

Straatsma, G. & Krisai-Greilhuber, I. (2003) Assemblage structure, species richness, abundance, and distribution of fungal fruit bodies in a seven-year plot-based survey near Vienna. *Mycological Research*, **107**, 632–640.

Straatsma, G., Ayer, F. & Egli, S. (2001) Species richness, abundance, and phenology of fungal fruit bodies over 21 years in a Swiss forest plot. *Mycological Research*, **105**, 515–523.

Supaart, S., Noritmitsu, S., Stephan, P. *et al.* (2003) Diversity of ectomycorrhizal fungi associated with Dipterocarpaceae in Malaysia. *Proceedings of International Conference on Forestry and Forest Products Research* (CFFPR 2001) Kepong: Forest Research Institute Malaysia, pp. 547–550.

Tanner, E. V. J. (1977) Four montane rain forests of Jamaica: a quantitative characterisation of the floristics, the soils and the foliar mineral levels, and a discussion of the inter-relationships. *Journal of Ecology*, **65**, 883–918.

Tanner, E. V. J., Kapos, V. & Franco, W. (1992) Nitrogen and phosphorus fertilization effects on Venezuelan montane forest trunk growth and litterfall. *Ecology*, **73**, 78–86.

Tanner, E. V. J., Vitousek, P. M. & Cuevas, E. (1998) Experimental investigation of nutrient limitation of forest growth on wet tropical mountains. *Ecology*, **79**, 10–22.

Tarafdar, J. C. & Marschner, H. (1994) Phosphatase-activity in the rhizosphere and hyphosphere of VA mycorrhizal wheat supplied with inorganic and organic phosphorus. *Soil Biology & Biochemistry*, **26**, 387–395.

Tawaraya, K., Takaya, Y., Turjaman, M. *et al.* (2003) Arbuscular mycorrhizal colonization of tree species grown in peat swamp forests of central Kalimantan, Indonesia. *Forest Ecology and Management*, **182**, 381–386.

Taylor, A. F. S. & Alexander, I. J. (1990) Demography and population-dynamics of ectomycorrhizas of Sitka spruce fertilized with N. *Agriculture Ecosystems & Environment*, **28**, 493–496.

Thomas, L., Mallesha, B. C. & Bagyaraj, D. J. (1994) Biological control of damping-off of cardamom by the VA mycorrhizal fungus, *Glomus fasciculatum*. *Microbiological Research*, **149**, 413–417.

Thomson, B. D., Grove, T. S., Malajczuk, N. & Hardy, G. E. S. J. (1994) The effectiveness of ectomycorrhizal fungi in increasing the growth of *Eucalyptus globulus* Labill in relation to root colonization and hyphal development in soil. *New Phytologist*, **126**, 517–524.

Tiessen, H., Cuevas, E. & Chacon, P. (1994) The role of soil organic-matter in sustaining soil fertility. *Nature*, **371**, 783–785.

Torti, S. D. & Coley, P. D. (1999) Tropical monodominance: a preliminary test of the ectomycorrhizal hypothesis. *Biotropica*, **31**, 220–228.

Torti, S. D., Coley, P. D. & Kursar, T. A. (2001) Causes and consequences of monodominance in tropical lowland forests. *American Naturalist*, **157**, 141–153.

Unestam, T. & Sun, Y. P. (1995) Extramatrical structures of hydrophobic and hydrophilic ectomycorrhizal fungi. *Mycorrhiza*, **5**, 301–311.

van der Heijden, M. G. A., Boller, T., Wiemken, A. & Sanders, I. R. (1998a) Different arbuscular mycorrhizal fungal species are potential determinants of plant community structure. *Ecology*, **79**, 2082–2091.

van der Heijden, M. G. A., Klironomos, J. N., Ursic, M. *et al.* (1998b) Mycorrhizal fungal diversity determines plant biodiversity, ecosystem variability and productivity. *Nature*, **396**, 69–72.

Vandenkoornhuyse, P., Husband, R., Daniell, T. J. *et al.* (2002) Arbuscular mycorrhizal community composition associated with two plant species in a grassland ecosystem. *Molecular Ecology* **11**, 1555–1564.

Vandenkoornhuyse, P., Ridgway, K. P., Watson, I. J., Fitter, A. H. & Young, J. P. W. (2003) Co-existing grass species have distinctive arbuscular mycorrhizal communities. *Molecular Ecology*, **12**, 3085–3095.

Vitousek, P. M. (1984) Litterfall, nutrient cycling, and nutrient limitation in tropical forests. *Ecology*, **65**, 285–298.

Vitousek, P. M. & Sanford, R. L. (1986) Nutrient cycling in moist tropical forest. *Annual Review of Ecology and Systematics*, **17**, 137–167.

Vitousek, P. M., Gerrish, G., Turner, D. R., Walker, L. R. & Mueller-Dombois, D. (1995) Literfall and nutrient cycling in four Hawaiian montane rainforests. *Journal of Tropical Ecology*, **11**, 189–203.

Whitbeck, J. L. (2001) Effects of light environment on vesicular-arbuscular mycorrhiza development in *Inga leiocalycina*, a tropical wet forest tree. *Biotropica*, **33**, 303–311.

Wright, S. J. (2002) Plant diversity in tropical forests: a review of mechanisms of species co-existence. *Oecologia*, **130**, 1–14.

Yazid, S. M., Lee, S. S. & Lapeyrie, F. (1994) Growth stimulation of *Hopea* spp. (Dipterocarpaceae) seedlings following ectomycorrhizal inoculation with an exotic strain of *Pisolithus tinctorius*. *Forest Ecology and Management*, **67**, 339–343.

Zangaro, W., Nisizaki, S. M. A., Domingos, J. C. B. & Nakano, E. M. (2003) Mycorrhizal response and successional status in 80 woody species from south Brazil. *Journal of Tropical Ecology*, **19**, 315–324.

CHAPTER EIGHT

An overview of arbuscular mycorrhizal fungal composition, distribution and host effects from a tropical moist forest

EDWARD ALLEN HERRE AND DAMOND KYLLO
Smithsonian Tropical Research Institute

SCOTT MANGAN
Smithsonian Tropical Research Institute and Indiana University

REBECCA HUSBAND
Smithsonian Tropical Research Institute and University of York

LUIS C. MEJIA
Smithsonian Tropical Research Institute

AHN-HEUM EOM
*Smithsonian Tropical Research Institute and
Korea National University of Education*

Introduction

Arbuscular mycorrhizal fungi (AMF) (Zygomycetes) are an ancient group, dating back to the invasion of land surfaces by plants. Currently, they are perhaps the most abundant soil fungi, and they form intimate relationships with the roots of the vast majority of terrestrial plant species across the planet. These fungal symbionts generally play a mutualistic role, aiding the host plant primarily by enhancing the acquisition of soil nutrients, particularly phosphorus (P). In addition, AMF species often affect plant hormone production/induction (Allen *et al.* 1980), resistance to root pathogens (Newsham *et al.* 1995); water uptake (Kyllo *et al.* 2003) and soil structure (Andrade *et al.* 1998; Rillig & Allen 1999). In return, all AMF species obligately depend on the host plant for photosynthetically fixed carbon. Given their obligate dependence, AMF are influenced by their hosts at essentially every phase in their life history – hyphal development, sporulation and spore germination (Hetrick & Bloom 1986; Sanders & Fitter 1992; Bever *et al.* 1996). On the other hand, the degree of mycorrhizal dependence often varies widely among the host plant species in a community (Janos 1980a; Azcon & Ocampo 1981; Hetrick *et al.* 1992; Kiers *et al.* 2000).

A central and still largely unanswered question is the degree to which host plant and AMF species influence each other's community composition in natural systems. Fundamentally, for community effects to occur, different combinations

Biotic Interactions in the Tropics: Their Role in the Maintenance of Species Diversity, ed. D. F. R. P. Burslem, M. A. Pinard and S. E. Hartley. Published by Cambridge University Press. © Cambridge University Press 2005.

of host and AMF species must produce different outcomes of survival and growth. Furthermore, given such differential effects, the potential for either host plant or AMF species to affect the other's community composition will depend largely on the identities and distributions of the associated species in a given habitat. The most conducive conditions for reciprocal effects would be that AMF species are heterogeneously distributed and differentially associate with, and affect, the growth and survival of different hosts. The form of the interactions between particular plant and AMF species (e.g. negative or positive feedbacks) will determine net effects on the diversities and distributions within plant and AMF communities (Bever *et al.* 1997; Bever 1999).

AMF have long been considered to be a relatively homogeneous group, both functionally and morphologically. Until recently, the diversity and composition of AMF species have been largely discounted as factors that significantly affect aboveground diversity for at least three reasons. First, if the number of rec-ognized AMF species (< 200) is roughly correct, then strict-sense specificity is impossible because there are least three orders of magnitude fewer mycorrhizal species than host plants they colonize. Second, it is clear that most AMF species are not host-specific colonizers (Janos 1980; Harley & Smith 1983; Clapp *et al.* 1995). Third, even at fairly small scales, AMF communities often contain a mix of many component species, and colonization of a given host's roots by multiple AMF species is both likely and observed (Allen 1996; Husband *et al.* 2002a, b). Thus, even if individual AMF species can produce different effects on particular hosts, one would expect colonization of roots by multiple AMF species to homog-enize and blur any differential effects on host plants, and thereby diminish any community-wide influences. These observations and considerations suggest that it is unlikely that AMF communities can have much effect on aboveground com-munity composition. However, it has also become increasingly obvious that this view is open to challenge.

More recently, many observations from temperate grasslands and microcosm experiments suggest that AMF are in fact likely to have a significant influ-ence on the distribution and diversity of plant communities (Grime *et al.* 1987; Gange *et al.* 1990; Bever 1994; Mills & Bever 1998; van der Heijden *et al.* 1998a; Hartnett & Wilson 1999; Olff *et al.* 2000; Klironomos 2002; Castelli & Casper 2003). For example, many studies indicate that different AMF species or mixes clearly invoke varying growth responses in different host plant species (Mosse 1972; Schenck & Smith 1982; Talukdar & Germida 1994; Streitwolf-Engel *et al.* 1997; van der Heijden *et al.* 1998a, b; Smith *et al.* 2000; Bever 2002; Helgason *et al.* 2002; Klironomos 2002). Likewise, the species identity of the plant host affects spore abundances of AMF species in the soil (Johnson *et al.* 1992; Sanders & Fitter 1992; Bever *et al.* 1997; Eom *et al.* 2000) or tendency to form hyphal asso-ciations with roots (Husband *et al.* 2002a, b). All of these observations suggest that AMF communities indeed influence the composition of host plants, at least in some temperate communities. But is this also true for more diverse tropical communities?

In contrast to most temperate and boreal tree species, which tend to form associations with ectomycorrhizae (Basidiomycetes), the vast majority of tropical tree species form associations with arbuscular mycorrhizae (Smith & Read 1997). However, basic biological information such as AMF floras (e.g. species lists based on spore morphology) is only beginning to become available for a few sites (e.g. Australia: Brundrett *et al.* 1999; Mexico: Guadarrama & Álvarez-Sánchez 1999; Nicaragua/Costa Rica: Picone 2000; Costa Rica: Lovelock *et al.* 2003). To date, studies examining the habitat associations of AMF communities in the neotropics have mostly concentrated on comparing AMF-spore compositions in soils of intact forests with those in adjacent disturbed soils such as pasture (Fischer *et al.* 1994; Johnson & Wedin 1997; Allen *et al.* 1998; Picone 2000; but see Lovelock *et al.* 2003). That shifts in AMF-spore communities are often detected across such starkly distinct habitats may come as little surprise (but see Picone 2000). Nonetheless, in contrast to some temperate studies documenting fine-scale differentiation (Bever *et al.* 1996; Pringle & Bever 2002), relatively little is known about the spatial scales in which changes in AMF communities can be detected within intact neotropical forests (Janos 1992; Husband *et al.* 2002a, b; Mangan *et al.* 2004; A. H. Eom *et al.*, unpublished results).

At best, the existing tropical species lists can only be considered as partially reflecting true AMF diversity (e.g. Bever *et al.* 2001; Helgason *et al.* 2002; Husband *et al.* 2002a, b; see below). Further, even in those sites for which there are partial floras, little information is available on the distributions of the component AMF species with respect to relevant ecological factors – space, time and host species (but see Lovelock *et al.* 2003). To make any advances in understanding tropical AMF–host interactions it is necessary to combine multiple approaches for charac-terizing the species composition and ecological properties of AMF communities at a series of sites. Further, having ecological and physiological characteriza-tions of at least the dominant components of the plant communities at these sites will aid the proper design and interpretation of experiments concerning interactions with the AMF community.

Here, after discussing limitations associated with each technique, we present results from a combination of spore- and molecular-based techniques to deter-mine the diversity and identities of AMF in the soils and roots (respectively) of such a tropical forest (Barro Colorado Island, Republic of Panama). We then assess AMF distribution among different sites, times and hosts. We present evidence that different components of the AMF community are functionally distinct, and cannot be considered as ecological equivalents. Further, we show that different host species are not ecological equivalents with respect to their associations with the AMF community. It thus appears that the AMF community at this site possesses all properties that are prerequisite for influencing the composition and distribution of the aboveground plant community.

Sources of information and methods
Morphological (spore-based) sampling and production of pure cultures

Traditionally, AMF identities have been determined by the morphology of spores, and AMF diversities and distributions in the field have been estimated on the basis of spore counts. However, this method for assaying AMF communities can be problematic. To begin with, the identification of AMF species based on spore morphology is a difficult enterprise, requiring a great deal of training and expertise in order to establish an acceptable level of consistency of identification. Ideally, therefore, taxonomic comparisons among different AMF communities should be conducted by the same researcher. Beyond that, the AMF community itself is difficult to characterize through spore abundances. AMF spore densities and species composition often change seasonally (Lee & Koske 1994; Allen *et al.* 1998; Schultz *et al.* 1999). Further, the rate of AMF spore production is affected by many factors (e.g. water, nutrient levels, light levels, host identity, etc.). For any given set of these conditions, such sporulation rates can vary among different AMF species (Morton *et al.* 1995; Bever *et al.* 1996; Eom *et al.* 2000). Therefore, the relative abundances of the spores of different AMF species in the soil may bear little relation to the relative abundances of AMF populations that have colonized roots (Clapp *et al.* 1995). In fact, it is likely that AMF diversity estimates based on spore morphology underestimate true species richness. For example, Bever and coworkers (2001) have demonstrated that AMF spore communities of temperate grasslands are much more diverse than would be expected from a single soil survey. By examining the fungal community at a single site through extensive soil-sampling and an assortment of subsequent trapping approaches over many years, they increased their initial estimate of 11 morphospecies to at least 37 different recognizable AMF morphospecies, one-third of which had not been previously described. Interestingly, this outcome suggests that the diversity of the AMF community at this temperate grassland site is roughly equal to that of the plants.

The spore identifications discussed in this chapter have all been performed by one person, Ahn-Heum Eom, and represent three types of collections from on and near Barro Colorado Island (BCI). The first involved the repeated collection of the AMF communities from the base of at least two individual trees for each of seven host species. The collection periods were: early February (early dry season), mid April (late dry season), mid June (early wet season) and mid December (late wet season) in order to characterize: (1) AMF species composition in the BCI community, (2) relative host associations and (3) the seasonal variation in spore abundance. The second method involved collections taken from four points around a 2-m perimeter of three adult host tree species growing in close proximity (each other's nearest neighbours) at each of four sites on BCI. This allowed us to establish the relative importance of location and host in affecting AMF spore community composition (A. H. Eom *et al.*, unpublished results). The third method involved structured sampling of AMF spores in soils of forested mainland and island sites in the vicinity of Gatun Lake, Republic of Panama. This sampling allowed us to better understand spatial structuring of AMF communities with particular attention to potential influences of island size (Mangan *et al.* 2004). The sampling consisted of individual plots, each with 16 soil sampling points arranged in a 9 × 9 m grid. Three sampling plots were located on small islands (less than 1 ha), three on medium-sized islands (2–4 ha) and three on adjacent mainland sites. At each plot we also conducted chemical analyses of the soil, as well as complete inventories of all vegetation greater than 0.5 m tall.

Molecular sampling

Although it requires appropriate training, assigning species names based on spore morphology is standard practice. Unfortunately, identifying AMF species within roots is nearly impossible because species-distinctive characters are generally lacking. However, advances in molecular techniques now make it feasible to identify directly the AMF within root tissues by using the polymerase chain reaction (PCR) to target specific AMF sequences. Several PCR-based methods have been developed over recent years, the majority of which target the ribosomal RNA genes (see Clapp *et al.* 2002 for review). Even so, the analysis of ribosomal genes is not without problems when applied to roots because any such approach is limited by the available genetic markers and is currently very time-consuming. Both Simon *et al.* (1992) and Helgason *et al.* (1998) designed primers (VANS1 and AM1 respectively) for the small subunit (SSU) rDNA intended to amplify all known glomalean fungi. Unfortunately, the VANS1 site is not well conserved in the Glomales (Clapp *et al.* 1999) and the AM1 site is absent from several highly diverged lineages (Redecker *et al.* 2000). Thus, these techniques are likely to give conservative estimates of AMF diversity *in planta*. Furthermore, the delineation of AMF species is ambiguous. In some cases, individual AMF spores have been found to contain multiple, genetically distinct nuclei (Kuhn *et al.* 2001; Sanders 2002), and give rise to subsequent, functionally different cultures (see Hart & Klironomos 2002). However, in contrast to patterns obtained with some more variable loci (Sanders *et al.* 1995; Lloyd-Macgilp *et al.* 1996; Clapp *et al.* 2001), differences between SSU sequences found within single spores are usually small compared with differences found among different AMF species (Clapp *et al.* 1999; Schussler *et al.* 2001; L. C. Mejia, personal observations). Also, in one of the few attempts to compare directly the morphological and molecular data for a host species (a forest herb in the United Kingdom), the results were broadly in agreement (see Merryweather & Fitter 1998; Helgason *et al.* 1999).

In the first application of molecular techniques to tropical host–AMF associations, Husband *et al.* (2002a, b) used AM1 primers to obtain SSU ribosomal gene sequences associated with seedlings and saplings of *Tetragastris panamensis* and *Faramea occidentalis*. First, newly emergent seedlings from cohorts of two host plant species were sampled from mixed seedling carpets at two different sites. Individuals from each cohort that survived to one year were then collected and all root systems were analysed. Second, at a third site, survivors from a cohort of *T. panamensis* were collected over two subsequent seasons, and, in the final collection, older (> 5 yr) saplings were also collected. Thus, one sample gave a picture of associated AMF of successful seedlings of two species at the beginning and end of their first year of survival at two sites (Husband *et al.* 2002a). The other gave a glimpse of AMF associations of seedlings of different ages sampled at the same time point, as well as AMF associated with the survivors in a cohort across two different years (Husband *et al.* 2002b).

Testing AMF effects on plant growth

In addition to sampling spores and roots to study the diversities and distributions of AMF, we have used two approaches for determining effects of different AMF inocula on the performance of host seedlings. Kiers *et al.* (2000) conducted three reciprocal inoculation experiments in the greenhouse using seedlings from six native tree species representing a range of life histories (early successional pioneers, a persistent understorey species and emergent species) typical of mature forest. Seeds were germinated in sterile soil and then either kept in sterile soil as controls or exposed to arbuscular mycorrhizal fungi in current association with naturally infected roots from adults of either the same or different species growing in intact forest. Using pure cultures of four AMF morphospecies and a complete AMF mixture, Eom *et al.* (unpublished results) inoculated two host plant species, *Luehea seemannii* and *Theobroma cacao* (a small-seeded pioneer and a large-seeded mid- to late-successional understorey tree species, respectively). The AMF species represented the three genera commonly found as spores at the site: *Glomus*, *Acaulospora* and *Scutellospora*.

Results

Tropical–temperate diversities

The factors that potentially affect the dramatic observed differences in aboveground plant diversity between temperate and tropical regions have been discussed extensively (Leigh 1999; Givnish 1999; Harms *et al.* 2000; Wright 2002). However, even the most basic estimates of belowground diversity are only now beginning to be obtained for sites in either region. Nevertheless, a comparison of both regions using the same survey techniques done by the same researcher or laboratory group is now possible for both morphological and molecular data. Eom has surveyed the spore communities of temperate grassland in Kansas (Eom *et al.* 2000) and tropical moist forest in Panama (unpublished results). In Panama, four forest sites on Barro Colorado Island (BCI) in which soils around three specific hosts were targeted, a total of 25 morphospecies were identified. In the additional study that included intensive sampling of grids and the inclusion of

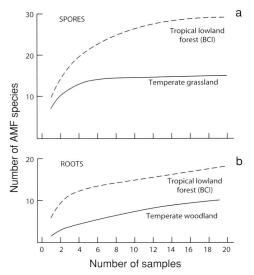

Figure 8.1 Comparison of BCI AMF species accumulation with accumulation in temperate sites where the same techniques have been used by the same researchers. (a) Morphological (spore-based) community surveys of BCI (each sample represents spores collected from a 10-g sample of soil) and a Kansas grassland (each sample represents spores collected from a 100-g sample of soil) from Eom *et al.* (2000), and (b) molecular survey of AMF species in roots (each sample represents AMF genotypes sampled from a single seedling root system; see Husband *et al.* 2002a). The curves for both the spore and root samples suggest conservatively that BCI has roughly twice the number of species of temperate regions where the same techniques have been applied.

several island and mainland sites around BCI, a total of 27 AMF morphospecies were encountered, with 17, 8, 1 and 1 from the genera *Glomus*, *Acaulospora*, *Sclerocystis* and *Scutellospora*, respectively (Mangan *et al.* 2004). Conservative comparisons of species accumulation curves from spore collections from both BCI and the temperate grassland (both sampled by A. H. Eom) suggest that the BCI AMF flora is roughly twice as large as that of the Kansas grassland (see Fig. 8.1a; Eom *et al.* 2000). This suggests that despite a higher overall AMF diversity at the BCI site, the ratio of AMF to plant species is much lower than at the temperate site.

The molecular surveys of roots from three host plant species show at least 30 different AMF types that exhibit genetic differences similar to those observed among closely related, named AMF species (see Husband *et al.* 2002a, b). A comparison of the AMF species accumulation curves from the same laboratory group using the same method for this tropical forest and temperate woodlands also suggests that the BCI AMF diversity is roughly twice as high (Fig. 8.1b). Further, the Shannon diversity index is appreciably higher for BCI ($H = 2.33$, $H_{max} = 3.135$, based on 48 roots from two host species) than for three temperate sites: a semi-natural woodland in England ($H = 1.44$, $H_{max} = 2.565$, based on 49 roots from five host species; Helgason *et al.* 1998), a temperate grassland ($H = 1.71$, $H_{max} = 2.890$, based on 47 roots from two host species; Vandenkoornhuyse *et al.* 2002), and temperate arable fields ($H = 1.16$, $H_{max} = 2.08$, based on 79 roots from four host species, calculated from Daniell *et al.* 2001).

Thus, both spores and molecular methods indicate that this tropical mycorrhizal community is relatively diverse compared with temperate sites where the similar procedures have been used to estimate AMF species diversity. However, unlike the results of temperate grasslands where the number of AMF and plant species are approximately equal, our current results from BCI suggest that, even with more extensive sampling, the AMF species diversity will not approach the magnitude of the aboveground diversity. Further, spore surveys offer no reason to suspect new genera or explosive proliferation of recognizable morphotypes. Nonetheless, we note that the molecular data have been predominantly collected from the seedlings of only two host species. More-extensive sampling of hosts is likely to increase the number of species identified using molecular methods, particularly if there are even modest degrees of differential host affinity (see below).

Interestingly, a phylogenetic tree combining molecular results from both temperate and tropical ecosystems indicates that there is no clear temperate–tropical differentiation of AMF floras. Neither is there a differentiation between Old World and New World (Husband *et al.* 2002a, b). It is remarkable to consider that only a few base substitutions separate the *G. mosseae* found in a deciduous English woodlot and that found in a Panamanian moist forest. Either there was an early and rapid radiation of AMF that pre-dated the successive continental breaks and was followed by a dramatic deceleration of rates of genetic change, or AMF have previously unsuspected and prodigious capacities for long-distance/intercontinental dispersal. Perhaps some combination of pirates, botanists and backpackers may have introduced English AMF to the New World tropics, or vice versa.

Spatial diversities

Landscape scale

Relatively extensive sampling has been conducted in two tropical sites of Central America: La Selva, Costa Rica (see McDade *et al.* 1994 for site description) and BCI, Republic of Panama (see Leigh 1999 for site description). In Costa Rica, Lovelock and coworkers (2003) examined the AMF spore community with respect to possible variation associated with host tree species, soil type, seasonality and rainfall. We have conducted similar sampling in a tropical moist forest of Panama (Mangan *et al.* and Eom *et al.* unpublished results). A comparison of the two distant sites (> 500 km apart) indicates surprising differences in basic AMF composition.

In the Costa Rican forest, over 90% of the AMF spore community is comprised of only two species (*Acaulospora morrowiae* and *A. mellea*). This contrasts with the findings from BCI in two ways. First, it is species of *Glomus*, not *Acaulospora*, that dominate the AMF community in BCI soils. Second, the degree of dominance by a few species is much less marked. On BCI, six *Glomus* species make up roughly 70% of the total spore volume (and spore number). Lovelock and coworkers suggest that dominance by *Acaulospora* may correlate with ecosystem type (pasture vs. forest), but this seems unlikely given that both the dominance

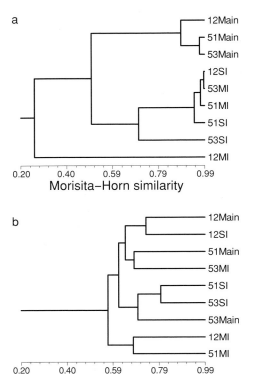

Figure 8.2 Similarities of AMF spore communities collected on mainland (Main), medium island (MI) and small island (SI) sites. The numbers refer to a geographical grouping of Main, MI and SI sites (see text). (a) Morisita–Horn similarities (based on relative abundances of AMF species). (b) Jaccard similarities (based on presence and absence of AMF species). Notice that analyses based on relative abundances clearly distinguish mainland (Main) from island (MI and SI) AMF floras, whereas the presence/absence analyses do not (see text).

by *Acaulospora* in Costa Rica and the dominance by *Glomus* in Panama occur in intact primary forest. Further, surveys of AMF spore communities from an intact Mexican seasonal forest also showed a dominance of *Glomus* (Allen *et al.* 1998). Possible explanations for this difference may be that La Selva has a greater total rainfall, and a less intense pattern of seasonality than BCI and the Mexican site. Another possibility is that, although the La Selva forest is more diverse overall, it is also more clearly dominated by a single tree species, *Pentaclethra macroloba* (Leguminosae) (McDade *et al.* 1994).

Intermediate and fine scale (mainland versus islands)
Within the vicinity of BCI, we found that the AMF spore community of any given mainland plot was more similar (Morisita–Horn index) to other distant (> 5 km) mainland plots than to nearby (within 0.7 km) island plots. This pattern reflects a more general pattern in all of our analyses of AMF distributions, and demonstrates the need to analyse relative abundance data in addition to presence/absence data when possible (see Fig. 8.2). Also, there was no decrease in AMF species richness (number of species, or Fisher's Alpha Index) either with decreasing forest size (size of the adjacent forest, mainland or island), or with decreasing species richness in the vegetation. In contrast, species richness of

vegetation did decrease with decreasing island size in a classical island biogeographical pattern that contrasted with the lack of such a decrease in the AMF diversity. Finally, within the smaller scale of the 9 × 9 m plots, we found no evidence for structuring of the AMF communities (no decay of Morisita–Horn community similarity with distance within plots; Mangan *et al.* 2004).

Differential host affinity

Spores

On BCI, we sampled AMF spore communities under the crowns of adults of three host tree species (*Luehea seemannii*, *Anacardium excelsum* and *Tetragastris panamensis*). These species were chosen because their life histories range from early pioneer to mature forest species, and because at each of four sites widely distributed across BCI, these trees were each other's closest neighbours. Each tree had four soil cores taken 2 m from the base for AMF community analysis. Both within and across sites, species composition of AMF spores was significantly influenced by the species of adult host (Eom *et al.*, unpublished results). Although most AMF species were present at most sites, and under most trees, the relative abundances varied significantly with host tree species.

Minimally, these results suggest that AMF species vary in their rates of sporulation, and/or in total underground biomass, in response to different hosts. This further suggests differential fungal/host affinity, an interpretation that is consistent with the observation that, although mycorrhizal fungi from all inocula were able to colonize the roots of all host species, the inoculum potential (the infectivity of an inoculum of a given concentration) and root colonization varied depending on the identity of the host seedling and the source of the inoculum (Kiers *et al.* 2000). In Costa Rica, host plant species also affected AMF community composition, and again the differences were primarily due to changes in relative abundances (Lovelock *et al.* 2003). However, in an interesting contrast, the differences in La Selva AMF communities were primarily due to changes in the relative abundance of the two dominant species. On BCI, the pattern was quite different. There were no significant differences in the relative abundances of the top AMF species of the two common genera, *Glomus* and *Acaulospora*. However, there were significant host effects on seven less-abundant AMF species.

Molecular analyses of roots

We collected roots from cohorts of *T. panamensis* and *Faramea occidentalis* seedlings from a series of mixed seedling carpets, and analysed the AMF sequences (Husband *et al.* 2002a, b). The two hosts have distinct life histories. *Tetragastris panamensis* is a mid-to-late successional species associated with mature forest, whereas *F. occidentalis* is a persistent understorey species. These mycorrhizal communities showed significant spatial heterogeneity and non-random associations with the different hosts. It appears that distinct AMF species preferentially colonize and

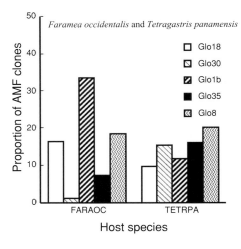

Figure 8.3 Comparison of the relative abundance of the five most common AMF species on six juvenile seedlings each of *Faramea occindentalis* (FARAOC) and *Tetragastris panamensis* (TETRPA) growing in a mixed-species stand at a single site on BCI. Notice that different AMF species dominate *Faramea* seedlings and *Tetragastris* seedlings (e.g. glo1b, glo30).

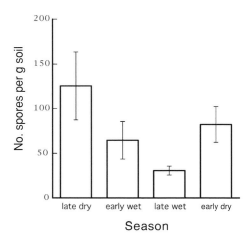

Figure 8.4 Density of spores at the base of at least two individuals for each of seven host tree species collected during four seasons on BCI. Notice that highest spore densities correspond to late dry season (mid-April), which immediately precedes the period of greatest seedling germination. This pattern reflects the seasonal pattern obtained at La Selva, in Costa Rica (see text, and see also Lovelock *et al.* 2003)

then differentially proliferate in roots of these two hosts (see Fig. 8.3; Kiers *et al.* 2000; Husband *et al.* 2002a, b).

Temporal patterns

Seasonal (spores)

At both BCI and La Selva, the overall density of spores was substantially higher at the end of the dry season than during the rainy season. Using four sampling periods on BCI, we find that overall spore density goes up just before the onset of the wet season (see Fig. 8.4; also see Mangan and Adler 2002). This corresponds with the period when the maximum germination of seeds is about to start (Garwood 1983). At both the BCI and La Selva sites, seasonality was also correlated with relative abundances of particular AMF species such that community composition would be different through time. For example, in La Selva,

A. morrowiae was the most abundant during the wet season, whereas *A. mellea* dominated during the drier season. However, season had no significant effect on community species diversity (Shannon's index) or richness (Lovelock *et al.* 2003). One possible explanation for these seasonal changes is that ecologically distinct AMF respond differently to changes in the abiotic environment, and as the environment changes over time, so do the dominant AMF. Such responses, plus changes in host phenology, are the implicit assumptions used previously to explain temporal variation in mycorrhizal communities (Lee & Koske 1994; Merryweather & Fitter 1998; Eom *et al.* 2000; Daniell *et al.* 2001; Mangan & Adler 2002). These possibilities require further examination.

Successional (molecular analyses of roots)

AMF species that dominate the roots of newly germinated seedlings are almost entirely replaced by previously rare species in the seedlings that survive a year (Husband *et al.* 2002a, b). In a second study, significantly different fungal populations were found to dominate 2-year-old seedlings and 5-year-old seedlings sampled at the same time point (Husband *et al.* 2002b). Both studies show a strong repeating pattern whereby the dominant mycorrhizal species are replaced by previously rare species in the surviving seedlings (see Fig. 8.5). Furthermore, both studies reveal a decrease in fungal evenness and diversity across plant age. Indeed, the repeated pattern both within host species and across sites suggests two non-mutually exclusive explanations. Either there is a succession of AMF types within a single host, possibly driven by differences in fungal life-history strategies (see Hart *et al.* 2001 for review); or individual AMF affect seedling recruitment, so that the most effective host–fungus combination has a higher probability of survival and is consequently enriched in the surviving population. The observation that the AMF combinations found in seedlings that survive for one or more years are not found in any of the earliest seedlings suggests that within-host succession of AMF plays the more important role (see Husband *et al.* 2002a, b). However, further experimental testing is needed to establish the relative importance of the roles of within-host succession and differential seedling survival depending on the identities of associated AMF.

Effects

Both pure culture and root inocula show that different AMF species or combinations generally produce different growth patterns (relative growth rates) in host seedlings, and that these effects vary interactively with AMF species (Kiers *et al.* 2000; Eom *et al.*, unpublished results; see Fig. 8.6). Further, both types of experiments also show that small-seeded pioneer plant species are more dependent on AMF for initial survival and growth (Kiers *et al.* 2000; D. Kyllo *et al.* unpublished results). Specifically, all of our experiments indicate that the tiny seedlings of small-seeded species soon die after germination in

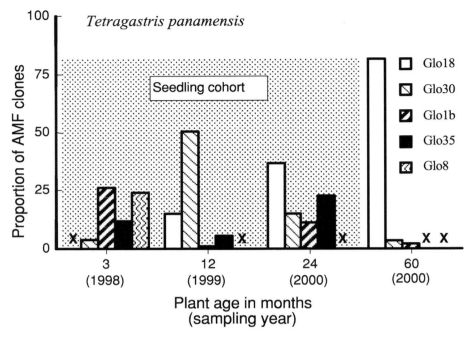

Figure 8.5 Sucession of AMF species associated with survivors of a cohort of *Tetragastris* seedlings between 1998 and 2000 (collected at 3 months, 12 months and 24 months). Further, during the 2000 collection, older saplings (older than 5 years) were also collected (60) for comparison of the AMF communities on the roots of the younger plants. Notice that the AMF species that dominate juvenile (3-month) seedlings (glo1b and glo8) decrease in abundance and are replaced by other species (e.g. glo18) in the older surviving seedlings and saplings. X indicates a zero count for the clone in the sample.

sterile soil. These are also the host species that show the most striking differences in growth when inoculated with different AMF sources. Consistent with these suggestions, recent greenhouse work with more than 80 woody species in Brazil clearly demonstrates that the early successional species generally show a much greater response to AMF than late successional species (Siqueira *et al.* 1998; Zangaro *et al.* 2003). Variation in response to different AMF species was greater in the host with greater mycorrhizal dependency, following a pattern suggested by van der Heijden (2002). It appears that having a relatively large seed provides some buffer against immediate dependence on AMF.

Two further interpretations are suggested by the combination of the effects on growth and the infectivity trials presented in Kiers *et al.* (2000). The experiments that use roots of adults as the source of inoculum for seedlings suggest that the AMF communities established in adult root systems do not necessarily optimize growth in conspecific seedlings (Kiers *et al.* 2000). If subsequent experimentation supports this pattern, then AMF associated with roots of adult

Growth rate (leaf area)—*3 months*

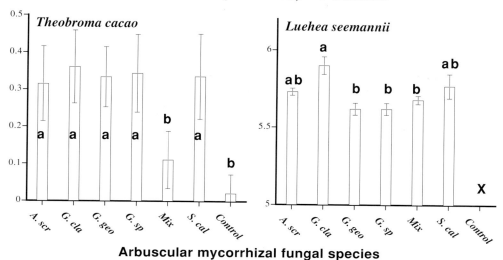

Arbuscular mycorrhizal fungal species

Figure 8.6 Growth responses of *Luehea seemannii* and *Theobroma cacao* seedlings in response to different AMF inocula. (Different letters refer to significantly different means $P < 0.05$). For both host species, AMF improved growth relative to controls, but the growth response was determined by the specific fungus–host combination. The pioneer species, *L. seemannii*, grew at varying rates depending on the AMF species and was obligately dependent on the mutualism for initial survival. Variation in response to different AMF species was greater in the host with greater mycorrhizal dependency following a pattern suggested by van der Heijden (2002). Although *T. cacao* showed no clear differences among different AMF species (and genera), the growth response of *T. cacao* was significantly lower for the mixed AMF species inoculum than single AMF species. The AMF species names are *Acaulospora scrobiculata* (A. scr), *Glomus clavisporum* (G. cla), *Glomus geosporum* (G. geo), *Glomus* sp. (G. sp.) and *Scutellospora calospora* (S. cal).

trees may contribute to the pervasive pattern of negative density dependence observed in the establishment of tropical seedlings (Augspurger 1984; Gilbert *et al.* 1994; Wills *et al.* 1997; Harms *et al.* 2000; Wright 2002). Finally, there is some suggestion that larger-seeded species show a higher level of differential affinity for certain AMF inocula (Kiers *et al.* 2000). Perhaps having a larger initial resource base also permits a seedling the luxury of being more 'choosy' concerning the AMF species with which associations are formed (see also Kitajima 2003). If so, we might expect differences in the tendency in larger-seeded species to form associations with particular AMF species that are more or less beneficial.

Discussion
Although preliminary, these results are nonetheless sufficient to begin to delineate the properties we might expect tropical mycorrhizal communities to

possess. For example, consistent with findings from other groups studying tropical AMF–host associations, it is now clear that different AMF species and mixes
of species produce different effects on host growth (Nemec 1978; Kiers *et al.*
2000). Further, different hosts react differently to any given set of AMF inocula
(Zangaro *et al.* 2003). Moreover, densities and distributions of AMF species are
not random with respect to space, time or host. Therefore, neither the AMF nor
the host plants are functional and ecological equivalents with respect to their
biotic interactions with each other. This demonstrates that AMF and host-plant
communities possess the prerequisite properties necessary for each to influence
the species composition and distribution of the other. As is the case with other
elements of the biotic environment (Augspurger 1984; Gilbert *et al.* 1994; Wills
et al. 1997; Harms *et al.* 2000; Wright 2002), we should not expect that AMF
communities provide a neutral background for the establishment and growth
of host plants (see also Connell & Lowman 1989). The preponderance of the evidence both from BCI and other sites suggests that tropical AMF communities
indeed affect aboveground community composition and distributions. However,
some important pieces are still missing from the puzzle.

How do AMF species in roots correspond to spores in the soils?
We have presented both morphological data collected from AMF spores in soils
and molecular data collected from AMF in association with roots. Both method-
ological approaches show higher AMF diversity in a tropical diverse forest than in
temperate ecosystems, when the same methods are used by the same researchers.
Sampling curves for both morphological and molecular data appear to approach
an asymptotic limit at roughly 30–40 species. Both approaches demonstrate clear
non-random AMF distributions with respect to space, time and host. But how do
spores relate to what is in roots? More specifically, how does the molecular information correspond to morphological information? In one of the few examples
to compare the morphological and molecular data directly for a host species,
the results were broadly in agreement (Merryweather & Fitter 1998; Helgason
et al. 1999). Is this also the case with the Panamanian samples?

Of the 30 AMF species that have been identified using the available sequences
taken from field-collected roots, only one shows a match with a sequence
obtained from the 12 spores from pure cultures of AMF species for which
sequences have been obtained. This is particularly striking because spores were
also collected from the sites where the roots were collected for molecular analyses. Therefore, our current understanding is that there are at least 41 genetically
distinct AMF species in BCI soils and roots. Nonetheless, if the AMF that are
present as spores and those that are present in roots represent samples from
the same population, we would expect the overlap to be higher. Minimally,
this suggests that the actual AMF diversity at this site estimated (at least 41) is
even higher than we had directly estimated on the basis of either soil or root

samples. It is possible that, with even more intensive sampling (e.g. Bever *et al.* 2001), some of the species currently found only as molecular signatures in roots might be encountered as spores. However, if there is no further overlap between the spores and roots, the total number of AMF species at this site would be at least 57, and it is noteworthy that only a relatively small number of individual seedlings from two species of host were used for the molecular sampling.

Further, the lack of overlap between root- and soil-collected sequences also suggests that these represent ecologically distinct groups of AMF in this forest. One group (those that dominate the root systems) tends to persist in active association with roots and tends not to produce large quantities of spores, while another (those that dominate the spore comunity in the soil) appears to be more transient in the association with roots, and tends to produce relatively large quantities of spores. In essence, the AMF groups are analogues of old-growth and pioneer tree species, respectively (Reader *et al.* 2001; also see Dalling & Hubbell 2002). If this is true, we might also expect that the different sets of mycorrhizae show different life-history strategies and/or fundamentally different sets of relationships with hosts. Specifically, the root-associated 'old-growth' AMF species (at least some of which apparently form much longer-standing associations) might provide greater benefits to the host than the 'pioneer' AMF species that might effectively be weeds, and provide less benefit for the hosts.

What is the relative importance of AMF spores and hyphae for colonizing germinating seedlings?

We know that spores derived from the different AMF pure cultures produce different growth effects in seedlings. We know that different host species respond differently to a given AMF inoculum. We know that the relative abundances of spores of different AMF species show relatively little variation over spatial scales that appear to correspond roughly with areas dominated by roots of a single canopy emergent tree (81 m^2). This suggests that at spatial scales of this order we can expect relatively homogeneous AMF spore communities that are likely to benefit the growth of some species more than others.

However, our preliminary results show that the composition of AMF species associated with surviving seedlings changes consistently through time, and that AMF communities in older seedlings and saplings tend to be dominated by relatively few species. Further, we know that few of the AMF species that we have identified from spores correspond genetically to AMF species that we have found associated with roots. Moreover, we do not know whether the spores or the AMF hyphae running throughout the adjacent soils from the roots of large individual plants are more important for colonization of seedlings. Are newly germinated seedlings colonized primarily from spores, only to have those spore-derived AMF displaced by subsequent colonization by hyphae? This scenario implies a competition for available seedlings by AMF that have access to widely different resource bases (spores vs. extensive hyphal systems connected to the carbon source of adult trees (see Grime *et al.* 1987; Kyllo 2001). A hypothesis to be tested is whether the AMF that tend to dominate older seedlings also correspond to those that dominate the roots of adults in the area. Specifically, the patterns revealed by our sampling thus far suggest that a forest floor consists of a patchwork of functionally distinct AMF communities that correspond roughly with the root systems of the different adult canopy trees. If this proves to be the case, and the results obtained by Kiers *et al.* (2000) from greenhouse experiments are relevant to the field, then the overall effect of AMF on host communities would be to maintain host diversities through negative density-dependence.

Ultimately, there are gaping holes in our view of AMF–host interactions in BCI that need to be filled. We need to develop an even more detailed view of what AMF species are dominating root systems of different host species of different ages. Are the AMF species associated with older seedlings and saplings largely a reflection of locally abundant AMF on existing, dominating root systems? Further, we only have pure cultures from about 12 of the genetically identified species available for experiments. Pure cultures will need to be produced from AMF species that currently have been found only in association with roots. These species will need to be characterized with respect to growth and survival effects on different hosts. Those effects will need to be placed in the context of the field samples that show how the various AMF species occur primarily as early or later associates in the roots of seedlings of particular host species, or predominately as spores. Nonetheless, despite the gaps in our knowledge, the preponderance of available evidence both from BCI and other sites suggests that neither hosts nor AMF species are functionally ecological equivalents, and that all prerequisites are fulfilled for AMF to influence the community composition and distribution of host plants. The possibilities are exciting. Studies of the ecological role of arbuscular mycorrhizal fungi in tropical forests are in their infancy.

Acknowledgements

We thank Rachel Gallery, Tanja Roehrich, Enith Rojas, Janneth Fabiola Santos, Zuleyka Maynard, Camila Pizano and Dora Alvarez for field and greenhouse assistance with several of the projects mentioned. We thank Peter Young and his York University laboratory for invaluable support of the molecular work. We thank Egbert Leigh, Joe Wright and Sunshine Van Bael for comments on early drafts of the manuscript. We thank the Andrew W. Mellon Foundation for most of the funding, with additional support from the American Cacao Research Institute, World Cacao Foundation, and the John Clapperton Fellowship of Mars Incorporated. We thank the Smithsonian Tropical Research Institute for making this work possible.

References

Allen, E. D., E. Rincon, M. F. Allen, A. Perez-Jimenez & P. P. Huante. 1998. Disturbance and seasonal dynamics of mycorrhizae in a tropical deciduous forest in Mexico. *Biotropica* 30: 261–274.

Allen, M. F. 1996. The ecology of arbuscular mycorrhizas: a look back into the 20th century and a peek into the 21st. *Mycological Research* 100: 769–782.

Allen, M. F., T. S. Moore & M. Christiensen. 1980. Phytohormone changes in *Bouteloua gracilis* infected by vesicular-arbuscular mycorrhizae: . Cytokinin increases in the host plant. *Canadian Journal of Botany* 58: 371–374.

Andrade, G., K. L. Mihara, R. G. Linderman & G. L. Bethlenfalvay. 1998. Soil aggregation status and rhizobacteria in the mycorrhizosphere. *Plant Soil* 202: 89–96.

Augspurger, C. K. 1984. Seedling survival of tropical tree species: interactions of dispersal distance, light-gaps, and pathogens. *Ecology* 65: 1705–1712.

Azcon, R. & J. A. Ocampo. 1981. Factors affecting the V-A infection and mycorrhizal dependency of thirteen wheat cultivars. *New Phytologist* 87: 677–685.

Bever, J. D. 1994. Feedback between plants and their soil communities in an old field community. *Ecology* 75: 1965–1977.

—— 1999. Dynamics within mutualism and the maintenance of diversity: inferences from a model of interguild frequency dependence. *Ecology Letters* 2: 52–62.

—— 2002. Negative feedback within a mutualism: host-specific growth of mycorrhizal fungi reduces plant benefit. *Proceedings of the Royal Society of London B* 269: 2595–2601.

Bever, J. D., J. B. Morton, J. Antonovics & P. A. Schultz. 1996. Host-dependent sporulation and species diversity of arbuscular mycorrhizal fungi in a mown grassland. *Journal of Ecology* 84: 71–82.

Bever, J. D., K. M. Westover & J. Antonovics. 1997. Incorporating the soil community into plant population dynamics: the utility of the feedback approach. *Journal of Ecology* 85: 561–573.

Bever, J. D., P. A. Schultz, A. Pringle & J. B. Morton. 2001. Arbuscular mycorrhizal fungi: more diverse than meets the eye, and the ecological tale of why. *Bioscience* 51: 923–931.

Brundrett, M. C., L. K. Abbott & D. A. Jasper. 1999. Glomalean mycorrhizal fungi from tropical Australia I. Comparison of the effectiveness and specificity of different isolation procedures. *Mycorrhiza* 8: 305–314.

Castelli, J. P. & B. B. Casper. 2003. Intraspecific AMF fungal variation contributes to plant–fungal feedback in a serpentine grassland. *Ecology* 84: 323–336.

Clapp, J. P., J. P. W. Young, J. W. Merryweather & A. H. Fitter. 1995. Diversity of fungal symbionts in arbuscular mycorrhizas from a natural community. *New Phytologist* 130: 259–265.

Clapp, J. P., A. H. Fitter & J. P. W. Young. 1999. Ribosomal small subunit sequence variation within spores of an arbuscular mycorrhizal fungus, *Scutellospora* sp. *Molecular Ecology* 8: 915–921.

Clapp, J. P., A. Rodriguez & J. C. Dodd. 2001. Inter- and intra-isolate rRNA large subunit variation in *Glomus coronatum* spores. *New Phytologist* 149: 539–554.

Clapp, J. P., T. Helgason, T. J. Daniell & J. P. W. Young. 2002. Genetic studies of the structure and diversity of arbuscular mycorrhizal communities. In *Mycorrhizal Ecology* (ed. M. G. A. van der Heijden & I. R. Sanders). Berlin: Springer-Verlag.

Connell, J. H. & M. B. Lowman. 1989. Low diversity tropical rain forests: some possible mechanisms for their existence. *American Naturalist* 134: 88–119.

Dalling, J. W. & S. P. Hubbell. 2002. Seed size, growth rate, and gap microsite conditions as determinants of recruitment success for pioneer species. *Journal of Ecology* **90**: 557–568.

Daniell, T. J., R. Husband, A. H. Fitter & J. P. W. Young. 2001. Molecular diversity of arbuscular mycorrhizal fungi colonising arable crops. *FEMS Microbiology Ecology* **36**: 203–209.

Eom, A. H., D. C. Hartnett & G. W. T. Wilson. 2000. Host plant species effects on arbuscular mycorrhizal fungal communities in tallgrass prairie. *Oecologia* **122**: 435–444.

Fischer, C. R., D. P. Janos, D. A. Perry, R. G. Linderman & P. Sollins. 1994. Mycorrhiza inoculum potentials in tropical secondary succession. *Biotropica* **26**: 369–377.

Gange, A. C., V. K. Brown & L. M. Farmer. 1990. A test of mycorrhizal benefit in an early successional plant community. *New Phytologist* **115**: 85–91.

Garwood, N. C. 1983. Seed germination in a seasonal tropical forest in Panama: a community study. *Ecological Monographs* **53**: 159–182.

Gilbert, G. S., S. P. Hubbell & R. B. Foster. 1994. Density and distance-to-adult effects of a canker disease of trees in a moist tropical forest. *Oecologia* **98**: 100–108.

Givnish, T. J. 1999. On the causes of gradients in tropical tree diversity. *Journal of Ecology* **87**(2): 193–210.

Grime, J. P., M. L. Mackey, S. H. Hillier & D. J. Read. 1987. Floristic diversity in a model system using experimental microcosms. *Nature* **328**: 420–422.

Guadarrama, P. & F. J. Álvarez-Sánchez. 1999. Abundance of arbuscular mycorrhizal fungi spores in different environments in a tropical rain forest, Veracruz, Mexico. *Mycorrhiza* **8**: 267–270.

Harley, J. L. & S. E. Smith. 1983. *Mycorrhizal Symbiosis*. London: Academic Press.

Harms, K. E., S. J. Wright, O. Calderón, A. Hernández & E. A. Herre. 2000. Pervasive density-dependent recruitment enhances seedling diversity in a tropical forest. *Nature* **404**: 493–496.

Hart, M. M. & J. N. Klironomos. 2002. Diversity of arbuscular mycorrhizal fungi and ecosystem functioning. In *Mycorrhizal Ecology* (ed. M. G. A. van der Heijden & I. R. Sanders). Berlin: Springer-Verlag.

Hart, M. M., R. J. Reader & J. N. Klironomos. 2001. Life-history strategies of arbuscular mycorrhizal fungi in relation to their successional dynamics. *Mycologia* **93**: 1186–1194.

Hartnett, D. C. & G. W. T. Wilson. 1999. Mycorrhizae influence plant community structure and diversity in tallgrass prairie. *Ecology* **80**: 1187–1195.

Helgason, T., T. J. Daniell, R. Husband, A. H. Fitter & J. P. W. Young. 1998. Ploughing up the wood-wide web? *Nature* **394**: 431.

Helgason, T., A. H. Fitter & J. P. W. Young. 1999. Molecular diversity of arbuscular mycorrhizal fungi colonising *Hyacinthoides non-scripta* (bluebell) in a seminatural woodland. *Molecular Ecology* **8**: 659–666.

Helgason, T., J. W. Merryweather & J. Denison. 2002. Selectivity and functional diversity in arbuscular mycorrhizas of co-occurring fungi and plants from a temperate deciduous woodland. *Journal of Ecology* **90**: 371–384.

Hetrick, B. A. & J. Bloom. 1986. The influence of host plant on production and colonization ability of vesicular-arbuscular mycorrhizal spores. *Mycologia* **78**: 32–36.

Hetrick, B. A. D., G. W. T. Wilson & T. C. Todd. 1992. Relationship of mycorrhizal symbiosis, rooting strategy and phenology among tall grass prairie forbs. *Canadian Journal of Botany* **70**: 1521–1528.

Husband, R., E. A. Herre, S. L. Turner, R. Gallery & J. P. W. Young. 2002a. Molecular diversity of arbuscular mycorrhizal fungi and patterns of host association over time and space in a tropical forest. *Molecular Ecology* 11: 2669–2678.

Husband, R., E. A. Herre & J. P. W. Young. 2002b. Temporal variation in the arbuscular mycorrhizal communities colonising seedlings in a tropical forest. *FEMS Microbiology Ecology* 42: 131–136.

Janos, D. P. 1980. Vesicular arbuscular mycorrhizae affect lowland tropical rain forest plant growth. *Ecology* 61: 151–162.

1992. Heterogeneity and scale in tropical vesicular-arbuscular mycorrhiza formation. In *Mycorrhizas in Ecosystems* (ed. D. J. Read, D. H. Lewis, A. H. Fitter & I. J. Alexander). Wallingford: CAB International, pp. 276–282.

Johnson, N. C. & D. A. Wedin. 1997. Soil carbon, nutrients, and mycorrhizae during conversion of dry tropical forest to grassland. *Ecological Applications* 7: 171–182.

Johnson, N. C., D. Tilman & D. Wedin. 1992. Plant and soil controls on mycorrhizal fungal communities. *Ecology* 73: 2034–2042.

Kiers, E. T., C. E. Lovelock, E. L. Krueger & E. A. Herre. 2000. Differential effects of tropical arbuscular mycorrhizal fungal inocula on root colonization and tree seedling growth: implications for tropical forest diversity. *Ecology Letters* 3: 106–113.

Kitajima, K. 2003. Impact of cotyledon and leaf removal on seedling survival in three tree species with contrasting cotyledon functions. *Biotropica* 35: 429–434.

Klironomos, J. N. 2002. Variation in plant response to native and exotic arbuscular mycorrhizal fungi. *Ecology* 84: 2292–2301.

Kuhn, G., M. Hijri & I. R. Sanders. 2001. Evidence for the evolution of multiple genomes in arbuscular mycorrhizal fungi. *Nature* 414: 745–748.

Kyllo, D. 2001. Effects of a common mycorrhizal network and light on growth and community structure of understorey shrubs, *Piper* & *Psychotria*, in a moist

neotropical forest. Unpublished Ph.D. thesis, University of Missouri – St Louis.

Kyllo, D., V. Velez & M. T. Tyree. 2003. Combined effects of arbuscular mycorrhizas and light on water uptake of the neotropical understory shrubs, *Piper* and *Psychotria*. *New Phytologist* 160: 443–454.

Lee, P. J. & R. E. Koske. 1994. *Gigaspora gigantea* – seasonal abundance and aging of spores in a sand dune. *Mycological Research* 98: 453–457.

Leigh, E. G. Jr. 1999. *Tropical forest ecology: a view from Barro Colorado Island*. New York: Oxford University Press.

Lloyd-Macgilp, S. A., S. M. Chambers & J. C. Dodd. 1996. Diversity of the ribosomal internal transcribed spacers within and among isolates of *Glomus mosseae* and related mycorrhizal fungi. *New Phytologist* 133: 103–111.

Lovelock, C. E., K. Andersen & J. B. Morton. 2003. Arbuscular mycorrhizal communities in tropical forests are affected by host tree species and environment. *Oecologia* 135(2): 268–279.

Mangan, S. A. & G. H. Adler (2002) Seasonal dispersal of arbuscular mycorrhizal fungi by spiny rats in a neotropical forest. *Oecologia* 131: 587–597.

Mangan, S. A., A.-H. Eom, G. H. Adler, J. B. Yavitt & E. A. Herre (2004) Diversity of arbuscular mycorrhizal fungi across a fragmented forest in Panama: insular spore communities differ from mainland communities. *Oecologia* 141: 687–700.

McDade, L. A., K. S. Bawa, H. A. Hespenheide & G. S. Hartshorn. 1994. *La Selva: Ecology and Natural History of a Neotropical Forest*. Chicago: Chicago University Press.

Merryweather, J. & A. Fitter. 1998. The arbuscular mycorrhizal fungi of *Hyacinthoides non-scripta*. Seasonal and spatial patterns of fungal populations. *New Phytologist* 138: 131–142.

Mills, K. M. & J. D. Bever. 1998. Maintenance of diversity within plant communities: soil pathogens as agents of feedback. *Ecology* **79**: 1595–1601.

Morton, J. B., S. P. Bentivenga & J. D. Bever. 1995. Discovery, measurement, and interpretation of diversity in arbuscular endomycorrhizal fungi (Glomales, Zygomycetes). *Canadian Journal of Botany* **73**: S25–S32.

Mosse, B. 1972. Effects of different *Endogone* strains on the growth of *Paspalum notatum*. *Nature* **239**: 221–223.

Nemec, S. 1978. Response of six citrus rootstocks to three species of *Glomus*, a mycorrhizal fungus. *Proceedings of the Florida State Horticulture Society* **91**: 10–14.

Newsham, K. K., A. H. Fitter & A. R. Watkinson. 1995. Arbuscular mycorrhizae protect an annual grass from root pathogenic fungi in the field. *Journal of Ecology* **83**: 991–1000.

Olff, H., B. Hoorens, R. G. M. de Goede, W. H. van der Putten & J. M. Gleichman. 2000. Small-scale shifting mosaics of two dominant grassland species: the possible role of soil-borne pathogens. *Oecologia* **125**: 45–54.

Picone, C. 2000. Diversity and abundance of arbuscular-mycorrhizal fungus spores in tropical forest and pasture. *Biotropica* **32**: 734–750.

Pringle, A. & J. D. Bever. 2002. Divergent phenologies may facilitate the coexistence of arbuscular mycorrhizal fungi in a North Carolina grassland. *American Journal of Botany* **89**: 1439–1446.

Reader, M. M., R. J. Klironomos & J. N. Klironomos. 2001. Life history strategies of arbuscular mycorrhizal fungi in relations to their successional dynamics. *Mycologia* **93**: 1186–1194.

Redecker, D., J. B. Morton & T. D. Bruns. 2000. Ancestral lineages of arbuscular mycorrhizal fungi (Glomales). *Molecular Phylogenetics and Evolution* **14**: 276–284.

Rillig, M. C. & M. F. Allen. 1999. What is the role of arbuscular mycorrhizal fungi in plant to ecosystem responses to elevated atmospheric CO_2. *Mycorrhiza* **9**: 1–8.

Sanders, I. R. 2002. Specificity in the arbuscular mycorrhizal symbiois. In *Mycorrhizal Ecology* (ed. M. G. A. van der Heijden & I. R. Sanders). Berlin: Springer-Verlag.

Sanders, I. R. & A. H. Fitter. 1992. Evidence for differential responses between host–fungus combinations of vesicular-arbuscular mycorrhizas from a grassland. *Mycological Research* **96**: 415–419.

Sanders, I. R., M. Alt, K. Groppe, T. Boller & A. Wiemken. 1995. Identification of ribosomal DNA polymorphisms among and within spores of the Glomales: application to studies on the genetic diversity of arbuscular mycorrhizal fungal communities. *New Phytologist* **130**: 419–427.

Schenck, N. C. & G. S. Smith. 1982. Responses of six species of vesicular-arbuscular mycorrhizal fungi and their effects on soybean at four soil temperatures. *New Phytologist* **92**: 193–201.

Schultz, P. A., J. D. Bever & J. Morton. 1999. *Acaulospora colossica* sp. nov. from an old field in North Carolina and morphological comparisons with similar species, *A. laevis* and *A. koskei*. *Mycologia* **91**: 676–683.

Schussler, A., H. Gehrig, D. Schwarzott & C. Walker. 2001. Analysis of partial Glomales SSU rRNA gene sequences: implications for primer design and phylogeny. *Mycological Research* **105**: 5–15.

Simon, L., M. Lalonde & T. D. Bruns. 1992. Specific amplification of 18S fungal ribosomal genes from vesicular-arbuscular endomycorrhizal fungi colonizing roots. *Applied and Environmental Microbiology* **58**: 291–295.

Siqueira, J. O., M. A. C. Carneiro, N. Curi, S. C. da Silva Rosado & A. C. Davide. 1998. Mycorrhizal colonization and mycotrophic growth of native woody species as related to successional groups in Southeastern Brazil. *Forest Ecology and Management* **107**: 241–252.

Smith, F. A., I. Jakobsen & S. E. Smith. 2000. Spatial differences in acquisition of soil phosphate between two arbuscular mycorrhizal fungi in symbiosis with *Medicago truncatula*. *New Phytologist* **147**: 357–366.

Smith, S. E. & D. J. Read. 1997. *Mycorrhizal Symbioses*, 2nd edn. London: Academic Press.

Streitwolf-Engel, R., T. Boller, A. Wiemken & I. R. Sanders. 1997. Clonal growth traits of two *Prunella* species are determined by co-occurring arbuscular mycorrhizal fungi from a calcareous grassland. *Journal of Ecology* **85**: 181–191.

Talukdar, N. C. & J. J. Germida. 1994. Growth and yield of lentil and wheat inoculated with 3 glomus isolates from Saskatchewan soils. *Mycorrhiza* **5**: 45–152.

van der Heijden, M. G. A. 2002. Arbuscular mycorrhizal fungi as a determinant of plant diversity: in search of underlying principles and mechanisms. In *Mycorrhizal Ecology* (ed. M. G. A. van der Heijden & I. R. Sanders). Berlin: Springer-Verlag.

van der Heijden, M. G. A., T. Boller, A. Wiemken & I. R. Sanders. 1998a. Different arbuscular mycorrhizal fungal species are potential determinants of plant community structure. *Ecology* **79**: 2082–2091.

van der Heijden, M. G. A., J. N. Klironomos, M. Ursic *et al.* 1998b. Mycorrhizal fungal diversity determines plant biodiversity, ecosystem variability and productivity. *Nature* **396**: 69–72.

Vandenkoornhuyse, P., R. Husband & T. J. Daniell. 2002. Arbuscular mycorrhizal community composition associated with two plant species in a grassland ecosystem. *Molecular Ecology* **11**: 1555–1564.

Wills, C., R. Condit, R. B. Foster & S. P. Hubbell. 1997. Strong density- and diversity-related effects help to maintain tree species diversity in a neotropical forest. *Proceedings of the National Academy of Science* **94**: 1252–1257.

Wright, S. J. 2002. Plant diversity in tropical forests: a review of mechanisms of species coexistence. *Oecologia* **130**: 1–14.

Zangaro, W., S. M. A. Nisizaki, J. C. B. Domingos & E. M. Nakano. 2003. Mycorrhizal response and successional status in 80 woody species from south Brazil. *Journal of Tropical Ecology* **19**: 315–324.

CHAPTER NINE

Tropical plants as chimera: some implications of foliar endophytic fungi for the study of host-plant defence, physiology and genetics

EDWARD ALLEN HERRE, SUNSHINE A. VAN BAEL,
ZULEYKA MAYNARD AND NANCY ROBBINS
Smithsonian Tropical Research Institute

JOSEPH BISCHOFF
Rutgers University

ANNE E. ARNOLD
Smithsonian Tropical Research Institute and Duke University

ENITH ROJAS, LUIS C. MEJIA, ROBERTO A. CORDERO,
CATHERINE WOODWARD AND DAMOND A. KYLLO
Smithsonian Tropical Research Institute

Introduction

Fungal endophytes are defined as those fungi that live inside plant tissues (e.g. roots, stems, leaves) without causing apparent harm to their host (see Wilson 1995). Although we will also mention stem-associated endophytes (see Evans *et al.* 2003) and endophytes associated with roots (mycorrhizae; see Herre *et al.*, this volume), throughout this chapter, we will focus primarily on the implications of recent studies of the endophytic fungi that live inside plant leaf tissue. These foliar endophytes are extremely diverse phylogenetically and have been documented in nearly all plants sampled (e.g. mosses, liverworts, ferns, conifers and angiosperms; Carroll 1988; Clay 1988; Petrini 1991; Schultess & Faeth 1998; Frohlich & Hyde 1999; Stone *et al.* 2000; Arnold *et al.* 2000; Arnold 2002; Arnold *et al.* 2003; Davis *et al.* 2003). Despite the growing recognition of their wide distribution across plant taxa, basic attributes of their biology are still poorly understood. Specifically, endophyte diversity, distributions, life cycles, interactions with hosts and other fungi, and their net chemical, physiological and ecological influences are only beginning to be appreciated and studied. This is particularly true in the extremely diverse tropics.

The best-studied endophytes are ascomycetes belonging to the family Clavicipitaceae. These fungi grow throughout the aboveground tissues of some temperate grass species (e.g. *Festuca arundinacea*, see Clay & Schardl 2002). Typically,

Biotic Interactions in the Tropics: Their Role in the Maintenance of Species Diversity, ed. D. F. R. P. Burslem,
M. A. Pinard and S. E. Hartley. Published by Cambridge University Press. © Cambridge University Press 2005.

in infected individuals, a single fungal genotype infects a single plant individual. In these grasses, endophytes appear to be predominantly asexual, and are transmitted vertically from maternal plants to their offspring via seeds (but see Schultess & Faeth 1998). The foliar endophytes that are associated with the grasses have often been shown to benefit their hosts through a variety of effects, including tolerance to heavy metals, increased drought resistance, reduced herbivory (due to chemicals, primarily alkaloids, produced by the endophytes), defence against pathogens, and enhanced growth and competitive ability (Carroll 1988; Clay *et al.* 1989; Ford & Kirkpatrick 1989; West *et al.* 1990; Gwinn & Gavin 1992; Welty *et al.* 1993; Saikkonen *et al.* 1998; Clay & Schardl 2002; but see Faeth 2002). These beneficial effects are consistent with much evolutionary theory that predicts that vertical transmission will tend to align symbiont interests with those of the host, and that this will tend to promote mutualistic interactions (reviewed in Herre *et al.* 1999).

However, the grass–claviciptaceous endophyte systems, characterized by vertical transmission of host-specific fungi that show low within-host fungal diversity, appear to present a special case that is not a useful general model for the vast majority of host–endophyte systems (Carroll 1988; Saikkonen *et al.* 1998; Faeth 2002). Specifically, most endophytes associated with foliage of woody plants (as well as most monocots; Schulthess & Faeth 1998; Frohlich & Hyde 1999) appear to be acquired from the environment. This sort of horizontal transmission is not generally considered to promote mutualistic interactions (Frank 1996; Herre *et al.* 1999), although many cases of horizontally transmitted mutualisms clearly exist (Herre 1999; Wilkinson 2001; Kiers *et al.* 2003; Arnold *et al.* 2003). Further, the community of endophytes even within a single leaf of a given host plant can be extremely diverse (Carroll 1988; Lodge *et al.* 1996; Bayman *et al.* 1998; Frohlich & Hyde 1999; Lebrón *et al.* 2001; Mejia *et al.* 2003; Arnold *et al.* 2003). Moreover, many endophytes of woody plants appear to be closely related to known pathogens (Carroll 1988; Freeman & Rodriguez 1993; Stone *et al.* 2000; Ortiz-Garcia *et al.* 2003). Therefore, it has been suggested that endophytes associated with leaves of woody angiosperms are unlikely to play protective or mutualistic roles in their host plants (Faeth 2002; Faeth & Fagan 2002).

Endophyte research in tropical areas has been primarily descriptive and restricted to dicot host plant species (e.g. Lodge *et al.* 1996; Bayman *et al.* 1998; Rajagopal *et al.* 2000; Gilbert *et al.* 2002b, Arnold & Herre 2003; Evans *et al.* 2003; Suryanarayanan *et al.* 2003). Recent work in Panama has demonstrated that tropical angiosperms host an extraordinary diversity of endophytes, which are horizontally transmitted and occur at very high densities in the leaf tissue of mature and old leaves (e.g. Arnold *et al.* 2000; Arnold *et al.* 2003). Moreover, this work has demonstrated that endophyte-free plants can be grown in the greenhouse and that single endophyte species or combinations can be experimentally reintroduced into plant tissue (Arnold & Herre 2003; Arnold *et al.* 2003; Mejia *et al.* 2003; see Table 9.1, Fig. 9.1). This technique allows for explicit experimental

Table 9.1 *Summary of major findings from studies of endophytic fungi (EF) in Panama*

Topic	Host plant species	Summary of findings
Diversity	*Theobroma cacao* (similar results in: *Ipomoea philomega* *I. squamata* *Merremia umbellata* *Witheringia solanacea*)	• A high diversity of endophytic fungi (EF) exists within a single host species. From 400 cm² leaf area ($n = 126$ leaves), 344 morphotaxa were isolated.[c] • Most EF morphotaxa are rare.[a–g] • Within a single leaf, the species diversity of EF increases after initial leaf flush and then decreases following leaf saturation.[e,f]
Transmission	*T. cacao* *Ipomoea philomega* *I. squamata* *Merremia umbellata* *Witheringia solanacea*	• EF transmission is horizontal: leaves are flushed free of EF, and EF are acquired from the habitat through time.[b,c,e,f,g] • Leaves appear to saturate in density of EF after 3–4 weeks.[e–g]
Spatial structure	*T. cacao*	• EF communities are very similar at small spatial scales (0–50 km).[c; but see a] • Similarity of EF communities declines dramatically with distances > 50 km.[c,e,f]
Host-affinity	*T. cacao* *Faramea occidentalis* *Heisteria concinna* *Ouratea lucens*	• EF exhibit differential host affinity.[a,c,g] • EF morphotaxa that dominate one host are markedly less common, rare or absent in other hosts.[a,b,f,g] • EF growth in vitro is affected by the inclusion of host-specific leaf extracts.[b,c] • EF that commonly occur in a given host tend to grow best in media with extracts of that host.[c,e,f]
Interactions	*T. cacao*	• In vitro interactions between two EF species range from indifference to overgrowth or active inhibition.[d,f] • EF species commonly found in a given host tend to dominate interaction trials with rare EF species when trials take place in media with extracts of the host.[d,e,f]
Experimental manipulation	*T. cacao*	• EF-free leaves can be experimentally produced and EF can be re-introduced into EF-free leaves.[b,c,d]
Pathogen resistance	*T. cacao*	• EF can enhance host defence against pathogens[c,d]

[a] Arnold *et al.* (2000)
[b] Arnold & Herre (2003)
[c] Arnold *et al.* (2003)
[d] Mejia *et al.* (2003)
[e] E. Rojas *et al.* (unpublished results)
[f] E. A. Herre *et al.* (unpublished results)
[g] S. A. van Bael *et al.* (unpublished results)

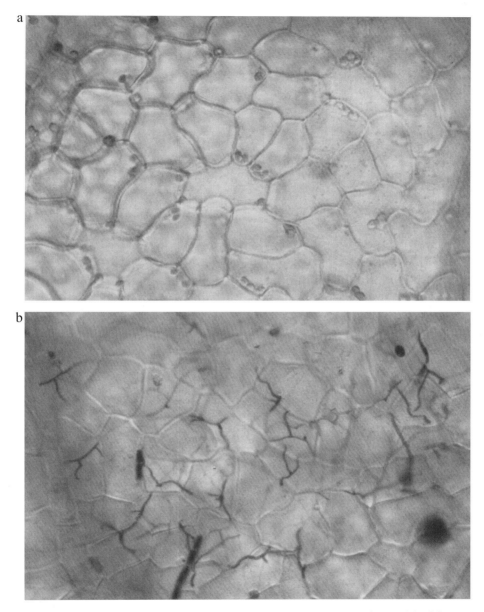

Figure 9.1 High magnification (400×) photograph of *Theobroma cacao* leaves: (a) without endophytic fungi, (b) with endophytic fungi (photos by L. C. Mejia).

comparisons of growth, physiology, defence, chemistry and genetic composition between plants (or their tissues) with and without endophytes. Such comparisons open a large number of possibilities that previously were only available in the study of grass–endophyte interactions. For example, experimental studies of plant–endophyte–herbivore interactions in these horizontally transmitted endophyte systems are now poised to complement the pioneering but primarily correlative work with adult oaks and other temperate trees (Carroll 1988; Wilson & Carroll 1994; Preszler *et al.* 1996; Wilson & Carroll 1997; Wilson & Faeth 2001; Faeth & Fagan 2002; Faeth & Hammon 1997).

In this chapter we discuss the implications of recent studies of the ecology of tropical foliar endophytes that use *Theobroma cacao* (Malvaceae) and other plant species as hosts in Panama (Arnold *et al.* 2000; Arnold & Herre 2003; Mejia *et al.* 2003; Arnold *et al.* 2003; findings are summarized in Table 9.1). Fundamentally, it is becoming increasingly clear that plants are effectively chimera: organisms composed of both plant and fungal tissues. As is the case with most examples in which simpler biological entities combine to form more complex associations (eukaryotes, multicellular organisms, social insects, etc.), the composite possesses emergent properties that neither party possesses separately (Maynard Smith & Szarthmary 1995; Margulis & Sagan 2002). Drawing on these and other studies, we will discuss some implications of the documented and potential influences of endophytic fungi for the study of tropical plant biology.

Life cycle and general natural history of endophytes

For many species that occur as endophytes, it appears that the portion of their life cycle that is involved with leaves begins as a taxonomically diverse assemblage of airborne spores that land on leaf surfaces (Carroll 1988). Generally, it appears that most leaves of tropical trees are flushed in a largely endophyte-free condition (Arnold 2002; Arnold & Herre 2003; Arnold *et al.* 2003; Mejia *et al.* 2003). After the wetting of the spore-laden leaf surfaces, a subset of spores germinate and are able to penetrate into the leaf tissue either through the stomata, or more directly through the cuticle (Arnold *et al.* 2003; Mejia *et al.* 2003). After penetration, a subset of the endophytes differentially proliferates within the leaf tissue (also see Deckert *et al.* 2001). During the lifetime of the leaf, there appear to be few, if any, recognizable symptoms of the presence of the endophytes. However, both tissue samples and microphotographs show that the plant tissue is indeed full of fungal hyphae (Fig. 9.1; Fig. 9.2; Arnold *et al.* 2000; Arnold *et al.* 2003; Mejia *et al.* 2003). Finally, many endophyte species appear to complete their life cycle (sporulate) on abscised leaves, effectively as saprophytes (see Fig. 9.2; J. Bischoff, L. C. Mejia and E. Rojas, personal observations).

The net effect from a plant's (or researcher's) perspective is that young leaf tissues (up to roughly 1 week following leaf flush) are relatively free of endophytic fungi. This is the time period during which secondary chemicals or DNA extracted from a leaf are most certain to be of exclusively plant origin (Arnold *et al.* 2003; Mejia *et al.* 2003, see below). During the first few weeks following leaf flush, the density of endophytic fungi in the leaf increases (as measured by the proportion of 2-mm^2 leaf punches that yield culturable endophytic fungi). At roughly 3–4 weeks the endophyte content of the leaf tissues appears to saturate (Arnold & Herre 2003; Arnold *et al.* 2003; Mejia *et al.* 2003). The trajectory of endophyte diversity within a leaf is less clear. However, it appears that diversity

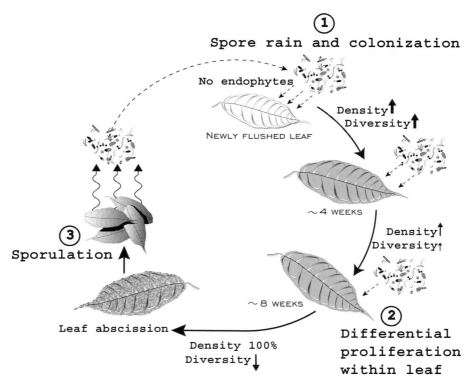

Figure 9.2 Inferred life cycle of many tropical foliar endophytic fungi based on observations made in *Theobroma cacao* (Arnold & Herre 2003; Arnold *et al.* 2003; Mejia *et al.* 2003; E. Rojas, J. Bischoff and A. E. Arnold, unpublished observations, see text; figure by D. A. Kyllo).

at first rises (as more of the leaf tissue becomes colonized by the diverse fungi that land on the leaf surface as spores, see Arnold *et al.* 2003). This phase (4–8 weeks following leaf flush) appears to be followed by a phase characterized by a reduction in overall diversity (as measured by the number of distinct morphospecies encountered per fungal isolate; E. Rojas *et al.* and E. A. Herre *et al.*, unpublished results). We interpret the observed decrease in diversity to reflect differential proliferation (or success in competition) of a subset of the colonizing fungi (see Arnold *et al.* 2003). In any event, whatever effects the endophytic fungi have on the host leaf are likely to become more pronounced with leaf age, as the leaf passes from an endophyte-free to an endophyte-saturated state (see below).

Effects on host-plant defence

By experimentally manipulating the presence or absence of foliar endophytic fungi among leaves within individual *Theobroma cacao* seedlings, Arnold *et al.* (2003) demonstrated that the presence of endophytes substantially reduced leaf loss and damage due to an oomycete pathogen, *Phytophthora* sp., commonly isolated from necrotic leaves. The endophytic fungi that were introduced into the endophyte-plus (E+) leaves had previously been isolated from healthy *T. cacao* leaves (Arnold *et al.* 2003; Mejia *et al.* 2003). Because E− (endophyte-free) and E+ leaves could be produced and compared within individual plants, it is possible to conclude that the benefit to the host of having endophytes was localized. Importantly, the fungal isolates used for inoculation were selected on the basis of a combination of extensive survey work that showed them to occur commonly in the leaves of healthy hosts, and in vitro experiments that showed them to possess anti-*Phytophthora* activity (Mejia *et al.* 2003; L. C. Mejia *et al.*, unpublished results). Given the observed in vitro activity of the inoculated endophytes against *Phytophthora* (Mejia *et al.* 2003), we suspect that direct inhibition of pathogens by resident endophytes contributes to the observed enhancement of host defence (Arnold *et al.* 2003; Mejia *et al.* 2003; E. A. Herre unpublished results).

That the endophytes would actively inhibit pathogens from invading and/or proliferating within the leaf tissues is also consistent with the observation that in the vast majority of cases, endophytes appear not to harm their host plant, and not to reproduce (sporulate) until after the host leaf has been abscised (J. Bischoff, E. Rojas, L. C. Mejia and A. E. Arnold, personal observation). We suggest that much of the observed host defence results from endophytes essentially 'guarding their turf' from potential usurpers. What we interpret as 'turf guarding' with respect to pathogens (and other endophytes) should also be expected to occur with respect to herbivores, as has been observed in the endophytes associated with grasses (Saikkonen *et al.* 1998; Clay & Schardl 2002). As is clearly true in the case of vertically transmitted endophytes in some grasses, it is in the interest of both the plant and endophyte for leaf tissues not to be lost to herbivores or pathogens. It is also certainly in the interest of the endophyte not to be displaced by other endophytes. Therefore, we can expect that the enhancement of host defences observed in *T. cacao* is not going to present an isolated case, and that endophytes will generally contribute a great deal that has previously been unappreciated to the defence of their hosts against pathogens and herbivores (see below).

Effects on host physiology, growth and the costs of endophytes

Endophytic fungi are obligately heterotrophic organisms. Further, direct observations (see Fig. 9.1) demonstrate abundant endophyte tissue within host leaves (Lodge *et al.* 1996; Bayman *et al.* 1998; Arnold *et al.* 2003). Therefore, we expect that endophytes derive their nutrition largely from their host plant, and that they should present at least a modest drain on host-plant resources. From the perspective of the defensive benefit they have been shown to provide, this hypothetical drain would constitute at least one principal 'cost' to the host plant. Obtaining accurate measurements of the 'cost' is a priority from a number of perspectives. Our efforts to measure such a cost have involved the comparison of growth rates and physiological parameters (see below) of E− and E+ plants. Current estimates suggest that the effects of endophytes on host growth and biomass accumulation are small, at least over a period of 23 weeks under greenhouse conditions.

However, a series of experiments comparing E+ and E− plants showed that maximum photosynthetic assimilation (A_{max}), stomatal conductance (g) and water-use efficiency were all significantly reduced in E+ seedlings relative to E− seedlings of *T. cacao* (R. A. Cordero *et al.*, unpublished results). We tentatively interpret these findings as reflecting a net drag on water movement through leaves due to the presence of the endophytes throughout the intercellular spaces (see Fig. 9.1). Given these decreases in water movement and A_{max}, we suspect that over longer periods than we have thus far examined, the effect of endophytes on reducing host biomass accumulation will become more clearly pronounced. Nonetheless, it is clear that foliar endophytes, like mycorrhizae, have the capacity to alter host physiological properties. A future research priority is to determine the degree to which different endophytic species induce different physiological responses in their hosts.

Effects on host chemistry and genetics

It can be taken as given that fungi are chemically distinct from plants. Therefore, it should not be surprising that E+ plant tissues have been found to have different chemical profiles from E− plant tissues (Bacon *et al.* 1977; Weber 1991; Petrini *et al.* 1992; Saikkonen *et al.* 1998; Yue *et al.* 2001; L. C. Mejia *et al.*, unpublished results). Further, in vitro studies give clear evidence that some endophyte species commonly encountered in *T. cacao* exude substances that inhibit the growth of other fungal species in the absence of physical contact between colonies (Mejia *et al.* 2003). Combined with the observation that endophytes in grasses are known to produce a suite of chemicals that deter herbivory by large grazing mammals, as well as insect herbivores (Clay & Schardl 2002), it is reasonable to expect that endophytes will contribute to the chemical mix extracted from a 'plant'. It follows that many chemicals that have previously been attributed to plants may actually be produced by the endophytes within them. This probability has a series of implications for studies of the role of 'plant' chemistry in host-plant defence, and in drug discovery programs.

Moreover, there are now several known instances in which researchers who have extracted DNA from a 'plant' have in fact isolated and amplified fungal DNA (Camacho *et al.* 1997; Chiang *et al.* 2001). Researchers conducting genetic studies of plants, particularly those studies using techniques that are not known to be specific for the host plant species (e.g. Rapids, AFLP, etc.), need to be conscious of this fact when collecting and interpreting their data (Chiang *et al.* 2003; C. Woodward *et al.*, unpublished results).

It is no longer a question of *whether* the endophytic fungi imbedded within host plant tissues (in leaves, in stems, or in roots) affect many properties that

Table 9.2 *Some observations and implications of the presence of endophytic fungi for the study of defence, physiology, and chemical and genetic composition of host plants*

Plant defences	• EF in *Theobroma cacao* can enhance host defence against pathogens (Arnold *et al.* 2003) • EF in *Festuca* sp. and other species also can enhance host defence against pathogens and herbivores (Carroll 1988; Saikkonen *et al.* 1998; Clay & Schardl 2002) ➤ We can expect that some portion of host-plant defences is actually due to endophytes, as opposed to being intrinsic to the host plant per se
Physiology	• EF in *Festuca* sp. and other species can increase drought and heavy-metal tolerance (Saikkonen *et al.* 1998; Clay & Schardl 2002) • EF in *Theobroma cacao* can affect levels of photosynthesis and hydraulic properties (Cordero *et al.*, unpublished results) • Mycorrhizae affect many physiological attributes of host plants ➤ We can expect that foliar endophytic fungi will influence at least some aspects of plant physiology
Chemical and genetic composition of leaf extracts	• The comparison of EF− and EF+ leaves shows that extracted secondary chemicals and genetic material can have an endophytic origin (Bacon *et al.* 1977; Weber 1991; Petrini *et al.* 1992; Saikkonen *et al.* 1998; Yue *et al.* 2001; Mejia *et al.*, unpublished results) ➤ Studies of plant chemistry and genetics must be designed with the likelihood of endophytic contribution/contamination in mind

researchers have long attributed to the plant (defence, growth, physiology, chemistry and genetic content, Table 9.2). Rather, the more appropriate questions are the degree to which those 'plant' properties are due to endophytic fungi, and the degree to which the identities of the endophytic fungi in a given host plant influence them. If the fungal effects are generally small, then viewing plants as 'just plants' is perfectly adequate. However, if the fungal effects on their hosts turn out to be large (as appears to be the case with plant defence in *T. cacao*; see Arnold *et al.* (2003), Mejia et al. (2003)), then much of how we go about studying and interpreting many seemingly familiar 'plant' characteristics may need to be reconsidered.

Acknowledgements

We thank two anonymous reviewers and David Burslem for constructive suggestions. We thank Greg Gilbert, Tom Gianfagna and Prakash Hebbar for essential technical advice and training. We thank the Smithsonian Institution, the Andrew W. Mellon Foundation, National Science Foundation (DEB 9902346 to L. McDade and A.E.A.), the American Cacao Research Institute, the World Cacao Foundation and the John Clapperton Fellowship of Mars Incorporated for financial support. Finally, we thank the Smithsonian Tropical Research Institute for providing the stimulating intellectual environment, infrastructure and logistical support that made this work possible.

References

Arnold, A. E. (2002) Neotropical fungal endophytes: diversity and ecology. Unpublished Ph.D. thesis, University of Arizona.

Arnold, A. E. & E. A. Herre (2003) Canopy cover and leaf age affect colonization by tropical fungal endophytes: ecological pattern and process in *Theobroma cacao* (Malvaceae). *Mycologia* **95**:388–398.

Arnold, A. E., Z. Maynard, G. S. Gilbert, P. D. Coley & T. A. Kursar (2000) Are tropical fungal endophytes hyperdiverse? *Ecology Letters* 3:267–274.

Arnold, A. E., L. C. Mejia, D. Kyllo *et al.* (2003) Fungal endophytes limit pathogen damage in a tropical tree. *Proceedings of the National Academy of Science* **100**:15649–15654.

Bacon, C. W., Porter J. K., Robbins, J. D. & Luttrell, E. S. (1977) *Epichloe typhina* from toxic tall fescue grasses. *Applied Environmental Microbiology* **34**:576–581.

Bayman, P., P. Angulo-Sandoval, Z. Baez-Ortiz & D. J. Lodge (1998) Distribution and dispersal of *Xylaria* endophytes in two tree species in Puerto Rico. *Mycological Research* **102**:944–948.

Camacho, F. J., D. S. Gernandt, A. Liston, J. K. Stone & A. S. Klein (1997) Endophytic fungal DNA, the source of contamination in spruce needle DNA. *Molecular Ecology* **6**:983–987.

Carroll, G. (1988) Fungal endophytes in stems and leaves: from latent pathogen to mutualistic symbiont. *Ecology* **69**:2–9.

Chiang, Y. C., C. H. Chou, P. R. Lee & T. Y. Chiang (2001) Detection of leaf associated fungi based on PCR and nucleotide sequence of the ribosomal internal transcribed spacer (ITS) in *Miscanthus*. *Botanical Bulletin of Academia Sinica* **42**:39–44.

Chiang, Y. C., C. H. Chou, S. Huang & T. Y. Chiang (2003) Possible consequences of fungal contamination on the RAPD fingerprinting in *Miscanthus* (Poaceae). *Australian Journal of Botany* **51**:197–201.

Clay, K. (1988) Fungal endophytes of grasses: a defensive mutualism between plants and fungi. *Ecology* **69**:10–16.

Clay, K. & C. Schardl (2002) Evolutionary origins and ecological consequences of endophyte symbiosis with grasses. *American Naturalist* **160**:S99–S127.

Clay, K., G. P. Cheplick & S. Marks (1989) Impact of the fungus *Balansia henningsiana* on *Panicum agrostoides*: frequency of infection, plant growth and reproduction, and resistance to pests. *Oecologia* **80**:374–380.

Davis, E. C., J. B. Franklin, A. J. Shaw & R. Vilgalys (2003) Endophytic *Xylaria* (Xylariaceae) among liverworts and angiosperms: phylogenetics, distribution, and symbiosis. *American Journal of Botany* **90**:1661–1667.

Deckert, R. J., L. H. Melville & R. L. Peterson (2001) Structural features of a Lophodermium endophyte during the cryptic life-cycle phase in the foliage of *Pinus strobes*. *Mycological Research* **105**:991–997.

Evans, H. C., K. A. Holmes & S. E. Thomas (2003) Endophytes and mycoparasites associated with an indigenous forest tree, *Theobroma gileri*, in Ecuador and a preliminary assessment of their potential as biocontrol agents of cacao diseases. *Mycological Progress* **2**:149–160.

Faeth, S. H. (2002) Are endophytic fungi defensive plant mutualists? *Oikos* **98**:25–36.

Faeth, S. H. & W. F. Fagan (2002) Fungal endophytes: common host plant symbionts but uncommon mutalists. *Integrative and Comparative Biology* **42**:360–368.

Faeth, S. H. & K. E. Hammon (1997) Fungal endophytes in oak trees: experimental analyses of interactions with leafminers. *Ecology* **78**:820–827.

Ford, V. L. & T. L. Kirkpatrick. (1989). Effects of *Acremonium coenophialum* in tall fescue on host disease and insect resistance and allelopathy to *Pinus taeda* seedlings. *Proceedings of the Arkansas Fescue Toxicosis Conference* **140**:29–34.

Frank, S. A. (1996) Host–symbiont conflict over the mixing of symbiotic lineages. *Proceedings of the Royal Society of London B* **263**:339–344.

Freeman, S. & R. J. Rodriguez (1993) Genetic conversion of a fungal plant pathogen to a nonpathogenic, endophytic mutualist. *Science* **260**:75–78.

Frohlich, J. & K. D. Hyde (1999) Biodiversity of palm fungi in the tropics: are global fungal diversity estimates realistic? *Biodiversity and Conservation* **8**:977–1004.

Gilbert, G. S., Mejía-Chang, M. & Rojas, E. (2002) Fungal diversity and plant disease in mangrove forests: salt excretion as a possible defense mechanism. *Oecologia* **132**:278–285.

Gwinn, K. D. & A. M. Gavin. (1992) Relationship between endophyte infection level of tall fescue seed lots and *Rhizoctonia zeae* seedling disease. *Plant Disease* **76**:911–914.

Herre, E. A. (1999) Laws governing species interactions? Encouragement and caution from figs and their associates. In L. Keller, ed. *Levels of Selection in Evolution*. Princeton: Princeton University Press, pp. 209–237.

Herre, E. A., N. Knowlton, U. G. Mueller & S. A. Rehner (1999) The evolution of mutualisms: exploring the paths between conflict and cooperation. *Trends in Ecology & Evolution* **14**:49–53.

Kiers, E. T., R. A. Rousseau, S. A. West & R. F. Denison (2003) Host sanctions and the legume-rhizobium mutualism. *Nature* **425**: 78–81.

Lebrón, L., D. J. Lodge, S. Laureano & P. Bayman (2001) Where is the gate to the party? *Phytopathology* **91**:116.

Lodge, D. J., P. J. Fisher & B. C. Sutton (1996) Endophytic fungi of *Manilkara bidentata* leaves in Puerto Rico. *Mycologia* **88**:733–738.

Margulis, L. & D. Sagan (2002) Acquiring genomes: a theory of the origins of species. New York: Basic Press.

Maynard Smith, J. & E. Szarthmary (1995) *The Major Transitions in Evolution*. New York: Freeman.

Mejia, L. C., E. Rojas, Z. Maynard *et al.* (2003) Inoculation of beneficial endophytic fungi into *Theobroma cacao* tissues. In *Proceedings of the 14th International Cocoa Research Conference*, Accra-Ghana.

Ortiz-Garcia, S., D. S. Gernandt, J. K. Stone & P. R. Johnson (2003) *Mycologia* **95**: 846–859.

Petrini, O. (1991) Fungal endophytes of tree leaves. In J. H. Andrews & S. S. Hirano, eds. *Microbial Ecology of Leaves*. New York: Springer-Verlag, pp. 179–197.

Petrini, O., T. N. Sieber, L. Toti & O. Viret (1992) Ecology, metabolite production and substrate utilization in endophytic fungi. *Natural Toxins* **1**:185–196.

Preszler, R. W., E. S. Gaylord & W. J. Boecklen (1996) Reduced parasitism of a leaf-mining moth on trees with high infection frequencies of an endophytic fungus. *Oecologia* **108**:159–166.

Rajagopal, K. & T. S. Suryanarayanan (2000) Isolation of endophytic fungi from leaves of neem (*Azadirachta indica* A. Juss.). *Current Science* **78**:1375–1378.

Saikkonen, K., S. H. Faeth, M. Helander & T. J. Sullivan (1998) Fungal endophytes: a continuum of interactions with host plants. *Annual Review of Ecology and Systematics* **29**:319–343.

Schulthess, F. M. & S. H. Faeth (1998) Distribution, abundances, and associations

of the endophytic fungal community of
Arizona fescue (*Festuca arizonica*). *Mycclogia*
90:569–578.

Stone, J. K., C. W. Bacon & J. R. White (2000) An
overview of endophytic microbes:
endophytism defined. In C. W. Bacon & J. F.
White, eds. *Microbial Endophytes*. New York:
Marcel Dekker, pp. 3–29.

Suryanarayanan, T. S., G. Venkatesan & T. S.
Murali (2003) Endophytic fungal
communities in leaves of tropical forest
trees: diversity and distribution patterns.
Current Science **85**:489–493.

Webber, J. (1981). A natural biological control of
Dutch elm disease. *Nature* **292**:449–450.

Welty, R. E., R. E. Barker & M. D. Azevedo.
(1993). Response of field-grown tall fescue
infected by *Acremonium coenophialum* to
Puccinia graminis ssp. *graminicola*. *Plant
Disease* **77**:574–575.

West, C. P., E. Izekor, R. T. Robbins, R. Gergerich
& T. Mahmood (1990). *Acremonium
coenophialum* effects on infestations of
barley yellow dwarf virus and soil-borne
nematodes and insects in tall fescue. In

S. S. Quisenberry & R. E. Joost, eds.
*Proceedings of the International Symposium of
Acremonium/ Grass Interactions*. Baton Rouge:
Louisiana Agricultural Experiment Station,
pp. 196– 198.

Wilson, D. (1995) Endophyte: the evolution of a
term, and clarification of its use and
definition. *Oikos* **73**:274–276.

Wilson, D. & G. C. Carroll (1994) Infection
studies of *Discula quercina*, an endophyte of
Quercus garryana. *Mycologia* **86**:635–647.

(1997) Avoidance of high-endophyte space by
gall-forming insects. *Ecology* **78**:2153–2163.

Wilson, D. & S. H. Faeth (2001) Do fungal
endophytes result in selection for leafminer
ovipositional preference? *Ecology*
82:1097–1111.

Wilkinson, D. M. (2001) Horizontally acquired
mutualisms, an unsolved problem in
ecology? *Oikos* **92**: 377–384.

Yue, Q., C. Wang, T. J. Gianfagna & W. A. Meyer
(2001) Volatile compounds of
endophyte-free and infected tall fescue
(*Festuca araundinacea* Schreb.) *Phytochemistry*
58: 935–941.

PART III

Plant–animal interactions

CHAPTER TEN

Implications of plant spatial distribution for pollination and seed production

JABOURY GHAZOUL

Imperial College London

Introduction

The production of offspring has long been thought to be predominantly limited by the availability of resources rather than mating opportunities, a view based on 'Bateman's principle' espoused by Bateman in 1948. Applied to vascular plants, this principle predicts that fruit production is limited by maternal resources rather than pollen transfer. However, in the past 20 years limitation of seed production by pollination has been reported in numerous studies, and reviews suggest that more than 50% of plants studied show increased fruit production following experimental pollen supplementation (Burd 1994). Of course, short-term pollen supplementation studies fail to capture the lifetime success of the whole plant, leading some critics to maintain that lifetime reproductive output remains resource-limited. Nevertheless, population-wide declines in reproductive success, at least in the short term, owing to reduced pollen availability or ineffectual pollination have now been recorded among a range of plant species and geographic locations.

The cause of the declining efficiency in pollination has often been traced to changes in the spatial distribution of plants in the population, which in turn has affected the abundance of pollinators or led to changes in their foraging behaviour. Plant spatial distribution has become increasingly relevant following rapid changes in land use and landscape structure driven by anthropogenic activities. Logging and land clearance for agriculture and development has caused degradation of forest habitats either through the partial removal of economically important species, or by wholesale clearance and effective fragmentation and isolation of remnant forest patches. In temperate regions heathlands and native grasslands have also been greatly fragmented. Thus, many plant populations have suffered a decline in density, effectively increasing distances between nearest neighbours, or subdivision of a once-continuous population into a series of relatively isolated subpopulations, and sometimes both.

Biotic Interactions in the Tropics: Their Role in the Maintenance of Species Diversity, ed. D. F. R. P. Burslem, M. A. Pinard and S. E. Hartley. Published by Cambridge University Press. © Cambridge University Press 2005.

The purpose of this chapter is to review how changes in the spatial distribution of plants affect their reproductive output. The basic premise is that as individuals within a population become increasingly isolated, reproductive processes such as pollination and seed dispersal function less efficiently. This may further lead to Allee effects whereby declining population density or abundance results in a disproportionate decline in reproductive output and population viability. The relevance of this to conservation is obvious – many isolated or depauperate plant populations, particularly in tropical regions, may not be ecologically viable, particularly when threats are compounded by invasive species or changing climate.

The extent of pollination limitation

How common is pollination limitation of fruit production among flowering plants? This question has central importance if spacing mechanisms acting through pollination are to affect fruit production. There remains great uncertainty about the relative importance of resource versus pollen limitation of fruit production among plants, and very few studies have examined both these factors simultaneously (Cunningham 2000). Bawa and Beach (1981) treated pollination as insignificant in limiting seed set and emphasized maternal resources instead. Yet Bawa himself reported elevated seed set following hand pollination in 26 of 34 trees examined in Costa Rica (Bawa 1974). Since then there has been an abundance of studies that appear to demonstrate at least some pollination limitation (see reviews by Bierzychudek 1981; Burd 1994; Larson & Barrett 2000). More recently, pollen limitation of seed set has been recorded from plants in arctic (Alatalo & Molau 2001; Elberling 2001); temperate (Ehrlen *et al.* 2002; Goodwillie 2001; Moody-Weis & Heywood 2001) and tropical regions (Aizen 2001; Ghazoul *et al.* 1998).

It is difficult to assess the relative frequency of pollen limitation and its distribution across taxa and life histories as negative results are less likely to be reported (but see Abe 2001; Albre *et al.* 2003; Bawa & Webb 1984; Whelan & Goldingay 1986; Zimmerman & Pyke 1988). However, Burd's (1994) survey of 258 species of flowering plant indicated that 62% were pollen-limited at least in some locations or times, and a comparative study of pollen-limited plants concluded that woody or tropical plants are more likely to be pollinator-limited than herbaceous or temperate ones, but only among self-incompatible species (Larson & Barrett 2000). Self-compatibility was not associated with pollen limitation presumably because a capacity for self-fertilization reduces dependence on pollen transfer by pollinators. Larson and Barrett (2000) suggest that herbaceous plants are less likely to be pollen-limited on account of the size of their floral displays relative to woody plants – although large displays attract more pollinators, the visitation rate per flower may be lower in woody plants. This may be correct, but it remains difficult to determine whether fruit set is pollen- or resource-limited in such circumstances. Similarly, the mechanism responsible for pollen-limitation differences among tropical and temperate species is uncertain, although tropical species tend to occur at lower densities and may have greater dependence on specialized pollinators than temperate species – but see Waser *et al.* (1996).

It now seems that pollen limitation is just as widespread among taxa, life forms and geographic localities as resource limitation, and indeed many species may be both pollen- and resource-limited across populations or times (Copland & Whelan 1989). Generally, self-compatible species are unlikely to be pollen-limited regardless of other life-history characteristics. Nevertheless, even self-pollinating plants can be pollen-limited if the effectiveness of self-pollination is reduced by lower pollinator activity or if there is a genetic load associated with selfing. As most tropical trees are self-incompatible (Bawa 1974; Bawa *et al.* 1985) they are predisposed to being pollen-limited, and it appears that many are. However, there is much uncertainty as to the actual occurrence of pollen limitation and whether observed occurrences reflect lifetime fitness or are short-lived. As most studies conducted on trees observe only a fraction of lifetime fitness the following discussion proceeds with this caveat in mind.

Plant spatial distribution and pollination

Pollinators and pollination may be affected to a greater or lesser extent by changes in the spatial distribution and density of plants that make up their habitat. Such changes have potential impacts on the effectiveness of the plant–pollinator interaction including reproductive failure (Bawa 1990; Jennersten 1988), reduced gene flow and increased selfing (Aizen & Feinsinger 1994a; Bawa 1990). In linking plant spatial distribution to reproductive ecology, studies have addressed the size, density and fragmentation of plant populations. It is important not to confound these population descriptors as although population size and density are frequently positively correlated (van Treuren *et al.* 1993) the mechanisms through which they affect reproduction are likely to differ. Indeed, there is evidence from both empirical and observational studies that the importance of plant size, density and fragmentation for pollination and seed set differs (Kunin 1997). In general, however, all these factors have been implicated in the depression of plant reproductive output, and a compilation of results from studies on 57 species shows only six that appear neutral in their response to increasing isolation or reduced population size (Table 10.1). While it is widely recognized that non-significant results are less likely to be published and reported, it appears that population size and spacing is an important factor in the determination of reproductive output for many plant species, ranging from annual herbs to large tropical trees.

Table 10.1 *A summary of impacts of plant population size, spacing and fragmentation on aspects of their reproductive ecology*

BS refers to breeding system as self-compatible (S) or self-incompatible (I). E/O refers to the type of study, as either experimental (E) or observational (O). Pollinators are listed when known.

Species	Family	BS	Life form	Cause	Response	Pollinator	E/O	Reference
Astragalus canadensis	Leguminosae	I	shrub/herb	density	fruit set	bees	E	Platt *et al.* (1974)
Diplotaxis erucoides	Brassicaceae	I	herb	density	fruit set; seed set per fruit; seed set per plant	small bees and flies	O	Kunin (1992)
Agalinis strictifolia	Scrophulariaceae	I	parasitic herb	density	no effect	honeybees and bumblebees	O	Dieringer (1992)
Silene dioica	Caryophyllaceae	S	herb	density	no effect	sphingid and noctuid moths	O	Pettersson (1997)
Echium vulgare	Boraginaceae	S	herb	density	no effect (pollinator visits)	bees	O	Klinkhamer & de Jong (1990)
Senecio integerrimus	Asteraceae	S	herb	density	no effect (pollinator visits, fruit set, seeds per plant)	bees and butterflies	E	Schmitt (1983)
Ceiba pentandra	Bombacaceae	I	tree	density	outcrossing	bats	O	Murawski & Hamrick (1992)
Cavanillesia platanifolia	Bombacaceae	S	tree	density	outcrossing	hawkmoths, hummingbirds, bees	O	Murawski *et al.* (1990)
Symphonia globulifera	Guttiferae	I	tree	density	outcrossing	hummingbirds	O	Aldrich & Hamrick (1998)
Senecio crassulus	Asteraceae	I	herb	density	pollen grains	bees	O	Thomson (1981)
Palicouria sp.	Rubiaceae	I	herb	density	pollen grains and seeds per fruit	hummingbirds	E	Feinsinger *et al.* (1991)
Besleria trifolia	Gesneriaceae	S	shrub/small tree	density	pollen grains; no effect on seeds per fruit	hummingbirds	E	Feinsinger *et al.* (1991)
Shorea siamensis	Dipterocarpaceae	PI	tree	density	pollination, seed set	small bees	O	Ghazoul *et al.* (1998)
Cynoglossum officinale	Boraginaceae	I	herb	density	pollinator visitation	bees	O	Klinkhamer *et al.* (1989)

Species	Family		Growth form	Pollinator	Variable	Response		Reference
Senecio jacobaea	Asteraceae	S	herb	bees and flies	density	pollinator visitation; no effect on fruit set or seeds per plant	E	Kunin (1997)
Neolitsia dealbata	Lauraceae	I	tree	small generalist insects	density	pollinator visitation; pollen dispersal; fruit set	O	House (1993)
Potentilla gracilis	Rosaceae		herb	bees and flies	density	pollinator visits	O	Thomson (1981)
Brassica kaber	Cruciferae	I	herb	honeybees and syrphids	density	pollinator visits and pollination quality	E	Kunin (1993)
Dipterocarpus obtusifolius	Dipterocarpaceae	I	tree	moths and butterflies	density	pollinator visits and pollination quality	O	Ghazoul (2002)
Taxus canadensis	Taxaceae	S	coniferous shrub	wind	density	seed set	O	Allison (1990)
Lesquerella fendleri	Brassicaceae	I	herb	small bees	density	seed set	O	Roll et al. (1997)
Cassia biflora	Caesalpiniaceae		shrub	bees	density	seed set	O	Silander (1978)
Salvia pratensis	Labiatae	S	herb	bumblebees	density and population size	outcrossing; no effect of population size on outcrossing	O and E	van Treuren et al. (1993)
Senecio integrifolius	Asteraceae		herb		flowering stems per plant	seed set	O	Widen (1993)
Paquira quinata	Bombacaceae		tree	bats	forest disturbance	outcrossing	O	Fuchs et al. (2003)
Pithecellobium elegans	Mimosoideae	I	tree		fragmentation	genetic diversity; fruit set	O	Hall et al. (1996)
Dinizia excelsa	Fabaceae	S	tree	exotic honeybees	fragmentation	increased selfing; larger genetic neighbourhoods	O	Dick et al. (2003)
Dianthus deltoides	Caryophyllaceae		herb	butterflies	fragmentation	lower seed set	O	Jennersten (1988)
Calyptrogyne ghiesbreghtiana	Palmaceae		palm	bats	fragmentation	pollen load	O	Cunningham (1996)
Celia grandiflora	Bombacaceae	I	tree	bats	fragmentation	pollination, fruit set	O	Quesada et al. (2003)
Acacia brachybotrya	Mimosoideae		tree		fragmentation	pollination, fruit set	O	Cunningham (2000)

(cont.)

Table 10.1 (cont.)

Species	Family	BS	Life form	Cause	Response	Pollinator	E/O	Reference
Eremophila glabra	Myoporaceae		tree	fragmentation	pollination, fruit set	bumblebees	O	Cunningham (2000)
Betonica officinalis	Labiatae		herb	fragmentation	visitation frequency; increased visitation time per patch; outcrossing		E	Goverde et al. (2002)
Lupinus sulphureus	Fabaceae	S	herb	fragmentation, density, population size	fruit set, inbreeding depression		O	Severns (2003)
Ipomoea purpurea	Polemoniaceae		herb	frequency dependence	outcrossing rate	bumblebees	O	Epperson & Clegg (1987)
Phlox pilosa	Polemoniaceae	I	herb	frequency dependence	seed set	butterflies		Levin (1972)
Delphinium nuttallianum	Ranunculaceae	I	herb	isolation	pollination, seed set	bumblebees and hummingbirds	E	Schulke & Waser (2001)
Anacardium excelsum	Anacardiaceae	I	tree	patch size	pollination, seed set	small bees, flies	O	Ghazoul & McLeish (2001)
Clarkia concinna	Onagraceae	S	herb	patch size and density	pollination, seed set	bees, butterflies and flies	E	Groom (1998)
Samanea saman	Leguminosae	I	tree	patch size and isolation	outcrossing, seedling vigour	sphingid moths	O	Cascante et al. (2002)
Nepeta cataria	Labiatae	S	herb	patch size and isolation	visitation rate	bees	O	Sih & Baltus (1987)
Brassica kaber	Cruciferae	I	herb	population size	no effect	bees and flies	E	Kunin (1997)
Leucochrysum albicans	Asteraceae	I	herb	population size	no effect		O	Costin et al. (2001)
Silene regia	Caryophyllaceae		herb	population size	outcrossing	hummingbirds	O	Menges (1991)
Phlox pilosa	Polemoniaceae	I	herb	population size	pollen deposition on stigmas	butterflies	O	Hendrix & Kyhl (2000)

Species	Family		Growth form	Predictor	Response	Pollinators	Status	Reference
Calypso bulbosa	Orchidaceae	I	herb	population size	pollen export negatively correlated with population size	bumblebees	O	Alexandersson & Ågren (1996)
Stellaria pubera	Caryophyllaceae	S	herb	population size	pollination success	bees and flies	E	Campbell (1985)
Epipactis helleborine	Orchidaceae	I	herb	population size	pollinia removal (pollinator visitation)	vespid wasps	O	Ehlers et al. (2002)
Orchis spitzelii	Orchidaceae	I	herb	population size	pollinia removal (pollinator visitation)	bumblebees	O	Fritz & Nilsson (1994)
Orchis palustris	Orchidaceae	I	herb	population size	pollinia removal (pollinator visitation)	bumblebees	O	Fritz & Nilsson (1994)
Anacamptis pyramidalis	Orchidaceae	I	herb	population size	pollinia removal (pollinator visitation)	butterflies	O	Fritz & Nilsson (1994)
Lythrum salicaria	Lythraceae	I	herb	population size	seed production per flower and per plant	bumblebees	NE	Ågren (1996)
Rutidosis leptorrhynchoides	Asteraceae	I	herb	population size	seed set	beetles, flies, moths	O	Morgan (1999)
Succisa pratensis	Dipsacaceae	S	herb	population size	seed set; seed viability	insects	O	Vergeer et al. (2003)
Arnica montana	Asteraceae	I	herb	population size	seed set; seedling size; number of flowering stems; total relative fitness	syrphid flies	O	Luijten et al. (2000)
Banksia goodii	Proteaceae	I	shrub	population size	total seed production of the population	honey-possums and birds	O	Lamont et al. (1993)
Primula elatior	Primulaceae	I	herb	population size and isolation	seed set; high fruit abortion at high density	bumblebees and flies	O	van Rossum (2002)

Plant density

Models of pollinator foraging and empirical field studies indicate that reproductive success in self-incompatible plants is likely to be strongly correlated
with local population density (Feinsinger et al. 1991; Kunin 1993, 1997). These
empirical studies are supported by field observations of reproductive success of
a variety of shrubs and trees each pollinated by agents as diverse as bats, bees,
flies, hummingbirds and wind (Allison 1990; Feinsinger et al. 1986; Ghazoul et
al. 1998; Heithaus et al. 1982; Kunin 1992; Platt et al. 1974). Plants growing at
low density may experience reproductive failure or decline owing to difficulties
in attracting pollinators from competing plants that occur at higher densities
(Ghazoul 2002; Kunin 1997). Pollinators are less likely to show floral constancy
(the propensity to visit the same type of flower as last visited irrespective of alternatives) on sparsely distributed patches, becoming effectively more generalist in
their selection of flowers (Kunin 1993). Pollinators may also visit more flowers
per plant at low plant densities (Bosch & Waser 1999; Ghazoul et al. 1998) or
within isolated patches (Schulke & Waser 2001), which is likely to increase self-
pollination.

For outcrossed pollen-limited species, seed set is a function of pollen transfer
between individuals, mediated through pollinator activity, and many pollinators (Ghazoul et al. 1998; Roll et al. 1997) spend proportionately less time moving between relatively isolated plants. Thus, increased spacing among flowering
conspecifics, be it through harvesting or habitat fragmentation, can reduce seed
set through lower pollinator visitation or pollen quality. The spatial scale over
which distance (and density) effects on reproductive output have been observed
can be as little as one metre, as observed in the desert mustard Lesquerella fendleri
pollinated by small bees (Roll et al. 1997).

Some authors have reported no reproductive disadvantage in spatial isolation
for some species (Dieringer 1992; Pettersson 1997). Relatively isolated Silene uniflora, pollinated by sphingid and noctuid moths, produced as many fruits and
seeds per flower as individuals growing in clumped groups (Pettersson 1997).
However, distances between plants were not recorded and isolation was assessed
instead by the number of neighbours within a 1-m radius of each plant. Thus
'isolated' plants may have been little more than a metre from their nearest
neighbours and well within foraging distance of the plants' pollinators. Similarly, Dieringer (1992) reported no relationship between density and seed production of Agalini strictifolia. In this case 'low'-density plots contained about five
plants per square metre, which again is well within the foraging capabilities of
the bumblebee and honeybee pollinators.

Reproductive decline associated with isolation might be expressed through
reduction in genetic variability and vigour of progeny rather than seed set.
Outcrossing has been reported to be a function of spatial isolation from other
flowering individuals (Murawski et al. 1990). For example, Cavanillesia platanifolia

trees belonging to clusters of flowering individuals are predominantly out-crossed relative to more isolated individuals (Murawski *et al.* 1990). Partially self-incompatible species may be particularly vulnerable to increasing genetic load (Ågren 1996; Aizen & Feinsinger 1994a; Kunin 1997). Isolated *Samanea saman* in Costa Rica showed only slightly depressed reproductive capacity compared with trees from large continuous populations, but the seeds were more inbred and seedling growth less vigorous (Cascante *et al.* 2002). Nevertheless, it should also be noted that at high plant densities (Ellstrand 1992) or large floral displays (Ohashi & Yahara 2001) pollinator flights can become shorter giving rise to low outcrossing.

Clearly there is a continuum of plant densities along which, at some point, pollination systems begin to break down through isolation by distance. The density at which pollination processes begin to fail is a function of the polli-nation vector, with powerfully flying agents likely to ensure effective pollina-tion even in comparatively widely spaced plant populations. However, as plants become increasingly isolated the rate of pollination breakdown is more likely to be related to changes in pollinator foraging behaviour, as pollinators switch to more profitable species to obtain better returns on energy invested in foraging, rather than physical capacity to cross resource-poor areas. Of course, in addi-tion to pollinator type there are several other factors that may contribute to a plant's susceptibility to reproductive decline in a changing landscape, and some of these are listed in Table 10.2.

Habitat fragmentation

There is concern that habitat fragmentation has caused the decline of pollinator populations and breakdown of plant–pollinator interactions (Didham *et al.* 1996; Kearns *et al.* 1998). Small habitat fragments may not support the same abundance or richness of pollinator species as continuous habitat (Aizen & Feinsinger 1994b; Jennersten 1988; Liow *et al.* 2001). Both experimental studies and field observa-tions have shown that increasing patch isolation results in lower seed set owing to a fall in pollinator numbers (Steffan-Dewenter & Tscharntke 1999) or changes in the foraging patterns of pollinators (Didham *et al.* 1996; Goverde *et al.* 2002) leading to reduced visits per flower (Jennersten 1988; Lamont *et al.* 1993; Schulke & Waser 2001) or smaller pollen loads per visit (Cunningham 2000). A combi-nation of these can result in both reduced seed set and increased inbreeding (Goverde *et al.* 2002).

Habitat fragmentation reduces local plant population sizes and isolates sub-populations, which may narrow the genetic neighbourhood and increase genetic load through inbreeding (Severns 2003). Such effects are, however, likely to be mediated by self-incompatibility mechanisms and pollinator foraging behaviour. Self-incompatible plants are less vulnerable to inbreeding depression than selfing species, although the corollary is declining seed set. Pollinators may

Table 10.2 *Plant, pollinator and community attributes that contribute to the susceptibility of plant reproductive failure in response to landscape changes*

Characteristics that provide resistance to reproductive decline	Characteristics that increase vulnerability to reproductive decline
Plant attributes	
Resource-limited seed set	Pollen-limited seed set
Clonal/chleistogamous reproduction	Obligate sexual reproduction
Self-compatible	Self-incompatible
Annual	Perennial
Synchronous flowering	Asynchronous flowering
Extended flowering period	Temporally constrained flowering
Monoecious	Dioecious
Pollinator type	
Diverse pollinator groups	Few pollinator species
Long-distance flight/movement	Short-distance flight
Large pollen carryover potential	Small pollen carryover potential
High constancy to host plant	Low constancy to host plant
Directional movement	Non-directional movement
Community attributes	
Substantial competition for pollinators	Little competition for pollinators
Low species diversity	High species diversity
Aggregated plant species distributions	Dispersed plant species distributions
Asynchronous flowering among species	Synchronous flowering among species
Continuous cover of favourable habitat	Fragmented habitat with hostile intervening matrix

exacerbate inbreeding in small patches depending on their behaviour – they may avoid moving between habitat patches and effectively reduce genetic neighbourhoods or, alternatively, increase outcrossing if they readily cross the intervening matrix between patches (Dick *et al.* 2003; Hendrix & Kyhl 2000). In moving between patches pollinators may disperse pollen greater distances than in contiguous habitat, and 'overdispersal' of pollen may be beneficial by countering genetic load, or detrimental if there is significant outbreeding depression. Introduced pollinators, such as African honeybees in Amazonian forests, may be effective in facilitating long-distance pollen flow in fragmented habitats thereby expanding genetic neighbourhoods and linking fragmented patches with continuous forest populations (Dick *et al.* 2003). Several studies indicate that while the distance that pollen is moved in a fragmented landscape can exceed that in a continuous habitat, the genetic diversity of progeny within a fragment tends to decrease owing to a reduction in the number of pollen donors (Aldrich & Hamrick 1998; Cascante *et al.* 2002; Dick *et al.* 2003).

If foraging ranges of pollinators are large, gene flow between fragments can be sustained and, generally, large pollinators have greater capacity to cross gaps between fragments. Steffan-Dewenter and Tscharntke (1999) found that pollinators visiting the most isolated fragments tended to have larger bodies. Large bees and hummingbirds can certainly traverse several hundred metres (Dick *et al.* 2003; Jennersten 1988; Schulke & Waser 2001) and bats several kilometres of resource-poor land (Law & Lean 1999). In a Mexican fragmented forest, flower visitation by two small nectarivorous bats was lower than in an adjacent undisturbed forest, but a larger nectarivorous bat showed no difference in activity among the sites (Quesada *et al.* 2003). The contrast between these bat pollinators might be explained by their home ranges, which is reported to be 2–4 ha for the smaller species (*Glossophaga soricina*; Fleming *et al.* 1993) and several square kilometres for the larger (*Leptonycteris curasoae*; Horner *et al.* 1998). The impact of habitat fragmentation on the reproductive neighbourhood may therefore be largely a function of the foraging behaviour of the pollinators and distances they traverse (Nielsen & Ims 2000).

Genetic research has also shown that pollen from tropical trees can travel far across fragmented forest landscapes (Aldrich & Hamrick 1998; Chase *et al.* 1996), and isolated trees may provide an important 'stepping stone' function for pollen movement thereby linking subpopulations. Movement of pollinators between flowering patches or forest fragments is therefore likely to be contingent on the presence and identity of floral resources in the intervening habitat matrix. In some cases trees isolated in a resource-poor matrix can even dominate population reproduction. Thus hummingbirds that would normally trapline (that is, visit sites in sequence over a wide area) adopt territorial behaviour at isolated *Symphonia globulifera* trees, securing pollination and elevated seed set albeit at the cost of increased proportion of selfed seeds (Aldrich & Hamrick 1998). A similar change in foraging behaviour has been observed at patches of *Betonica officinalis* by bumblebees that visited more inflorescences and flew longer total distances in small patches, although they also tended to remain longer (Goverde *et al.* 2002).

Pollen from wind-pollinated trees has usually been assumed to be abundant and to be carried over long distances. Recent genetic analyses have concluded that, on the contrary, wind-pollinated trees may be pollen-limited and that pollen movement by wind might be spatially very restricted (Knapp *et al.* 2001; Koenig & Ashley 2003; Sork 1993). Increased fragmentation could, therefore, lead to reproductive failure in some wind-pollinated species.

Population size

A relationship between population size and seed set due to pollination effectiveness has been observed in self-incompatible animal-pollinated species (Ågren 1996; Sih & Baltus 1987) and wind-pollinated plants (Nilsson & Wastljung 1987).

Small populations are likely to be less attractive or less apparent to pollinators than large populations (Ågren 1996; Jennersten & Nilsson 1993; Sih & Baltus 1987) leading not only to a decline in pollinator visits but also to poor pollen quality (Silander 1978; and see above). Fruit set of *Banksia goodii*, for example, declined to zero as population size decreased (Lamont *et al.* 1993). In this case the decline was precipitous as the bird pollinators avoided populations that were below a certain threshold size. Other studies have found no correlation between population size and reproductive success (Sowig 1989; van Treuren *et al.* 1993), although van Treuren *et al.* (1993) did note a positive effect of population density on outcrossing.

For wind-pollinated species pollination is a function of pollen production, which is relatively low for small populations. For example, populations of yew that had been reduced in size by browsing deer produced less pollen and had lower levels of pollination success and seed set than larger populations (Allison 1990).

In contrast to flowers that reward pollinators with pollen or nectar, pollination success and seed set of non-rewarding species declines as population size increases. Most of these species achieve pollination by deceiving pollinators, and, as non-rewarding plants become increasingly common, pollinators learn to avoid them (Alexandersson & Ågren 1996; Fritz & Nilsson 1994).

Population or patch size and individual isolation can interact such that isolated plants in small populations have the greatest risk of being pollen-limited (Groom 1998). Thus for *Clarkia concinna* seed set as a function of isolation is more pronounced in the smallest populations, and once population size exceeded 50 individuals no clear relation between seed set or pollen receipt and isolation could be detected (Groom 2001). The widely reported correlation of population size with density may therefore lead to pollination failure through density and abundance mechanisms.

Pollinator behaviour

As agents of pollen transfer, the foraging behaviour of pollinators is crucially important to pollination success and seed set in pollen-limited plants. The constancy with which a pollinator visits a particular plant species is relevant to the quality of pollen load received by the flowers, with generalist pollinators more likely to carry a diverse array of pollen types of which only a small proportion is compatible to each visited flower. For rare plants this can be exacerbated as generalist pollinators are often presumed to forage in a frequency-dependent manner with disproportionate costs for rare species. Furthermore, most plants are generalist in terms of their pollinators (Waser *et al.* 1996) and different pollinator species forage in different ways and to varying degrees of effectiveness. Tropical hummingbirds and euglossine bees commonly visit plants sequentially over a wide geographic area (traplining) and will return to the same plants in the same sequence repeatedly (Janzen 1971; Linhart & Feinsinger 1980). Other pollinators are territorial and tend to move pollen among one or a few adjacent plants (Feinsinger *et al.* 1986; Willmer *et al.* 1994).

Likewise, over 70 bee species have been found on flowers of *Andira inermis* in Costa Rica, but cross-pollination is due to foraging by only six *Centris* species (Frankie *et al.* 1976). Some pollinators (honeybees and butterflies) may readily move between patches while others (beeflies) typically return to a profitable patch if they fail to encounter another within a few metres (Groom 1998). Even congenerics can differ markedly in their foraging behaviour – of two species of *Trigona* bee in Costa Rica one forages in large groups on dense flowering patches while the other forages individually or in small groups visiting widely spaced plants (Johnson & Hubbell 1975). Given the variety of pollinator behaviours under differing circumstances, determining the impacts of plant spatial distribution on the pollination biology of plants is exceedingly complex.

Pollinator foraging behaviour is not fixed and may change in response to changes in the spatial distribution of plants. Bumblebees foraging in fragmented landscapes visit more flowers more intensively, and consequently spend more time, in small fragments than in large fragments or continuous habitat (Rasmussen & Brodsgaard 1992; Cresswell 1997, 2000; Goverde *et al.* 2002). The directionality of bumblebee flight also decreases with increased plant density (Cresswell 1997), while edge plants provoke changes in direction or reversals (Rasmussen & Brodsgaard 1992). Directionality in flight decreases self-pollination by decreasing the probability of encountering previously visited plants. Flight directionality is also influenced by the richness of the rewards, and small patches with rapidly depleted floral rewards can *increase* directionality (Goverde *et al.* 2002).

Pollinators that visit a wide array of plants may forage in a frequency-dependent way showing greater constancy to common species or floral morphs (Epperson & Clegg 1987; Smithson & Macnair 1996; Smithson & Macnair 1997). For example, flower visitors may move from the canopy to the understorey or vice versa depending on the relative availability of floral resources (Ghazoul 2002). Common flowers are probably more apparent to pollinators but, additionally, pollinators may improve foraging efficiency by specializing on the most common flower types, leading to a disproportionate number of visits to common flowers at the expense of rare species or morphs. Other studies have been unable to detect frequency-dependent foraging (Waser 1982), and Smithson and Macnair (1996) argue that positive frequency-dependent selection is only likely to be important to reproductive success if pollinators are rare. Where frequency-dependent visitation does occur rare species could be disadvantaged where fitness is correlated with number of pollinator visits received. This may be further compounded by poor pollen quality as most pollen carried by pollinators is likely to be from the more common species (Silander 1978). Selfing may compensate for few pollinator visits although this may bring disadvantages through inbreeding depression and reduced paternal function (Epperson & Clegg 1987; Levin & Kerster 1971). Alternatively, infrequently visited rare flowers could represent rich and untapped sources of nectar and thereby attract specialist pollinators. Birds, for example, visit flowers of *Dipterocarpus* trees predominantly in disturbed areas where an abundance of understorey flowers diverts the attention of the trees' primary pollinators (Ghazoul 2002, 2004).

Interactions among pollinators at plants can further affect the dynamics of pollination. Increasing pollinator activity during periods of high flowering density can lead to an increase in aggressive interactions among territorial bees (Frankie *et al.* 1976). In consequence inter-tree movements by bees, and therefore outcrossing, is increased. It is possible that predators of pollinators may also be attracted to high densities of mass-flowering trees and that this promotes movement of pollinators to other trees to avoid predation.

Pollination by beetles, flies, butterflies and moths is particularly common in tropical regions. Beetles are important pollinators of many tropical palms and some canopy trees (Consiglio & Bourne 2001; Knudsen *et al.* 2001; Sakai *et al.* 1999), although little is known about their flight abilities and capacity for promoting cross-pollination over large distances. Flies carry substantial pollen loads (Kearns 2001) but move relatively short distances (up to a few metres) between plants (Olesen & Warncke 1989; Widen & Widen 1990). Moths can also have large pollen loads (Willmott & Burquez 1996) and may fly considerable distances between successively visited plants, even, in some cases, bypassing neighbouring plants (Richter & Weis 1998). However, capacity for long flights does not necessarily lead to extensive pollen dispersal (Marr *et al.* 2000). Butterflies can disperse pollen over similar distances to moths but are thought to carry only relatively small amounts of pollen (Murawski & Gilbert 1986). Thus non-hymenopteran pollinators can be effective in promoting long-distance gene flow for many plants. Additionally floral constancy, and hence improved pollen quality, has also been reported for beetles, flies and lepidopterans. Of course, the flight range of pollinators does not necessarily reflect pollen transfer distances. Forest honeybees for example may show fidelity on a daily basis to very small patches no more than a few tens of metres wide even though the location of the patch may be several kilometres from the nest (Visscher & Seeley 1982). Similarly, large euglossine bees show local site fidelity in most circumstances despite being capable of covering large areas in a 'traplining' fashion (Ackerman *et al.* 1982). In general, however, little is known in detail about the foraging behaviour of insect pollinator groups other than social bees, and the effects of changes in plant spatial distributions on the effectiveness of this suite of insects as pollinators remains largely unknown.

Pollinator abundance

Declining abundance or density of floral resources can disproportionately affect pollinator populations by reducing resource availability (Aizen & Feinsinger 1994a, b; Cunningham 2000; Jennersten 1988; Sih & Baltus 1987). When several species of plants that share pollinators occur together at low density they may have a facilitatory effect in attracting pollinators, although the risk remains that pollinators in such conditions will carry a multispecies pollen and hence low quality of transferred pollen. The scale over which a facilitatory effect occurs will vary with the mobility of pollinators. Pollinator behaviour can also change at different densities, with some species (e.g. bumblebees and hummingbirds) becoming less discriminate at low densities. Thus rare plants may benefit initially as similar flowering species become more abundant but this may change rapidly as pollinators specialize on the more abundant species (Kunin 1997), resulting in both lower visitation of rare plants and increased likelihood of transfer of non-specific pollen. On the other hand, as flower density increases pollinators may become more abundant, but also more sedentary or territorial (Willmer et al. 1994) leading to reduced gene flow and possibly inbreeding effects.

In the extreme case of specialized and obligate plant–pollinator interactions, such as figs and fig wasps, each mutualistic partner is wholly dependent on the other for its persistence. Where pollinators are short-lived and have no diapause, as is the case with fig wasps, population persistence is dependent on the continuous and asynchronous availability of flowers. Reduction in plant density and abundance could result in gaps in the continuous flowering sequence that will inevitably lead to a collapse of pollinator populations if the duration of the gaps exceeds the adult insect's lifespan. As the lifespan of adult fig wasps is very brief, even a flower-free period of a few days may be sufficient to cause collapse of the fig wasp population and consequently reproductive failure of figs. Increasing length of the fig sexual phase will lessen the likelihood of non-flowering phases arising, as will an increase in the dispersal distance of the fig wasps. The resilience of the fig reproductive system to environmental change may be due to enormous fig-wasp dispersal (Chase et al. 1996), which has been recorded at over 80 km from the pollen source (S. Compton, personal communication). Indeed, anthropogenically driven changes in natural habitats have rarely led to local pollinator extinctions (but see Washitani 1996). Bee communities may be more resistant to habitat fragmentation than previously thought (Cane 2001) and a more likely outcome of fragmentation may be more-subtle changes in pollinator foraging patterns.

Vertebrate pollinators with high energy demands, such as hummingbirds and glossophagine bats, can only persist in a habitat that has sufficient spatial density of nectar resources. Although empirical or field data are lacking, it is expected that habitat fragmentation or reduction in plant population abundance and density will reduce the availability of profitable foraging locations

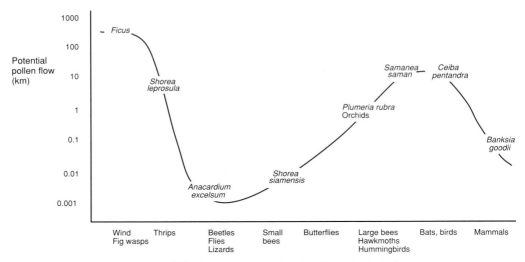

Figure 10.1 A conceptual framework for potential pollen flow distances by different pollinator groups arranged by body size.

resulting in the population decline of bats and birds. Indeed, flower-visiting bats in Mexican dry forests visited *Ceiba grandiflora* flowers less frequently in disturbed than undisturbed forests, resulting in marked reduction in seed set in disturbed sites (Quesada *et al.* 2003). However, nectarivorous bats have also been noted to be highly mobile in fragmented landscapes, flying as much as 5 km across cleared land and encompassing home ranges of up to 1796 hectares (Law & Lean 1999). For the common blossom bat *Syconycteris australis* such mobility results in no difference in population densities among small and large fragments and continuous forest (Law & Lean 1999).

A conceptual framework for pollinator responses to distance

We can begin to develop a conceptual framework that links pollinator types to plant reproductive vulnerability using body size as an indicator of likely response (Fig. 10.1). At the smallest body sizes insects such as fig wasps and thrips have poor capacity for long-distance flight but nevertheless traverse exceptional distances through wind-assisted movement (see above). Slightly larger insects, including small flies, beetles and bees, have relatively poor powers of long-distance flight and are too large for wind-assisted movement. Plants pollinated by these insects are most vulnerable to pollination failure associated with density reduction and consequently increasing distances between neighbouring flowering trees. As pollinator body size and power of flight continues to increase, the distance between neighbouring plants becomes less important as a factor contributing to reproductive success or failure. The largest-bodied pollinators, terrestrial mammals, are incapable of flight and consequently plants pollinated by rodents, lemurs and possums are likely to be more vulnerable to distance effects.

A further consideration is the relative importance of local density and habitat patchiness, which is likely to change along this continuum of pollinator body sizes. At the smallest body sizes wind-assisted travel ensures that plants are relatively secure from spatial effects acting on pollination. Slightly larger insects may respond behaviourally to changes in the density of flowering trees, but as a group are expected to be relatively insensitive to fragmentation as populations may persist with little change in all but the very smallest of fragments. By contrast, larger vertebrate pollinators may easily cross large distances between neighbouring flowering trees in continuous forest cover, but may be susceptible to population-related declines in fragmented landscapes through habitat loss, increased predation or hunting. The susceptibility of plants pollinated by vertebrates to pollination failure is likely to be a function of abundance as well as the behavioural response of the pollinators to landscape structure.

Seed dispersal

Dispersal of pollen is only one mechanism by which gene flow is achieved by plants, the other being seed movement. Genetic analyses have shown that gene flow through pollen movement can be subject to Allee effects (e.g. Murawski *et al.* 1994, and see above) but also that pollen can be dispersed considerable distances in fragmented landscapes (Chase *et al.* 1996; Nason *et al.* 1998). Few studies, however, have examined the impacts of habitat fragmentation or population size reduction on seed dispersal as a mechanism for gene flow. Yet gene flow through seeds accounts for two-thirds of the total genetic neighbourhood size (pollen is haploid) and seed dispersal is equally likely to be affected by habitat fragmentation and degradation as pollen dispersal. Seed dispersal of trees is generally leptokurtic leading to acute genetic differentiation of adjacent populations (Hamilton 1999), and in a fragmented landscape small patches may exchange seeds infrequently, limiting the size of the genetic neighbourhood (Hamilton 1999).

Most tropical trees have seeds that are dispersed by animals, so the effects of habitat fragmentation on the behaviour and abundance of animal seed dispersers is of paramount importance. In a recent study using artificial fruits the activity of frugivorous birds in Brazilian forest fragments (ranging in size from 251 to 36,000 hectares) was monitored. Small fragments had lower frugivore activity than large fragments although there was also a marked edge effect with increased frugivory at forest edges (Galetti *et al.* 2003). Similarly, dispersal of juniper seed declined dramatically in fragmented forests owing to the local decline of thrushes, the main seed dispersers (Santos *et al.* 1999). *Dipteryx panamensis* seed dispersal was numerically greater in continuous than fragmented forests in Costa Rica, but the reverse was true for another canopy tree, *Carapa guianensis*, indicating that fragmentation effects on seed fate may be species-specific and a function of the dispersal agents involved (Guariguata *et al.* 2002). In Tanzania, recruitment of 31 animal-dispersed trees was three times as high in large fragments (exceeding 30 ha in size) as in small fragments, whereas recruitment of eight wind- or gravity-dispersed trees showed no difference, the result being linked to the declining abundance of frugivores in small forest fragments (Cordeiro & Howe 2001).

Perhaps the most vulnerable species in terms of seed dispersal are plants that have few large fruits and attract few species of large dispersers, mainly mammals and birds. It is these animals that are most vulnerable to hunting, fragmentation and habitat loss, which may have knock-on effects on plant recruitment (Corlett 1998).

Thresholds

Models of Allee effects show extinction likelihoods increasing dramatically with declining population size, and often a density or population-size threshold can be identified below which extinction is almost inevitable (Kunin & Iwasa 1996; Veit & Lewis 1996). It has been more difficult to verify these theoretical results with real population data, although extinction thresholds have been suggested in some studies. Seed set of *Clarkia concinna*, for example, dropped to zero when isolation distance exceeded 26 to 104 m depending on patch size (Groom 1998). Likewise, small remnant populations of *Banksia goodii* had complete reproductive failure below a certain patch size (Lamont *et al.* 1993).

The existence of thresholds remains uncertain for most species but where they occur they will probably be mediated by the type of pollinator. Density thresholds below which there is a precipitous decline in pollination success are likely to be much lower for plants pollinated by agents that move very short distances. Unfortunately little is known about pollinator movement and behaviour under differing circumstances, and most studies include only a superficial treatment of pollinator foraging and instead make assumptions about what the animals do. Pollinator foraging models do give some insight into the contribution of pollinator behaviour to the establishment of thresholds. These models predict, for example, that constancy to a particular floral type will break down at low flower density leading to a sudden acceleration of pollination failure. However, interactions among plants and flower visitors are complex and form a highly connected web (Memmott 1999) and although plants are often visited by specialized insects they often attract more generalized flower visitors too. Recent work by Cotton (1998) has showed that, contrary to previously accepted beliefs, hermit hummingbirds are as generalized as other hummingbirds, and in tropical lowland forest many pollinators are both generalized and opportunistic in their flower-visiting behaviours (Momose *et al.* 1998). Additionally, pollinator visitation may not be the best indicator of seed set as Olsen (1997) discovered for

Heterotheca subaxillaris, an Asteraceae in Texas, whose most effective pollinator in terms of seed set was the least abundant flower visitor.

Allee effects and extinction

Are Allee effects alone enough to cause precipitous population declines leading to extinction? To date there are few documented examples of localised plant extinctions that can be attributed to Allee effects. Small populations have been observed to produce zero seed (e.g. *Banksia goodii*, Lamont *et al.* 1993), but this does not necessarily reflect the lifetime fitness of individuals. Groom (1998) reported 28 *Clarkia concinna* patch extinctions from a total of 211 patches over a 5-year period. Catastrophic disturbance accounted for more than half of these extinctions, with 9 of the remaining 12 being populations known to have low reproductive success (cause of extinction for the remaining three populations could not be established). All recorded extinctions were of small and isolated populations. These data demonstrate that such patches have greater risk of extinction but at best provide weak support for pollination failure as the principal causal mechanism.

Although there is mounting evidence from experimental and field studies of the importance of plant spacing to pollination success and seed set, many of these studies are conducted over only a fraction of the lifetime of the plants concerned. Conflicting results can be obtained if populations are studied over several years or from different sites (Costin *et al.* 2001; Hendrix & Kyhl 2000; Morgan 1999; Robertson *et al.* 1999). Weather conditions can affect pollinator activity (Hendrix & Kyhl 2000; Morgan 1999) and pollinator populations can undergo marked fluctuations from one year to the next (Roubik 2001).

Conclusions

Pollination studies have usually been conducted on populations at high density, presumably owing to the ease of study, and as a result early reviews emphasized the role of resource limitation in reproduction. More recently, attention has shifted to fragmented or sparse populations, anthropogenically derived and natural, where pollen-limited seed set subject to Allee effects has been widely noted. Declining reproductive efficiency associated with reduced densities may lead to positive feedbacks that drive sparse populations to extinction. Annuals are particularly vulnerable as failure to achieve full reproductive potential cannot be turned into increased investment in vegetative tissue. It is unlikely that long-lived perennials could be driven to extinction by Allee effects alone, but population recovery could be slowed considerably. Furthermore, consecutive years of low seed production present an opportunity for competitors and exotics to establish with long-term consequences for community composition. As many plant species in the tropics are undergoing rapid population decline, Allee effects on seed production are likely to become increasingly relevant to conservation of plant species.

References

Abe, T. (2001) Flowering phenology, display size, and fruit set in an understorey dioecious shrub, *Aucuba japonica* (Cornaceae). *American Journal of Botany*, **88**, 455–461.

Ackerman, J. D., Mesler, M. R., Lu, K. L. & Montalvo, A. M. (1982) Food-foraging behavior of male *Euglossini* (Hymenoptera: Apidae): vagabonds or trapliners? *Biotropica*, **14**, 241–248.

Ågren, J. (1996) Population size, pollinator limitation, and seed set in the self-incompatible herb *Lythrum salicaria*. *Ecology*, **77**, 1779–1790.

Aizen, M. A. (2001) Flower sex ratio, pollinator abundance, and the seasonal pollination dynamics of a protandrous plant. *Ecology*, **82**, 127–144.

Aizen, M. A. & Feinsinger, P. (1994a) Forest fragmentation, pollination and plant reproduction in a Chaco dry forest, Argentina. *Ecology*, **75**, 330–351.

(1994b) Habitat fragmentation, native insect pollinators, and feral honey bees in Argentine 'chaco serrano'. *Ecological Applications*, **4**, 378–392.

Alatalo, J. M. & Molau, U. (2001) Pollen viability and limitation of seed production in a population of the circumpolar cushion plant, *Silene acaulis* (Caryophyllaceae). *Nordic Journal of Botany*, **21**, 365–372.

Albre, J., Quilichini, A. & Gibernau, M. (2003) Pollination ecology of *Arum italicum* (Araceae). *Botanical Journal of the Linnean Society*, **141**, 205–214.

Aldrich, P. R. & Hamrick, J. L. (1998) Reproductive dominance of pasture trees in a fragmented tropical forest. *Science*, **281**, 103–105.

Alexandersson, R. & Ågren, J. (1996) Population size, pollinator visitation and fruit production in the deceptive orchid *Calypso bulbosa*. *Oecologia*, **107**, 533–540.

Allison, T. D. (1990) Pollen production and plant density affect pollination and seed production in *Taxus canadensis*. *Ecology*, **71**, 516–522.

Bawa, K. S. (1974) Breeding systems of tree species of a lowland tropical community. *Evolution*, **28**, 85–92.

(1990) Plant–pollinator interactions in tropical rain forests. *Annual Review of Ecology and Systematics*, **21**, 399–422.

Bawa, K. S. & Beach, J. H. (1981) Evolution of sexual systems in flowering plants. *Annals of the Missouri Botanical Garden*, **68**, 254–274.

Bawa, K. S. & Webb, C. J. (1984) Flower, fruit and seed abortion in tropical trees: implications for the evolution of paternal and maternal reproductive patterns. *American Journal of Botany*, **71**, 736–751.

Bawa, K. S., Bullock, S. H., Perry, D. R., Coville, R. E. & Grayum, M. H. (1985) Reproductive biology of tropical lowland rain forest trees. II. Pollination systems. *American Journal of Botany*, **72**, 346–356.

Bierzychudek, P. (1981) Pollinator limitation of plant reproductive effort. *American Naturalist*, **117**, 838–840.

Bosch, M. & Waser, N. M. (1999) Effects of local density on pollination and reproduction in *Delphinium nuttallianum* and *Aconitum columbianum* (Ranunculaceae). *American Journal of Botany*, **86**, 871–879.

Burd, M. (1994) Bateman's principle and plant reproduction: the role of pollen limitation in fruit and seed set. *The Botanical Review*, **60**, 83–139.

Campbell, D. R. (1985) Pollen and gene dispersal: the influences of competition for pollination. *Evolution*, **39**, 418–431.

Cane, J. H. (2001) Habitat fragmentation and native bees: a premature verdict? *Conservation Ecology*, **5**, 3 (online). http://www.consecol.org/vol5/iss1/art3/

Cascante, A., Quesada, M., Lobo, J. J. & Fuchs, E. A. (2002) Effects of dry tropical forest fragmentation on the reproductive success and genetic structure of the tree *Samanea saman*. *Conservation Biology*, **16**, 137–147.

Chase, M. R., Moller, C., Kesseli, R. & Bawa, K. S. (1996) Distant gene flow in tropical trees. *Nature*, **383**, 398–399.

Consiglio, T. K. & Bourne, G. R. (2001) Pollination and breeding system of a neotropical palm *Astrocaryum vulgare* in Guyana: a test of the predictability of syndromes. *Journal of Tropical Ecology*, **17**, 577–592.

Copland, B. J. & Whelan, R. J. (1989) Seasonal variation in flowering intensity and pollination limitation of fruit set in four co-occurring *Banksia* species. *Journal of Ecology*, **77**, 509–523.

Cordeiro, N. J. & Howe, H. F. (2001) Low recruitment of trees dispersed by animals in African forest fragments. *Conservation Biology*, **15**, 1733–1741.

Corlett, R. T. (1998) Frugivory and seed dispersal by vertebrates in the Oriental (Indomalayan) region. *Biological Reviews of the Cambridge Philosophical Society*, **73**, 413–448.

Costin, B. J., Morgan, J. W. & Young, A. G. (2001) Reproductive success does not decline in fragmented populations of *Leucochrysum albicans* subsp. *albicans* var. *tricolor* (Asteraceae). *Biological Conservation*, **98**, 273–284.

Cotton, P. A. (1998) The hummingbird community of a lowland Amazonian rainforest. *Ibis*, **140**, 512–521.

Cresswell, J. E. (1997) Spatial heterogeneity, pollinator behaviour and pollinator-mediated gene flow: bumblebee movements in variously aggregated rows of oil-seed rape. *Oikos*, **78**, 546–556.

(2000) A comparison of bumblebees' movements in uniform and aggregated distributions of their forage plant. *Ecological Entomology*, **25**, 19–25.

Cunningham, S. A. (1996) Pollen supply limits fruit initiation by a rain forest understorey palm. *Journal of Ecology*, **84**, 185–194.

(2000) Depressed pollination in habitat fragments causes low fruit set. *Proceedings of the Royal Society of London B*, **267**, 1149–1152.

Dick, C. W., Etchelecu, G. & Austerlitz, F. (2003) Pollen dispersal of tropical trees (*Dinizia excelsa*: Fabaceae) by native insects and African honeybees in pristine and fragmented Amazonian rainforest. *Molecular Ecology*, **12**, 753–764.

Didham, R. K., Ghazoul, J., Stork, N. E. & Davis, A. J. (1996) Insects in fragmented forests: a functional approach. *Trends in Ecology & Evolution*, **11**, 255–260.

Dieringer, G. (1992) Pollinator effectiveness and seed set in populations of *Agalinis strictifolia* (Scrophulariaceae). *American Journal of Botany*, **79**, 1018–1023.

Ehlers, B. K., Olesen, J. M. & Ågren, J. (2002) Floral morphology and reproductive success in the orchid *Epipactis helleborine*: regional and local across-habitat variation. *Plant Systematics and Evolution*, **236**, 19–32.

Ehrlen, J., Kack, S. & Ågren, J. (2002) Pollen limitation, seed predation and scape length in *Primula farinosa*. *Oikos*, **97**, 45–51.

Elberling, H. (2001) Pollen limitation of reproduction in a subarctic-alpine population of *Diapensia lapponica* (Diapensiaceae). *Nordic Journal of Botany*, **21**, 277–282.

Ellstrand, N. C. (1992) Gene flow by pollen: implications for plant conservation genetics. *Oikos*, **63**, 77–86.

Epperson, B. K. & Clegg, M. T. (1987) Frequency-dependent variation for outcrossing rate among flower color morphs of *Ipomoea purpurea*. *Evolution*, **41**, 1302–1311.

Feinsinger, P., Murray, K. G., Kinsman, S. & Busby, W. H. (1986) Floral neighborhood and pollination success in four hummingbird-pollinated cloud forest plant species. *Ecology*, **67**, 449–464.

Feinsinger, P., Tiebout, H. M. & Young, B. E. (1991) Do tropical bird-pollinated plants exhibit density-dependent interactions? Field experiments. *Ecology*, **72**, 1953–1963.

Fleming, T. H., Nunez, R. A. & Sternberg, L. S. (1993) Seasonal changes in the diets of migrant and non-migrant nectarivorous bats as revealed by carbon stable isotope analysis. *Oecologia*, **94**, 72–75.

Frankie, G. W., Opler, P. A. & Bawa, K. S. (1976) Foraging behaviour of solitary bees: implications for outcrossing of a neotropical forest tree species. *Journal of Ecology*, **64**, 1049–1058.

Fritz, A.-L. & Nilsson, L. A. (1994) How pollinator-mediated mating varies with population size in plants. *Oecologia*, **100**, 451–462.

Fuchs, E. J., Lobo, J. A. & Quesada, M. (2003) Effects of forest fragmentation and flowering phenology on the reproductive success and mating patterns of the tropical dry forest tree *Pachira quinata*. *Conservation Biology*, **17**, 149–157.

Galetti, M., Alves-Costa, C. P. & Cazetta, E. (2003) Effects of forest fragmentation, anthropogenic edges and fruit colour on the consumption of ornithocoric fruits. *Biological Conservation*, **111**, 269–273.

Ghazoul, J. (2002) Flowers at the front line of invasion? *Ecological Entomology*, **27**, 638–640.

(2004) Alien abduction: disruption of native plant–pollinator interactions by invasive species. *Biotropica*, **36**, 156–164.

Ghazoul, J. & McLeish, M. (2001) Reproductive ecology of tropical forest trees in logged and fragmented habitats in Thailand and Costa Rica. *Plant Ecology*, **153**, 335–345.

Ghazoul, J., Liston, K. A. & Boyle, T. J. B. (1998) Disturbance-induced density-dependent seed set in *Shorea siamensis* (Dipterocarpaceae), a tropical forest tree. *Journal of Ecology*, **86**, 462–473.

Goodwillie, C. (2001) Pollen limitation and the evolution of self-compatibility in *Linanthus* (Polemoniaceae). *International Journal of Plant Sciences*, **162**, 1283–1292.

Goverde, M., Schweizer, K., Baur, B. & Erhardt, A. (2002) Small-scale habitat fragmentation effects on pollinator behaviour: experimental evidence from the bumblebee *Bombus veteranus* on calcareous grasslands. *Biological Conservation*, **104**, 293–299.

Groom, M. J. (1998) Allee effects limit population viability of an annual plant. *American Naturalist*, **151**, 487–496.

(2001) Consequences of subpopulation isolation for pollination, herbivory, and population growth in *Clarkia concinna concinna* (Onagraceae). *Biological Conservation*, **100**, 55–63.

Guariguata, M. R., Arias-Le Claire, H. & Jones, G. (2002) Tree seed fate in a logged and fragmented forest landscape, northeastern Costa Rica. *Biotropica*, **34**, 405–415.

Hall, P., Walker, S. & Bawa, K. (1996) Effect of forest fragmentation on genetic diversity and mating system in a tropical tree, *Pithecellobium elegans*. *Conservation Biology*, **10**, 757–768.

Hamilton, M. B. (1999) Tropical tree gene flow and seed dispersal. *Nature*, **401**, 129–130.

Heithaus, E. R., Stashko, E. & Anderson, P. K. (1982) Cumulative effects of plant–animal interactions on seed production by *Bauhinia ungulata*, a neotropical legume. *Ecology*, **63**, 1294–1302.

Hendrix, S. D. & Kyhl, J. F. (2000) Population size and reproduction in *Phlox pilosa*. *Conservation Biology*, **14**, 304–313.

Horner, M. A., Fleming, T. H. & Sahley, C. T. (1998) Foraging behaviour and energetics of a nectar-feeding bat, *Leptonycteris curasoae* (Chiroptera: Phyllostomidae). *Journal of Zoology*, **244**, 575–586.

House, S. M. (1993) Pollination success in a population of dioecious rain forest trees. *Oecologia*, **96**, 555–561.

Janzen, D. H. (1971) Euglossine bees as long-distance pollinators of tropical plants. *Science*, **171**, 203–205.

Jennersten, O. (1988) Pollination in *Dianthus deltoides* (Caryophyllaceae): effects of habitat fragmentation on visitation and seed set. *Conservation Biology*, **2**, 359–366.

Jennersten, O. & Nilsson, S. G. (1993) Insect flower visitation frequency and seed production in relation to patch size of *Viscaria vulgaris* (Caryophyllaceae). *Oikos*, **68**, 283–292.

Johnson, L. K. & Hubbell, S. P. (1975) Contrasting foraging strategies and coexistence of two bee species on a single resorce. *Ecology*, **56**, 1398–1406.

Kearns, C. A. (2001) North American dipteran pollinators: assessing their value and conservation status. *Conservation Ecology*, **5** [online] http://www.consecol.org/vol5/iss1/art5.

Kearns, C. A., Inouye, D. W. & Waser, N. W. (1998) Endangered mutualisms: the conservation of plant–pollinator interactions. *Annual Review of Ecology and Systematics*, **29**, 83–112.

Klinkhamer, P. G. L. & de Jong, T. J. (1990) Effects of plant size, plant density and sex differential nectar reward on pollinator visitation in the protandrous *Echium vulgare* (Boraginaceae). *Oikos*, **57**, 399–405.

Klinkhamer, P. G. L., de Jong, T. J. & de Bruyn, G.-J. (1989) Plant size and pollinator visitation in *Cynoglossum officinale*. *Oikos*, **54**, 201–204.

Knapp, E. E., Goedde, M. A. & Rice, K. J. (2001) Pollen-limited reproduction in blue oak: implications for wind pollination in fragmented populations. *Oecologia*, **128**, 48–55.

Knudsen, J. T., Tollsten, L. & Ervik, F. (2001) Flower scent and pollination in selected neotropical palms. *Plant Biology*, **3**, 642–653.

Koenig, W. D. & Ashley, M. V. (2003) Is pollen limited? The answer is blowin' in the wind. *Trends in Ecology & Evolution*, **18**, 157–159.

Kunin, W. & Iwasa, Y. (1996) Pollinator foraging strategies in mixed floral arrays: density effects and floral constancy. *Theoretical Population Biology*, **49**, 232–263.

Kunin, W. E. (1992) Density and reproductive success in wild populations of *Diplotaxis erucodes* (Brassicaceae). *Oecologia*, **91**, 129–133.

(1993) Sex and the single mustard: population density and pollinator behavior effects on seed-set. *Ecology*, **74**, 2145–2160.

(1997) Population size and density effects in pollination: pollinator foraging and plant reproductive success in experimental arrays of *Brassica kaber*. *Journal of Ecology*, **85**, 225–234.

Lamont, B. B., Klinkhamer, P. G. L. & Witkowski, E. T. F. (1993) Population fragmentation may reduce fertility to zero in *Banksia goodii* – a demonstration of the Allee effect. *Oecologia*, **94**, 446–450.

Larson, B. M. H. & Barrett, S. C. H. (2000) A comparative analysis of pollen limitation in flowering plants. *Biological Journal of the Linnean Society*, **69**, 503–520.

Law, B. S. & Lean, M. (1999) Common blossom bats (*Syconycteris australis*) as pollinators in fragmented Australian tropical rainforest. *Biological Conservation*, **91**, 201–212.

Levin, D. A. (1972) Pollen exchange as a function of species proximity in *Phlox*. *Evolution*, **26**, 251–258.

Levin, D. A. & Kerster, H. W. (1971) Neighborhood structure of plants under diverse reproductive methods. *American Naturalist*, **105**, 345–354.

Linhart, Y. B. & Feinsinger, P. (1980) Plant–hummingbird interactions: effects of island size and degree of specialization on pollination. *Journal of Ecology*, **68**, 745–760.

Liow, L. H., Sodhi, N. S. & Elmqvist, T. (2001) Bee diversity along a disturbance gradient in tropical lowland forests of Southeast Asia. *Journal of Applied Ecology*, **38**, 180–192.

Luijten, S. H., Dierick, A., Oostermeijer, J. G. B., Raijmann, L. E. L. & Den Nijs, H. C. M. (2000) Population size, genetic variation, and reproductive success in a rapidly declining, self-incompatible perennial (*Arnica montana*) in The Netherlands. *Conservation Biology*, **14**, 1776–1787.

Marr, D. L., Leebens-Mack, J., Elms, L. & Pellmyr, O. (2000) Pollen dispersal in *Yucca filamentosa* (Agavaceae): the paradox of self-pollination behavior by *Tegeticula yuccasella* (Prodoxidae). *American Journal of Botany*, **87**, 670–677.

Memmott, J. (1999) The structure of a plant–pollinator food web. *Ecology Letters*, **2**, 276–280.

Menges, E. (1991) Seed germination percentage increases with population size in a fragmented prairie species. *Conservation Biology*, **5**, 158–164.

Momose, K., Yumoto, T., Nagamitsu, T. *et al.* (1998) Pollination biology in a lowland dipterocarp forest in Sarawak, Malaysia. I. Characteristics of the plant–pollinator community in a lowland dipterocarp forest. *American Journal of Botany*, **85**, 1477–1501.

Moody-Weis, J. M. & Heywood, J. S. (2001) Pollination limitation to reproductive success in the Missouri evening primrose, *Oenothera macrocarpa* (Onagraceae). *American Journal of Botany*, **88**, 1615–1622.

Morgan, J. W. (1999) Effects of population size on seed production and germinability in an endangered, fragmented grassland plant. *Conservation Biology*, **13**, 266–273.

Murawski, D. A. & Gilbert, L. E. (1986) Pollen flow in *Psiguria warscewiczii*: a comparison of *Heliconius* butterflies and hummingbirds. *Oecologia*, **68**, 161–167.

Murawski, D. A. & Hamrick, J. L. (1992) The mating system of *Cavanillesia platanifolia* under extremes of flowering-tree density: a test of predictions. *Biotropica*, **24**, 99–101.

Murawski, D. A., Hamrick, J. L., Hubbell, S. P. & Foster, R. B. (1990) Mating systems of two bombacaceous trees of a neotropical moist forest. *Oecologia*, **82**, 501–506.

Murawski, D. A., Gunatilleke, I. A. U. N. & Bawa, K. S. (1994) The effects of selective logging on inbreeding in *Shorea megistophylla* (Dipterocarpaceae) from Sri Lanka. *Conservation Biology*, **8**, 997–1002.

Nason, J. D., Herre, E. A. & Hamrick, J. L. (1998) The breeding structure of a tropical keystone plant resource. *Nature*, **391**, 685–687.

Nielsen, A. & Ims, R. A. (2000) Bumble bee pollination of the sticky catchfly in a fragmented agricultural landscape. *Ecoscience*, **7**, 157–165.

Nilsson, S. G. & Wastljung, U. (1987) Seed predation and cross-pollination in mast-seeding beech (*Fagus sylvatica*) patches. *Ecology*, **68**, 260–265.

Olesen, J. M. & Warncke, E. (1989) Temporal changes in pollen flow and neighborhood-structure in a population of *Saxifraga hirculus* L. *Oecologia*, **79**, 205–211.

Olsen, K. M. (1997) Pollination effectiveness and pollinator importance in a population of *Heterotheca subaxillaris* (Asteraceae). *Oecologia*, **109**, 114–121.

Pettersson, M. W. (1997) Solitary plants do as well as clumped ones in *Silene uniflora* (Caryophyllaceae). *Ecography*, **20**, 375–382.

Platt, W. J., Hill, G. R. & Clark, S. (1974) Seed production in a prairie legume (*Astragalus canadensis* L). *Oecologia*, **17**, 55–63.

Quesada, M., Stoner, K. E., Rosas-Guerrero, V., Palacios-Guevara, C. & Lobo, J. A. (2003) Effects of habitat disruption on the activity of nectarivorous bats (Chiroptera: Phyllostomidae) in a dry tropical forest: implications for the reproductive success of the neotropical tree *Ceiba grandiflora*. *Oecologia*, **135**, 400–406.

Rasmussen, I. R. & Brodsgaard, B. (1992) Gene flow inferred from seed dispersal and pollinator behavior compared to DNA analysis of restriction site variation in a patchy population of *Lotus corniculatus* L. *Oecologia*, **89**, 277–283.

Richter, K. S. & Weis, A. E. (1998) Inbreeding and outcrossing in *Yucca whipplei*: consequences for the reproductive success of plant and pollinator. *Ecology Letters*, **1**, 21–24.

Robertson, A. W., Kelly, D., Ladley, J. J. & Sparrow, A. D. (1999) Effects of pollinator loss on endemic New Zealand mistletoes (Loranthaceae). *Conservation Biology*, **13**, 499–508.

Roll, J., Mitchell, R. J., Cabin, R. J. & Marshall, D. L. (1997) Reproductive success increases with local density of conspecifics in a desert mustard (*Lesquerella fendleri*). *Conservation Biology*, **11**, 738–746.

Roubik, D. W. (2001) Ups and downs in pollinator populations: when is there a decline? *Conservation Ecology*, **5** [online] http://www.consecol.org/vol5/iss1/art2/

Sakai, S., Momose, K., Yumoto, T., Kato, M. & Inoue, T. (1999) Beetle pollination of *Shorea parvifolia* (section Mutica, Dipterocarpaceae) in a general flowering period in Sarawak, Malaysia. *American Journal of Botany*, **86**, 62–69.

Santos, T., Telleria, J. L. & Virgos, E. (1999) Dispersal of Spanish juniper *Juniperus thurifera* by birds and mammals in a fragmented landscape. *Ecography*, **22**, 193–204.

Schmitt, J. (1983) Flowering plant density and pollinator visitation in *Senecio*. *Oecologia*, **60**, 97–102.

Schulke, B. & Waser, N. M. (2001) Long-distance pollinator flights and pollen dispersal between populations of *Delphinium nuttallianum*. *Oecologia*, **127**, 239–245.

Severns, P. (2003) Inbreeding and small population size reduce seed set in a threatened and fragmented plant species, *Lupinus sulphureus* ssp. *kincaidii* (Fabaceae). *Biological Conservation*, **110**, 221–229.

Sih, A. & Baltus, M. S. (1987) Patch size, pollinator behavior, and pollinator limitation in catnip. *Ecology*, **68**, 1679–1690.

Silander, J. A. (1978) Density-dependent control of reproductive success in *Cassia biflora*. *Biotropica*, **10**, 292–296.

Smithson, A. & Macnair, M. R. (1996) Frequency-dependent selection by pollinators: mechanisms and consequences with regard to behaviour of bumblebees *Bombus terrestris* (L) (Hymenoptera: Apidae). *Journal of Evolutionary Biology*, **9**, 571–588.

(1997) Density-dependent and frequency-dependent selection by bumblebees *Bombus terrestris* (L) (Hymenoptera: Apidae). *Biological Journal of the Linnean Society*, **60**, 401–417.

Sork, V. L. (1993) Evolutionary ecology of mast-seeding in temperate and tropical oaks (*Quercus* spp.). *Vegetatio*, **107/108**, 133–147.

Sowig, P. (1989) Effects of flowering plant's patch size on species composition of pollinator communities, foraging strategies, and resource partitioning in bumblebees (Hymenoptera: Apidae). *Oecologia*, **78**, 550–558.

Steffan-Dewenter, I. & Tscharntke, T. (1999) Effects of habitat isolation on pollinator communities and seed set. *Oecologia*, **121**, 432–440.

Thomson, J. D. (1981) Spatial and temporal components of resource assessment by flower-feeding insects. *Journal of Animal Ecology*, **50**, 49–59.

van Rossum, F., Echchgadda, G., Szabadi, I. & Triest, L. (2002) Commonness and long-term survival in fragmented habitats: *Primula elatior* as a study case. *Conservation Biology*, **16**, 1286–1295.

van Treuren, R., Bijlsma, R., Ouborg, N. J. & van Delman, W. (1993) The effects of population size and plant density on outcrossing rates in locally endangered *Salvia pratensis*. *Evolution*, **47**, 1094–1104.

Veit, R. R. & Lewis, M. A. (1996) Dispersal, population growth, and the Allee effect: dynamics of the house finch invasion of eastern North America. *American Naturalist*, **148**, 255–274.

Vergeer, P., Rengelink, R., Copal, A. & Ouborg, N. J. (2003) The interacting effects of genetic variation, habitat quality and population size on performance of *Succisa pratensis*. *Journal of Ecology*, **91**, 18–26.

Visscher, P. K. & Seeley, T. D. (1982) Foraging strategy of honeybee colonies in a temperate deciduous forest. *Ecology*, **63**, 1790–1801.

Waser, N. M. (1982) A comparison of distances flown by different visitors to flowers of the same species. *Oecologia*, **55**, 251–257.

Waser, N. M., Chittka, L., Price, M. V., Williams, N. M. & Ollerton, J. (1996) Generalization in pollination systems, and why it matters. *Ecology*, **77**, 1043–1060.

Washitani, I. (1996) Predicted genetic consequences of strong fertility selection due to pollinator loss in an isolated population of *Primula sieboldii*. *Conservation Biology*, **10**, 59–64.

Whelan, R. J. & Goldingay, R. L. (1986) Do pollinators influence seed-set in *Banksia paludosa* Sm. and *Banksia spinulosa* R. Br.? *Australian Journal of Ecology*, **11**, 181–186.

Widen, B. (1993) Demographic and genetic effects on reproduction as related to population size in a rare, perennial herb, *Senecio integrifolius* (Asteraceae). *Biological Journal of the Linnean Society*, **50**, 179–195.

Widen, B. & Widen, M. (1990) Pollen limitation and distance-dependent fecundity in females of the clonal gynodioecious herb *Glechoma hederacea* (Lamiaceae). *Oecologia*, **83**, 191–196.

Willmer, P., Gilbert, F., Ghazoul, J., Zalat, S. & Semida, F. (1994) A novel form of territoriality: daily paternal investment in an anthophorid bee. *Animal Behaviour*, **48**, 535–549.

Willmott, A. P. & Burquez, A. (1996) The pollination of *Merremia palmeri* (Convolvulaceae): can hawk moths be trusted? *American Journal of Botany*, **83**, 1050–1056.

Zimmerman, M. & Pyke, G. H. (1988) Reproduction in *Polemonium*: assessing the factors limiting seed set. *American Naturalist*, **131**, 723–738.

CHAPTER ELEVEN

Seed dispersal of woody plants in tropical forests: concepts, examples and future directions

HELENE C. MULLER-LANDAU
University of Minnesota
BRITTA DENISE HARDESTY
University of Georgia

Introduction

Understanding seed dispersal is critical to understanding plant population and community dynamics (Nathan & Muller-Landau 2000), especially in tropical forests where seed rain of virtually all plant species is sparse and patchy (Hubbell *et al.* 1999; Muller-Landau *et al.* 2002). Seed rain determines potential population growth rates and spatial patterns, as well as the relative influences of post-dispersal processes such as seed predation (e.g. Wright *et al.* 2000), microhabitat requirements for establishment (e.g. Svenning 1999) and density-dependent survival (e.g. Harms *et al.* 2000). Despite its importance, we know very little about seed dispersal of tropical trees, because it has been studied in only a tiny proportion of the many tropical tree species and seed dispersers, and because the patterns that have been observed have largely eluded easy generalization.

Just as there is a greater diversity of plant species and animal species in the tropics than in other regions, there is also a greater diversity of seed-dispersal strategies and patterns. Seed dispersal by animals predominates – it is the main strategy of 70%–90% of tropical forest plant species (Willson *et al.* 1989) – and involves a tremendous diversity of animal species and behaviours. Birds, bats, arboreal and terrestrial mammals (everything from mice to elephants), ants, dung beetles, even fish can disperse seeds (Levey *et al.* 1994). Animals may consume fruit and drop, spit or defecate the seeds, carry seeds in their coats or scatter-hoard seeds for later consumption. Abiotic strategies such as wind, water and ballistic dispersal form the main mode of seed movement for the remaining 10%–30% of tropical tree species (Willson *et al.* 1989). The seeds of many species may experience both primary dispersal (movement before the seed reaches the ground) and secondary dispersal (subsequent movement) usually by different modes or different animal species (e.g. movement by ballistic dispersal and then

Biotic Interactions in the Tropics: Their Role in the Maintenance of Species Diversity, ed. D. F. R. P. Burslem, M. A. Pinard and S. E. Hartley. Published by Cambridge University Press. © Cambridge University Press 2005.

by ants; movement by primate frugivores and then by rodent scatter-hoarders). Thus, understanding the complete seed rain of any given plant species may require understanding seed movement by a wide range of processes, each influenced by many different factors.

In this chapter, we begin by considering the relative importance of different vectors and animal groups to seed removal and the degree of specialization in plant–frugivore interactions in tropical forests. Next, we turn to seed deposition, where we examine patterns with respect to distance from the parent plant, habitat and clumping, in turn. Most seed-dispersal research has focused on patterns of seed deposition with respect to distance from the parent and/or distance to the nearest conspecific. We argue that the scales and magnitude of clumping of seed rain, and any biases in where seeds arrive by habitat, are at least as important. For each type of pattern, we consider the selective forces involved, and provide examples of variation in observed patterns and recommendations for researchers. Throughout, we draw extensively on studies from Barro Colorado Island, Panama (BCI), one of the best-studied tropical forests in the world (Leigh *et al.* 1996), both because we know this forest best, and because the multitude of studies at this site makes more comprehensive analyses possible. In the last section, we discuss exciting new developments in technologies and analytical methods and their promise for studies of seed dispersal, as well as areas most urgently in need of further study, including conservation implications.

Seed removal

The relative importance of different vectors

Overall, the proportion of plant species that are fleshy-fruited and thus presumed to be animal-dispersed varies between 70% and 100% in tropical rain forests around the world, with the proportion of tree species consistently above 65% (Willson *et al.* 1989). As precipitation decreases and dry-season length increases, animal dispersal becomes less common and wind dispersal more common (Bullock 1995; Gentry 1982; Willson *et al.* 1989). Water dispersal is, unsurprisingly, more common on wetter sites (Gentry 1983). There appears to be no relationship between edaphic conditions and dispersal mode (Gentry 1983). Within all forests, dispersal mode varies with life form. Wind dispersal is most common in lianas (often > 50% species), less common in large trees, and very rare in the understorey of tropical forests (Bullock 1995; Gentry 1982). Studies inferring whether plants are bird- or mammal-dispersed based on congener information or fruit morphology further suggest that mammal dispersal is more common than bird dispersal among taller trees while bird dispersal is relatively more common in lower-statured trees (Gentry 1982).

Most studies of the frequency of different dispersal syndromes among plant species in tropical forests infer dispersal syndrome from seed and fruit morphology. It is relatively straightforward to classify species as primarily wind-dispersed, ballistically dispersed, ectozoochorously dispersed (on the coats of animals) or endozoochorously dispersed (by fruit consumption). It is more difficult to go further and infer which group of animals is the principal disperser of endo-zoochorous fruits. Janson (1983) proposed that fruits in a Peruvian forest could be identified as predominantly bird- or mammal-dispersed on the basis of their colour, size and degree of protection. In particular, he classified as bird-dispersed fruits that were unprotected (no husk), relatively small (< 1.4 cm in the smallest dimension) and red, black, white, blue, purple or multi-coloured. Fruits that were protected, large and coloured orange, yellow, brown or green were classi-fied as mammal-dispersed. This is similar to classifications used by other authors (Gautier-Hion *et al.* 1985; Gentry 1982; Willson *et al.* 1989), although Gautier-Hion *et al.* (1985) find that in an African forest this division corresponds not to disper-sal by birds versus mammals, but rather birds and monkeys versus ruminants, rodents and elephants.

The accuracy of the Janson classification of fruits as mainly bird- or non-volant mammal-dispersed (Janson 1983) was tested on BCI (S. J. Wright and O. Calderón, unpublished data), another neotropical forest with a relatively similar animal community (Leigh 1999; Terborgh & Wright 1994). Of 78 com-mon, animal-dispersed, woody plant species, only 37 (47%) fit neatly into one of Janson's two categories. Of the 20 species thus classified as bird fruits, all are indeed known to be dispersed by birds, and all 17 species classed as mammal species are known to be dispersed by mammals. However, these species are also dispersed by other means (all 20 'bird' fruits are dispersed by nonvolant mam-mals, and 14 of 17 'mammal' fruits by birds). Of the 41 unclassified species, all are known to be dispersed by at least two of the three vertebrate groups (birds, bats and nonvolant mammals). Overall, it appears that inferences from fruit morphology do not reliably identify fruit consumers of a given plant species in this moderately diverse tropical forest (Howe 1986).

Given the diversity of plants and animals in tropical forests, there are very few sites at which we can identify the major animal fruit consumers (and generally seed dispersers) of many plant species from observations rather than inferences. In Central Panama, such information has been compiled for most plant species by S. J. Wright and O. Calderón from their own observations and numerous studies of frugivorous animals (e.g. Forget & Milleron 1991; Kalko *et al.* 1996b; Milton 1980; Poulin *et al.* 1999; Wehncke *et al.* 2003) and animal-dispersed plants (e.g. Croat 1978; Martin 1985). Furthermore, long-term studies of forest dynam-ics provide complementary information on the adult stature and abundances of free-standing woody plant species, allowing us to examine not only the pro-portion of species but also the proportion of basal area within various life-form classes that are dispersed in different ways (Condit *et al.* 1996; Hubbell & Foster 1983).

On BCI, primary dispersal of 73% of the woody plant species (encompassing 71% of the basal area of free-standing woody plants over 1 cm in diameter within the 50 ha Forest Dynamics Plot) is exclusively by animals, and of 26% (25%) is

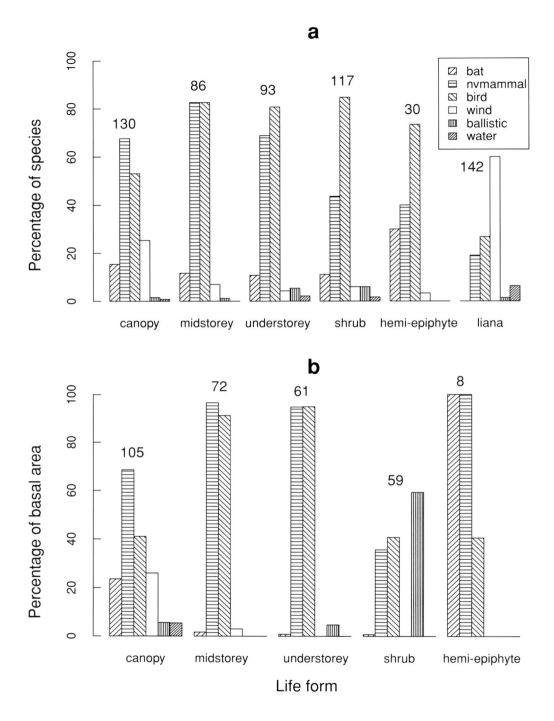

exclusively abiotic by wind, water, and ballistic means (Fig. 11.1). Wind dispersal is very common among lianas (60% of species) and is found in a quarter of canopy trees, but is rare among smaller-statured trees and hemi-epiphytes. Explosive dispersal is generally rare, but is the strategy employed by several of the most common shrub species (especially *Hybanthus prunifolius*), and thus 60% of shrub basal area is accounted for by ballistically dispersed species. Bird dispersal is most common in midstorey trees, understorey trees and shrub species. Dispersal by bats appears to be rare, although this probably reflects a relative paucity of knowledge even at this well-studied site, where 72 species of bats (E. Kalko, personal communication), including 25 frugivorous species, are known to occur (Kalko *et al.* 1996a). Water dispersal is very rare, and is found most often among liana species, 6% of which are thought to be mainly water-dispersed. The proportion of species secondarily dispersed by ants, dung beetles and other invertebrates is not known on BCI, but is likely to be high.

An alternative way to assess the importance of different animal groups to seed removal is to calculate the amount of fruit and seed biomass consumed by each. This requires information for each species on population density, feeding rates (grams dry weight consumed per day) and the proportion of fruit in the diet. Feeding rates can be estimated from body mass using allometric relationships (Nagy 1987). To truly understand the implications for seed dispersal requires further information on seed treatment and subsequent viability – disperser quality as well as quantity (Schupp 1993). In the absence of this information, and, more generally, in the absence of information distinguishing fruit and seed consumption, the estimates that we can make include both seed dispersal and seed predation.

Although birds and nonvolant mammals are similarly likely to participate in dispersing seeds of a particular species on BCI (Fig. 11.1), nonvolant mammals

---

Figure 11.1 For each life form, the percentages of (a) all species on BCI and (b) species weighted by their basal area in the BCI 50-ha Forest Dynamics Plot whose seeds are dispersed by different groups of animals (nvmammal = nonvolant mammals) and/or different abiotic methods. Species that are dispersed in multiple ways are accordingly counted multiple times, and thus the totals in each category often exceed 100%. All animal-disperser assignments are based on actual observations. Species for which no dispersal syndrome information was available were omitted. Free-standing woody plant life forms are defined by mature height: canopy trees as > 20 m, midstorey trees as 10–20 m, understorey treelets as 4–10 m, and shrubs as 1–4 m. Note that no basal area data is available for lianas, and that the basal area data for hemi-epiphytes includes only free-standing individuals. The number of species included in each life-form category in each panel is given above the corresponding bars. Dispersal syndrome data courtesy of S. J. Wright and O. Calderon; basal area data courtesy of the Center for Tropical Forest Science, Smithsonian Institution, Washington DC.

consume many times more fruit and seed biomass than birds and bats combined (Tables 11.1, 11.2). Two-thirds of this consumption is by animals that are known to be both seed dispersers and seed predators – mostly by terrestrial mammals. Even if we exclude species that act exclusively or in part as seed predators, however, nonvolant mammals still dominate in the biomass of fruit consumed. Seed predation is much less common among birds (Willis 1990) and has rarely been recorded among bats (but see Flannery 1995; Nogueira & Peracchi 2003). Each of these groups of animals includes many species on BCI: 16 species of nonvolant mammals (Leigh 1999), 25 species of bats (Kalko *et al.* 1996a) and 79 species of birds (Willis 1990) are known to consume fruits and/or seeds. Fruit and seed biomass consumption varies tremendously within groups, with most species and families contributing relatively little (Table 11.1). One species, *Alouatta palliata* (mantled howler monkey), accounts for over half the total for nonvolant mammals that are exclusive seed dispersers, and one species of bat, *Artibeus jamaicensis* (Jamaican fruit bat), similarly accounts for over half of all fruit biomass consumed by bats. Among birds, the most important species, *Tinamus major* (great tinamou), accounts for only 11% of the fruit and seed biomass consumption.

Animal communities differ among tropical forests, and thus the relative importance of different animal groups to fruit and seed consumption, and seed dispersal, also vary. We expect that most other tropical sites, with the exception of some monodominant forests, will have frugivore assemblages as diverse as or more diverse than that of BCI. For example, a family-level analysis suggests that the frugivore community is more diverse at La Selva, an everwet site in Costa Rica (Levey *et al.* 1994). In more-diverse frugivore communities, the most important species are likely to account for a smaller share of total fruit and seed consumption than on BCI. A pantropical comparison of primarily frugivorous birds by Snow (1981) finds ecologically similar though taxonomically disparate groups of avian frugivores in each region, except that Africa has comparatively few species of both specialized avian frugivores and plants belonging to families that such birds prefer. Levey *et al.* (1994) compare frugivore communities across the tropics for selected groups of birds and mammals, and find many parallels, but also some intriguing differences among the mammals. Calculations by Leigh (1999) suggest that mammals eat several times more fruit on the ground in BCI than they do in a climatically similar tropical forest in Parque Manu in Amazonian Peru, while consumption in trees by nonvolant mammals and birds is similar. Corlett (1998) reviews fruit and seed consumption by vertebrates in the Asian tropics, and finds frugivory to be very common among both mammals and birds, while seed predation is restricted to fewer species. Overall, few quantitative data are available for comparisons of frugivore communites, beyond examinations of species lists and general dietary information, and this is an obvious avenue for further research.

Table 11.1 *The densities and estimated fruit and seed consumption of birds and mammals whose diets include a substantial proportion of seeds and fruits (>10%) on BCI, by family*

Family	Status	Frugivorous species	Frugivore density	Estimated fruit and seed consumption
			(km^{-2})	$(kg\ ha^{-1}\ yr^{-1})$
Mammals				
Cebidae (cebid monkeys)	disperser	4	104	57
Procyonidae (raccoons)	disperser	3	44	19
Phyllostomidae (leaf-nosed bats)	disperser	25	346	8.9
Didelphidae (opossums)	disperser	5	107	6.7
Cervidae (deer)	disperser	1	3	2.8–6.3
Megalonychidae (two-toed sloths)	disperser	1	50	0–2.9
Mustelidae (weasels)	disperser	1	2	0.80–1.4
Callitrichidae (marmosets and tamarins)	disperser	1	3	0.36
Dasyproctidae (agoutis)	disperser/predator	1	100	59
Agoutidae (pacas)	disperser/predator	1	40	50–56
Tayassuidae (peccaries)	disperser/predator	1	10	17–30
Echimyidae (spiny and tree rats)	disperser/predator	1	180	20
Sciuridae (squirrels)	disperser/predator	1	180	14
Tapiridae (tapirs)	disperser/predator	1	0.5	1.0–6.1
Muridae (mice)	disperser/predator	1	133	3.3–3.7
Birds				
Ramphastidae (toucans)	disperser	3	60	0.7–6.2
Tyrannidae (tyrant-flycatchers)	disperser	14	239	0.3–3.0
Cotingidae (cotingas)	disperser	6	55.3	0.17
Thraupidae (tanagers)	disperser	11	149	0.2–2.1
Trogonidae (trogons)	disperser	5	47	0.2–1.7
Cracidae (curassows, guans and chachalacas)	disperser	2	4.7	0.9
Cathartidae (American vultures)	disperser	1	4.7	0.1–1.1
Pipridae (manakins)	disperser	3	77	0.11–0.96
Motmotidae (motmots)	disperser	1	13	0.09–0.80
Muscicapidae/ Turdidae (solitaires, thrushes and allies)	disperser	5	15	0.04–0.32
Picidae (woodpeckers)	disperser	1	10	0.03–0.30

(*cont.*)

Table 11.1 (*cont.*)

Family	Status	Frugivorous species	Frugivore density	Estimated fruit and seed consumption
Icteridae (American orioles and blackbirds)	disperser	5	6	0.02–0.20
Parulidae (wood-warblers)	disperser	2	4	0.005–0.042
Vireonidae (vireos)	disperser	2	3.7	0.005–0.042
Corvidae (jays, magpies and crows)	disperser	1	0.1	0.0005–0.0042
Columbidae (Pigeons)	disperser/predator	7	68	4.2
Tinamidae (tinamous)	disperser/predator	1	13	2.8
Emberizidae (New World sparrows and Old World buntings)	disperser/predator	9	137	0.2–2.1
Fringillidae (fringilline and cardueline finches and allies)	disperser/predator	1	2	0.021
Psittacidae (parrots)	predator	4	67	5.1
Rallidae (rails, gallinules and coots)	predator	2	6.7	0.04–0.38

Families are ordered by decreasing fruit and seed consumption within each category (mammalian dispersers, mammalian dispersers and predators, etc.). Estimated fruit and seed consumption for each species was calculated as the product of animal density (per square kilometre), feeding rates per animal (kg dry mass per year), and the proportion of the diet composed of fruits and seeds. Densities and estimated fruit and seed consumption were summed over fruit- and seed-consuming species to arrive at totals for each family. Feeding rates for each species were calculated from body mass using allometric equations given in Nagy (1987). Densities, body masses and diets for nonvolant mammals from Leigh (1999), Gompper (1996) and J. Giacalone Willis (personal communication); for birds from Willis (1990); and for bats from Kalko *et al.* (1996a). The complete species list and species-level data are available from the authors upon request.

Specialization and generalization among plant–disperser interactions

Researchers have long debated whether there are strong coevolutionary relationships between plants and their animal dispersers (Howe 1993). While there is ample evidence of mutual benefits in these relationships in tropical forests, neither theory nor data support the idea of tight coevolution, such as that observed between figs and fig wasps (Wheelwright & Orians 1982). Individual plant species are seldom adapted to dispersal by one particular frugivore species (Levey *et al.* 1994). Tight coevolutionary relationships are the exception rather than the rule.

Most plant species are dispersed by multiple dispersers, and most dispersers disseminate seeds from multiple plant species. The interaction may vary with site, seasonality, resource availability, nutrient reward of the plant and dietary

Table 11.2 *The estimated dry biomass of seeds and fruits consumed*

	Exclusive disperser	Both disperser and predator	Exclusive predator
Consumption in trees			
by nonvolant mammals	77.8 (13.5)	7.2 (0.5)	
by bats	8.94 (25)		
by birds	13.14 (66.5)	1.13 (4)	5.16 (7)
Consumption on ground			
by nonvolant mammals	11.83 (2.5)	168.38 (6.5)	
by birds	2.92 (5.5)	2.85 (2)	0.17 (1)

Values are in kg per ha per year on Barro Colorado Island and (in parenthesis) the number of species involved, for various categories of seed dispersers and predators (see Table 11.1 for families included; sums were taken at the species level). Species that consume substantial amounts of fruits and/or seeds on the ground as well as in trees were counted as half in each list. The complete species list and species-level data are available from the authors upon request.

requirements of the frugivore. In general, dietary requirements and fruit availability shift seasonally and regionally, with many species of large frugivorous animals tracking different fruit resources as they become available throughout the year (Blake & Loiselle 1991; Whitney & Smith 1998). The disperser assemblage of a given tree species can vary with habitat, as Howe and Vande Kerckhove (1979) found for *Casearia carymbosa* at wet- and dry-forest sites in Panama.

Despite the lack of clear specialization, there is abundant evidence that animals differ in their effectiveness as seed-dispersers (Schupp 1993), and that plants differ in the rewards they offer and the costs to animals of obtaining these rewards (Gautier-Hion *et al.* 1985). The next section discusses differences among animals in their seed deposition patterns, which affect both the quality and effectiveness of seed dispersal (Schupp 1993). Also of concern when evaluating dispersal is how the frugivore treats the fruit or seed ingested, since this determines whether the frugivore acts as a disperser or seed predator. A meta-analysis of gut treatment effects found that, on average, bats and birds had strong positive impacts on germination, while nonvolant mammals had slight positive effects (Traveset & Verdú 2002). However, such generalities mask important variation: ecologically similar species may exert very different effects on fruits, and the same animal species may have very different effects on different plant species. Small seeds pass intact through the guts of guenons (*Cercopithecus* sp.), Old World monkeys, whereas larger seeds are increasingly likely to be destroyed, and no seeds above 2 mm in diameter survive gut passage (Gautier-Hion 1984). However, large seeds may be effectively dispersed by guenons if they are carried away from the parent in cheek pouches and then spat out rather than swallowed (Rowell & Mitchell 1991). In contrast, seeds of most species pass intact through the gut of the ecologically analogous New World capuchins (*Cebus* sp.): 31 of 35 plant species tested in Panama germinated after gut passage (Rowell & Mitchell 1991).

Differences among animals in dispersal effectiveness and food preferences should provide scope for selection on plant species for traits that attract or repel potential fruit consumers based on their dispersal effectiveness and related contributions to plant fitness. Fruit syndromes, discussed previously, certainly suggest that these differences have been important for fruit-trait evolution (Levey *et al.* 1994). Cases of directed deterrence provide even stronger evidence for such selection. The capsaicin found in the fruits of wild chiles repels mammals, which tend to be seed predators of chiles, while having no effect on birds, thus ensuring that virtually all fruit removal and subsequent seed dispersal is by birds (Tewksbury & Nabhan 2001).

Once again, we turn to the extensive history of research on BCI to assess the actual degree of specialization and generality by both animals and plants with respect to fruit consumption and seed dispersal, and to test for patterns in variation in specialization among taxa. Such comparisons among species are best done in the same forest, as the diversity of plants and animals varies among forests, and the diversity of available partners is expected to affect the diversity of interactions for most species. A valuable recent review of the diet diversity of frugivorous vertebrates in another tropical site, the Guiana Shield, is provided by Forget and Hammond (2005).

Among the 13 species of frugivorous mammals whose diets have been intensively studied on BCI, the number of plant species whose fruits or seeds are consumed varies between 8 and 166. Monkeys have particularly broad diets: *Ateles geoffroyi* (Central American spider monkey) has been observed consuming the fruit of 130 plant species over one year (C. Campbell, personal communication), *Alouatta palliata* (mantled howler monkey) 103 species (Milton 1980), and *Cebus capucinus* (white-throated capuchin monkey) 95 species in just a 4-month period (Wehncke *et al.* 2003). Other nonvolant mammals have somewhat more restricted diets: *Potos flavus* (kinkajou) has been observed feeding on 74 species (Kays 1999), *Sciurus granatensis* (red-tailed squirrel) on 58 (Glanz *et al.* 1982), *Nasua narica* (white-nosed coati) on 51 (Gompper 1994), and *Dasyprocta punctata* (Central American agouti) on 38 (Smythe *et al.* 1982). Among bats, all those studied seem to have relatively narrow diets: only 17 plant species have been recorded in observations and facial samples of the dominant frugivorous bat *Artibeus jamaicensis* over many years of study (Handley *et al.* 1991), and only 18 and 19 species are consumed by the much smaller bats *Carollia castanea* and *C. perspicillata*, respectively (Thies 1998).

An understorey mist-netting study near BCI in Central Panama provides information on the diversity of bird diets, which seems generally to be lower than that of mammals – but this may in part reflect more-limited sampling (Poulin *et al.* 1999). Of the four bird species for which more than 50 diet samples were obtained (via regurgitation or defecation), samples from *Pipra mentalis* (red-capped manakin) included fruit material from 45 plant species, *Pipra coronata* (blue-crowned manakin) 26 species, *Mionectes oleagina* (ochre-bellied flycatcher) 19 species, and *Phaethornis superciliosis* (long-tailed hermit) just three species (Poulin *et al.* 1999). (*Pipra coronata* and *Mionectes* are not found on BCI, and *Phaethornis* is not listed as a fruit consumer there by Willis 1990.)

In-depth studies of fruit removal from individual plant species on BCI paint a similarly varied picture of the number of fruit- and seed-consuming species per plant species – although overall there are fewer fruit-consuming species per plant species than fruit food-source species per animal species. Martin (1985) summarized results of avian frugivore visitation studies in seven canopy tree species in Panama and found that between 7 and 46 bird species visit and pre-sumably consume fruit of a particular tree species, with larger-fruited species receiving fewer visitors. In the forest understorey, the diversity of animal species recorded as taking fruit of a given plant species seems to be lower than in the canopy, and the relationship between fruit size and number of avian visitors (Martin 1985) appears to break down. Two species of manakins account for 62% of all *Miconia* fruit found in diet samples of more than 2000 individuals of 103 bird species in central Panama (Poulin *et al.* 1999). Similarly, a mere six species of birds, three resident manakins and three migrant thrushes, account for 97% of all *Psychotria* fruits found in the same bird-diet samples. Focal watches of indi-vidual shrubs of *Piper dilatum* revealed that 49% of all fruits are taken by bats – probably almost entirely by the two *Carollia* species mentioned earlier (Thies 1998).

Records of feeding observations on BCI overall show that most plant species have their fruit consumed not only by many different animal species, but by species from multiple major animal groups. Of the 431 BCI woody plant species that are biotically dispersed, 40% are thought to be dispersed almost exclusively by one of three major groups of vertebrates: birds, bats or nonvolant mammals (S. J. Wright and O. Calderón, unpublished data). A full 60% are known to be dispersed by species belonging to two or more of these groups (S. J. Wright and O. Calderón, unpublished data). If we consider basal area, we find an even higher proportion with multiple disperser groups: 93% of the basal area of species that are dispersed by vertebrates are dispersed by species belonging to two or more groups of vertebrates; the proportion for shrubs alone is 89%. The abundant species that dominate basal area are the ones whose fruit consumption is best-known; thus, we expect that a similar proportion of rare species will prove to be consumed by multiple vertebrate groups as more data accumulate.

Not only are some fruits of a species taken by some animals and other fruits by other animal species, but the same fruit and seed may itself be handled by multiple animal species. Many species on BCI – as elsewhere – experience primary dispersal by birds, bats or arboreal mammals, and then secondary dispersal by ground-dwelling rodents or other animals, including ants and dung beetles (Feer & Forget 2002). For example, leafcutter ants disperse seeds of *Simarouba amara*, whose fruits are also consumed by birds and nonvolant mammals. The impact of ant dispersal activity is evident in the exceptionally high *Simarouba* seedling densities observed near leafcutter-refuse piles located near reproductive trees, in comparison to other sites without ant nests nearby (B. D. Hardesty, unpublished data). While ants and dung beetles move only small seeds, scatter-hoarding rodents such as agoutis are important as seed dispersers, as well as seed predators, of many particularly large-seeded plant species (Forget & Milleron 1991). In addition to their effect on seed movement, seed treatment by secondary dispersers can be important to seed survival; in particular, seed burial by agoutis, ants or dung beetles can help seeds escape subsequent predation by other insects and rodents (Wright & Duber 2001).

While all these data indicate a tremendous diversity of frugivores for any given plant species, a small proportion of these partners are responsible for a disproportionate share of fruit consumption and associated seed-dispersal services. For example, closer examination of focal tree watch data discussed earlier shows that in all cases the majority (> 50%) of visits are by just 1–3 animal species, and 90% by at most 10 species, and seed-removal rates are even more skewed (e.g. Howe 1977; Howe 1980; Russo 2003b). Indeed, if we calculate a Shannon diversity index of animal visitors instead of merely examining species richness, we find that values for visits range from 2.5 to 3.5, and values for actual seed removal range from 1 to 1.5. If we could also incorporate information on ultimate seed fate, we might well find that the diversity of animal species that provide effective seed-dispersal services is even lower.

Similarly, relative specialization of animals on particular fruits is more apparent when we examine the numbers of fruits consumed for each species, rather than simply the number of species, and when we look at somewhat broader taxonomic groups. On BCI, there are several animal species whose diet is dominated by one genus of plants: for example, 92% of the occurrences of fruit or seed material in *Carollia castanea* fecal samples were of *Piper* species (as were 58% of those of *C. perspicillata*) (Thies 1998). Dietary overlap among different animal groups is also much lower when fruit species are weighted by their proportional abundance in the diet, as exemplified by a study of overlap between hornbill and primate diets in Cameroon (Poulsen *et al.* 2002). This disparity between occurrence and quantitative importance could potentially explain the mismatch between fruit-dispersal syndromes that seem designed to attract a particular species or group of dispersers and the wider diversity of species that actually consume fruit. To evaluate these possibilities fully, we need more such quantitative data on fruit removal and consumption, and ultimate seed fate.

Neither complete specialization nor complete generalization is found among either plants or their seed dispersers in tropical forests. Instead, we find more complicated intermediate patterns. At their most specific, plant–disperser inter-actions seem to involve particular genera that account for a majority – but never all – of each other's fruit removal and diet, respectively. Sets of fruit traits that appear designed to attract birds over mammals, or mammals over birds, do not reliably predict the array of fruit consumers of a species. But it is possible that they may be better predictors of the dominant consumers, or alternatively, of those that are most effective at dispersing seeds – of the animal species whose actions are really determining plant fitness, and thus selection on plant traits. In the next section, we examine patterns of seed deposition, and how they differ in favourability among animals and other dispersal vectors.

Seed deposition

In the last section we examined seed removal from parent plants. Now we turn to examine what we know about where seeds are deposited. We can divide con-siderations of seed deposition into three general, somewhat overlapping areas: dispersal distance, habitat-specificity and clumping. All three aspects contribute to determining the quality and effectiveness of seed dispersal (*sensu* Schupp 1993), which varies among plant species and dispersal vectors. To understand the importance of each aspect, we must consider not only how it affects the probability of success of individual seeds, but also how it affects parent-plant fitness – especially since it is largely maternal genes that determine fruit mor-phology and other traits affecting seed dispersal (Ronce *et al.* 2001).

The most obvious impacts of seed-deposition patterns are on seed success, and these impacts have been the principal focus of many studies evaluating the advantages of particular dispersal strategies. Seed survival, establishment proba-bility and subsequent recruitment success may be improved by dispersal to sites with better environmental conditions and lower numbers of natural enemies. Because natural enemies are thought to concentrate near conspecific adults and in areas of high conspecific density (Connell 1971; Janzen 1970), and seed suc-cess is strongly negatively density-dependent in tropical forests (Harms *et al.* 2000), much research has focused on how longer seed-dispersal distances and reduced clumping enhance seed survival (described as the 'escape' hypothesis by Howe & Smallwood 1982). In addition, abiotic conditions for establishment may be more favourable at sites beyond the shade of the parent crown. Dis-proportionate dispersal to particularly favourable types of regeneration sites, or 'directed dispersal' (Howe & Smallwood 1982), has thus been another focus of study (nonetheless, its importance may still be underestimated: see Wenny 2001).

More generally, however, seed dispersal away from parents will be favoured even if individual seeds actually have a lower probability of succeeding if dis-persed (Levin *et al.* 2003). This is because dispersal reduces kin competition and allows for bet-hedging over environmental variation, and thus enhances reproductive success of the parent plant and its genes favouring dispersal. If all seeds stay beneath the parent, then they compete intensively with their siblings, severely constraining the number that could possibly survive to adulthood. As a result, a genotype that has even a minute chance of successfully dispersing and succeeding in another site will always outcompete one that never disperses and merely retains the same site (Hamilton & May 1977). If the environment varies in space and time, dispersal away from parent plants and from siblings may be selected because it allows for bet-hedging over spatial and temporal uncertainty, even if it means that some individual seeds are dispersed to habitats where they have only a very low probability of success (Cohen & Levin 1991; Howe & Smallwood 1982).

Here we consider patterns of seed deposition with respect to distances from parent trees and habitat, as well as clumping that is due to other factors. In each case, we begin by drawing attention to the selective forces on these patterns, noting especially cases in which they are conflicting. We briefly summarize how these patterns are measured, and review typical examples. Finally, we make suggestions for standard methods.

Distances

Dispersal distances are the most frequently studied aspect of seed deposition. Studies have examined both the true dispersal distance from the mother tree, and the 'effective dispersal distance' – the distance of a seed from the nearest conspecific adult (Bustamente & Canals 1995). Effective dispersal distances are often easier to measure, and they too have ecological and evolutionary relevance. In general, longer true dispersal distances make it possible for the offspring of a single parent to sample a larger area, thus reducing kin competition and effecting bet-hedging over a greater number and variety of regeneration environments (although severe clumping of seed deposition can reduce or even eliminate these benefits). On the population level, longer dispersal distances generally reduce seed limitation by reducing dispersal limitation (Clark *et al.* 1998; Nathan & Muller-Landau, 2000). On population fronts, longer dispersal distances also aid the colonization of new habitats free of conspecific competitors (dispersal distances determine rates of population advance). However, if there are environmental gradients or the habitat is patchy, longer dispersal distances may also be more likely to land seeds in unfavourable environments. Longer effective dispersal distances are expected to reduce exposure to natural enemies and intraspecific competition, but they may also reduce the likelihood of encountering specialized mutualists or tapping into parental resources via mycorrhizal networks (Fitter *et al.* 1998).

Seed-dispersal distances can be measured or estimated via a number of methods. Each method has different advantages, disadvantages and biases. Comparing dispersal distances for different dispersal vectors is fraught with difficulties because different methods are more appropriate and more often used for some vectors than others, and because each method has its own distinct inherent biases. Dispersal by scatter-hoarding rodents is typically investigated with seed mark-recapture studies, dispersal by wind with seed-traps or seedling mapping studies, and dispersal by arboreal and volant animals by following animals and predicting dispersal distances from movement rates and gut passage time. Thus, while we report dispersal distances for a range of studies, we generally restrict our comparisons to studies using similar methods.

Patterns

A rare study that used the same methods to assess primary dispersal of multiple species dispersed by multiple vectors took place on BCI. Seed shadows of 81 individual tree species were estimated from a seed-trap study (Wright *et al.* 1999) within a mapped plot (Hubbell & Foster 1983) using inverse modelling (Muller-Landau 2001). Species' mean dispersal distances ranged from 2.8 to 152 m, with considerable variation within each dispersal syndrome. The identity of the most important dispersal vector (wind, water, ballistic, small bird, large bird, bat or nonvolant mammal) did not explain variation among species in mean dispersal distances, except that ballistically dispersed species had shorter mean dispersal distances (Muller-Landau 2001).

Spatial genetic structure can be used to infer the spatial scales of gene flow, which includes pollen as well as seed movement. If pollen movement modes or distances can be controlled for, or are uncorrelated with seed dispersal modes, then comparisons of spatial genetic structure among seed dispersal modes should provide information on seed-dispersal distances. Hamrick and Loveless (1986) compare genetic diversity within and among populations of eight tree and shrub species on BCI dispersed variously by birds, by bats and ballistically. They found that the relationship between seed-dispersal mode and the distribution of genetic variation was not as significant as they had anticipated. A review combining temperate and tropical studies found that overall, gene flow rates were lowest for gravity- and explosively dispersed species, and highest for species dispersed by wind or seed ingestion by animals (Hamrick *et al.* 1995). No studies presenting genetic data on actual seed-movement patterns in the tropics are published at this time, though there are several studies currently under way (B. D. Hardesty, unpublished data; F. A. Jones, personal communication; C. Woodward, personal communication).

Biogeographic patterns can also provide some insight into dispersal distances. All other things being equal, longer dispersal distances are expected to result in larger range sizes (Chave *et al.* 2002). Gentry (1983) notes that range sizes and geographic distributions are largest for wind-dispersed species, intermediate for species with characteristics suggesting bird dispersal, and smallest for species apparently mammal-dispersed. This implies that seed dispersal by wind results in the longest dispersal distances, seed dispersal by birds intermediate, and seed dispersal by mammals the shortest distances. Of course, dispersal syndrome may be correlated with other factors that also influence range size, such as life-history strategy or habitat specialization. Information on long-distance dispersal rates can be gleaned from an examination of floras of isolated islands, since the arrival of a species there demonstrates a capacity for long-distance dispersal. Island floras tend to have many bird-dispersed species, some wind-dispersed species and sometimes a few bat-dispersed species, but no species dispersed by terrestrial and arboreal mammals (Gentry 1983; Yockteng & Cavelier 1998). It is possible, however, that dispersal distances by arboreal and terrestrial mammals can be quite long on contiguous land, even if these animals cannot generally move seeds across long expanses of water.

For animal seed dispersers, movement rates, home-range size and gut-passage time together should predict dispersal distances, and as these vary among dis-perser groups, so should dispersal distances. Comparing among studies, mean dispersal distances for animal groups seem to be shortest for ground-dwelling rodents and small birds, intermediate for bats and monkeys, and longest for large canopy birds. The longest measured mean dispersal distances are reported by Holbrook and Smith (2000), who estimate that the mean seed-dispersal dis-tances for two species of African hornbills (*Ceratogymna atrata* and *C. cylindricus*) range from 1127 to 1947 m, depending on seed size and disperser species. In a later study, they record large-scale movements of hornbills up to 290 km, sug-gesting this species also has the potential to occasionally provide super-long-distance dispersal (Holbrook *et al.* 2002). Thies (1998) found that mean seed dispersal distances by a neotropical fruit bat (*Carollia castanea*) on BCI were 106 m for females and 280 m for males, with a range from 15 to 1400 m and modal dispersal distances of 100–200 m. Wehncke *et al.* (2003) estimated that modal dispersal distances by white-faced monkeys (*Cebus capucinus*) on BCI were also in the 100–200 m range, with a mean dispersal distance of 216 m and a range of 20–844 m. For a small tropical tyrant flycatcher, *Mionectes oleaginous* in Costa Rica, seed-dispersal distances estimated from the mean gut-passage time and the mean movement in that time (which will generally not be equal to the mean dispersal distances) ranged from 21 to 100 m, depending on the plant species (Westcott and Graham 2000). In comparison, a seed mark-recapture study of seed dispersal by the red acouchi (*Myoprocta exilis*) in French Guiana found seeds cached at distances of 0 to 124 m from their original locations (Jansen *et al.* 2002).

Within each of these broad categories of animal disperser there is consider-able variation in dispersal distances. Some of this variation seems to be related to body size, which is hardly surprising since metabolic rates, feeding rates and home ranges all scale with body size (Brown 1995). Westcott and Graham (2000) show that there is a positive, almost linear, relationship between body mass and median dispersal distance among eight tropical bird species. Kalko *et al.*

(1996b) report that home-range size scales positively with body size among fig-eating bat species on BCI, and suggest that seed-dispersal distances will be correspondingly larger for larger bats. It has also been suggested that the larger primate species provide longer-distance seed-dispersal services than smaller ones (Peres & van Roosmalen 2002). Additional research is needed to assess these patterns. Nevertheless, there are certain to be exceptions; for example, large, but less-mobile arboreal birds such as guans and turacaos may not move seeds as far as smaller avian dispersers.

The same animal species may move seeds of different plant species different distances (Westcott & Graham 2000). Gut-passage times typically increase with seed size (Holbrook & Smith 2000), and also vary among seeds of similar size (Murray 1988), possibly because of differences in fruit chemical composition. Because gut-passage times for larger-seeded species tend to be longer, we might expect a positive relationship between seed size and dispersal distances among endozoochorously dispersed species. Mean estimated dispersal distances of large-seeded species by hornbills are larger than those for small-seeded species (Holbrook & Smith 2000). To evaluate whether overall mean dispersal distances are higher for large-seeded plants, we must further consider changes in which species take fruit. Because larger-seeded fruits are typically eaten by larger animal species (Kalko *et al.* 1996b; Peres & van Roosmalen 2002), and because these larger animals should have larger home ranges, we might expect a positive relationship to hold up overall as well. Among 44 animal-dispersed species on BCI, however, there was no relationship between seed mass and mean dispersal distance (Muller-Landau 2001). Theory suggests that dispersal distances by scatter-hoarding rodents should increase with seed size for a different reason – because larger seeds offer more reward for the effort of caching (Jansen *et al.* 2002). Several studies have found such an increasing relationship between seed size and caching distance (Forget *et al.* 1998; Jansen *et al.* 2002).

Among wind-dispersed species, mechanistic models predict that dispersal distance will decrease with increasing seed size, because heavier seeds tend to fall more quickly (Augspurger 1986). Estimated mean dispersal distances did indeed decrease with seed mass among 16 wind-dispersed species on BCI (Muller-Landau 2001).

Rare, long-distance seed dispersal events may be particularly important for plant populations (Cain *et al.* 2000). The magnitude and frequency of long-distance dispersal events will not necessarily vary in parallel with mean or typical dispersal distances among species (Muller-Landau *et al.* 2003). For example, while bats usually have fast food processing and/or gut-passage times and correspondingly short to moderate dispersal distances, Shilton *et al.* (1999) find that some small seeds are ingested and may remain in the gut of the Old World fruit bat *Cynopterus sphinx* for more than 12 hours, resulting in potential dispersal distances of kilometres or even tens or hundreds of kilometres. Unfortunately, such rare, long-distance dispersal events are particularly difficult to measure (Nathan *et al.* 2003), although inferences using genetic techniques make their study much more feasible (Cain *et al.* 2000).

Recommendations

Comparisons of distributions of dispersal distances among studies are complicated not only by differences in methodology but also by the fact that dispersal distance distributions vary in their shapes and that different statistics on these distributions are reported by different authors. Because of the variation in shapes, a distribution with a higher mean dispersal distance may not also have a higher median, much less a higher number of seeds going beyond 100 m, a longer 90th percentile distance, etc. No single statistic can adequately summarize any distribution, and for this reason we encourage authors to report multiple statistics. In particular, we recommend reporting the mean as well as the 10th, 25th, 50th (median), 75th and 90th percentiles of the distribution. Minimum and maximum dispersal distances are also interesting, but are particularly sensitive to sample size (which should, of course, also be reported). Many papers present histograms of observed or estimated dispersal distances, which provide valuable information on the entire distance distribution. Future syntheses and comparisons would be greatly aided if authors also report the entire distribution of dispersal distances in electronic appendices.

Habitat-specific deposition

In addition to variation with distance from tree, seed-deposition rates may also vary with habitat (Nathan & Muller-Landau 2000), for example between gaps and understorey sites (Schupp *et al.* 1989). Our definition of a habitat worthy of consideration is one that makes a difference for the recruitment of the plant species under consideration – whether positive or negative – for reasons beyond simply the distance from parent trees (treated in the previous section) or the density of seeds deposited (treated in the next section, on clumping). Disproportionate deposition of seeds to particularly favourable sites for recruitment via directed dispersal (Howe 1986; Howe & Smallwood 1982) is one example of habitat-specific deposition. The other possibility, of course, is that seeds are disproportionately deposited in sites that are less favourable for recruitment – a possibility that has rarely been considered in the literature. Directed dispersal has the obvious selective advantage that it increases the odds of success of individual seeds. Countering this advantage are the kin-selection and bet-hedging benefits of spreading seeds over a larger area, including less-favourable sites, as long as there is some probability of successful recruitment at these sites. Thus, even were it possible, dispersal exclusively to the best regeneration habitat may not be the optimal strategy, although higher-than-average seed deposition in these areas should always be favoured over random deposition with respect to habitat.

Patterns

Disproportionate dispersal to gaps is undoubtedly the most frequently studied type of habitat-specific deposition (Schupp *et al.* 1989; Wenny 2001; Wenny & Levey 1998). Gaps are a favourable recruitment environment for many species (Brokaw & Busing 2000), and are the only environment in which regeneration can occur for some species (Dalling & Hubbell 2002). The most spectacular cases of directed dispersal to gaps involve birds that display or lek in gaps (e.g. Wenny & Levey 1998). Several studies have also suggested that wind-dispersed seeds may arrive in gaps at higher densities than in adjacent forests due to small-scale turbulence, but evidence remains equivocal (e.g. Augspurger & Franson 1988; Dalling *et al.* 2002). Much less noted is the opposite pattern – disproportionate dispersal to understorey sites. Such habitat-specific deposition is likely to be common for seeds dispersed by arboreal animals that rarely or never come to the ground. If the species thus dispersed regenerate better in the understorey, then this too would qualify as directed dispersal; for most species, however, it is likely to constitute the opposite.

Numerous studies have also documented patterns of tropical forest seed deposition (Duncan & Chapman 2002) and subsequent recruitment (Alrich & Hamrick 1998; Sezen *et al.* 2005) in pastures and fields near forests. Such fields and pastures offer a very different regeneration environment from forests, one that is considerably more hostile to most tropical forest plant species (Holl *et al.* 2000). Seed deposition by arboreal and terrestrial species typically drops off sharply at forest edges, while seed deposition by wind declines slowly with distance beyond the forest boundary, and seed deposition by birds and bats decreases but continues in a patchy fashion (e.g. Gorchov *et al.* 1993). Some bird and bat species move over and into open areas, where they deposit seeds mostly, but not exclusively, beneath isolated trees or other emergent structures (Duncan & Chapman 2002; Slocum & Horvitz 2000). Because abiotic conditions for seedling regeneration are better under isolated trees (Toh *et al.* 1999), relatively high seed deposition in these areas can also be considered directed dispersal.

Recommendations

Few studies have examined, or at least reported, disproportionate dispersal to any habitats other than gaps, fields or abandoned pastures. This is surprising, because there are numerous other habitat divisions upon which many tropical forest plants are specialized (e.g. Clark *et al.* 1995; Svenning 1999), and because many animal species show uneven patterns of habitat use (e.g. McShea *et al.* 2001). We believe that biases in seed-deposition with respect to habitats other than gaps are likely to be widespread, and deserve greater attention. More studies should examine whether seed-deposition rates vary with habitats distinguished by their topography, soils and/or water availability. In addition, habitats may also differ in abiotic ways as a result of animal activity. For example, trails used frequently by animals, including humans, may present a difficult regeneration microhabitat in which seedling densities are lower than in surrounding areas (Voysey *et al.* 1999). Such microhabitat differences deserve further attention.

The habitats and microhabitats worth studying will depend on the plant species involved and the factors that affect its regeneration success. When detailed information on such factors is lacking (as it usually is), we recommend tests for biases with respect to multiple classes of habitat, distinguished not only by understorey light availability but also by topography, soils and water availability, when possible. Better methods of recording locations of seed deposition (e.g. use of the global positioning system, GPS), of combining this information with available data on landscape characteristics (e.g. from satellite images) and of analysing the resulting associations or lack thereof (via geographic information systems, GIS) will greatly facilitate such studies. It is important that even negative results be reported. Finally, such studies will be greatly enriched if they are paired with parallel investigations of seed- and seedling-recruitment success in different habitats.

Clumping

While seed-dispersal distances have received much more theoretical and empirical attention, the degree and scale of clumping of seed deposition are important for seed success and parental fitness for most of the same reasons. When seed rain is highly clumped, seeds experience higher density-dependent mortality, reducing seed success (Harms *et al.* 2000, but see Fragoso 1997). Further, clumping increases kin competition and allows the offspring of a tree to sample fewer sites, thus limiting bet-hedging over spatiotemporal variation. Clumping increases seed limitation of populations because fewer sites are reached by seeds, an effect Schupp *et al.* (2002) refer to as dissemination limitation. Like short dispersal distances, clumping reduces interspecific competition, slows competitive exclusion and can potentially help to maintain nonequilibrium diversity (Chave *et al.* 2002; Hurtt & Pacala 1995).

Clumping of seed rain can occur at multiple spatial scales (area over which seeds are distributed), temporal scales (time over which seeds are deposited) and intensities (numbers of seeds deposited). An example from BCI illustrates how clumping at small spatial scales is evident over multiple temporal scales in the seed rain of two tree species, one animal- and one wind-dispersed. Looking only at traps that are more than 15 m from a conspecific adult, an examination of all weekly seedfall records over 15 years shows that most records have very few seeds, while some have very high seedfall – evidence of clumping of seed rain among traps (Fig. 11.2). If we look by year, some traps continue to have much higher seedfall than others, because high-seedfall weeks are nonrandomly distributed among traps. Similarly, the 15-year record continues to show clumping of seed rain, with high concentrations in a few traps – evidence of even longer-timescale

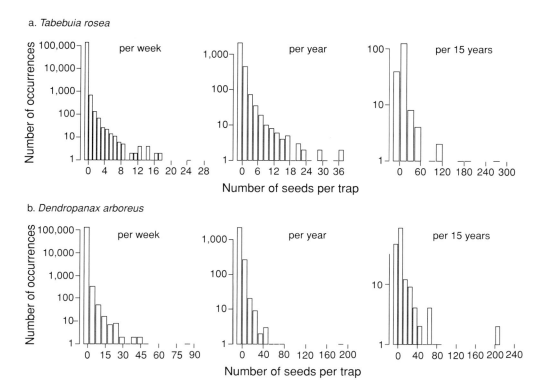

Figure 11.2 Number of occurrences of different numbers of seeds of (a) the wind-dispersed tree *Tabebuia rosea* and (b) the animal-dispersed tree *Dendropanax arboreus* falling into traps in a given week, year, or 15 years on Barro Colorado Island, Panama. This includes only those traps more than 15 m from conspecific adults. A horizontal mark along the horizontal axis represents a single trap characterized by that combination of seed number and time interval. Each seed-trap is 0.5 m^2. Seed-trap data courtesy of S. J. Wright, O. Calderón, and the Environmental Science Program of the Smithsonian Institution.

spatial autocorrelation in seed rain (Fig. 11.2). Some of the nonrandomness of the distribution of seeds among traps is due to differences in distances to, number of, and fecundity of nearby reproductive adults. However, the magnitude of clumping exceeds what would be expected from distance-dependence alone (H. C. Muller-Landau *et al.*, unpublished analyses), and it is not associated with any identified differences in regeneration habitat (Dalling *et al.* 2002). It appears that this clumping is the result of other processes that result in clustered deposition of seeds, and repeated deposition of seeds in some sites over others.

Patterns

Seed dispersal by animals is particularly likely to result in clumping, which varies depending on the dispersing animal and its behaviour (Schupp 1993;

Schupp *et al.* 2002). Clumping of seed dispersal by wind is much less pronounced. When seed shadows that varied only with distance from source trees were fitted to seed-rain data for animal and wind-dispersed species on BCI, animal-dispersed species showed much higher variances around the best-fit expected values than did wind-dispersed species (H. C. Muller-Landau *et al.*, unpublished analyses; Dalling *et al.* 2002). However, wind-dispersed species still deviated considerably from the expectation if all seeds were independently dispersed and their deposition depended only on distance from source trees – suggesting that seed deposition by wind is also significantly clumped.

At the shortest temporal scales (seconds to minutes), clumping among animal-dispersed species typically reflects simultaneous or near-simultaneous deposition of multiple seeds by a single animal or a group of animals. For example, individual defecations of white-faced monkeys (*Cebus capucinus*) on BCI almost always contain multiple seeds of the same species of plant (Wehncke *et al.* 2003). Depending on the plant species, the mean number of seeds per dropping varies from 4.3 (*Cordia bicolor*) to 1430 (*Cecropia insignis*). These seeds are all generally deposited within several metres of each other. For social animals, larger-scale clumping (tens of metres) may also be generated by defecation at around the same time (within half an hour) by multiple members of a group, spread out over the crown of one tree or a few neighbouring trees (Christina Campbell, personal communication).

At longer temporal scales, it is the repeated use of particular trails or sites that induce most clumping in seed deposition by animals. On scales of hours to days, clumping on short spatial scales (one to a few metres) can result from the repeated use of the same feeding roosts by bats, who return there to process fruits and drop or defecate multiple seeds beneath the roost (Fleming & Heithaus 1981). Over the course of one fruiting season of a given plant species, additional clumping can accumulate at medium spatial scales (tens of metres) because of continued or repeated use of nest sites by birds (Kinnaird 1998), sleep or nest trees by primates (Voysey *et al.* 1999), latrines by tapirs (Fragoso 1997) and display sites by birds (Krijger *et al.* 1997). Indeed, some of these sites may be used repeatedly for multiple years, inducing clumping over yet longer timescales. J. M. Fragoso (personal communication) has observed some tapir latrine sites used consistently for 12 years, and Théry and Larpin (1993) suggest that the same leks may be used by bird species for multiple generations. Unfortunately, few studies are long enough to document such patterns.

Different animal species and groups vary in the clumping of their seed deposition depending on their food-handling, gut physiology, social system and other traits. Howe (1989) suggested that animal species could be classified as either scatter or clump dispersers, based on their seed-deposition behaviour. We argue that it is important to characterize clumping more precisely, because it can occur at many different scales, with potentially very different implications. In general, the consensus is that the ranking of primary disperser groups, from most clumped to most scattered, is primates to terrestrial mammals to bats to birds. Of course, the degree of clumping varies further among species within groups, and among individuals and over time within species (e.g. Thies 1998; Wenny 2000).

Consideration of clumping highlights the importance of secondary dispersers such as rodents, ants and dung beetles. Secondary dispersers rarely move seeds as far as primary dispersers and their movements will mostly be at random with respect to seed-bearing trees; therefore they can often be ignored when dispersal-distance distributions are calculated. However, secondary dispersers often substantially change the clumping of seed distributions by their activities. Scatter-hoarding rodents generally reduce seed-clumping, as they remove individual seeds from tapir latrines (Fragoso 1997), bat feeding roosts and other sites of concentrated seed deposition and then most often cache seeds individually (Forget & Milleron 1991). Dung beetles also generally move one or a few seeds at a time from the clumps found in defecations of primates and other mammals, and thus spread those originally concentrated seeds over a greater area through their activities (Andresen 2001). On the other hand, central-place foragers such as ants may increase clumping of seed deposition by collecting seeds from a wide area and depositing most of them within the relatively smaller area of the nest or its refuse piles (Passos & Oliveira 2002).

Clumping of seed deposition by wind has received very little attention. One obvious source of variation in seed deposition beyond that due to distance from source trees is wind direction. For example, Augspurger (1984) noted that seed shadows of wind-dispersed trees on BCI were displaced to the south of fruiting trees, reflecting the prevailing northerly wind direction during the dry season when these species released their seeds. Yet while this process certainly increases variance in seed densities, it does not result in the same intensity or short spatial scales of clumping evident among animal-dispersed species. Two other processes, affecting primary and secondary dispersal by wind, respectively, have the potential to produce such more obvious clumping. Variation in canopy geometry and topography may entrain winds and generate turbulence that results in seeds disproportionately dropping in some areas rather than others. Variation in ground cover (e.g. mud vs. dry leaf litter) may affect whether wind-dispersed seeds stick where they first land, or are potentially picked up and redispersed by subsequent gusts. These possibilities deserve further investigation. Overall, however, seed deposition by wind is likely to result in relatively little clumping, particularly in comparison with vertebrate-dispersed plant species.

Recommendations

Better quantification of the intensity and scale of clumping of seed deposition is needed if we are to improve our understanding of this phenomenon and its implications. Studies of animal seed dispersers would ideally take data on all the information necessary to recreate the patterns of clumping produced. The required information would vary depending on the species. For a social primate, for instance, we would want data on the number of conspecific seeds per defecation, the spatial area covered by defecations of individual animals, the likelihood that multiple group members defecate at the same time, the area over which individuals in a group are typically distributed, the proportion of defecations that are under sleep trees, the average area into which defecations occur beneath a sleep tree, the number of sleep trees used during the fruiting period of a plant species, etc. (Russo 2003a). Knowledge of the natural history of an animal species will be invaluable in determining what sort of data will be useful.

When seed-deposition data are collected from seed traps or censuses rather than by following animals, the appropriate scales of data collection may be more difficult to determine. Ideally, all seeds would be mapped, so that spatial patterns at all scales could be detected, and censuses would be nearly continuous, so that patterns at all temporal scales could also be assessed. More often, data consist of counts of seeds in seed traps or sample plots of a fixed size with a fixed census interval, and data can only be analysed at a few spatial and temporal scales (e.g. Fig. 11.2). In this case, the best that can be done is to try to choose spatial and temporal scales that are most biologically relevant, given what is known about the plants and their dispersers (the scales at which clumping is likely to occur, and the scales at which it is likely to matter). Additionally, using a nested (or even fractally nested) spatial or temporal sampling design may make it possible to analyse processes at multiple spatial or temporal scales (Hardesty & Parker 2003), even if the fundamental sampling unit has a fixed size. Just as for dispersal distances, no single statistic can capture the clumping of seed deposition. Again, we recommend authors report multiple statistics, for multiple spatial and temporal scales. In particular, for studies that involve monitoring seed densities in traps or plots of fixed dimensions and fixed duration, we recommend reporting the proportion with seeds, the mean, variance and skewness of seed densities, and the 10th, 25th, 50th, 75th and 90th percentiles of the numbers of seeds per trap (or sample) among traps (samples) having seeds. Finally, we encourage authors to make as much of their original data as possible available in electronic appendices or other formats, so that the results of their studies can continue to enlighten us beyond the lifetimes of individual research projects or even investigators.

Future directions

Here we highlight both the most promising new techniques and the most pressing needs for research into the dispersal of tropical forest seed. Recent advances in technology and analytical methods have great potential for enhancing our understanding of seed dispersal. Particularly promising are new developments in genetic techniques (Ouborg *et al.* 1999), methods for collecting and analysing spatially explicit data (Hunsaker *et al.* 2001), and modelling abilities and tools (Turchin 1998). More research is also urgently needed on applied issues relating to disruption of plant–disperser interactions, and on understudied taxonomic and functional groups.

Genetic methods

In the past, most studies using genetic methods have focused on overall spatial patterns of genetic variation, gene flow, and relatedness (e.g. Nason *et al.* 1998). When gene flow is mainly by seed dispersal, dispersal distances can be inferred from genetic relatedness. For example, Kinlan and Gaines (2003) estimate dispersal distances from genetic isolation-by-distance slopes for numerous marine species in order to compare dispersal among taxonomic and trophic groups. Spatial variation in plants reflects both seed and pollen movement. Because pollen movement tends to occur over longer distances and thus dominate large-scale patterns of genetic variation, most studies of genetic variation in plants have focused on gene flow via pollen (e.g. Dick *et al.* 2003). In plants, we can also estimate historical dispersion patterns by combining information on nuclear markers with uniparentally inherited chloroplast or mitochondrial markers (Ouborg *et al.* 1999).

Molecular techniques make it possible not only to determine overall relatedness but positively to identify parents for any given seed or seedling (Ouborg *et al.* 1999). Matching highly variable multilocus genotypes (using microsatellites) from known maternal individuals to embryo tissue of seeds permits unambiguous identification of the mother tree, and thereby direct estimates of seed movement (Godoy & Jordano 2001). With highly variable microsatellite markers, direct multilocus genotype-matching or likelihood techniques both can be used to assign mother trees to established offspring. In contrast, allozymes generally show more limited variability among individuals – especially at local spatial scales. Thus, allozyme data often require use of more loci to achieve high parental-exclusion probabilities, and most studies allow only probabilistic linkage of offspring to parents using likelihood techniques (Schnabel *et al.* 1998). Allozymes have the advantage that they are inexpensive and much less time-consuming to develop than costly microsatellite markers, although the latter are rapidly becoming more accessible. For any method selected for parentage assignment, the number of potential parents in the study population as well as the number and evenness of alleles or loci in the population play a substantial role in determining exclusion probabilities. Numerous freeware genetic analysis programs (e.g. CERVUS, Marshall *et al.* 1998) are now available to aid in parentage analysis to measure seed-dispersal distances, evaluate relative reproductive success of parents in a population, and calculate differences in gene movement and hence the relative contribution to spatial genetic structure of both pollen and seed movement. A number of recent reviews evaluate DNA-based techniques (Campbell *et al.* 2003; Ouborg *et al.* 1999) and compare methods of kinship and pedigree analysis (Jones & Arden 2003).

The ability to identify mother trees of seeds or seedlings genetically makes it possible to 'view' seed dispersal events in a whole new way. For example, genetic studies have shown that the conspecific seeds concentrated under fruiting trees of *Prunus mahaleb* are not, as often presumed, all the offspring of that tree; instead, a significant fraction have other mothers and apparently were deposited there by animal dispersers that move from one fruiting tree to another (Jordano & Godoy 2002). Molecular markers allow us to understand better the types of clumping created by animal seed dispersers in particular. We can determine whether clumped individuals are full siblings, half sibs, or unrelated, and thus, the likelihood that they were dispersed together or by an animal making repeated visits to a perching site after ingestion. These different kinds of clumping of animal-generated seed shadows will all be reflected in population genetic structuring, which will subsequently influence gene movement and recruitment patterns (Jordano & Godoy 2002). Using genetic markers, we can measure dispersal distances at both the seed and seedling stages, and use comparisons of the two to determine whether seeds that were dispersed to different sites (e.g. farther from the parent) have different probabilities of successful recruitment. These and related genetic analyses may have important implications for the evolutionary consequences of dispersal and may also provide insights to the ecological and genetic basis of local differentiation in rainforest tree populations.

Spatial measurements and analyses

Because dispersal is inherently a spatial process, advances in collecting and analysing spatial data are particularly exciting. Small, lightweight handheld global positioning systems (GPS) that work within the forest make mapping of trees, seeds, disperser movements and seed-deposition events much easier and faster. Our ability to track movements of animals via attached transmitters has also improved greatly in recent years. Technological advances in automated radiotelemetry systems (Briner *et al.* 2003) hold great promise for constant, fine-scale tracking of frugivore movement patterns of multiple individuals and species, arboreal and terrestrial, without the labour-intensive fieldwork required for obtaining locations using handheld telemetry. Increasing miniaturization of transmitters makes it possible to track ever smaller animals (Bradshaw & Bradshaw 2002) – and even seeds (Sone & Kohno 1996) – for ever-longer periods. Furthermore, new transmitters make it possible to track not only location but also additional information such as the amount of activity (Kenward 2001), physiological data such as intestinal movements (Meile & Zittel 2002), or even sound recordings (Roland Kays, unpublished data). Finally, GPS and satellite tracking permit very large-scale monitoring of movements (Weimerskirch *et al.* 2002).

Advances in remote sensing also enhance our ability to collect spatial data on terrain, vegetation and other habitat variables that may affect seed and disperser movements. Excellent satellite images of many types are now available (Foody 2003), sometimes at little or no cost. These images can be used to classify habitats and thus inform analyses of animal movement (McShea *et al.* 2001) and potential seed rain with respect to habitat. Inexpensive ultralight aeroplanes also bring an older remote-sensing technology – aerial photographs – within easier reach. High-resolution aerial photographs can be used to delineate crowns and potentially identify trees to functional groups (Trichon 2001) or even species in some cases (S. Bohlman, personal communication). Particularly for tree species whose fruits are apparent in the upper canopy, it may also be possible to estimate fruit production of individual trees from overflights or aerial photographs.

All these spatial data can be analysed more easily and in ever greater detail using Geographic Information Systems (Hunsaker *et al.* 2001). Numerous methods and software tools have been developed specifically to analyse the home ranges and movement patterns of animals (Kenward 2001; Worton 1987). Such analyses can incorporate information on habitat features and investigate preferential use of different habitats by animals (McShea *et al.* 2001), thus improving analyses of the potential for habitat-specificity and large-scale clumping of seed deposition. Further, these more precise data on animal-movement patterns can allow better construction and tests of mechanistic models of animal behaviour. Moorcroft *et al.* (1999) use a mechanistic model of animal behaviour to predict home ranges of wolves in Yellowstone. Such models offer the promise of greater generality than phenomenological descriptions alone can provide.

Modelling

Both mechanistic and phenomenological models have proved useful to the study of seed dispersal (Nathan & Muller-Landau, 2000). Phenomenological models are fitted to data on, for example, densities of seeds at different distances from an isolated tree. Their main utility is to serve as a concise description of those data (e.g. the distances from the tree may be said to follow a roughly exponential distribution with a mean of 8 m) and as a basis for interpolating and extrapolating from those data (Ribbens *et al.* 1994). Mechanistic models, in contrast, are parameterized based on independent data – for example, models of dispersal distances may be based on data on disperser movements and gut-passage times for seed dispersal by animals (Murray 1988), or on wind speed, tree height and diaspore terminal velocity for seed dispersal by wind (Nathan *et al.* 2002a). Such models allow us to predict patterns from processes, and to understand better how variation or changes in processes could alter the resulting patterns.

In the past, models have been used mainly to describe or predict the distribution of dispersal distances. But models can be, should be and increasingly are used to describe and predict other patterns as well, including clumping of seeds at different scales, differential seed deposition by habitat (LePage *et al.* 2000) and the overall distribution of seeds over the whole landscape (Dalling *et al.* 2002). Both mechanistic and phenomenological models of increasing complexity and realism are easier than ever to construct and implement, and can serve as tools to help us better describe, investigate and understand seed dispersal in tropical forests.

The simplest mechanistic models of dispersal by animals predict seed-dispersal distances from data on how much time elapses between an animal taking a fruit and depositing seeds and data on how far an animal moves in that time (Murray 1988; Wehncke *et al.* 2003). For endozoochorous dispersal, this involves measuring gut-passage time and animal movement while foraging. Because seed deposition by animals may be highly clumped or diffuse depending on the disperser, and may vary with habitat, such models are generally insufficient to predict patterns of seed rain over a landscape. The next step is to incorporate information that produces clumping at different scales, including number of seeds per defecation, animals per group, proportion of seeds deposited under sleep trees, number of sleep trees, etc., as well as data on the actual spatial distribution of fruiting trees. Russo (2003a) has done exactly this in a study of dispersal of *Virola calophylla* seeds by *Ateles paniscus* (spider monkeys) in Manu, Peru. In the future, such models might also include information on how seed production and seed removal vary among trees and/or patch sites, and may reflect increasingly sophisticated models of fine-scale animal-movement patterns. To fully predict seed rain of any given animal-dispersed plant species would, of course, require constructing such mechanistic models for multiple species of seed-dispersing animals, and incorporating information on the proportion of seeds removed by each.

Recently, there have been tremendous advances in mechanistic models of seed dispersal by wind (Nathan *et al.* 2002b; Tackenberg 2003). These new models incorporate spatiotemporal variation in windspeed and thus can reproduce seed uplifting by turbulent updrafts – a necessary and sufficient condition for long-distance seed dispersal by wind (Nathan *et al.* 2002b). In contrast, early pioneering work by Augspurger (1986) simply used the height of seed release and seed terminal velocity to predict how far seeds might go at different windspeeds if vertical windspeed was zero. Thus far, the new models have been applied only in temperate forests (Nathan *et al.* 2002b) and grasslands (Tackenberg 2003). However, the same principles apply everywhere, and given sufficiently precise data on windspeed, similar models can be constructed for seed dispersal by wind in tropical forests. Recent theoretical work has also improved our understanding of how differently shaped wind-dispersed diaspores move through the air

(Minami & Azuma 2003). This is a necessary precursor for understanding seed dispersal by wind in tropical forests, where diaspores vary tremendously in their size and shape (Augspurger 1986), and thus in their behaviour in different wind environments (Minami & Azuma 2003).

We know of no mechanistic models of seed dispersal by water, and this seems an obvious frontier for future research. Hydrochory is particularly important in flooded forests in the Amazon and in mangroves the world over. Hydrological models of water movement in various river systems (e.g. Coe et al. 2002) and oceanographic models of currents are already available, as are increasingly good remote-sensing data on water levels and flooding extent (Campos et al. 2001). In combination with information on fruit flotation characteristics (e.g. Williamson & Costa 2000) and the phenology of fruit production, hydrologic and oceanographic models should allow for prediction of seed movement via water dispersal. Indeed, oceanographic models have already been incorporated into mechanistic models of larval fish dispersal (Cowen et al. 2000). The increasing availability of data on climate (especially precipitation), topography and soils also aids construction of hydrological models, where none are yet available. Given the long distances that seeds may disperse in rivers, along floodplains and in oceans, studies of genetic relatedness may be the best way to test and validate models of seed dispersal by water. In these systems, seed movement is likely to be over longer distances than pollen movement, making large-scale genetic structure an accurate reflection of seed movement. Extensive data on the genetic structure of mangroves are already available, and have been used to test hypotheses concerning historical distributions and gene flow (Dodd & Rafii 2002; Duke et al. 2002).

Like mechanistic models, phenomenological models have focused mainly on the distribution of seed-dispersal distances. Early models relied on data collected around isolated trees, but techniques now exist for fitting dispersal-distance distributions from seed or seedling data collected in areas of overlapping seed shadows of multiple trees (Clark et al. 1998; Ribbens et al. 1994). Collection of more seed data at multiple spatial scales should allow fits not only of dispersal distances but also of the clumping and patchiness of seed deposition. Given additional information on the microhabitat of all the areas where seed data were collected, models could also incorporate habitat-specific deposition, much as in LePage et al. (2000). They examined the combined effects of seed-dispersal distances and habitat on seedling recruitment. As with all modelling efforts, additional parameters can be reliably fitted only given sufficient data, and fitting of more parameters will generally require more data.

Mechanistic models of seed-dispersal processes could be used to constrain phenomenological models, reducing the number of parameters that need to be fitted and narrowing their confidence intervals. For example, mechanistic models of seed dispersal by wind might reveal that in general, seed shadows of wind-dispersed trees have a characteristic functional form. In this case, new data on seed distributions of a wind-dispersed tree (in an area where windspeeds or other information necessary for constructing a mechanistic model are lacking) might be fitted with a phenomenological model having this form. Similarly, data on seed production of individual trees might be used in inverse modelling of seed shadows, thus allowing for higher confidence in dispersal distances and other purely fitted parameters.

Understudied areas

A review of the tropical forest seed-dispersal literature shows that some woody plant groups are relatively well studied (although not necessarily yet well understood), while others have been studied very little, if at all.

Seed dispersal of lianas appears never to have been examined – virtually the only work is on their general dispersal syndromes and as information on consumption of liana fruits when diets of animal frugivores are studied. This parallels the general understudy of lianas – a situation which should be remedied since they are a major component of tropical forests (Schnitzer & Bongers 2002) and seem to be increasing in abundance with global change (Phillips *et al.* 2002; Wright *et al.* in press). The lack of work on seed dispersal by lianas may be due in part to the difficulty of precisely mapping the spread of individual lianas in the canopy, and thus the locations of fruits, by observers on the ground (S. Schnitzer, personal communication). Lianas can be distinguished on aerial photographs with a resolution of 1:6000 (S. Bohlman, personal communication), and this sort of remote sensing may aid future studies of seed dispersal by lianas.

Seed dispersal by wind, water and explosion are understudied relative to their prevalence among species and basal area in tropical forests. In our literature searches of seed-dispersal studies on BCI in particular and in tropical forests in general, approximately 6% concerned seed dispersal by wind. In contrast, between 10% and 25% of trees and over 50% of lianas in tropical forests are dispersed primarily by wind (Bullock 1995; Gentry 1982). Water dispersal and ballistic dispersal, while considerably less common in tropical forests, are even rarer in the literature. Water dispersal is an important component of Amazonian floodplain forests, and ballistic dispersal appears to be a successful strategy for a number of dominant shrub species. Both deserve additional attention.

Studies of seed dispersal by animals are much more numerous, but among these, attention has focused on some animal groups more than others. Naturally enough for diurnal animals such as ourselves, we have investigated diurnal dispersers and dispersal activities much more than nocturnal ones. Bats – although they generally comprise more than half the mammalian fruit-consuming species in tropical forests – have been studied at only a few sites (BCI being a notable example: see Kalko *et al.* 1996a). Other nocturnal frugivores such as kinkajous and olingos have received hardly any attention (but see Kays 1999). Terrestrial herbivores such as deer, peccaries and tapirs, which are known to consume fruits and seeds, have also rarely been studied in the context of seed dispersal (but see Fragoso 1997) – in general, we do not even know whether they serve primarily as seed predators or seed dispersers. Overall, monkeys, large birds and the large caviomorph rodents have received much more study than bats, small birds, ants and dung beetles.

Conservation implications

Severe disruptions of plant–disperser interactions are widespread in tropical forests, and their ultimate consequences for seed dispersal and plant communities remain unknown. Hunting and forest fragmentation are changing animal abundances in many if not most forests (Laurance & Bierregaard 1997; Redford & Robinson 1987). Population declines or even local extirpation are particularly likely for certain taxonomic and functional groups: among important seed dispersers, large birds and primates are at particularly high risk (Redford & Robinson 1987). Because for many plant species these animals are the most important, or perhaps only, dispersers observed in intact tropical forests, ecologists have long speculated that their loss could catastrophically reduce seed dispersal and recruitment in affected plant species (e.g. Bond 1994; Corlett 1998). This argument alone is not persuasive, however, because it is possible that fruits will be consumed and seeds successfully dispersed by the remaining taxa, which are likely to experience compensatory increases in abundance (Wright 2003). Indeed, Hawthorne and Parren (2000) find that even though forest elephants are known to disperse seeds of many species in the rain forests of Upper Guinea, an examination of forest inventory data suggests as few as two plant species have historically depended exclusively on elephants for seed dispersal and germination-promoting seed scarification, and even those can regenerate occasionally without elephants.

Only a few studies have compared seed-dispersal processes in areas of perturbed and intact animal communities, although the number of such studies is increasing. Pizo (1997) found that the number of bird species taking fruits of the neotropical tree *Cabralea canjerana* decreased from 35 in contiguous forest to 14 in a 250-ha fragment. The species identities of the most important fruit consumers changed, with associated changes in seed-deposition behaviour including an increase in fruit dropped beneath crowns in the fragmented forest (Pizo 1997). In a study in central Panama, Wright *et al.* (2000) found that the proportion of seeds of two palm species left under conspecific crowns increased from 15% at protected sites to greater than 90% at highly hunted sites (Wright *et al.* 2000), while seed-dispersal distances decreased (Wright & Duber 2001). Cordeiro and Howe (2003) present compelling evidence of significantly reduced seedling recruitment for an endemic bird-dispersed tropical tree in Tanzania in fragmented versus intact forest. This study, at a large spatial scale and using forest fragments over 70 years old, provides an excellent example of the effect of vertebrate dispersers on local forest diversity. Two other studies in Gabon and Tanzania have demonstrated decreased seedling recruitment in fragmented forests (Chapman & Onderdonk 1998; Cordeiro & Howe 2001), but these studies cannot rule out the possibility that changes in seed predation rather than seed dispersal are responsible.

Evidence of the longer-term impact of changes in disperser communities can be gleaned from studies of areas that lost dispersers in the last few centuries or millennia. McConkey and Drake (2002) examine seed dispersal on the Polynesian island of Tonga, where the largest avian seed dispersers became extinct following human colonization approximately 3000 years ago. They conclude that even those species whose fruits have the usual features associated with consumption and dispersal by birds and whose seeds are too large to be dispersed by the extant avian fauna are currently being consumed by bats, whose diet appears to be more varied on these islands than on the mainland (McConkey & Drake 2002). Corlett (2002) reports that in Singapore, where the animal species that would usually disperse seeds of large-fruited plants are absent, large-fruited animal-dispersed species as well as large-fruited wind-dispersed species are absent from secondary forests. Again, however, it is possible that this is due not to changes in seed dispersal but to changes in seed predation and/or the seedling recruitment environment; experiments are needed to resolve this question. An examination of the palaeological evidence might also reveal whether the extinction of Pleistocene megafauna precipitated further extinctions of plants for which they may have served as obligate dispersers (Howe 1985; Janzen & Martin 1982).

Overall, it appears that changes in animal communities following hunting and forest fragmentation will have significant qualitative effects on seed dispersal for some plant species. The identities of fruit consumers are likely to change, and with them seed-dispersal distances and deposition patterns. Whether these changes will result in population declines or disappearance of any plant species remains an open question. At present, it appears that no extinctions, local or otherwise, of tropical forest tree species can unambiguously be attributed to loss of seed dispersers. More studies are urgently needed to assess how seed dispersal changes in disturbed forests, and how altered dispersal affects plant populations and communities.

Knowledge of seed dispersal would also be useful in enhancing seed input to degraded lands that are open for restoration and reforestation. While wind-dispersed seeds of tropical forest plants typically disperse beyond forest edges as well as, or better than, they do within forests, animal-mediated seed rain generally decreases quickly beyond the forest boundary (e.g. Ingle 2003). Studies have shown that seed rain is much greater beneath emergent structures such as isolated trees and snags (Duncan & Chapman 2002; Slocum & Horvitz 2000), because birds and bats are more likely to visit there (da Silva et al. 1996). Seedling densities and diversities also tend to be higher in these areas (Toh *et al.* 1999). This suggests that planting trees or adding artificial perches or other structures could enhance tree seed input and thus forest regeneration in degraded tropical areas, just as they do in some temperate systems (McClanahan & Wolfe 1993). However, an experiment in which artificial perches were added to abandoned pasture found no difference in recruitment beneath them relative to nearby open areas (Holl *et al.* 2000). It appears that microclimate differences beneath isolated trees are at least as important for tree regeneration as seed input (Toh *et al.* 1999). As a result, establishment of some sort of preliminary woody cover, via plantations or other means, may be necessary to enhance both seed rain and seedling establishment of tropical forest species on degraded lands, and thereby jumpstart succession (Holl *et al.* 2000; Wunderle 1997).

Conclusions

The plants and animals involved in the biotic interactions that result in seed dispersal have very different selective forces acting upon them. Plants seek to have their seeds well-distributed and deposited intact in favourable environments at the lowest cost in animal rewards or deterrents, while animals seek nutritional rewards at their lowest cost in energy expenditure and risk. There is often a parallel disparity between the plant and animal foci of different dispersal studies (Howe 1993). Animal ecologists are most often interested in who is taking what and from where, and such studies may not report the results – from the plant perspective – of the fruit removal, ingestion and subsequent deposition. Plant ecologists often overlook the relative importance of who is moving the seeds in favour of where seeds are surviving. Ideally, studies should integrate both perspectives, investigating both ultimate seed fate and the processes that produce that fate. Happily, increasing numbers of investigators are bridging this gap (e.g. Balcomb & Chapman 2003), and the resulting studies, by contributing to our understanding of both seed-dispersal processes and their consequences, will ultimately provide the best basis for generalization.

To grasp the larger picture of seed dispersal requires integration – not only of different methods, but also of results over multiple studies. Such integration is greatly facilitated by the Internet, and by the many databases that are increasingly available there. For example, digital images of herbarium specimens and electronic florulas make it easier to assess many traits for many plant species, from any place on the globe that has Internet access. Electronic archives of old and new journal issues make it easier to do meta-analyses of previous studies, although there are still important gaps, as few tropical natural history journals are thus available. However, the true potential of the web is just beginning to be realized in ecology. In molecular biology, it is now expected that investigators deposit essentially all their data in publicly available databases at the time of journal publication. A similar effort in ecology could revolutionize our ability to make comparisons among sites and species, assess the generality of patterns, test our models and generally improve our understanding. While the deposition of data in electronic appendices to journals is a start, it would be far more useful if data were deposited in topic-specific archives, such as those provided by the Interaction Web Database at the National Center for Ecological Analysis and Synthesis (NCEAS) (http://www.nceas.ucsb.edu/interactionweb/), in a standardized format that made comparisons easier. Increasing use of such databases offers great promise for advances in our ability to integrate results across studies.

Fundamentally, we need not only better studies and better tools for bringing together their results but also simply more studies, and especially more geographically comparative or at least easily comparable studies. We know animal assemblages (as well as plant communities) vary among tropical forests, especially between neotropical and palaeotropical forests (Cristoffer & Peres 2003). We do not know how these differences affect seed dispersal and thereby plant regeneration. Long-term studies replicated at multiple sites across the tropics would be the ideal means to document the spatial and temporal variability of seed-deposition patterns and their consequences among tropical forests. Even without explicit coordination or replication, convergence on standard methods and analyses will make subsequent comparisons and integration more feasible.

Essentially every generalization or rule that has been or could be advanced regarding tropical forest seed dispersal has exceptions. Syndromes of fruit traits do not necessarily predict the identity of fruit consumers and seed dispersers; many plant–disperser pairs are neither very specialized nor very generalized; etc. This is hardly surprising, given the tremendous diversity of tropical plants and animals. Such outcomes should be seen not as failures that indicate that generalization is impossible and each plant–disperser system must be studied anew, but as partial successes and opportunities to fine-tune our understanding and better represent reality.

Acknowledgements

We thank Richard Corlett, Pierre-Michel Forget and Joe Wright for helpful comments that improved the manuscript. Joe Wright and Osvaldo Calderón provided data on seed-dispersal syndromes and the clumping of seed deposition on BCI, work supported by the Environmental Sciences Program of the Smithsonian. The Center for Tropical Forest Science provided data on woody plant life forms and relative abundances on BCI, work supported by the National Science Foundation, the John D. and Catherine T. MacArthur Foundation, and the Smithsonian Tropical Research Institute. H.C.M. gratefully acknowledges the support of a postdoctoral fellowship at the National Center for Ecological Analysis and Synthesis, a Center funded by the National Science Foundation award DEB-0072909, the University of California, and the Santa Barbara campus.

References

Alrich, P. & Hamrick, J. L. (1998) Reproductive dominance of pasture trees in a fragmented tropical forest mosaic. *Science*, **281**, 103–105.

Andresen, E. (2001) Effects of dung presence, dung amount and secondary dispersal by dung beetles on the fate of *Micropholis guyanensis* (Sapotaceae) seeds in central Amazonia. *Journal of Tropical Ecology*, **17**, 61–78.

Augspurger, C. K. (1984) Seedling survival of tropical tree species: interactions of dispersal distance, light-gaps, and pathogens. *Ecology*, **65**, 1705–1712.

(1986) Morphology and dispersal potential of wind-dispersed diaspores of neotropical trees. *American Journal of Botany*, **73**, 353–363.

Augspurger, C. K. & Franson, S. E. (1988) Input of wind-dispersed seeds into light-gaps and forest sites in a neotropical forest. *Journal of Tropical Ecology*, **4**, 239–252.

Balcomb, S. R. & Chapman, C. A. (2003) Bridging the gap: influence of seed deposition on seedling recruitment in a primate–tree interaction. *Ecological Monographs*, **73**, 625–642.

Blake, J. G. & Loiselle, B. A. (1991) Variation in resource abundance affects capture rates of birds in 3 lowland habitats in Costa Rica. *Auk*, **108**, 114–130.

Bond, W. J. (1994) Do mutualisms matter: assessing the impact of pollinator and disperser disruption on plant extinction. *Philosophical Transactions of the Royal Society of London B*, **344**, 83–90.

Bradshaw, S. D. & Bradshaw, F. J. (2002) Short-term movements and habitat use of the marsupial honey possum (*Tarsipes rostratus*). *Journal of Zoology*, **258**, 343–348.

Briner, T., Airoldi, J.-P., Dellsperger, F., Eggimann, S. & Nentwig, W. (2003) A new system for automatic radiotracking of small mammals. *Journal of Mammalogy*, **84**, 571–578.

Brokaw, N. & Busing, R. T. (2000) Niche versus chance and tree diversity in forest gaps. *Trends in Ecology and Evolution*, **15**, 183–188.

Brown, J. H. (1995) *Macroecology*. Chicago: University of Chicago Press.

Bullock, S. H. (1995) Plant reproduction in neotropical dry forests. In *Seasonally Dry Tropical Forests* (ed. S. H. Bullock, H. A. Mooney & E. Medina). Cambridge: Cambridge University Press, pp. 277–303.

Bustamente, R. O. & Canals, L. M. (1995) Dispersal quality in plants: how to measure efficiency and effectiveness of a seed disperser. *Oikos*, **73**, 133–136.

Cain, M. L., Milligan, B. G. & Strand, A. E. (2000) Long-distance seed dispersal in plant populations. *American Journal of Botany*, **87**, 1217–1227.

Campbell, D., Duchesne, P. & Bernatchez, L. (2003) AFLPn utility for population assignment studies: analytical investigation and empirical comparison with microsatellites. *Molecular Ecology*, **12**, 1979–1991.

Campos, I. D., Mercier, F., Maheu, C. *et al.* (2001) Temporal variations of river basin waters from Topex/Poseidon satellite altimetry. Application to the Amazon basin. *Comptes Rendus de l'Academie des Sciences IIa: Sciences de la terre et des Planetes*, **333**, 633–643.

Chapman, C. A. & Onderdonk, D. A. (1998) Forests without primates: primate/plant codependency. *American Journal of Primatology*, **45**, 127–141.

Chave, J., Muller-Landau, H. C. & Levin, S. A. (2002) Comparing classical community models: theoretical consequences for patterns of diversity. *American Naturalist*, **159**, 1–23.

Clark, D. A., Clark, D. B., Sandoval, M. R. & Castro, C. M. V. (1995) Edaphic and human effects on landscape-scale distributions of tropical rain forest palms. *Ecology*, **76**, 2581–2594.

Clark, J. S., Macklin, E. & Wood, L. (1998) Stages and spatial scales of recruitment limitation in Southern Appalachian forests. *Ecological Monographs*, **68**, 213–235.

Coe, M. T., Costa, M. H., Botta, A. & Birkett, C. (2002) Long-term simulations of discharge and floods in the Amazon Basin. *Journal of Geophysical Research: Atmospheres*, **107**.

Cohen, D. & Levin, S. A. (1991) Dispersal in patchy environments: the effects of temporal and spatial structure. *Theoretical Population Biology*, **39**, 63–99.

Condit, R., Hubbell, S. P. & Foster, R. B. (1996) Changes in tree species abundance in a neotropical forest: impact of climate change. *Journal of Tropical Ecology*, **12**, 231–256.

Connell, J. H. (1971) On the roles of natural enemies in preventing competitive exclusion in some marine animals and in rain forest trees. In *Dynamics of Populations, Proceedings of the Advanced Study Institute on Dynamics of Numbers in Populations, Oosterbeek 1970* (ed. P. J. den Boer & G. R. Gradwell). Wageningen: Centre for Agricultural Publishing and Documentation, pp. 298–312.

Cordeiro, N. J. & Howe, H. F. (2001) Low recruitment of trees dispersed by animals in African forest fragments. *Conservation Biology*, **15**, 1733–1741.

(2003) Forest fragmentation servers mutualism between seed dispersers and an endemic African tree. *Proceedings of the National Academy of Sciences*, **100**, 14052–14056.

Corlett, R. T. (1998) Frugivory and seed dispersal by vertebrates in the Oriental (Indomalayan) region. *Biological Reviews of the Cambridge Philosophical Society*, **73**, 413–448.

(2002) Frugivory and seed dispersal in degraded tropical East Asian landscapes. In *Seed Dispersal and Frugivory: Ecology, Evolution and Conservation* (ed. D. J. Levey, W. R. Silva

& M. Galetti). Wallingford, UK: CAB International, pp. 451–465.

Cowen, R. K., Lwiza, K. M. M., Sponaugle, S., Paris, C. B. & Olson, D. B. (2000) Connectivity of marine populations: open or closed? *Science*, **287**, 857–859.

Cristoffer, C. & Peres, C. A. (2003) Elephants versus butterflies: the ecological role of large herbivores in the evolutionary history of two tropical worlds. *Journal of Biogeography*, **30**, 1357–1380.

Croat, T. B. (1978) *Flora of Barro Colorado Island*. Stanford, CA: Stanford University Press.

da Silva, J. M. C., Uhl, C. & Murray, G. (1996) Plant succession, landscape management, and the ecology of frugivorous birds in abandoned Amazonian pastures. *Conservation Biology*, **10**, 491–503.

Dalling, J. W. & Hubbell, S. P. (2002) Seed size, growth rate and gap microsite conditions as determinants of recruitment success for pioneer species. *Journal of Ecology*, **90**, 557–568.

Dalling, J. W., Muller-Landau, H. C., Wright, S. J. & Hubbell, S. P. (2002) Role of dispersal in the recruitment limitation of neotropical pioneer species. *Journal of Ecology*, **90**, 714–727.

Dick, C. W., Etchelecu, G. & Austerlitz, F. (2003) Pollen dispersal of tropical trees (*Dinizia excelsa*: Fabaceae) by native insects and African honeybees in pristine and fragmented Amazonian rainforest. *Molecular Ecology*, **12**, 753–764.

Dodd, R. S. & Rafii, Z. A. (2002) Evolutionary genetics of mangroves: continental drift to recent climate change. *Trees: Structure and Function*, **16**, 80–86.

Duke, N. C., Lo, E. Y. Y. & Sun, M. (2002) Global distribution and genetic discontinuities of mangroves: emerging patterns in the evolution of Rhizophora. *Trees: Structure and Function*, **16**, 65–79.

Duncan, R. S. & Chapman, C. A. (2002) Limitations of animal seed dispersal for enhancing forest succession on degraded lands. In *Seed Dispersal and Frugivory: Ecology, Evolution and Conservation* (ed. D. J. Levey, W. R. Silva & M. Galetti). Wallingford, UK: CAB International, pp. 437–450.

Feer, F. & Forget, P. M. (2002) Spatio-temporal variation in post-dispersal seed fate. *Biotropica*, **34**, 555–566.

Fitter, A. H., Graves, J. D., Watkins, N. K., Robinson, D. & Scrimgeour, C. (1998) Carbon transfer between plants and its control in networks of arbuscular mycorrhizas. *Functional Ecology*, **12**, 406–412.

Flannery, T. F. (1995) *Mammals of the South-West Pacific & Moluccan Islands.* New York: Cornell University Press.

Fleming, T. H. & Heithaus, E. R. (1981) Frugivorous bats, seed shadows, and the structure of tropical forests. *Biotropica*, **13**, Supplement, 45–53.

Foody, G. M. (2003) Remote sensing of tropical forest environments: towards the monitoring of environmental resources for sustainable development. *International Journal of Remote Sensing*, **24**, 4035–4046.

Forget, P.-M. & Hammond, D. S. (2005). Rainforest vertebrates and food plant diversity in the Guiana Shield. In *Tropical Rainforests of the Guiana Shield* (ed. D. S. Hammond). Wallingford, UK: CAB International, 233–294.

Forget, P.-M. & Milleron, T. (1991) Evidence for secondary seed dispersal by rodents in Panama. *Oecologia*, **87**, 596–599.

Forget, P.-M., Milleron, T. & Feer, F. (1998) Patterns in post-dispersal seed removal by neotropical rodents and seed fate in relation to seed size. In *Dynamics of Tropical Communities* (ed. D. M. Newbery, H. H. T. Prins & N. D. Brown). Oxford: Blackwell Scientific, pp. 25–49.

Fragoso, J. M. V. (1997) Tapir-generated seed shadows: scale-dependent patchiness in the Amazon rain forest. *Journal of Ecology*, **85**, 519–529.

Gautier-Hion, A. (1984) Seed dispersal by African forest cercopithecines. *Revue d'Ecologie: La Terre et la Vie*, **39**, 159–165.

Gautier-Hion, A., Duplantier, J.-M., Quris, R. *et al.* (1985) Fruit characters as a basis of fruit choice and seed dispersal in a tropical forest vertebrate community. *Oecologia*, **65**, 324–337.

Gentry, A. H. (1982) Patterns of neotropical plant species diversity. In *Evolutionary Biology* (ed. M. K. Hecht, B. Wallace & G. T. Prance), Vol. **15**. New York: Plenum Press, pp. 1–84.

(1983) Dispersal ecology and diversity in neotropical forest communities. In *Dispersal and Distribution: An International Symposium* (ed. K. Kubitzki). Hamburg: Paul Parey.

Glanz, W. E., Thorington, R. W. Jr, Giacalone-Madden A. J. & Heaney, L. R. (1982) Seasonal food use and demographic trends in *Sciurus granatensis*. In *Ecology of a Tropical Forest: Seasonal Rhythms and Long-Term Changes* (ed. E. G. Leigh, Jr, A. S. Rand & D. M. Windsor). Washington DC: Smithsonian Institution Press, pp. 239–252.

Godoy, J. A. & Jordano, P. (2001) Seed dispersal by animals: exact identification of source trees with endocarp DNA microsatellites. *Molecular Ecology*, **10**, 2275–2283.

Gompper, M. E. (1994) The behavioral ecology and genetics of a coati (*Nasua narica*) population in Panama. Unpublished Ph.D. dissertation, University of Tennessee.

(1996) Sociality and asociality in white-nosed coatis (*Nasua narica*): foraging costs and benefits. *Behavioral Ecology*, **7**, 254–263.

Gorchov, D. L., Cornejo, F., Ascorra, C. & Jaramillo, M. (1993) The role of seed dispersal in the natural regeneration of rain forest after strip-cutting in the Peruvian Amazon. *Vegetatio*, **107/108**, 339–349.

Hamilton, W. D. & May, R. M. (1977) Dispersal in stable habitats. *Nature*, **269**, 578–581.

Hamrick, J. L., Godt, M. J. W. & Sherman-Broyles, S. L. (1995) Gene flow among plant populations: evidence from genetic markers. In *Experimental and Molecular Approaches to Plant Biosystematics* (ed. P. C. Hoch & A. G. Stephenson). St Louis, MO: Missouri Botanical Garden, pp. 215–232.

Hamrick, J. L. & Loveless, M. D. (1986) The influence of seed dispersal mechanisms on the genetic structure of plant populations. In *Frugivores and Seed Dispersal* (ed. A. Estrada & T. H. Felming). Dordrecht: Dr W. Junk Publishers, pp. 211–223.

Handley, J., Charles O., Wilson, D. E. & Gardner, A. L. (eds.) (1991) *Demography and Natural History of the Common Fruit Bat, Artibeus jamaicensis, on Barro Colorado Island, Panamá*. Washington, DC: Smithsonian Institution, pp. 1–173.

Hardesty, B. D. & Parker, V. T. (2003) Community seed rain patterns and a comparison to adult community structure in a West African tropical forest. *Plant Ecology*, **164**, 49–64.

Harms, K. E., Wright, S. J., Calderón, O., Hernández, A. & Herre, E. A. (2000) Pervasive density-dependent recruitment enhances seedling diversity in a tropical forest. *Nature*, **404**, 493–495.

Hawthorne, W. D. & Parren, M. P. E. (2000) How important are forest elephants to the survival of woody plant species in Upper Guinean forests? *Journal of Tropical Ecology*, **16**, 133–150.

Holbrook, K. M. & Smith, T. B. (2000) Seed dispersal and movement patterns in two species of Ceratogymna hornbills in a West African tropical lowland forest. *Oecologia*, **125**, 249–257.

Holbrook, K. M., Smith, T. B. & Hardesty, B. D. (2002) Implications of long-distance movements of frugivorous rain forest hornbills. *Ecography*, **25**, 745–749.

Holl, K. D., Loik, M. E., Lin, E. H. V. & Samuels, I. A. (2000) Tropical montane forest restoration in Costa Rica: overcoming barriers to dispersal and establishment. *Restoration Ecology*, **8**, 339–349.

Howe, H. F. (1977) Bird activity and seed dispersal of a tropical wet forest tree. *Ecology*, **58**, 539–550.

(1980) Monkey dispersal and waste of a neotropical fruit. *Ecology*, **61**, 944–959.

(1985) Gomphothere fruits: a critique. *The American Naturalist*, **125**, 853–865.

(1986). Seed dispersal by fruit-eating birds and mammals. In *Seed Dispersal* (ed. D. R. Murray). San Diego: Academic Press, pp. 123–190.

(1989) Scatter and clump-dispersal and seedling demography: hypothesis and implications. *Oecologia*, **79**, 417–426.

(1993) Specialized and generalized dispersal systems: where does 'the paradigm' stand? *Vegetatio*, **107/108**, 3–13.

Howe, H. F. & Smallwood, J. (1982) Ecology of seed dispersal. *Annual Review of Ecology and Systematics*, **13**, 201–228.

Howe, H. F. & vande Kerckhove, G. A. (1979) Fecundity and seed dispersal of a tropical tree. *Ecology*, **60**, 180–189.

Hubbell, S. P. & Foster, R. B. (1983) Diversity of canopy trees in a neotropical forest and implications for conservation. In *Tropical Rain Forest: Ecology and Management* (ed. S. L. Sutton, T. C. Whitmore & A. C. Chadwick). Oxford: Blackwell Scientific Publications, pp. 25–41.

Hubbell, S. P., Foster, R. B., O'Brien, S. T. *et al.* (1999) Light-gap disturbances, recruitment limitation, and tree diversity in a neotropical forest. *Science*, **283**, 554–557.

Hunsaker, C. T., Goodchild, M. F., Friedl, M. A. & Case, T. J., eds. (2001) *Spatial Uncertainty in Ecology: Implications for Remote Sensing and GIS Applications*. New York: Springer Verlag, p. 440.

Hurtt, G. C. & Pacala, S. W. (1995) The consequences of recruitment limitation: reconciling chance, history and competitive differences between plants. *Journal of Theoretical Biology*, **176**, 1–12.

Ingle, N. R. (2003) Seed dispersal by wind, birds, and bats between Philippine montane rainforest and successional vegetation. *Oecologia*, **134**, 251–261.

Jansen, P. A., Bartholomeus, M., Bongers, F., Elzinga, J. A., den Ouden, J. & van Wieren, S. E. (2002) The role of seed size in dispersal by a scatter-hoarding rodent. In *Seed Dispersal and Frugivory: Ecology, Evolution and Conservation* (ed. D. J. Levey, W. R. Silva & M. Galetti). Wallingford, UK: CAB International, pp. 209–225.

Janson, C. H. (1983) Adaptation of fruit morphology to dispersal agents in a neotropical forest. *Science*, **219**, 187–189.

Janzen, D. H. (1970) Herbivores and the number of tree species in tropical forests. *American Naturalist*, **104**, 501–528.

Janzen, D. H. & Martin, P. S. (1982) Neotropical anachronisms: the fruits the gomphotheres ate. *Science*, **215**, 19–27.

Jones, A. G. & Arden, W. R. (2003) Methods of parentage analysis in natural populations. *Molecular Ecology*, **12**, 2511–2523.

Jordano, P. & Godoy, J. A. (2002) Frugivore-generated seed shadows: a landscape view of demographic and genetic effects. In *Seed Dispersal and Frugivory: Ecology, Evolution and Conservation* (ed. D. J. Levey, W. R. Silva & M. Galetti). Wallingford, UK: CAB International, pp. 305–321.

Kalko, E. K. V., Handley, C. O. Jr & Handley, D. (1996a) Organization, diversity, and long-term dynamics of a neotropical bat community. In *Long-Term Studies of Vertebrate Communities* (ed. M. L. Cody & J. A. Smallwood). San Diego: Academic Press, pp. 503–553.

Kalko, E. K. V., Herre, E. A. & Handley, C. O. (1996b) Relation of fig fruit characteristics to fruit-eating bats in the New and Old World tropics. *Journal of Biogeography*, **23**, 565–576.

Kays, R. W. (1999) Food preferences of kinkajous (*Potos flavus*): a frugivorous carnivore. *Journal of Mammalogy*, **80**, 589–599.

Kenward, R. E. (2001) *A Manual for Wildlife Radio Tagging*. London: Academic Press.

Kinlan, B. P. & Gaines, S. D. (2003) Propagule dispersal in marine and terrestrial environments: a community perspective. *Ecology*, **84**, 2007–2020.

Kinnaird, M. F. (1998) Evidence for effective seed dispersal by the Sulawesi red-knobbed hornbill, *Aceros cassidix*. *Biotropica*, **30**, 50–55.

Krijger, C. L., Opdam, M., Théry, M. & Bongers, F. (1997) Courtship behaviour of manakins and seed bank composition in a French Guianan rain forest. *Oecologia*, **107**, 347–355.

Laurance, W. F. & Bierregaard, R. O., eds. (1997) *Tropical Forest Remnants: Ecology, Management, and Conservation of Fragmented Communities*. Chicago: University of Chicago Press.

Leigh, E. G. Jr (1999) *Tropical Forest Ecology: A View from Barro Colorado Island*. Oxford: Oxford University Press.

Leigh, J., Egbert, G., Rand, S. & Windsor, D. M., eds. (1996) *The Ecology of a Tropical Forest: Seasonal Rhythms and Long-Term Changes*, 2nd edn. Washington DC: Smithsonian Institution Press.

LePage, P. T., Canham, C. D., Coates, K. D. & Bartemucci, P. (2000) Seed abundance versus substrate limitation of seedling recruitment in northern temperate forests of British Columbia. *Canadian Journal of Forest Research*, **30**, 415–427.

Levey, D. J., Moermond, T. C. & Denslow, J. S. (1994) Frugivory: an overview. In *La Selva: Ecology and Natural History of a Tropical Rain Forest* (ed. L. A. McDade, K. S. Bawa, H. A. Hespenheide & G. S. Hartshorn). Chicago: University of Chicago Press, pp. 282–294.

Levin, S. A., Muller-Landau, H. C., Nathan, R. & Chave, J. (2003) The ecology and evolution of seed dispersal: a theoretical perspective. *Annual Review of Ecology and Systematics*, **34**, 575–604.

Marshall, T. C., Slate, J., Kruuk L. E. B. & Pemberton Z. J. M. (1998) Statistical confidence for likelihood-based paternity inference in natural populations. *Molecular Ecology*, **7**, 639–655.

Martin, T. E. (1985) Resource selection by tropical frugivorous birds: integrating multiple interactions. *Oecologia*, **66**, 563–573.

McClanahan, T. R. & Wolfe, R. W. (1993) Accelerating forest succession in a fragmented landscape: the role of birds and perches. *Conservation Biology*, **7**, 279–388.

McConkey, K. R. & Drake, D. R. (2002) Extinct pigeons and declining bat populations: are large seeds still being dispersed in the tropical Pacific? In *Seed Dispersal and Frugivory: Ecology, Evolution and Conservation* (ed. D. J. Levey, W. R. Silva & M. Galetti). Wallingford, UK: CAB International, pp. 381–395.

McShea, W. J., Aung, M., Poszig, D., Wemmer, C. & Monfort, S. (2001) Forage, habitat use, and sexual segregation by a tropical deer (*Cervus eldi thamin*) in a dipterocarp forest. *Journal of Mammalogy*, **82**, 848–857.

Meile, T. & Zittel, T. T. (2002) Telemetric small intestinal motility recording in awake rats: a novel approach. *European Surgical Research*, **34**, 271–274.

Milton, K. (1980) *The Foraging Strategy of Howler Monkeys: A Study in Primate Economics*. New York: Columbia University Press.

Minami, S. & Azuma, A. (2003) Various flying modes of wind-dispersal seeds. *Journal of Theoretical Biology*, **225**, 1–14.

Moorcroft, P. R., Lewis, M. A. & Crabtree, R. L. (1999) Home range analysis using a mechanistic home range model. *Ecology*, **80**, 1656–1665.

Muller-Landau, H. C. (2001) Seed dispersal in a tropical forest: empirical patterns, their origins and their consequences for forest dynamics. Unpublished Ph.D. dissertation, Princeton University.

Muller-Landau, H. C., Wright, S. J., Calderón, O., Hubbell, S. P. & Foster, R. B. (2002) Assessing recruitment limitation: concepts, methods and examples for tropical forest trees. In *Seed Dispersal and Frugivory: Ecology, Evolution and Conservation* (ed. D. J. Levey, W. R. Silva & M. Galetti). Wallingford, UK: CAB International, 35–53.

Muller-Landau, H. C., Levin, S. A. & Keymer, J. E. (2003) Theoretical perspectives on evolution of long-distance dispersal and the example of specialized pests. *Ecology*, **84**, 1957–1967.

Murray, K. G. (1988) Avian seed dispersal of three neotropical gap-dependent plants. *Ecological Monographs*, **58**, 271–298.

Nagy, K. A. (1987) Field metabolic rate and food requirement scaling in mammals and birds. *Ecological Monographs*, **57**, 111–128.

Nason, J. D., Herre, E. A. & Hamrick, J. L. (1998) The breeding structure of a tropical keystone plant resource. *Nature*, **391**, 685–687.

Nathan, R., Horn, H. S., Chave, J. & Levin, S. A. (2002a) Mechanistic models for tree seed dispersal by wind in dense forests and open landscapes. In *Seed Dispersal and Frugivory: Ecology, Evolution and Conservation* (ed. D. J. Levey, W. R. Silva & M. Galetti). Wallingford, UK: CAB International, pp. 69–82.

Nathan, R., Katul, G. G., Horn, H. S. *et al.* (2002b) Mechanisms of long-distance dispersal of seeds by wind. *Nature*, **418**, 409–413.

Nathan, R. & Muller-Landau, H. C. (2000) Spatial patterns of seed dispersal, their determinants and consequences for recruitment. *Trends in Ecology and Evolution*, **15**, 278–285.

Nathan, R., Perry, G., Cronin, J. T., Strand, A. E. & Cain, M. L. (2003) Methods for estimating long-distance dispersal. *Oikos*, **103**, 261–273.

Nogueira, M. R. & Peracchi, A. L. (2003) Fig-seed predation by two species of Chiroderma: discovery of a new feeding strategy in bats. *Journal of Mammalogy*, **84**, 225–233.

Ouborg, N. J., Piquot, Y. & Groenendael, V. (1999) Population genetics, molecular markers and the study of dispersal in plants. *Journal of Ecology*, **87**, 551–568.

Passos, L. & Oliveira, P. S. (2002) Ants affect the distribution and performance of seedlings of *Clusia criuva*, a primarily bird-dispersed rain forest tree. *Journal of Ecology*, **90**, 517–528.

Peres, C. A. & van Roosmalen, M. (2002) Primate frugivory in two species-rich neotropical forests: implications for the demography of large-seeded plants in overhunted areas. In *Seed Dispersal and Frugivory: Ecology, Evolution and Conservation* (ed. D. J. Levey, W. R. Silva & M. Galetti). Wallingford, UK: CAB International, pp. 407–421.

Phillips, O. L., Martinez, R. V., Arroyo, L. *et al.* (2002) Increasing dominance of large lianas in Amazonian forests. *Nature*, **418**, 770–774.

Pizo, M. A. (1997) Seed dispersal and predation in two popuolations of *Cabralea canjerana* (Meliaceae) in the Atlantic forest of southeastern Brazil. *Journal of Tropical Ecology*, **13**, 559–578.

Poulin, B., Wright, S. J., Lefebrve, G. & Calderon, O. (1999) Interspecific synchrony and asynchrony in the fruiting phenologies of congeneric bird-dispersed plants in Panama. *Journal of Tropical Ecology*, **15**, 213–227.

Poulsen, J. R., Clark, C. J., Connor, E. F. & Smith, T. B. (2002) Differential resource use by primates and hornbills: implications for seed dispersal. *Ecology*, **83**, 228–240.

Redford, K. H. & Robinson, J. G. (1987) The game of choice: patterns of Indian and colonist hunting in the neotropics. *American Anthropologist*, **89**, 650–667.

Ribbens, E., Silander, J. A. Jr & Pacala, S. W. (1994) Seedling recruitment in forests: calibrating models to predict patterns of tree seedling dispersion. *Ecology*, **75**, 1794–1806.

Ronce, O., Olivieri, I., Clobert, J. & Danchin, E. (2001) Perspectives on the study of dispersal evolution. In *Dispersal* (ed. J. Clobert, E. Danchin, A. A. Dhondt & J. D. Nichols). Oxford: Oxford University Press, pp. 341–357.

Rowell, T. E. & Mitchell, B. J. (1991) Comparison of seed dispersal by guenons in Kenya and capuchins in Panama. *Journal of Tropical Ecology*, **7**, 269–274.

Russo, S. E. (2003a) Linking spatial patterns of seed dispersal and plant recruitment in a neotropical tree, *Virola calophylla* (Myristicaceae). Unpublished Ph.D. dissertation, University of Illinois.

(2003b) Responses of dispersal agents to tree and fruit traits in *Virola calophylla* (Myristicaceae): implications for selection. *Oecologia*, **136**, 80–87.

Schnabel, A., Nason, J. D. & Hamrick, J. L. (1998) Understanding the population genetic structure of *Gleditsia triacanthos* L.: seed dispersal and variation in female reproductive success. *Molecular Ecology*, **7**, 819–832.

Schnitzer, S. A. & Bongers, F. (2002) The ecology of lianas and their role in forests. *Trends in Ecology and Evolution*, **17**, 223–230.

Schupp, E. W. (1993) Quantity, quality and the effectiveness of seed dispersal by animals. *Vegetatio*, **107/108**, 15–29.

Schupp, E. W., Howe, H. F., Augspurger, C. K. & Levey, D. J. (1989) Arrival and survival in tropical treefall gaps. *Ecology*, **70**, 562–564.

Schupp, E. W., Milleron, T. & Russo, S. E. (2002) Dissemination limitation and the origin and maintenance of species-rich tropical forests. In *Seed Dispersal and Frugivory: Ecology, Evolution and Conservation* (ed. D. J. Levey, W. R. Silva & M. Galetti). Wallingford, UK: CAB International, pp. 19–33.

Sezen, U. U., Chazdon, R. L. & Holsinger, K. E. (2005) Genetic consequences of tropical second-growth forest regeneration. *Science*, **307**, 891.

Shilton, L. A., Altringham, J. D., Compton, S. G. & Whittaker, R. J. (1999) Old World fruit bats can be long-distance seed dispersers through extended retention of viable seeds in the gut. *Proceedings of the Royal Society of London B*, **266**, 219–223.

Slocum, M. G. & Horvitz, C. G. (2000) Seed arrival under different genera of trees in a neotropical pasture. *Plant Ecology*, **149**, 51–62.

Smythe, N., Glanz, W. E., Leigh, J. & Egbert G. (1982) Population regulation in some terrestrial frugivores. In *The Ecology of a Tropical Forest: Seasonal Rhythms and Long-term Changes* (ed. E. G. Leight Jr, A. S. Rand & D. M. Windsor). Washington DC: Smithsonian Institution Press, pp. 227–238.

Snow, D. W. (1981) Tropical frugivorous birds and their food plants: a world survey. *Biotropica*, **13**, 1–14.

Sone, K. & Kohno, A. (1996) Application of radiotelemetry to the survey of acorn dispersal by Apodemus mice. *Ecological Research*, **11**, 187–192.

Svenning, J.-C. (1999) Microhabitat specialization in a species-rich palm community in Amazonian Ecuador. *Journal of Ecology*, **87**, 55–65.

Tackenberg, O. (2003) Modeling long distance dispersal of plant diaspores by wind. *Ecological Monographs*, **73**, 173–189.

Terborgh, J. & Wright, S. J. (1994) Effects of mammalian herbivores on plant recruitment in two neotropical forests. *Ecology*, **75**, 1829–1833.

Tewksbury, J. J. & Nabhan, G. P. (2001) Seed dispersal: directed deterrence by capsaicin in chillies. *Nature*, **412**, 403–404.

Théry, M. & Larpin, D. (1993) Seed dispersal and vegetation dynamics at a cock-of-the-rock's lek in the tropical forest of French Guiana. *Journal of Tropical Ecology*, **9**, 109–116.

Thies, W. (1998) Resource and habitat use in two frugivorous bat species (Phyllostomidae: *Carollia perspicillata* and *C. castanea*) in Panama: mechanisms of coexistence. Unpublished Ph.D. thesis, University of Tübingen.

Toh, I., Gillespie, M. & Lamb, D. (1999) The role of isolated trees in facilitating tree seedling recruitment at a degraded sub-tropical rainforest site. *Restoration Ecology*, **7**, 288–297.

Traveset, A. & Verdú, M. (2002) A meta-analysis of the effect of gut treatment on seed germination. In *Seed Dispersal and Frugivory: Ecology, Evolution and Conservation* (ed. D. J. Levey, W. R. Silva & M. Galetti). Wallingford, UK: CAB International, pp. 339–350.

Trichon, V. (2001) Crown typology and the identification of rain forest trees on large-scale aerial photographs. *Plant Ecology*, **153**, 301–312.

Turchin, P. (1998) *Quantitative Analysis of Movement: Measuring and Modeling Population Redistribution in Animals and Plants.* Sunderland, MA: Sinauer.

Voysey, B. C., McDonald, K. E., Rogers, M. E., Tutin, C. E. G. & Parnell, R. J. (1999) Gorillas and seed dispersal in the Lope Reserve, Gabon. II: Survival and growth of seedlings. *Journal of Tropical Ecology*, **15**, 39–60.

Wehncke, E. V., Hubbell, S. P., Foster, R. B. & Dalling, J. W. (2003) Seed dispersal patterns produced by white-faced monkeys: implications for the dispersal limitation of neotropical tree species. *Journal of Ecology*, **91**, 677–685.

Weimerskirch, H., Bonadonna, F., Bailleul, F., Mabille, G., Dell'Omo, G. & Lipp, H. P. (2002) GPS tracking of foraging albatrosses. *Science*, **295**, 1259–1259.

Wenny, D. G. (2000) Seed dispersal, seed predation, and seedling recruitment of a neotropical montane tree. *Ecological Monographs*, **70**, 331–351.

(2001) Advantages of seed dispersal: a re-evaluation of directed dispersal. *Evolutionary Ecology Research*, **3**, 51–74.

Wenny, D. G. & Levey, D. J. (1998) Directed seed dispersal by bellbirds in a tropical cloud forest. *Proceedings of the National Academy of Sciences*, **95**, 6204–6207.

Westcott, D. A. & Graham, D. L. (2000) Patterns of movement and seed dispersal of a tropical frugivore. *Oecologia*, **122**, 249–257.

Wheelwright, N. T. & Orians, G. H. (1982) Seed dispersal by animals: contrasts with pollen dispersal, problems of terminology, and constraints on coevolution. *The American Naturalist*, **119**, 402–413.

Whitney, K. D. & Smith, T. B. (1998) Habitat use and resource tracking by African Ceratogymna hornbills: implications for seed dispersal and forest conservation. *Animal Conservation*, **1**, 107–117.

Williamson, G. B. & Costa, F. (2000) Dispersal of Amazonian trees: hydrochory in *Pentaclethra macroloba*. *Biotropica*, **32**, 548–552.

Willis, E. O. (1990) Ecological roles of migratory and resident birds on Barro Colorado Island, Panama. In *Migrant Birds in the Neotropics: Ecology, Behavior, Distribution, and Conservation* (ed. A. Keast & E. S. Morton). Washington DC: Smithsonian Institution Press, pp. 205–225.

Willson, M. F., Irvine, A. K. & Walsh, N. G. (1989) Vertebrate dispersal syndromes in some Australian and New Zealand plant-communities, with geographic comparisons. *Biotropica*, **21**, 133–147.

Worton, B. J. (1987) A review of models of home range for animal movement. *Ecological Modelling*, **38**, 277–298.

Wright, S. J. (2003) The myriad consequences of hunting for vertebrates and plants in tropical forests. *Perspectives in Plant Ecology, Evolution and Systematics*, **6**, 73–86.

Wright, S. J. & Duber, H. C. (2001) Poachers and forest fragmentation alter seed dispersal, seed survival, and seedling recruitment in the palm *Attalea butyracea*, with implications for tropical tree diversity. *Biotropica*, **33**, 583–595.

Wright, S. J., Caldwell, M. M., Hernandéz, A. & Paton, S. (2004) Are lianas increasing in importance in tropical forests? A 16-year record from Barro Colorado Island, Panamá. *Ecology*, 85, 484–489.

Wright, S. J., Carrasco, C., Calderón, O. & Paton, S. (1999) The El Niño Southern Oscillation, variable fruit production and famine in a tropical forest. *Ecology*, **80**, 1632–1647.

Wright, S. J., Zeballos, H., Dominguez, I., Gallardo, M. M., Moreno, M. C. & Ibanez, R. (2000) Poachers alter mammal abundance, seed dispersal, and seed predation in a neotropical forest. *Conservation Biology*, **14**, 227–239.

Wunderle, J. M. (1997) The role of animal seed dispersal in accelerating native forest regeneration on degraded tropical lands. *Forest Ecology and Management*, **99**, 223–235.

Yockteng, R. & Cavelier, J. (1998) Diversity and dispersal mechanisms of trees in Isla Gorgona and of the tropical moist forests of the Colombian–Equatorian Pacific. *Revista de Biologia Tropical*, **46**, 45–53.

CHAPTER TWELVE

The role of trophic interactions in community initiation, maintenance and degradation

JOSÉ MANUEL VIEIRA FRAGOSO

University of Hawaii

The generality of the role of trophic cascades in creating trophic structure is still in debate (Strong 1992; Polis *et al.* 2000; Holt 2000). In part this is because spatial (van Noughys & Hanski 2002) and temporal effects (Dyer & Coley 2002; Dyer & Letourneau 2003) make it difficult to pinpoint patterns in the impact of intertrophic effects and cascades on community structure. For example, predators may exert effects at some times but not others (Sinclair 2003), and in some locations but not others (van Noughys & Hanski 2002). Sinclair (2003) and Sinclair and Krebs (2002), working with vertebrates (hares, lynx, predatory birds and others), found that predators regulated their prey populations at high but not low densities in the boreal forest of Canada, and that this effect did not cascade down to plant populations.

To date, support for the existence of trophic cascades and their role in influencing trophic structure comes primarily from biologically simple ecosystems with two or three interacting organisms and trophic levels (see the recent volume by Tscharntke and Hawkins 2002). Although there are suggestions in the literature (Tscharntke & Hawkins 2002) that it will be difficult to find evidence in support of strong trophic cascades in high-biodiversity terrestrial systems with complex food webs, Terborgh and co-workers have found correlations that support their occurrence in predator–herbivore–plant systems involving both vertebrates and invertebrates (Terborgh 1992, Terborgh *et al.* 2001). Letourneau and Dyer (1998) also experimentally reproduced top-down effects for four trophic levels (soil nutrients, a plant, insect herbivores and insect predator) in Costa Rica. Recognition of strong multitrophic interactions in high-diversity terrestrial systems may also have been limited because work to date has not included appropriately scaled studies with large vertebrate herbivores and their predators (Persson 1999). Furthermore, most studies have simplified complex food webs into simpler food chains in their attempt to understand the nature of the system (Pace *et al.* 1999).

Biotic Interactions in the Tropics: Their Role in the Maintenance of Species Diversity, ed. D. F. R. P. Burslem, M. A. Pinard and S. E. Hartley. Published by Cambridge University Press. © Cambridge University Press 2005.

To explore how spatial and temporal scales and long-term dynamics may affect both the functioning of multitrophic interactions and our ability to detect their impact on community structure, I consider one multi-species trophic interaction and trophic cascade in a complex food web in a terrestrial tropical system. As documented to date (Fragoso 1994; 1997a; Fragoso *et al.* 2003; Silvius 2002; Silvius & Fragoso 2002; 2003), the system consists of the palm *Attalea maripa* (formerly *Maximiliana maripa*, revised by Henderson 1995), its fruits and seeds, two obligate invertebrate herbivores of *A. maripa* seeds, a wasp parasitoid of at least one and probably both of the invertebrate herbivores, a large vertebrate herbivore, and a number of omnivorous vertebrate species that feed on the fruits and seeds and that prey on the invertebrate herbivores. I describe the key role of seed dispersal by tapirs (*Tapirus terrestris*) in allowing *Attalea* to escape its predators, and subsequent local regulation by a multi-level trophic system, in which the ecological importance of each interaction shifts through time. I further consider how the interactions between organisms in the same and different trophic levels can affect the diversity and abundance of different species, and speculate on how these interactions can initiate community formation, biodiversity patterns, trophic organization, community degradation and death. Because the interactions are focused on seed dispersal and seed predation, they help set patterns of diversity among tropical forest tree species, and may represent an alternative or complementary scenario to explanations based on canopy disturbance and gap dynamics. I compiled and correlated data from studies conducted over 16 years at Maracá Island Ecological Reserve in tropical rainforest in the state of Roraima, Amazonian Brazil (Fragoso 1994; 1997a, b; 1998a, b; 1999; Fragoso & Huffman 2000; Fragoso *et al.* 2003; Silvius 1999; 2002; Silvius & Fragoso 2002; 2003).

The system
Producer trophic level
Attalea maripa

This palm forms the producer trophic level of the system. It grows to 30 m in height and reaches the canopy or sub-canopy of old-growth forests (Fragoso 1997a). It occurs in a clumped distribution with high-density patches containing 12–24 trees per 0.25 ha (Fragoso 1998b). Within patches *Attalea* co-occurs with many other tree species (Milliken & Ratter 1989). *Attalea* also occurs in patches in savannas, riverine forests and secondary forests (Fragoso *et al.* 2003). Within the forest matrix, palm patches range from less than a hectare to about 10 hectares in extent, although most are probably between half a hectare and 2 hectares; patches are separated by hundreds to thousands of metres of palm-free forest (Fragoso *et al.* 2003). Adult *Attalea* also occur as solitaires. *Attalea* fruiting is synchronized within climatic regions (J. M. V. Fragoso, personal observations). In the northern part of Roraima most trees flower from January to March and ripe fruit falls en masse from April (peaking in June) to October (Fragoso 1998a). Adult trees produce one to three infructescences each containing from a few hundred to more than 2500 fruits (Fragoso 1998b; Silvius & Fragoso 2002). Mature fruits are yellowish-green, oval-shaped and 4 to 7 cm long. A slightly sweet soapy-tasting fleshy pulp with some fibres surrounds a thick woody endocarp (3 to 6 cm length) that generally contains one and occasionally two or three seeds (Fragoso 1994; Silvius 2002). Ripe fruits drop naturally below parent trees, or are dropped or spat out by arboreal herbivores (frugivores; e.g. primates, parrots, macaws) after pulp has been consumed (Fragoso 1997a). Pulp-free seeds can remain dormant for at least 14 months (J. M. V. Fragoso, unpublished data).

Herbivore trophic level

Pachymerus cardo *(Fåhraeus) and* Speciomerus giganteus *(Chevrolat)*

These bruchid beetles specialize on *Attalea* seeds, although they are not confined to the genus (Delobelle *et al.* 1995; Silvius 1999). They appear to be the dominant representatives of the herbivorous trophic level supported directly by *Attalea* fruits and seeds (Fragoso 1997a; Fragoso *et al.* 2003; Silvius 2003; Silvius & Fragoso 2002). Signs of vertebrate or insect herbivory on the leaves of seedlings or adults are not obvious (personal observations). It is during the larval stage that beetles feed on *Attalea*. Females can lay 1–30 eggs on an endocarp, and oviposit preferentially on endocarps whose pulp has been incompletely consumed by frugivores (Silvius & Fragoso 2002). However, they eventually lay eggs on all fruits and endocarps on the soil surface within palm patches (Fragoso 1997a). Upon hatching (6–7 days), first-instar larvae burrow through the remaining pulp and woody endocarp and take up residence in the endosperm (Silvius 1999). They consume the endosperm over a 3- to 12-month period (possibly up to 2 years; K. Silvius, personal communication), after which one adult beetle (occasionally two or three, in multiple-seeded endocarps) emerges from each endocarp, although many first-instar larvae may have entered each seed. Up to 97% of seeds remaining on soil surfaces within 30 m of parent trees are consumed by beetles (Fragoso 1997a), as are almost all seeds within a palm patch (K. Silvius, personal communication). Adult beetles oviposit on endocarps at night (J. M. V. Fragoso, unpublished data). In forest, eggs are laid only on fallen fruits.

Tapirus terrestris *(Linnaeus) and other vertebrate herbivores*

Lowland tapirs (*Tapirus terrestris*) eat the pulp of intact freshly fallen *Attalea* fruits; although they ingest the seeds, they do not consume them and the seeds pass intact and viable through the digestive tract (Fragoso 1997a; Fragoso & Huffman 2000). Tapirs thus belong to the herbivore trophic level, which includes animals that feed only on the pulp of the fruits as well as those that feed on seeds. Tapirs are large (250 kg) and can ingest many fruits at one time; up to 210 fruits have been found in one faecal pile (Fragoso & Huffman 2000). Thousands of viable seeds become aggregated at tapir latrines in upland and wetland areas

(Fragoso 1997a). The immersion of endocarps in faeces prevents beetles from accessing and killing seeds (Fragoso *et al.* 2003). The long-distance movement of endocarps several kilometres beyond *Attalea* patches by tapirs makes it difficult if not impossible for beetles to find these seeds (Fragoso *et al.* 2003). Nothing is known about how far beetles disperse, but they appear to be weak flyers with restricted movements, as they are never captured in light traps or by other capture methods that intercept insect movements (Fragoso *et al.* 2003). Thus, seed-handling by tapirs ensures the survival of seeds that would otherwise have been killed by beetles. This interaction is best described as a mutualism between *Attalea* and tapirs, and exemplifies what Bronstein and Barbosa (2002) classify as a protective mutualism, where a species deters other species in the same trophic level from negatively affecting the producer species. Endocarps are secondarily dispersed from latrines by rodents and many of these seeds eventually germinate and create high-density patches of seedlings, saplings and probably adult palms (Fragoso 1997a). Tapirs mediate the interaction between *Attalea* palms, beetles and rodents, and are responsible for initiating new *Attalea* patches.

Other vertebrate species in this trophic level include at least five primate, three macaw, two parrot and four ungulate species and potentially a number of marsupials, armadillos and carnivores (Fragoso 1997a, and personal observations); however, these animals, with the possible exception of macaws and parrots, move seeds only one metre to tens of metres from parent trees (Fragoso 1997a). Because the seeds remain within a conspecific patch they are accessible to beetles for egg-laying and consumption (Fragoso 1997a; Silvius & Fragoso 2002). The scarce amounts of faecal matter produced per defecation by other fruit eaters that swallow fruits and defecate seeds, such as spider monkeys (*Ateles belzebuth* É. Geoffroy; K. Silvius, personal communication), are insufficient to cover and protect the large *Attalea* endocarps.

Predator trophic level one (exclusively)
Parasitoid
A potentially important member of the first predator trophic level is an unidentified braconid wasp whose larvae parasitize beetle larvae (Silvius, 1999; 2002). This parasitic wasp may be an obligate predator of *Pachymerus* and *Speciomerus* beetle larvae. Adult wasps lay their eggs inside *Attalea* endocarps that contain beetle larva. As wasp larvae mature they consume their beetle larval host. Once the beetle larva is consumed the wasp larva pupates inside the endocarp and two to five or more adult wasps eventually emerge (K. Silvius, unpublished data).

Omnivores: trophic levels predator one and two, and herbivore
Tayassu pecari *(Link)*
One of the more important vertebrates in the system may be the white-lipped peccary (white-lips, *Tayassu pecari*). These large ungulates (up to 50 kg) occur in herds of 40–400 and perhaps thousands of individuals (Fragoso 1998a; Kiltie &

Terborgh 1983; Mayer & Brandt 1982). Although omnivorous, white-lipped peccaries forage extensively on *Attalea* endocarps (Fragoso 1999), consuming bruchid larvae in the endocarps (Silvius 2002), and also occasionally consume endosperm (Fragoso 1994; 1999). These animals simultaneously belong to the herbivore and first predator trophic levels. To the extent that they encounter and consume wasp larvae and adult wasps, white-lipped peccaries (and the other omnivores described below) also constitute a second predator trophic level. White-lipped peccary populations fluctuate from very high to very low densities at intervals of 10 to 15 years (Fragoso 1997b, 2004). At low densities white-lips open few *Attalea* endocarps and usually only those encountered incidentally in wetland tapir latrines when they are foraging for other items (Fragoso 1994; and unpublished data). As the population increases they begin foraging on endocarps around a few to eventually all parent trees and patches (K. Silvius, unpublished data). They also uproot and kill hundreds of seedlings (Kiltie & Terborgh 1983; Silman *et al.* 2003; Fragoso, unpublished data).

Rodents

Agoutis (*Dasyprocta leporina* Linnaeus) are large rodents (3 to 5 kg) that feed on endosperm and beetle larvae as do Amazon red squirrels (*Sciurus igniventris* Wagner) and smaller rodents such as spiny rats (*Proechimys* sp.; Silvius 2002). These animals also simultaneously belong to the herbivore and first predator trophic levels. Squirrels may be the only vertebrate species that specializes on *Attalea* and other palm endocarps in this system; however, they regularly consume both endosperm and beetle larvae (Silvius 2002). As is the case for white-lipped peccaries, these rodents may also consume wasp larvae and thus influence wasp population dynamics as level-two predators. White-lipped and collared (*Tayassu tajacu* Linnaeus) peccaries, agoutis, squirrels, smaller rodents, deer and other terrestrial herbivores also consume the pulp from fallen fruits and are thus also members of the herbivore trophic level (Fragoso 1997a). The multiplicity of trophic levels to which these vertebrates belong indicates that they probably mediate interactions between one another in various ways.

Next I consider the nature of the interactions between *Attalea* palms herbivorous beetles, parasitic wasps, predatory white-lipped peccaries and rodents across time and space in an attempt to understand how multi-trophic interactions may contribute to community initiation, structure and dynamics. I also consider how these interactions influence one another under a scenario of *Attalea* patches transitioning from formation through to maintenance, degradation and death. Figure 12.1 depicts the type and direction of interaction and the trophic-level placement of the species under consideration.

To determine whether the multitrophic interactions described in Figure 12.1 remain constant during all life phases of an *Attalea* patch, I classified *Attalea* patches into different stages (phases) of patch development: birth, growth,

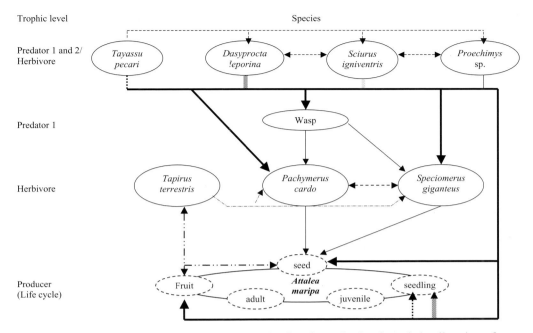

Figure 12.1 Trophic levels of nine species forming a food web, and the direction of consumption, in an Amazonian tropical forest. Additional frugivores (not shown) consume the fruit pulp. Legend: predation (one species) —, predation (several species) —, interference —·—, cooperation —··— and competition ———. The arrowhead indicates the direction of the interaction; two arrowheads pointing in opposite directions on a single line indicate mutualism or competition. The point of influence of an animal species in the top row of the figure occurs where its line-pattern departs from the thick solid line that almost encircles the figure. When the exit line is also black and thick it indicates that all the top row animal species affect a taxon or plant part.

expansion, decline and death, based on long-term work by Fragoso (1994; 1997a; 1998a; 1999), Fragoso and Huffman (2000) and Fragoso *et al.* (2003).

Phases in *Attalea* patch development
Phase 1
All phases and constituent organisms can be viewed in Figure 12.2. In Phase 1, tapirs disperse *Attalea* seeds into an area without adult *Attalea*, dominated by adults of other tree species (Fragoso *et al.* 2003). Rodents re-disperse these endocarps a few to hundreds of metres beyond the arrival points (Forget & Milleron 1991; Forget *et al.* 1998; Fragoso 1997a; Wenny 1999). Because some *Attalea* seed predators (e.g. white-lipped peccaries) locate their prey by searching for adult palms (Fragoso 1997a), and others such as agoutis (Silvius & Fragoso 2003) are territorial, there will be few seed predators at the arrival points at this time (Fragoso *et al.* 2003). Without predators almost all seeds survive (Fragoso *et al.* 2003). Many

Phase 1 (seed dispersal)

Phase 1: Seedlings

Phase 2: First Fruit

Phase 3:

Phase 4:

Phase 5:

Figure 12.2 An illustration of the organisms involved in five phases of *Attalea maripa* patch formation, maintenance and death. The number of individuals of a plant type represents their numerical importance in that phase relative to other plants, while individual size (largest) indicates a dominant or driver animal species. Species in order of appearance are: dicotyledonous trees, tapir (*Tapirus terrestris*), *Attalea maripa* seeds, agouti (*Dasyprocta leporine*), *Attalea* seedlings and adults, bruchid beetle (*Pachymerus cardo* and *Speciomerus giganteus*), rat (*Proechimys* sp.), squirrel (*Sciurus igniventris*), braconid wasp sp., and white-lipped peccary (*Tayassu pecari*).

germinate, establish seedlings and reach maturity (Fragoso 1997a). At this time the *Attalea* patch should exhibit a truncated age distribution because of an abundance of seedlings and saplings, and a lack of adult palms. High recruitment levels should continue until slightly beyond the appearance of the first fruit-producing adult trees. This may take 10 to 20 years based on *Attalea* recruitment from seeds to adults following deforestation at nearby ranches (Fragoso, personal observation).

At this stage the *Attalea* population is most likely to be controlled by soil fertility, light and water levels in conjunction with positive interactions with herbivores (tapirs and rodents). Since herbivorous beetles and other seed-eating organisms do not occur in the area or occur at very low densities, they have little or no influence on *Attalea* (Fragoso 1997a; Fragoso *et al.* 2003). Similarly, with no fruit-producing palms, white-lipped peccaries, the other potential herbivorous influence on seeds, are unlikely to be visiting the area (Fragoso 1998a), unless some other tree species is producing fruit or seeds (Silvius 2002). This *Attalea* population is regulated by bottom-up, pre-producer-level trophic elements.

Phase 2

The first *Attalea* palms produce fruit. Assuming that intraspecific competition is stronger than interspecific competition, and that *Attalea* trees affect soil nutrient levels differently from dicot species, fruit production per tree should be high, as the soil nutrient levels will be higher than for sites where *Attalea* has been long established. Both assumptions need to be tested, but support comes from the observations that (1) *Attalea* seedlings and saplings establish and grow under the closed forest canopy in the absence of disturbance, and (2) palms release large amounts of slow-decaying litter that physically prevents nearby seedling establishment and may contribute few nutrients to the soil. Seed survival and recruitment into all age classes, including adults, continues to be high (Fragoso 1997a). However, demographically the population is still dominated by younger plants (a reverse J curve with very few adults). A few bruchids will have discovered the first *Attalea* fruits and they will have infested a few endocarps. These beetles will show high egg-laying rates per female because of low intraspecific competition for seeds (Wright 1983; 1990). Bruchid egg, hatching, larval survivorship and recruitment rates to the adult stage will be very high, although their density remains low. Low beetle-larvae densities mean few to no parasitoids can be supported, and this species has yet to colonize the area. The system is still regulated by bottom-up trophic forces, meaning that *Attalea* productivity is dependent on soil nutrients and light levels. Competition with other palm species is probably low, since the other arborescent species in the area, *Astrocaryum vulgare* (Giseke) and *Oenocarpus bacaba* (Martius), occur at low densities (Fragoso, unpublished data). The population density of bruchid beetles is still too low to affect seed survivorship and population recruitment.

The community structure of the area will have changed as new frugivorous, granivorous and grubivorous (larvae feeders; as per Silvius 2002) species arrived or increased in density in response to the new abundance of *Attalea* fruits, seeds and beetle larvae. White-lipped peccaries, as well as various primate, bird and invertebrate species that feed on fruit pulp or seeds may now have located the new palm patch; however, their visits to the area should still be infrequent, especially for white-lipped peccaries since they tend to forage in traditional feeding areas dispersed over a home range 100 to 200 km² in area (Fragoso 1998a; 1999; Carrillo *et al.* 2002; Keuroghlian 2003). Hence, at this stage the palm and the bruchid beetle populations are relatively unaffected by a predatory trophic level (white-lipped peccaries). Rodents such as agoutis, squirrels and smaller rats (e.g. *Proechimys*) should now be more common in the area, be feeding on *Attalea* fruit and seeds, and have begun feeding on beetle larvae. However, these animals continue to occur at low densities and this coupled with their territorial behaviour suggests that they would have a weak effect or none on the *Attalea* population. Squirrels will probably not establish many territories in the area, because of the lack of a year-round source of palm endocarps, and agoutis will still prefer to feed on alternative food sources, as *Attalea* is not a preferred food (Silvius 2002; Silvius & Fragoso 2003). The *Attalea*-dominated producer trophic level now regulates the diversity and abundance of the animal species in higher trophic levels.

Phase 3

Attalea fruit and seed production per tree and patch continues to be high, and I posit that the seed population is in a state of dynamic equilibrium. Thus, the number of seeds consumed by predators and recruiting into older plants is approximately equal. At this stage the *Attalea* population and animal species at higher trophic levels continue to be regulated by *Attalea*-dominated bottom-up relationships but the system is beginning to flip into another state (*sensu* Holling 1986), from one controlled by the producer trophic level, to one regulated by a dominant herbivore, the bruchid beetle. In this phase female beetles may lay up to 30 eggs per endocarp; most of these hatch and many larvae penetrate individual endocarps (Silvius & Fragoso 2002). The endocarp population attains a beetle larvae infestation rate of 90% or more, and most multiple-seeded endocarps support more than one larva. The number of eggs laid by wasps per beetle larva per endocarp is now high but wasp infestation of the endocarp/beetle larvae population remains at medium levels; however, wasp survival and recruitment rates into adulthood should be very high. Beetle recruitment rates between life stages are at medium levels owing to high predation by vertebrates, wasps and perhaps fungus (Silvius 2002; Fragoso unpublished data). High densities of larvae-infested seeds have led to more visits by white-lipped peccaries to the patch (Fragoso 1998a; 1999), and concomitant higher rates of beetle-larvae consumption.

White-lipped peccaries, members of the second predator trophic level, have begun to exert a regulating influence on the larvae population and are competing with wasps for the larva resource, potentially limiting wasp populations. The various rodent species may also have experienced population increases.

The local adult tree community has shifted from one with high species diversity to a lower-diversity system dominated by adult *Attalea* palms (Milliken & Ratter 1989; Fragoso *et al.* 2003). The species diversity and abundance of vertebrates and invertebrates in the community is high, owing to the diversity and abundance of food types available (*Attalea* fruit pulp and seeds, beetle eggs on seeds and fruits, and bruchid larvae, pupae and wasp larvae in endocarps). A diversity of insect species feed and/or lay their eggs on *Attalea* pulp, and insect predators hunt in the fruit patches (Silvius & Fragoso 2002; K. Silvius, personal communication); because of the sheer abundance of the resource, many individuals and species can be supported; because the resource is ephemeral, it is unlikely that one species can dominate it. At this point tapirs ingest fruits at the patch and initiate the creation of new *Attalea* patches far from the present patch. Seed survival within the parent palm patch is low but sufficient to maintain its structure and extent. The palm population now exhibits a reverse J age distribution curve.

Phase 4

The numerical dominance of *Attalea* trees is declining as adults senesce and recruitment falls, and diversity of other tree species in the understorey is increasing. *Attalea* trees are still producing fruit, but seed predation by beetles (before and after short-distance dispersal) is very high (Fragoso 1997a; Fragoso *et al.* 2003; Silvius & Fragoso 2002). Seedling mortality rates are also high because of intense soil-rooting and trampling by peccaries (Cintra 1997; Fragoso unpublished data; Silman *et al.* 2003). Primary short-distance seed dispersal by vertebrates is low and too few seeds are surviving to maintain the recruitment rates necessary for patch maintenance. The population has an age structure with more adults than saplings and seedlings. This phase probably lasts for a long time, determined by the lifespan of the palm.

The bruchid beetle, a member of the first predator trophic level, now dominates the herbivorous trophic level and controls the palm. Female bruchid beetles continue to lay up to 30 eggs per seed; however, many of these are consumed by ants before hatching (K. Silvius, personal communication). Egg-survival rates per seed are still high enough to maintain 95% to 100% seed-infestation rates (Fragoso 1997a). A high density of beetle larvae continues to attract a diversity and abundance of grubivorous species such as parasitic wasps, white-lipped peccaries, agoutis and other rodents (Fragoso 1997a; Silvius 2002; Silvius & Fragoso 2002). Wasp-larvae infestation of beetle larva is at its highest levels, with a high number of wasp eggs laid per beetle larva and over the larvae-endocarp population. The vertebrate community attains its highest diversity and abundance, and the frequency of visits by white-lipped peccaries to the area is at its peak. Consumption of larvae by these organisms is at its highest level (Silvius 2002; Silvius & Fragoso 2002). The herbivore trophic level (beetle larvae) is now strongly influencing the dynamics of adjacent trophic levels (up and down), and may eventually cause the collapse of the *Attalea* population. However, members of the predator trophic levels, especially white-lipped peccaries and wasps (first predator level), exert a strong negative influence on the beetle population (Fragoso & Silvius unpublished data), and probably introduce non-linear dynamics into the system, in contrast to the linearity we have described up to this point. For example, if the white-lipped peccary population is at a high or increasing point in its cycle (Fragoso 1997b; Fragoso, 2004), they may exert enough negative pressure on the beetle population to negate or eliminate their effect on the *Attalea* population. However, if the white-lips are at a low point in their population cycle, wasps are released from competition with and predation by white-lips and may exert control of the beetle population. Thus, community assembly and disassembly depends not only on internal factors, but on an externally controlled population fluctuation – two patches will not follow the same trajectory if they are established at different times in the population shift of white-lipped peccaries. If predation pressure is too low to reduce the beetle population, the *Attalea*-population size begins to decline. At this point the community of species supported by beetles has peaked in diversity and abundance. The seeds that are surviving at high rates are those that were swallowed by tapirs and dispersed to their latrines.

Phase 5

Rodents, organisms belonging to the first and second predator trophic levels, are now driving the *Attalea* system. Adult *Attalea* palms are now rare or non-existent in the area. If a few palms occur they appear as solitaires. The area is now dominated by other tree species. *Attalea* fruit and seed production is non-existent or a tiny fraction of that in phases 2 to 4. If a few fruits and seeds are available rodents quickly find and eat them, or eat the fruit pulp and quickly cache the seed for later consumption and before beetles can lay eggs on them. Most if not all the cached seeds will be found and eaten. Quick seed removal translates into few if any beetle eggs being laid, and few larvae attaining adulthood. The few beetles that do lay eggs should lay high numbers per endocarp, but predation by ants (Silvius & Fragoso 2002) and death due to handling by rodents should also be very high, negating the effects of high levels of egg-laying. Low to no egg survival results in proportionally low endocarp-infestation rates. The low survival rate translates into very few to no adult beetles and eventually local beetle extinction. The few adult beetles to emerge from endocarps will either die or disperse long distances at low frequencies in search of fruit-producing

Attalea patches in phases 2 to 4. This dispersal probably occurs through self-propelled flight.

With few beetles, the wasps and ants die out, or the ants switch to alternate foods. The tapirs, primates, peccaries, deer, macaws, parrots, rodents and other vertebrates stop visiting the area, or if they continue they do so because of another tree species whose fruit, seeds or grubs colonized the region. The timing of white-lipped-peccary foraging visits to the area will now coincide with the fruiting phenology of the new dominant tree species. The area may also now support a more diverse adult tree community. Rodents thus deliver the final death-blow to the *Attalea* and beetle population. Some evidence indicates prey population collapse occurs when a species is a secondary prey item for a predator sustained by alternative food items (Sinclair & Krebs 2002), as larvae are for white-lipped peccaries, and as they become for rodents, but not for wasps. How long the area remains uninhabited by *Attalea* is unknown. This may be as short a time as it takes for a new generation of adult palms to arise from the local seed bank (*Attalea* seeds may remain dormant for over 14 months) or to be moved into the area by tapirs from an *Attalea* population in an earlier phase of development.

Patterning in tropical multitrophic interactions

The model I present above is based on empirical evidence collected during 16 years of study at Maracá Island Ecological Reserve, Amazonian Brazil, and on some speculation. Important components of the model are colonization and extinction dynamics driven by interactions between organisms belonging to different trophic levels. Initially, spatial expansion and numerical growth of a new plant (producer) population is limited by physical parameters. However, this situation is temporary. When individuals in the producer trophic level begin reproducing and attain high group reproductive output, herbivorous species (beetle larvae, rodents, white-lipped peccaries and other seed- and fruit-eating organisms) are attracted to the region. The upper trophic levels of the system are regulated by the producer trophic level at this stage. The system eventually flips to one regulated by the herbivore (beetles) trophic level owing to the growth of the herbivore population, a top-down effect. An increased density of herbivores attracts species in the next trophic level up, predators (white-lipped peccaries; predators with respect to this system, but omnivores in terms of overall diet), and increases the density of the resident omnivore population (rodents: omnivores with respect to this system, frugivore-granivores in terms of overall diet). The rate of visits by one member of the top predator trophic level (white-lipped peccaries), discovery of the beetle population by strict parasitoids, and an increase in the resident population of omnivores (rodents) lead to control of the system by the first and second predator trophic levels, so that the system is now controlled by top predators. Population and community regulation of the system is fluid, however, and the organism or trophic level that drives it depends on the population dynamics of and interactions among other organisms. The succession in dominance appears to be linear with succeeding species taking control in a time-lagged fashion. The speed at which this happens will depend on the rate at which different species locate the palm patches, which will in turn depend on their searching behaviour (for large species like white-lipped pec-caries and tapirs) and their dispersal pattern (for smaller or territorial species such as insects and rodents). The species driving the system at any one time is dependent on the maturity of the plants, the abundance of the herbivores, and the abundance and diversity of the predators.

Non-linear dynamics are introduced to the system when the population dynamics of a driver species are not linked to the system. For example, the population fluctuations of white-lipped peccaries appear to be driven by fac-tors operating at larger spatial scales than *Attalea*-patch dynamics and at longer time scales than those of animals restricted to *Attalea* patches. This component of white-lipped peccary ecology may introduce non-linearity into the succession-dominance pattern of the *Attalea* community. Short-term environmental fluctu-ations can be tracked through rapid reproduction by small-bodied species; how-ever, large-bodied, long-lived vertebrate species such as peccaries cannot track these changes, making their populations less responsive to the environment (Sinclair 2003). White-lipped peccaries are long-lived mammals (Walker 1999). White-lipped peccars populations respond to factors unrelated to the dynamics of the *Attalea* community (Fragoso 1997b; 2004). As their population nears a high point in its cycle, more beetle larvae and/or *Attalea* endosperms will be eaten; therefore white-lips will have a severe negative effect on the larval population. This effect eventually cascades down and up to other trophic levels, potentially increasing seed survivorship in following years or reducing the population of parasitic wasps. Synergy with palm reproductive phenology, another non-linear, decoupled factor in the system, can intensify the effect of the white-lips: if a year of high white-lip predation is followed by a year of low fruit production, then local beetle populations may be strongly impacted. However, when white-lipped peccaries are at low or medium population densities, their influence on the system is minor (Fragoso 1994, unpublished data). At this time most endocarps beneath parent trees remain untouched by white-lipped peccaries, and larvae continue to drive the system. Under this scenario most endocarps produce at least one adult beetle, and over 90% of endocarps in palm patches produce beetles. In this case the beetle-larvae population has outgrown any regulatory effect that wasps may have, and most larvae become adult beetles. Since the wasp-parasitoid generation time is probably much shorter than that of white-lipped peccaries their populations can react more quickly to this abundance of food, eventually growing to a level that begins reducing the beetle population (van Noughys & Hanski 2002). Even in the absence of white-lipped peccaries, non-linearity can be introduced by fluctuations in fruit production.

White-lipped-peccary population density appears to fluctuate at intervals of approximately 20–30 years between high or low points (Fragoso 1997b; Fragoso, 2004; Silman *et al.* 2003). During these intervals, a few highs and lows in beetle and wasp populations may have occurred, but we do not know how long individual beetles and wasps can live. In captivity white-lipped peccaries can live 12 to 15 years (Eisenberg 1989), begin reproducing at about 1 year of age, and bear one litter of 1–4 young (normally two). We also know that agoutis live 2–3 years, reproduce once to twice per year and may bear two young but only one survives (Silvius & Fragoso 2003; Smythe 1978). We do not know how long an adult bruchid beetle survives, but we do know that eggs laid on an endocarp take approximately 6–7 days to hatch, that first-instar larvae probably take a few days to penetrate to the endosperm, and the beetles remain as larvae from 3 months to a year (and sometimes up to 2 years) before pupating within the endocarp and emerging as adults (Silvius 2002). Assuming that like many beetles the adults begin searching for mates soon after emergence and reproduce within the same year, then we see that the beetles have a very short generation time, and in comparison to the other organisms described above, an extremely high reproductive output. Following a similar logic for a small invertebrate parasitoid of the beetles, we would predict them to have very high reproductive outputs; for example, we know that 4 to 19 wasps may emerge from one endocarp (J. M. V. Fragoso, personal observations), suggesting that a female lays many eggs at one time. Interrelating this information leads to the prediction that wasp population growth tracks the beetle larvae in a time-lagged way. However, the wasps may eventually exert a strong predation pressure on the beetle population, which begins to decline. Since white-lips are decoupled from the system, their population dynamics are unaffected by those of *Attalea*, beetles and wasps, but they can strongly influence the beetles and wasps during high points in their population fluctuations. These population-level interactions produce time-lagged responses from the various trophic levels, and these may be linear or non-linear depending on whether a strongly interacting organism is coupled or decoupled from the system.

Van Noughys and Hanski (2002), working with a more limited set of organisms, no vertebrates and fewer trophic levels, observed that populations of two parasitoid species of a herbivorous insect fluctuated in response to the population of their prey species. They also found that the dynamics of one interacting pool of organisms were decoupled from other spatially distinct pools. The parasitoids depressed the population of their prey, eventually causing the local extinction of the herbivore. These observations are very similar to what we predict to occur in the *Attalea* system for beetle and wasps.

Large mammals because of their size can have major impacts on the physical structure of habitats, rates of ecosystem processes and the diversity of communities (Sinclair 2003). For this reason many researchers have suggested that large mammals can exert significant influence on other elements of the communities to which they belong (see Terborgh *et al.* 2001). There is debate as to whether top predators are capable of exerting strong negative influences on their prey species (Wright *et al.* 1994; Terborgh *et al.* 2001); but several studies show that rodent populations in areas without predators can exist at very high densities (e.g. Adler 1994; 1998; Terborgh *et al.* 2001). On islands without predators rodent populations were 35 times as great as those on adjacent mainland areas of the Guri impoundment in Venezuela (Terborgh *et al.* 2001). On islands without predators and with high populations of terrestrial herbivorous vertebrates, the herbivores' plant-food species declined significantly, and the plant community transitioned into one dominated by species that were not preferred by the herbivores. Terborgh *et al.* (2001) observed a correlation between a lack of predators and possible cascading effects down into lower trophic levels. They did not consider interactions between invertebrates, between vertebrates and invertebrates, and between producers and all of the others. Nor did they consider how the suite of interactions and their outcomes could change with time. The assumption of their study is that the effect of predators, and the other organisms, is constant through time and space, with the only important variable being the presence or absence of a 'key' organism, in their case one that produced significant 'top-down effects'.

Terborgh and colleagues' model differs from the one presented here in that the *Attalea* model is dynamic and the strength or type of outcomes from interaction between species and trophic levels exhibits fluidity in both linear and non-linear fashions. In other words, a key species at one time (e.g. white-lipped peccaries) may be a passive 'rider' or even completely lacking at another time (or place). This temporarily missing or weakly present element may not be unusual in such systems, as long as their influence or a similar one returns to the system before it flips completely into another ecological state. Thus, I predict that the strength of top-down or bottom-up effects is a function of sampling time and spatial scale and location of the study. In effect, one should consider the stage of assembly of the local community when isolation occurred; differences in these stages, and the organisms that were most strongly affecting community processes, may be partially responsible for the sharp differences in community structure found on the Guri islands after several decades of isolation. Also important is whether the population is viewed from the context of metapopulations or single populations (van Noughys & Hanski 2002). In non-island situations, where populations are not separated from one another by impermeable barriers (e.g. extensive water bodies for terrestrial species), long-distance dispersal eventually occurs between areas, and this re-sets relationships and the importance of different trophic levels. Thus there can be no constancy in top-down or bottom-up effects even within one location.

In conclusion, the discussion of whether cascades occur in high-diversity terrestrial systems, and whether they affect community structure, may benefit from a shift in our perspective on trophic interactions. If we view them as short-lived, variable and dynamic, we may see that cascades often exist at one location and time, but that they may be different from those occurring at the same location at a different time, or a different location at the same time, still with the same organisms. The pattern of interactions is driven by different species depending on who is at a population high, and on the scale of measurement of the study. For example, white-lipped-peccary population highs occur at intervals of 20–30 years, and thus they drive the system at this scale. Beetles drive the system at the 8-year scale and wasps at the 6-year scale, based on a logic of longer lifespans described above. Because agoutis and squirrels can switch their diet their populations are unlikely to be driving the system. But if their populations explode (as in the Guri dam case) then the potential is there for them to be drivers.

Acknowledgements

This paper is dedicated to the memory of John F. Eisenberg, whose insatiable curiosity stimulated much of the research and speculation about both plants and animals contained in this paper, and to C. S. 'Buzz' Holling, who introduced me to large-scale, lumpy approaches to ecology. I thank Kirsten M. Silvius for letting me know when my logic was unsound, and for critical reviews, thoughtful suggestions and insight into the functioning of the *Attalea* sytem. Without Michelle Pinard's graciousness this paper would not have been published. I also thank Jerome Chave and Dan Wenny for their quick and meaningful reviews.

References

Adler, G. A. (1994) Tropical forest fragmentation and isolation promote asynchrony among populations of a frugivorous rodent. *Journal of Animal Ecology*, **63**, 903–911.

(1998) Impacts of resource abundance on populations of a tropical forest rodent. *Ecology*, **79**, 242–254.

Bronstein, J. L. & Barbosa, P. (2002) *Multitrophic/Multispecies Mutualistic Interactions: The Role of Non-mutualists in Shaping and Mediating Mutualisms* (ed. T. Tscharntke & B. A. Hawkins). New York: Cambridge University Press, pp. 44–66.

Carrillo, E., Saenz, J. C. & Fuller, T. K. (2002) Movements and activities of white-lipped peccaries in Corcovado National Park, Costa Rica. *Biological Conservation*, **108**, 317–324.

Cintra, R. (1997) Leaf litter effects on seed and seedling predation of the palm *Astrocaryum*

murumuru and the legume tree *Dipteryx micrantha* in Amazonian forest. *Journal of Tropical Ecology*, **13**, 709–725.

Delobelle, A., Couturier, G., Kahn, F. & Nilsson, J. A. (1995) Trophic relationships between palms and bruchids (Coleoptera: Bruchidae: Pachymerini) in Peruvian Amazonia. *Amazoniana*, **8**, 209–219.

Dyer, L. A. & Coley, P. D. (2002) *Tritrophic Interactions in Tropical versus Temperate Communities* (ed. T. Tscharntke & B. A. Hawkins). New York: Cambridge University Press, pp. 67–88.

Dyer, L. A. & Letourneau, D. (2003) Top down and bottom up diversity cascades in detrital versus living food webs. *Ecology Letters*, **6**, 60–68.

Eisenberg, J. F. (1989) *Mammals of the Neotropics: The Northern Neotropics*. Vol. I. Chicago: University of Chicago Press.

Forget, P. M. & Milleron, T. (1991) Evidence for secondary dispersal by rodents in Panama. *Oecologia*, **87**, 596–599.

Forget, P. M., Milleron, T. & Feer, F. (1998) *Patterns in Post-dispersal Seed Removal by Neotropical Rodents and Seed Fate in Relation to Seed Size* (ed. D. M. Newbery, H. T. Prins & N. D. Brown). Oxford: Blackwell Scientific, pp. 25–49.

Fragoso, J. M. V. (1994) Large mammals and the community dynamics of an Amazonian rain forest. Unpublished Ph.D. thesis, University of Florida.

(1997a) Tapir-generated seed shadows: scale-dependent patchiness in the Amazon rain forest. *Journal of Ecology*, **85**, 519–529.

(1997b) *Desapariciones Locales del Báquiro Labiado* (Tayassu pecari*) en la Amazonia: Migración, Sobre-cosecha, o Epidemia?* (ed. T. G. Fang, R. E. Bodmer, R. Aquino & M. H. Valqui). La Paz: Instituto de Ecología, pp. 309–312.

(1998a) Home range and movement patterns of white-lipped peccary (*Tayassu pecari*) herds in the northern Brazilian Amazon. *Biotropica*, **30**, 458–469.

(1998b) *White-lipped Peccaries and Palms on the Ilha de Maracá* (ed. W. Milliken & J. A. Ratter). Chichester: Wiley.

(1999) Perception of scale and resource partitioning by peccaries: behavioral causes and ecological implications. *Journal of Mammalogy*, **80**, 993–1003.

(2004) *A Long-Term Study of White-lipped Peccary* (Tayassu pecari) *Population Fluctuations in Northern Amazonia: Anthropogenic vs. 'Natural' Causes* (ed. K. Silvius, R. E. Bodmer & J. M. V. Fragoso). New York: Columbia University Press.

Fragoso, J. M. V. & Huffman, J. (2000) Seed-dispersal and seedling recruitment patterns by the last neotropical megafaunal element in Amazonia, the tapir. *Journal of Tropical Ecology*, **16**, 369–385.

Fragoso, J. M. V., Silvius, K. M. & Correa, J. A. (2003) Long distance seed dispersal by tapirs increases seed survival and aggregates tropical trees. *Ecology*, **84**, 1998–2006.

Henderson, A. (1995) *The Palms of the Amazon*. New York: Oxford University Press.

Holling, C. S. (1986) *Resilience of Ecosystems: Local Surprise and Global Change* (ed. C. W. Clark & R. E. Munn). New York: Cambridge University Press, pp. 292–317.

Holt, R. D. (2000) Trophic cascades in terrestrial ecosystems. Reflections on Polis *et al*. *Trends in Ecology and Evolution*, **11**, 444–445.

Keuroghlian, A. (2003) The response of peccaries to seasonal fluctuations in an isolated patch of tropical forest. Unpublished Ph.D. thesis, University of Nevada.

Kiltie, R. A. & Terborgh, J. (1983) Observations on the behaviour of rain forest peccaries in Peru: why do white-lipped peccaries form herds? *Zeitschrift für Tierzuchtung und Zuchtungbiologie*, **62**, 241–255.

Letourneau, D. K. & Dyer, L. A. (1998) Experimental test in a lowland tropical forest shows top-down effects through four trophic levels. *Ecology*, **79**, 1678–1687.

Mayer, J. J. & Brandt, P. N. (1982) *Identity, Distribution and Natural History of the Peccaries, Tayassuidae* (ed. M. A. Mares & H. H. Genoways). Pennsylvania: University of Pittsburgh, pp. 433–455.

Milliken, W. & Ratter, J. A. (1989) *The Vegetation of the Ilha de Maracá: First Report of the Vegetation Survey of the Maracá Rain Forest Project*. Edinburgh: Royal Botanical Garden.

Pace, M. L., Cole, J. J., Carpenter, S. R. & Kitchell, J. F. (1999) Trophic cascades revealed in diverse ecosystems. *Trends in Ecology and Evolution*, **14**, 483–488.

Polis, G. A., Sears, A. L. W., Huxel, G. R., Strong, D. R. & Maron, J. (2000) When is a trophic cascade a trophic cascade? *Trends in Ecology and Evolution*, **11**, 473–475.

Persson, L. (1999) Trophic cascades: abiding heterogeneity and the trophic level concept at the end of the road. *Oikos*, **85**, 385–397.

Silman, M. R., Terborgh, J. W. & Kiltie, R. A. (2003) Population regulation of a dominant rain forest tree by a major seed predator. *Ecology*, **84**, 431–438.

Silvius, K. M. (1999) Interactions among *Attalea* palms, bruchid beetles, and neotropical terrestrial fruit-eating mammals: implications for the evolution of frugivory. Unpublished Ph.D. thesis, University of Florida.

(2002) Spatio-temporal patterns of palm endocarp use by three Amazonian forest mammals: granivory or 'grubivory'? *Journal of Tropical Ecology*, **90**, 1024–1032.

Silvius, K. M. & Fragoso, J. M. V. (2002) Pulp handling by vertebrate seed dispersers increases palm seed predation by bruchid beetles in the northern Amazon. *Journal of Ecology*, **90**, 1024–1032.

(2003) Agouti (*Dasyprocta agouti*) home range size, habitat use and diet in Amazonian moist tropical forest. *Biotropica*, **35**, 74–83.

Sinclair, A. R. E. (2003) Mammal population regulation, keystone processes and ecosystem dynamics. *Philosophical Transactions of the Royal Society of London B*, **358**, 1729–1740.

Sinclair, A. R. E. & Krebs, C. J. (2002) Complex numerical responses to top-down and bottom-up processes in vertebrate populations. *Philosophical Transactions of the Royal Society of London B*, **357**, 1221–1231.

Smythe, N. (1978) *The Natural History of the Central American Agouti* (Dasyprocta punctata). Smithsonian Contributions to Zoology No. 157. Washington DC: Smithsonian Institution Press.

Strong, D. R. (1992) Are trophic cascades all wet? Differentiation and donor control in speciose ecosystems. *Ecology*, **73**, 747–754.

Terborgh, J. (1992) Maintenance of diversity in tropical forests. *Biotropica*, **24**, 283–292.

Terborgh, J., Lopez, L., V. Nunez, P. *et al.* (2001) Ecological meltdown in a predator-free forest fragment. *Science*, **294**, 1923–1925.

Tscharntke, T. & Hawkins, B. A. (2002) *Multitrophic Level Interactions*. New York: Cambridge University Press.

van Noughys, S. & Hanski, I. (2002) *Multitrophic Interactions in Space: Metacommunity Dynamics in Fragmented Landscapes* (ed. T. Tscharntke & B. A. Hawkins). New York: Cambridge University Press, pp. 124–149.

Walker, R. M. (1999) *Walker's Mammals of the World,* 6th edn. Baltimore: John Hopkins University Press.

Wenny, D. G. (1999) Two-stage dispersal of *Guarea glabra* and *G. kunthiana* (Meliaceae) in Monteverde, Costa Rica. *Journal of Tropical Ecology*, **15**, 481–496.

Wright, S. J. (1983) The dispersion of eggs by a bruchid beetle among *Scheelea* palm seeds and the effect of distance to the parent palm. *Ecology*, **64**, 1116–1021.

(1990) Cumulative satiation of seed predator over the fruiting season of its host. *Oikos*, **58**, 272–276.

Wright, S. J., Gompper, M. E. & Deleon, B. (1994) Are large predators keystone species in neotropical forests: the evidence from Barro Colorado Island. *Oikos*, **71**, 279–294.

CHAPTER THIRTEEN

Impacts of herbivores on tropical plant diversity

ROBERT J. MARQUIS

University of Missouri – St Louis

Introduction

Herbivore attack is hypothesized to contribute to the high level of plant species richness common in many tropical terrestrial communities (Janzen 1970; Connell 1971). Janzen and Connell independently proposed that seed and seedling predation in tropical forests would prevent any one highly competitive plant species from excluding other species. Two sets of observations served as the basis for the initial hypothesis. Firstly, high seed and seedling predation is common for many tropical tree species, potentially influencing the population dynamics of those species (Janzen 1970; Connell 1971). Secondly, predation on a superior competitor in an adjacent lower trophic level in temperate marine systems was shown to reduce the likelihood of competitive exclusion (Connell 1971) or actually to reduce exclusion (Paine 1966). Since this hypothesis was first proposed (Janzen 1970; Connell 1971), herbivory and seed predation, at least for temperate systems, have been shown to influence the richness and diversity of plant species (Dirzo 1984; Augustine & McNaughton 1998; Howe & Brown 1999), vegetation structure (Brown & Heske 1990) and plant succession (Brown & Gange 1992). In turn, herbivory (Rausher & Feeny 1980; Doak 1992; Ehrlen 1995) and predispersal seed predation (Louda & Potvin 1995; Ehrlen 1996; Kelly & Dyer 2002) have been demonstrated to influence plant population dynamics in temperate systems.

Support for the impacts of herbivores on plant diversity in tropical systems is much less complete. At least five hypotheses have been proposed. Janzen (1970) and Connell (1971) first proposed that herbivores in tropical forests might maintain plant species richness through high levels of postdispersal seed and seedling predation near adult plants and lower levels far away. As any one species becomes more abundant, there would be fewer and fewer safe sites for establishment in the forest, and thus its population growth rate would diminish, preventing competitive exclusion. McNaughton (1985) suggested that grazing by ungulates on the canopy grass species in the Serengeti may reduce competition, resulting in increased availability of resources for poorer competitors in the subcanopy, and

Biotic Interactions in the Tropics: Their Role in the Maintenance of Species Diversity, ed. D. F. R. P. Burslem, M. A. Pinard and S. E. Hartley. Published by Cambridge University Press. © Cambridge University Press 2005.

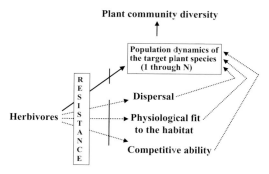

Figure 13.1 The potential pathways by which herbivores may affect the population dynamics of a plant species. Solid lines indicate mechanisms by which herbivores directly influence plant population dynamics, and dashed lines indicate indirect pathways. Resistance (here defined as the combined effects of escape, defence and tolerance) determines the impact of herbivores on each of the four potential mechanisms that influence the population dynamics of the plant species in the community. Whole plant death is most relevant for direct effects on plant population dynamics. Summed across all plant species, these pathways will then determine the impact of herbivores on the number of plant species that occur in a given location.

in that way reduce the loss of species through competitive exclusion. Connell and Lowman (1989) proposed that a lack of herbivore attack on the dominant species is one potential explanation for monospecific tropical forests. Langenheim and Stubblebine (1983) hypothesized that if herbivores specifically attack certain chemical phenotypes within a plant species, then conspecific seedlings could establish around parent plants if they are sufficiently different in chemical profile from their parent. Such a pattern of attack could hasten competitive exclusion despite Janzen–Connell patterns of seed predation, but also would lead to maintenance of high genetic diversity within the target plant species. Finally, Connell (1971) proposed that natural-enemy impact (in this case that of herbivores) will be greatest, and competitive exclusion among plant species will be least likely, in environments in which there are few extremes of temperature and rainfall throughout the year, i.e. lowland wet forests.

In all of these hypotheses, herbivores influence local plant diversity by reducing the likelihood of species extinction through competitive exclusion. It is important, however, to consider the impact of herbivores in a broader context to understand the mechanisms by which they might influence plant diversity in tropical systems (Fig. 13.1). Their impact on competitive interactions is only one such mechanism. Instead, herbivores may directly influence the occurrence of a plant species in a given location by reducing its local abundance or entirely eliminating it, independent of its competitive ability. More-indirect effects include three pathways. First, herbivores could determine whether a species occurs at a given site by influencing its dispersal capability, and thus the probability that it could arrive there. Second, once a seed has arrived at a site, herbivores may influence the physiological capability of the plant to establish, grow and reproduce in the face of the abiotic conditions found there. Third, herbivores could influence whether a species occurs in a given location by reducing its competitive ability. It is this latter mechanism that has received the most attention.

The influence of herbivores on the number of species and genotypes that occur in a community will depend on the sum of these impacts across all species. Attack can and does occur at all stages of the life cycle of plants: seed (both pre- and postdispersal), seedling, juvenile, and adult, both vegetative (including below ground) and reproductive plant parts. Further, to predict the relative impact of herbivores on a given plant species, we need knowledge of the resistance components that influence the impact of herbivores on plant fitness and plant population dynamics. Resistance to herbivore attack has three components (Dirzo 1984): (1) escape from herbivores, (2) defence (as manifested in the amount of damage caused once an herbivore has found a potential host plant) and (3) tolerance (the ability to regrow and reproduce following attack). For tropical systems, Janzen (1970) and Connell (1971) emphasized escape. In contrast, Coley (1983) and others following her lead (Filip *et al.* 1995; Gross *et al.* 2000; Marquis *et al.* 2001) have emphasized defence as a resistance mechanism in tropical plants, demonstrating that species vary in traits associated with leaf quality, and that at least some of this variation influences interspecific variation in levels of folivory. There are few studies of the relative tolerance of tropical plant species to herbivory (Augustine & McNaughton 1998; Ribeiro 1999). There are apparently no systems, tropical or extra-tropical, for which all three components of resistance have been estimated. This is an important point because tradeoffs between resistance components (e.g. defence vs. tolerance) could lead to maintenance of species. Finally, the natural enemies of herbivores may reduce herbivore impact both at the level of the individual plant species (Dyer & Letourneau 1999) and potentially at the community level (Terborgh 1992; Coley & Barone 1996).

In this chapter, I review the evidence available for each of the outlined mechanisms (Fig. 13.1). Evidence reviewed is from tropical terrestrial systems, although recent studies in neotropical streams demonstrate that fish have controlling impacts on algal biomass and community composition (Flecker *et al.* 2002 and references therein). Available studies are strongly biased geographically. For example, 87% of the tests of the Janzen–Connell hypothesis reviewed by Hyatt *et al.* (2003) were conducted in the neotropics, while most studies of the impacts of megaherbivores have been conducted in African savanna and grasslands. When pertinent, I draw upon findings from extra-tropical systems, particularly when data from tropical systems are lacking. Finally, I discuss directions for future research. Diversity here is broadly defined to include not only plant species richness and evenness, but also diversity of plant genotypes within a species and the physiognomy (vegetation structure) of the associated community.

Direct impacts of herbivores

Herbivores potentially can have direct impacts on the population dynamics of their host plants, irrespective of their competitive relationships with co-occurring plant species (Louda 1982). Actual demonstration of such impacts is relatively limited (see Chapter 14), even for temperate species (Rausher & Feeny 1980; Louda 1982; Doak 1992; Ehrlen 1995; 1996; Louda & Potvin 1995; Kelly & Dyer 2002). There appears to be only one study of the effects of herbivory on the population dynamics of a tropical plant species: Ticktin (2003) demonstrated that collared peccaries (*Tayassu tajuca*) decreased the population growth rate of *Aechmea magdalenae*, a terrestrial bromeliad, during an El Niño year, when the preferred food of the peccaries was reduced in abundance. In particular, this study shows that herbivory may be episodic in its impacts (see also Bebber *et al.* 2002).

Experimental studies that exclude large herbivores in tropical forests demonstrate significant impacts on seedling density and diversity at the community level. Where *Gecarcinus quadratus* crabs occur naturally, plant diversity is lower than in the naturally crab-free zone along a Costa Rican beach. When crabs were excluded, diversity of surviving seedlings mirrored that in the crab-free zone, suggesting that crabs are the cause of lower species richness of trees where they occur (Sherman 2002). Exclusion of wild pigs (*Sus scofa*) resulted in higher seedling density and diversity in Malaysia, presumably because of their effects as seed predators and through trampling of seedlings (Ickes *et al.* 2001). Excluding vertebrates increased community-wide seedling recruitment over two years on both Barro Colorado Island, Panama, and at Cocha Cashu, Peru (Terborgh & Wright 1994). Finally, seedling density was also higher (but species richness and evenness lower) in a Mexican forest in which large mammals have been extirpated by hunting than in a forest in which the mammalian fauna remains intact (Dirzo & Miranda 1991).

Perhaps the most dramatic effects of herbivores in tropical systems are seen in Africa. Elephants, giraffes, wildebeest and antelope, particularly at high densities, all have been shown to have major effects on vegetation structure, with the capacity to turn woodlands into grasslands (Augustine & McNaughton 1998). Elephants (*Loxodonta africana*) are often species-selective in their feeding (e.g. Holdo 2003). They suppress recruitment of preferred tree species into the canopy (Holdo 2003), reduce average canopy height and tree density (e.g. Mapaure & Campbell 2002) by toppling and debarking trees, and reduce regeneration through seedling-feeding (Barnes 2001). Impala browsing can prevent seedling establishment of woody plants (Prins & Van der Juegd 1993). Decline in wildebeest, buffalo and giraffe populations associated with the introduction of rinderpest led to the conversion of large expanses of savanna to woodlands in the Serengeti-Mara region (Dublin 1995; Sinclair 1995). Subsequent control of rinderpest in the southern Serengeti increased herb dominance in grasslands and increased woody plant abundance in woodlands, because the trampling and grazing of grass decreased the frequency of fire, further favouring woody plants over grasses (Sinclair 1995). In the northern Serengeti, however, high elephant densities resulted in conversion of woodlands to grasslands (Dublin 1995; see also Mapaure and Campbell 2002 for an example from Zimbabwe). Thus, African megaherbivores can determine the vegetation physiognomy of a location. Surprisingly, there are no studies of effects of herbivory on the population dynamics of any African plant species.

Impacts of herbivores on seed dispersal

Dispersal can have a major influence on the structure of tropical plant communities (see Chapters 10 and 11). Mangrove community zonation is in part controlled by the dispersal characteristics of mangrove propagules (Rabinowitz 1978). In turn, dispersal limitation appears to influence local colonization of light gaps in a Panamanian forest (Dalling *et al.* 2002).

Janzen (1970) hypothesized that predispersal seed predation would reduce the average dispersal distance of seeds, thus reducing escape from postdispersal seed and seedling predators. This hypothesis can be extended to include the other potential benefits of seed dispersal (Howe & Smallwood 1982) as they affect the likelihood that a species might be maintained in a community: these are directed dispersal to safe sites (Wenny & Levey 1998) and colonization of vacant sites. The impact of predispersal seed predation on seed shadows apparently has not been quantified directly. In a description of the traits that contribute to the characteristics of seed shadows, however, Thiede and Augspurger (1996) showed that architectural traits (plant height, crown width and branch number), crop size and infructescence length in a wind-dispersed species all influenced the nature of seed shadows, including the mean distance seeds were dispersed, the standard deviation of dispersal distance, and skewness and kurtosis of the seed shadows. Thus, predispersal seed predation could influence the nature of seed shadows by reducing total crop size, and damage to meristems and branches can influence the relevant plant architectural traits (Caraglio *et al.* 2001). The total number of seeds could also be reduced by herbivore attack to foliage, stems and roots before reproduction (Marquis 1992).

Impacts of herbivores on physiological fit to microhabitat

Herbivores might maintain species and genotype diversity in a given site composed of multiple microhabitats (Lubchenco & Cubit 1980) or in a given region along an environmental gradient (Janzen 1974; Lubchenco 1980; Louda *et al.* 1990) by determining or reinforcing associations of plants with particular microhabitats. Tropical plant species (and genotypes: Silvera *et al.* 2003) are often distributed non-randomly within a site (or across a landscape gradient: Baker *et al.* 2003) with regard to soil type (e.g. Clark *et al.* 1999), topography (Tuomisto *et al.* 2002) and disturbance size (Silvera *et al.* 2003), and among microhabitats within disturbances (Brandani *et al.* 1988). Invariably, these habitat preferences are suggested to be due to differences in the physiological capability of a plant species to survive and grow under the given levels of light availability, soil fertility and soil moisture among these microhabitats or along these gradients.

The contribution of herbivores to such patterns is rarely tested, despite their demonstrated role in some temperate systems (e.g. Lubchenco 1980; Lubchenco & Cubit 1980; Louda 1982). There are three noted exceptions. Firstly, studies by King (2003) suggest that seedling-clipping by mammalian herbivores contributes to the apparent flooding-tolerance of the tree *Calophyllum brasiliense* (Clusiaceae) along a tributary of the Amazon River. Seedlings of this species are more abundant in areas of low topography and in early succession forests, because flooding in these habitats reduces the number of mammalian herbivores and increases light levels by preventing establishment of flooding-intolerant plant species. Second, Smith *et al.* (1989) found that propagule predation in the mangroves *Avicennia germinans* and *Rhizophora apiculata* was least where these mangrove species are dominant and highest where they are subdominant, suggesting that seed predation contributes to mangrove zonation. Third, Fine *et al.* (2004) reciprocally transplanted seedlings of 20 species of trees, either adapted to grow on high nutrient, clay soils or on nearby low nutrient, white-sand soils. When protected from herbivores, clay-soil specialists grew faster on both soil types. In contrast, when exposed to herbivores, white-sand soil specialists outgrew clay specialists on white-sand soil. These results provide strong evidence that herbivores are contributing to habitat specialization in this system.

Herbivores also may influence plant genotype distribution within tropical species. In the understorey wet-forest shrub *Piper arieianum* in Costa Rica, different genotypes suffer different levels of leaf-area loss (Marquis 1984), and specific herbivore species show preferences for different genotypes (Marquis 1990). Such genotype-specific patterns of attack could maintain genetic diversity within a species at a given location depending on fluctuations in the relative abundance of the relevant herbivore species and tolerance of the genotypes to leaf-area loss. In the Serengeti, non-grazing genotypes of grasses increase in abundance in fenced exclosures relative to grazing genotypes (McNaughton 1979), and increase in abundance along a geographic gradient of decreasing grazing intensity (McNaughton 1984). Finally, plant species vary geographically in defence traits depending on the presence of seed predators (Janzen 1975) or protective ant species (Janzen 1973; Koptur 1985).

Impacts of herbivores on competitive ability

Herbivores affect plant species (and genotype) diversity via competition by changing the relative competitive abilities among those species (or genotypes). Competition (negative/negative interactions among two species) has been classified into 'real' competition for a shared resource and apparent competition through a shared intermediary (Holt 1977; Connell 1990; Louda *et al.* 1990).

Real competition can be broken down further into interference competition and exploitation competition. In plants, interference competition is essentially competition for space, either through dispersal to and occupation of a safe site (pre-emptive competition), litterfall, shading and overgrowth of upright plants by vines (overgrowth competition), or allelopathy (chemical competition). Exploitation competition, in contrast, occurs when the shared resource is nutrients and water, or for shared mutualists (pollinators and seed dispersers).

Hypotheses about the impact of herbivores on plant diversity (see above) have a similar mechanism in common. Any given species is prevented from driving other species to local extinction because of density-dependent, plant-species-specific attack by herbivores. As a species increases in abundance, density-dependent mortality by host-specific herbivores prevents competitive exclusion. The Janzen–Connell model (Janzen 1970; Connell 1971) incorporates a spatial component in the process. One of the presented hypotheses (McNaughton 1985) also implies that there is a negative tradeoff between competitive ability and resistance to attack by herbivores. This tradeoff has served as the basis for predicting changes in the species richness with the addition of herbivores in temperate systems (Paine 1966; Lubchenco 1978). When there is a negative tradeoff, addition of herbivores will increase species richness. If instead a positive correlation exists, herbivores in essence will enhance the competitive ability of the superior competitors and reduce species richness. Predictions about the impacts of addition or exclusion of herbivores on resulting plant species richness have been borne out for temperate intertidal communities (snails as herbivores) and temperate grasslands (vertebrate herbivores). Such tradeoffs have been investigated in only one tropical system. In the Serengeti grass *Themada triandra*, there are intraspecific tradeoffs between grazing and non-grazing genotypes (Hartvigsen and McNaughton 1995): the former are short and fast-growing while the latter are tall and slow-growing.

Given this mechanism, we would need to know the following to predict the outcome of herbivore-influenced competition on plant species diversity in a single habitat: (1) the relative resistance of plant species to herbivores, and (2) the relative competitive abilities of plants in the presence and absence of herbivores. Both are essentially unknown for tropical systems (but see McNaughton 1983 for competition studies in the absence of vertebrate herbivores). Regarding the latter, in a meta-analysis of competition and predation field experiments, Gurevitch *et al.* (2000) list no tropical studies.

Real competition: interference competition
Pre-emptive competition
The Janzen–Connell model (Janzen 1970; Connell 1971) describes pre-emptive competition based on plant traits that contribute to escape. Species are prevented from pre-empting potential establishment sites near to parent trees

because of high seed and seedling predation. Attack decreases with distance, so the farther a seed is dispersed from the parent, the greater chance it has of escaping its herbivores. As a result, sites beneath and near the canopy of adults are available for colonization by seeds and seedlings of other tree species. Herbivores will prevent any one species (or group of few species) from dominating a site because as those species become common, there will be fewer places far away from all parent trees that represent safe sites.

Three recent summaries are available of the current evidence regarding the Janzen–Connell model. Hammond and Brown (1998) reviewed tests of the model at the species level: 15 out of 19 populations attacked by insects showed survivorship was greater farther away from a conspecific parent, while this was true for only 2 out of 27 populations attacked by vertebrates. Thus, distance-related mortality is common, and the agents of this mortality are generally insect herbivores and not vertebrate herbivores. More recently, Hyatt et al. (2003), in a meta-analysis of 40 different studies involving 75 unique plant species, found that seedling predation decreased with distance, while distance had no overall effect on seed predation.

Wright (2002) reviewed the results of community-level evaluations. Each case looked for evidence of distance-dependent or density-dependent survival and/or growth, inferring that such distance and density effects would be supportive evidence for the Janzen–Connell hypothesis. Four studies show significant negative distance effects (Connell et al. 1984; Hubbell & Foster 1990; Condit et al. 2000) or density effects (Harms et al. 2000) for a majority but not all species. Contradictory evidence comes from Condit et al. (1992) who showed that survivorship over a 5-year period was lowest near conspecifics for only 15 of 80 species tested.

Manipulation of the herbivore fauna can provide an additional test of the Janzen–Connell model. When predators of vertebrate herbivores are hunted out of a region, recruitment curves were shallower (there was decreased survival close to parent trees as herbivore attack increased) (Wright & Duber 2001). When a major herbivore, the white-lipped peccary Tayassu pecari, disappeared from the Cocha Cashu region, recruitment curves became steeper (there was increased survival closer to parent trees) (Silman et al. 2003).

The Janzen–Connell model implies that the loss or absence of herbivores will lead eventually to a forest dominated by a few or a single tree species, the so-called monodominant forests (Connell & Lowman 1989). A number of tests of this hypothesis are available, providing both supportive and contradictory evidence. Firstly, if escape from herbivores is driving monodominance then damage should be lower on the dominant species. In only one case (Makana et al. 1998) has herbivore attack on a monodominant species in monodominant stands been found to be lower than on the same species in neighbouring more-diverse stands (Hart 1995; Gross et al. 2000; Nascimento & Proctor 2001). Secondly, outbreaks of defoliating insects should be rare or non-existent. This corollary is difficult to test given that negative evidence is not likely to be reported. There are, however, two reports of widespread defoliation on monodominant trees (Anderson 1961; Nascimento & Proctor 1994). Finally, we would predict that damage should be lower on dominant species than for closely related, non-dominant species. Anecdotal evidence suggests that very low seed-predation levels for *Pentaclethra macroloba* in the Caribbean lowlands of Costa Rica may be responsible for its high abundance there (Janzen 1970). Its African counterpart, *P. macrophylla*, is low in abundance and has very high levels of seed predation (M. Lieberman, personal communication).

The Janzen–Connell effect may be manifested differently at the subspecific level. Langenheim and Stubblebine (1983) proposed that only seedlings with phenotypes dissimilar in anti-herbivore chemistry from that of their parent would be able to survive under parent trees. Sánchez-Hidalgo *et al.* (1999) found that chemical phenotypes (and presumably genotypes) of *Nectandra ambigens* similar to parental phenotypes were only growing outside the canopy of parents while all seedling phenotypes growing under the canopy were chemically distinct from those parents. Langenheim and Stubblebine (1983) found similar results for a number of species of *Hymenaea*.

Overgrowth competition
Vines growing on trees have been shown to reduce the growth of those trees (e.g. Clark & Clark 1990). Davidson *et al.* (1988) suggested that protective ants of ant-plants might prevent such colonization and its detrimental effects by clipping the tips of invading vines. No studies are available that test the impacts of herbivores on vine colonization of their support hosts.

Defoliation of the canopy by insect herbivores can release species otherwise stunted in growth, resulting in increased species diversity, as shown for an old field in North America (Carson & Root 2000). Almost all tropical forest plants growing in the understorey respond positively to canopy openings. The likelihood of release of suppressed understorey plants in tropical forests following defoliation may vary with rainfall: insect outbreaks and the resultant high level of defoliation appear to be most common in tropical dry forests (Janzen 1988) and almost absent in wet forests (Marquis & Braker 1994). In the Serengeti, excluding vertebrate herbivores reduced evenness among plant species and caused an almost complete change in species composition inside exclosures, but no change in species richness (McNaughton 1979, 1983). The mechanism of release appears to be through overtopping competition, given that grazing favours short-statured species. Hand removal of plant species increased abundance of the remaining species, demonstrating that these species compete (McNaughton 1983).

Another mechanism by which herbivores may influence overtopping competition is through their impact on litterfall (Carson & Root 2000). Folivore attack, particularly by galling species (e.g. Faeth *et al.* 1981) and heavy defoliation of single leaves (R. J. Marquis and J. T. Lill, in preparation) can lead to early leaf abscission, increasing litterfall. Litter has been shown to reduce survivorship and growth of seedlings in tropical forests (e.g. Molofsky & Augspurger 1992) and can actually increase attack by herbivores and pathogens (García-Guzmán & Benítez-Malvido 2003).

Chemical competition

Although allelopathy has been suggested as a potential mechanism for density-dependent mortality of juvenile trees on BCI (Hubbell & Foster 1990), there apparently is only one study demonstrating allelopathy in a tropical tree (Webb *et al.* 1967).

Real competition: exploitation competition

Resource: nutrients and water

A few trenching studies have been conducted in the tropics demonstrating that competition for underground resources influences growth (Coomes & Grubb 2000; see also Chapters 2 and 3, this volume). Apparently none have been conducted both in the presence and absence of herbivores. One intriguing result is that of Denslow *et al.* (1991), who grew seedlings of two species of *Inga* either under or away from the canopy of understorey palms and cyclanths in lowland wet forest. Seedlings of both species grew less under the shade of monocots, but trenching showed that the effect was not due to root competition. Instead, both species suffered higher levels of stem damage due to herbivore attack under monocot shade than away from monocots. This result points to the need for competition experiments that incorporate a herbivore treatment.

Canopy defoliation also may alter nutrient schedules in the understorey through frass input from above (Rinker *et al.* 2001). Such nutrient release in combination with higher light levels may increase the survivorship and growth of otherwise-suppressed understorey individuals, decreasing the probability of competitive exclusion. Again, this phenomenon is more likely to occur in seasonally dry forests than in wet forests.

Resource: pollinators

Competition for shared mutualists may also be important. Plants in tropical forests have been hypothesized to have flowering times that are displaced to avoid competition for shared pollinators (Gentry 1974; Stiles 1978). Herbivore damage to foliage may decrease flower production and nectar production (Aizen & Raffaele 1996) while damage to flowers may decrease their attractiveness to pollinators (Mothershead & Marquis 2000; Malo *et al.* 2001). All three would decrease the ability of a plant to attract pollinators. Such herbivore attack would tend to decrease plant species richness.

Resource: seed dispersers

Plants may also compete for disperser services. Howe and Vande Kerchove (1979) demonstrated that individual trees of *Virola surinamensis* compete for dispersers, with the aril/seed weight ratio determining the proportion of seeds dispersed. Foliage damage and predispersal seed predation will tend to decrease crop size, and therefore decrease the ability of an individual to compete for dispersers with neighbours. Foliage damage may also decrease propagule size and relationships of reward to seed size, which also may influence its attractiveness to dispersers. These relationships are almost entirely unexplored (but see Christensen & Whitham 1993).

Apparent competition

Apparent competition arises when two members of the same trophic level share natural enemies, and natural-enemy attack increases when the two species co-occur. As a result, each species is negatively affected by the presence of the other but through predation by a shared enemy (Holt 1977; see also Connell 1990 for a case in which the intermediate species is a member of the same trophic level). Dyer and Letourneau (1999) showed that other plant species, as well as conspecifics, growing in the neighbourhood of the ant-plant *Piper cenocladum* suffered higher foliar damage when protective ant abundance on focal *P. cenocladum* was decreased by addition of the clerid beetle predator of the ants (Chapter 15). The proposed mechanism is that many of the folivores of *P. cenocladum* are generalists; when the abundance of these folivorous species increased on focal *P. cenocladum* shrubs, they spilled over onto neighbouring plants. The impact of variation in abundance of this beetle on plant diversity would depend on the relative attack levels of the herbivores among plant species and the relative tolerance of each species to damage.

Synthesis

In exclosure studies, vertebrate herbivores and land crabs decrease survivorship and density of seedling populations in tropical forests. Impacts of herbivores on species richness vary among studies, either decreasing it (Sherman 2002; Ickes *et al.* 2003) or increasing it (Dirzo & Miranda 1991). The degree to which these differential impacts on species richness are related to tradeoffs between competitive abilities and resistance to attack is unknown. Indeed, whether seedlings are actually competing in the early stages of establishment is unclear. There have been no studies in tropical systems in which the mechanism of competition has been tested (*sensu* Goldberg 1990) in the presence and absence of herbivores. By doing so, we could begin to identify the traits that contribute to competitive ability and resistance to herbivores. Such an approach would increase our ability to predict the impact of herbivore attack on the outcome of competition.

In contrast to the impacts of large herbivores in tropical forests as assessed by exclosures, insect seed predators and folivores account for the majority of distance effects in tests of the Janzen–Connell model. Relative and interactive effects of vertebrate and insect herbivores have rarely been studied (Hart 1995; Musgrove & Compton 1997). The contribution of insect herbivores relative to the impacts of vertebrate exclusion on seedling dynamics awaits testing. The challenge will be to establish long-term experiments. Large exclosures maintained for numbers of years are required to determine the impact of herbivores on diversity in tropical forests (Gilbert *et al.* 2001). A number of the mechanisms, however, can be addressed with short-term experiments under natural conditions or in common gardens.

Our best information on herbivore impacts on diversity comes from long-term exclosures in the Serengeti (McNaughton 1979). There, evenness decreased under herbivore pressure, species composition changed dramatically, but the number of species was unaffected. In the Serengeti, however, plant lifespan is much shorter than that of forest trees, making study of such effects more feasible. Aerial photographs and ground-sampling have demonstrated that even more dramatic effects occur when populations of large vertebrates are high in Africa. At high densities these megaherbivores can convert woodlands to savanna, but the direction of change for any given location is determined by complex interactions among herbivore density, fire and soil type (McNaughton 1983; Dublin 1995; Sinclair 1995). The relative impact of these animals as herbivores and ecosystem engineers (*sensu* Jones *et al.* 1997) is not known. Besides actually eating plants, large vertebrate herbivores trample them, deposit faeces and urine, and cut plants for shelter (McNaughton 1983; Ickes *et al.* 2003). Studies that follow individual plants and their fate would help to resolve this issue. Along this line, conspicuously absent are studies of the impacts of leaf-cutting ants on vegetation diversity, although their role as ecosystem engineers has been investigated (Moutinho *et al.* 2003).

Of the five early hypotheses concerning the impacts of herbivores on tropical plant diversity, all but that of Connell (1971) appear to hold true. When controlling for soil fertility, damage levels due to insect folivory are actually higher at intermediate levels of rainfall in deciduous forests (Marquis *et al.* 2002), rather than in wet forests as Connell proposed. Coley and Barone (1996) hypothesize that natural-enemy attack on the herbivores themselves is highest in lowland wet forests. Connell (1971) anticipated the potential complicating influence of the third trophic level in developing his original hypothesis. Further studies of the impacts of herbivores on plant diversity should not ignore the potential influence of the third trophic level (Terborgh 1992).

It is important to point out that in the short run, herbivore presence may be detrimental to only a few species, the least resistant. When competition for mutualists is the mechanism by which herbivores influence diversity, loss of a few species through herbivore attack may have a compounding negative effect on the remaining species through near or complete collapse of the associated mutualist community. This may occur when multiple plant species are necessary to maintain a community of mutualists (Howe 1977; Gilbert 1979). Thus, understanding the mechanism of competition will aid our ability to predict the effect of herbivores on tropical plant diversity. Even more broadly, the role of many of the potential mechanisms of herbivore impact on tropical plant diversity, as outlined here, essentially remains unstudied. In considering the interaction of soil type and herbivores on seedling growth and survival, the study by Fine *et al.* (2004) represents an important first step in this quest.

The examples available clearly demonstrate that the impact of herbivory must be considered in the context of other environmental factors, both biotic (pollinators, plant competitors, natural enemies of herbivores, and mycorrhiza) and abiotic (fire, soil nutrients, and soil water availability). Other factors, including environmental heterogeneity and dispersal limitation, are likely to contribute to species maintenance. The relative impact of herbivory and these other factors, and how they interact, is yet to be explored.

Acknowledgements

I thank Michelle Pinard and David Burslem for the opportunity to participate in the symposium, and Beatriz Baker, Karina Boege, Rebecca Forkner, June Jeffries, Carol Kelly, Alejandro Masís, Peter Van Zandt and two anonymous reviewers for their comments on earlier versions of this manuscript.

References

Aizen, M. A. & Raffaele, E. (1996) Nectar production and pollination in *Alstroemeria aurea*: response to level and pattern of flowering shoot defoliation. *Oikos*, **76**, 312–322.

Anderson, J. A. R. (1961) The destruction of *Shorea albida* forest by an unidentified insect. *Empire Forest Review*, **40**, 19–29.

Augustine, D. J. & McNaughton, S. J. (1998) Ungulate effects on the functional species composition of plant communities: herbivore selectivity and plant tolerance. *Journal of Wildlife Management*, **62**, 1165–1183.

Baker, T. R., Burslem, D. F. R. P. & Swaine, M. D. (2003) Associations between tree growth, soil fertility and water availability at local and regional scales in Ghanaian tropical rain forest. *Journal of Tropical Ecology*, **19**, 109–125.

Barnes, M. E. (2001) Effects of large herbivores and fire on the regeneration of *Acacia erioloba* woodlands in Chobe National Park, Botswana. *African Journal of Ecology*, **39**, 340–350.

Bebber, D., Brown, N. & Speight, M. (2002) Drought and root herbivory in understorey *Parashorea* Kurz (Dipterocarpaceae) seedlings in Borneo. *Journal of Tropical Ecology*, **18**, 795–804.

Brandani, A., Hartshorn, G. S. & Orians, G. H. (1988) Internal heterogeneity of gaps and species richness in Costa Rican tropical wet forest. *Journal of Tropical Ecology*, **4**, 99–119.

Brown, J. H. & Heske, E. J. (1990) Control of a desert–grassland transition by a keystone rodent guild. *Science*, **250**, 1705–1707.

Brown, V. C. & Gange, A. C. (1992) Secondary plant succession: how is it modified by insect herbivory? *Vegetatio*, **101**, 3–13.

Caraglio, Y., Nicolini, E. & Petronelli, P. (2001) Observations on the links between the architecture of a tree (*Dicorynia guianensis* Amshoff) and Cerambycidae activity in French Guiana. *Journal of Tropical Ecology*, **17**, 459–463.

Carson, W. P. & Root, R. B. (2000) Herbivory and plant species coexistence: community regulation by an outbreaking phytophagous insect. *Ecological Monographs*, **70**, 73–99.

Christensen, K. M. & Whitham, T. G. (1993) Impact of insect herbivores on competition between birds and mammals for pinyon pine seeds. *Ecology*, **74**, 2270–2278.

Clark, D. B. & Clark, D. A. (1990) Distribution and effects on tree growth for lianas and woody hemiepiphytes in a Costa Rican tropical wet forest. *Journal of Tropical Ecology*, **6**, 321–331.

Clark, D. B., Palmer, M. W. & Clark, D. A. (1999) Edaphic factors and the landscape-scale distributions of tropical rain forest trees. *Ecology*, **80**, 2662–2675.

Coley, P. D. (1983) Herbivory and defensive characteristics of tree species in a lowland tropical forest. *Ecological Monographs*, **53**, 209–233.

Coley, P. D. & Barone, J. A. (1996) Herbivory and plant defenses in tropical forests. *Annual Review of Ecology and Systematics*, **27**, 305–335.

Condit, R., Hubbell, S. P. & Foster, R. B. (1992) Recruitment near conspecific adults and the maintenance of tree and shrub diversity in a neotropical forest. *American Naturalist*, **140**, 261–286.

Condit, R., Ashton, P. S., Baker, P. *et al.* (2000) Spatial patterns in the distribution of tropical tree species. *Science* **288**, 1414–1418.

Connell, J. H. (1971) On the role of natural enemies in preventing competitive exclusion in some marine animals and rain forest trees. In *Dynamics of Populations*

(ed. P. J. den Boer & G. R. Gradwell). Wageningen: Centre for Agricultural Publishing and Documentation, pp. 298–312.

(1990) Apparent versus "real" competition in plants. *Perspective on Plant Competition* (ed. J. B. Grace & D. Tilman). New York: Academic Press, pp. 9–26.

Connell, J. H. & Lowman, M. D. (1989) Low-diversity tropical rain forests: some possible mechanisms for their existence. *American Naturalist*, **134**, 88–119.

Connell, J. H., Tracy, J. G. & Webb, L. J. (1984) Compensatory recruitment, growth, and mortality as factors maintaining rain forest diversity. *Ecological Monographs*, **54**, 41–164.

Coomes, D. A. & Grubb, P. J. (2000) Impacts of root competition in forests and woodlands: a theoretical framework and review of experiments. *Ecological Monographs*, **70**, 171–207.

Dalling, J. W., Muller-Landau, H. C., Wright, S. J. & Hubbell, S. P. (2002) Role of dispersal in the recruitment limitation of neotropical pioneer species. *Journal of Ecology*, **90**, 714–727.

Davidson, D. W., Longino, J. T. & Snelling, R. R. (1988) Pruning of host plant neighbors by ants: an experimental approach. *Ecology*, **69**, 801–808.

Denslow, J. S., Newell, E. & Ellison, A. M. (1991) The effects of understory palms and cyclanths on the growth and survival of *Inga* seedlings. *Biotropica*, **23**, 225–234.

Dirzo, R. (1984) Herbivory: a phytocentric overview. In *Perspectives on Plant Population Ecology* (ed. R. Dirzo & J. Sarukhan). Sunderland, MA: Sinauer, pp. 141–165.

Dirzo, R. & Miranda, A. (1991) Altered patterns of herbivory and diversity in the forest understory: a case study of the possible consequences of contemporary defaunation. In *Plant–Animal Interactions, Evolutionary Ecology in Tropical and Temperate Regions* (ed. P. W. Price, T. M. Lewinsohn,

G. W. Fernandes & W. W. Benson). New York: Wiley, pp. 273–287.

Doak, D. K. (1992) Lifetime impacts of herbivory for a perennial plant. *Ecology*, **73**, 2086–2099.

Dublin, H. T. (1995) Vegetation dynamics in the Serengeti-Mara ecosystem: the role of elephants, fire, and other factors. In *Serengeti II, Dynamics, Management, and Conservation of an Ecosystem* (ed. A. R. E. Sinclair & P. Arecese). Chicago: University of Chicago Press, pp. 71–90.

Dyer, L. A. & Letourneau, D. K. (1999) Trophic cascades in a complex terrestrial community. *Proceedings of the National Academy of Sciences*, **96**, 5072–5076.

Ehrlen, J. (1995) Demography of the perennial herb *Lathyrus vernus*. II. Herbivory and population dynamics. *Journal of Ecology*, **83**, 297–308.

(1996) Spatiotemporal variation in predispersal seed predation intensity. *Oecologia*, **108**, 708–713.

Faeth, S. H., Connor, E. F. & Simberloff, D. (1981) Early leaf abscission: a neglected source of mortality for folivores. *American Naturalist*, **117**, 409–415.

Filip, V., Dirzo, R., Maass, J. M. & Sarukhan, J. (1995) Within- and among-year variation in the levels of herbivory on the foliage of trees from a Mexican tropical deciduous forest. *Biotropica*, **27**, 78–86.

Fine, P. V. A., Mesones, I. & Coley, P. D. (2004) Herbivores promote habitat specialization by trees in Amazonian forests. *Science*, **305**, 663–665.

Flecker, A. S., Taylor, B. W., Bernhardt, E. S. *et al.* (2002) Interactions between herbivorous fishes and limiting nutrients in a tropical stream ecosystem. *Ecology*, **83**, 1831–1844.

García-Guzmán, G. & Benítez-Malvido, J. (2003) Effect of litter on the incidence of leaf-fungal pathogens and herbivory in seedlings of the tropical tree *Nectandra*

ambigens. *Journal of Tropical Ecology*, **19**, 171–177.

Gentry, A. H. (1974) Flowering phenology and diversity in tropical Bignoniaceae. *Biotropica*, **6**, 64–68.

Gilbert, G. S., Harms, K. E., Hamill, D. N. & Hubbell, S. P. (2001) Effects of seedling size, El Niño drought, seedling density, and distance to nearest conspecific adult on 6-year survival of *Ocotea whitei* seedlings in Panama. *Oecologia*, **127**, 509–516.

Gilbert, L. E. (1980) Food web organization and conservation of neotropical diversity. In *Conservation Biology* (ed. M. E. Soule & B. A. Wilcox). Sunderland, MA: Sinauer Associates, pp. 11–34.

Goldberg, D. E. (1990) Components of resource competition in plant communities. In *Perspective on Plant Competition* (ed. J. B. Grace & D. Tilman). San Diego: Academic Press, pp. 27–49.

Gross, N. D., Torti, S. D., Feener, D. H. Jr & Coley, P. D. (2000) Monodominance in an African rain forest: is reduced herbivory important? *Biotropica*, **32**, 430–439.

Gurevitch, J., Morrison, J. A. & Hedges, L. V. (2000) The interaction between competition and predation: a meta-analysis of field experiments. *American Naturalist*, **155**, 435–453.

Hammond, D. S. & Brown, V. K. (1998) Disturbance, phenology and life-history characteristics factors influencing distance/density-dependent attack on tropical seeds and seedlings. *Dynamics of Tropical Communities* (ed. D. M. Newbery, H. H. T. Prins & N. D. Brown). Oxford: Blackwell, pp. 51–78.

Harms, K. E., Wright, S. J., Calderon, O., Hernandez, A. & Herre, E. A. (2000) Pervasive density-dependent recruitment enhances seedling diversity in a tropical forest. *Nature*, **404**, 493–495.

Hart, T. B. (1995) Seed, seedling, and sub-canopy survival in monodominant and mixed forests of the Ituri Forest, Africa. *Journal of Tropical Ecology*, **11**, 443–459.

Hartvigsen, G. & McNaughton, S. J. (1995) Tradeoff between height and relative growth rate in a dominant grass from the Serengeti ecosystem. *Oecologia*, **102**, 273–276.

Holdo, R. M. (2003) Woody plant damage by African elephants in relation to leaf nutrients in western Zimbabwe. *Journal of Tropical Ecology*, **19**, 189–196.

Holt, R. D. (1977) Predation, apparent competition and the structure of prey communities. *Theoretical Population Biology*, **12**, 197–299.

Howe, H. F. (1977) Bird activity and seed dispersal of a tropical wet forest tree. *Ecology*, **58**, 539–550.

Howe, H. F. & Brown, J. S. (1999) Effects of birds and rodents on synthetic tallgrass communities. *Ecology*, **80**, 1776–1781.

Howe, H. F. & Smallwood, J. (1982) Ecology of seed dispersal. *Annual Review of Ecology and Systematics*, **13**, 201–228.

Howe, H. F. & Vande Kerckhove, G. A. (1979) Nutmeg dispersal by tropical birds. *Science*, **210**, 925–927.

Hubbell, S. P. & Foster, R. B. (1990) The fate of juvenile trees in a neotropical forest: implications for the natural maintenance of tropical tree diversity. In *Reproductive Ecology of Tropical Forest Plants* (ed. K. S. Bawa & M. Hadley), Vol. 7. Paris: The Parthenon Publishing Group, pp. 317–341.

Hyatt, L. A., Rosenberg, M. S., Howard, T. G. *et al.* (2003) The distance dependence predation of the Janzen–Connell hypothesis: a meta-analysis. *Oikos*, **103**, 590–602.

Ickes, K., Dewalt, S. J. & Appanah, S. (2001) Effects of native pigs (*Sus scrofa*) on woody understorey vegetation in a Malaysian lowland rain forest. *Journal of Tropical Ecology*, **17**, 191–206.

Ickes, K., Dewalt, S. J. & Thomas, S. C. (2003) Resprouting of woody saplings following stem snap by wild pigs in a Malaysian rain forest. *Journal of Ecology*, **91**, 222–233.

Janzen, D. H. (1970) Herbivores and the number of tree species in tropical forests. *American Naturalist*, **104**, 501–528.

(1973) Dissolution of a mutualism between *Cecropia* and its *Azteca* ants. *Biotropica*, **5**, 15–28.

(1974) Tropical black water rivers, animals, and mast-fruiting by Dipterocarpaceae. *Biotropica*, **6**, 69–103.

(1975) Behavior of *Hymenaea courbaril* when its predispersal seed predator is absent. *Science*, **189**, 145–147.

(1988) Ecological characterization of a Costa Rican dry forest caterpillar fauna. *Biotropica*, **20**, 120–135.

Jones, C. G., Lawton, J. H. & Shachak, M. (1997) Positive and negative effects of organisms as physical ecosystem engineers. *Ecology*, **78**, 1946–1957.

Kelly, C. A. & Dyer, R. J. (2002) Demographic consequences of inflorescence-feeding insects for *Lyatris cylindracea*, an iteroparous perennial. *Oecologia*, **132**, 350–360.

King, R. T. (2003) Succession and microelevation effects on seedling establishment of *Calophyllum brasiliense* Camb. (Clusiaceae) in an Amazonian river meander forest. *Biotropica*, **35**, 462–471.

Koptur, S. (1985) Alternative defenses against herbivores in *Inga* (Fabaceae: Mimosoideae) over an elevational gradient. *Ecology*, **66**, 1639–1650.

Langenheim, J. H. & Stubblebine, W. H. (1983) Variation in leaf resin composition between parent tree and progeny in *Hymenaea*: implications for herbivory in the humid tropics. *Biochemistry Systematics and Ecology*, **11**, 97–106.

Louda, S. M. (1982) Distribution ecology: variation in plant recruitment over a gradient in relation to insect seed predation. *Ecological Monographs*, **52**, 25–42.

Louda, S. M., Keeler, K. H. & Holt, R. D. (1990) Herbivore influences on plant performance and competitive interactions. *Perspective on Plant Competition* (ed. J. B. Grace & D. Tilman). New York: Academic Press, pp. 413–444.

Louda, S. M. & Potvin, M. A. (1995) Effect of inflorescence-feeding insects on the demography and fitness of a native plant. *Ecology*, **76**, 229–245.

Lubchenco, J. (1978) Plant species diversity in a marine intertidal community: importance of herbivore food preferences and algal competitive abilities. *American Naturalist*, **112**, 23–39.

(1980) Algal zonation in the New England Rocky intertidal community: an experimental analysis. *Ecology*, **61**, 333–344.

Lubchenco, J. & Cubit, J. (1980) Heteromorphic life histories of certain marine algae as adaptations to variations in herbivory. *Ecology*, **61**, 676–687.

Makana, J.-R., Hart, T. B. & Hart, J. A. (1998) Forest structure and diversity of lianas and understory treelets in monodominant and mixed stands in the Ituri Forest, Zaire. In *Forest Biodiversity Research, Monitoring and Modeling. Conceptual Background and Old World Case Studies* (ed. F. Dallmeier & J. A. Comiskey). New York: The Parthenon Publishing Group, pp. 429–446.

Malo, J. E., Leirana-Alcocer, J. & Parra-Tabla, V. (2001) Population fragmentation, florivory, and the effects of flower morphology alterations on the pollination success of *Myrmecophila tibicinis* (Orchidaceae). *Biotropica*, **33**, 529–534.

Mapaure, I. N. & Campbell, B. M. (2002) Changes in miombo woodland cover in and around Sengwa Wildlife Research Area, Zimbabwe, in relation to elephants and fire. *African Journal of Ecology*, **40**, 212–219.

Marquis, R. J. (1984) Leaf herbivores decrease fitness in a tropical plant. *Science*, **226**, 537–539.

(1990) Genotypic variation in leaf damage in *Piper arieianum* (Piperaceae) by a multi-species assemblage of herbivores. *Evolution*, **44**, 104–120.

(1992) Selective impact of herbivores. In *Ecology and Plant Resistance to Herbivores and Pathogens* (ed. R. S. Fritz & E. L. Simms). Chicago: University of Chicago Press, pp. 301–325.

Marquis, R. J. & Braker, H. E. (1994) Plant–herbivore interactions at La Selva: diversity, specificity, and impact. In *La Selva: Ecology and Natural History of a Neotropical Rainforest* (ed. L. McDade, G. S. Hartshorn, H. A. Hespenheide & K. S. Bawa). Chicago: University of Chicago Press, pp. 261–281.

Marquis, R. J., Diniz, I. R. & Morais, H. C. (2001) Patterns and correlates of interspecific variation in foliar insect herbivory and pathogen attack in Brazilian cerrado. *Journal of Tropical Ecology*, **17**, 1–23.

Marquis, R. J., Morais, H. C. & Diniz, I. R. (2002) Interactions among cerrado plants and their herbivores: unique or typical? In *The Cerrados of Brazil* (ed. P. S. Oliveira & R. J. Marquis). New York: Columbia University Press, pp. 203–223.

McNaughton, S. J. (1979) Grassland–herbivore dynamics. In *Serengeti, Dynamics of an Ecosystem* (ed. A. R. E. Sinclair & M. Norton-Griffiths). Chicago: University of Chicago Press, pp. 46–81.

(1983) Serengeti grassland ecology: the role of composite environmental factors and contingency in community organization. *Ecological Monographs*, **53**, 291–320.

(1984) Grazing lawns: animals in herds, plant form, and coevolution. *American Naturalist*, **124**, 863–886.

(1985) Ecology of a grazing ecosystem: the Serengeti. *Ecological Monographs*, **55**, 259–294.

Molofsky, J. & Augspurger, C. K. (1992) The effect of leaf litter on early seeding establishment in a tropical forest. *Ecology*, **73**, 68–77.

Mothershead, K. & Marquis, R. J. (2000) Indirect effects of leaf herbivores on plant–pollinator interactions in *Oenothera macrocarpa* (Onagraceae). *Ecology*, **81**, 30–40.

Moutinho, P., Nepstad, D. C. & Davidson, E. A. (2003) Influence of leaf-cutting ant nests on secondary forest growth and soil properties in Amazonia. *Ecology*, **84**, 1265–1276.

Musgrove, M. K. & Compton, S. G. (1997) Effects of elephant damage to vegetation on the abundance of phytophagous insects. *Journal of Tropical Ecology*, **35**, 370–373.

Nascimento, M. T. & Proctor, J. (1994) Insect defoliation of a monodominant Amazonian rainforest. *Journal of Tropical Ecology*, **10**, 633–636.

(2001) Leaf herbivory on three tree species in a monodominant and two other Terra Firme forests on Maracá Island, Brazil. *Acta Amazonica*, **31**, 27–38.

Paine, R. T. (1966) Food web complexity and species diversity. *American Naturalist*, **100**, 65–75.

Prins, H. H. T. & van der Juegd, H. P. (1993) Herbivore population crashes and woodland structure in East Africa. *Journal of Ecology*, **81**, 305–314.

Rabinowitz, D. (1978) Dispersal properties of mangrove propagules. *Biotropica*, **10**, 47–57.

Rausher, M. D. & Feeny, P. (1980) Herbivory, plant density, and plant reproductive success: the effect of *Battus philenor* on *Aristolochia reticulata*. *Ecology*, **61**, 905–917.

Ribeiro, S. P. (1999) The role of herbivory in *Tabebuia* spp. life history and evolution. Unpublished Ph.D. thesis, Imperial College at Silwood Park, Ascot.

Rinker, H. B., Lowman, M. D., Hunter, M. D., Schowalter, T. D. & Fonte, S. J. (2001) Literature review: canopy herbivory and soil ecology, the top-down impact of forest processes. *Selbyana*, **22**, 225–231.

Sánchez-Hidalgo, M. E., Martinez-Ramos, M. & Espinosa-Garcia, F. J. (1999) Chemical differentiation between leaves of seedlings and spatially close adult trees from the tropical rainforest species *Nectandra ambigens* (Lauraceae): an alternative test of the Janzen–Connell model. *Functional Ecology*, **13**, 725–732.

Sherman, P. M. (2002) Effects of land crabs on seedling densities and distributions in a mainland neotropical rain forest. *Journal of Tropical Ecology*, **18**, 67–89.

Silman, M. R., Terborgh, J. W. & Kiltie, R. A. (2003) Population regulation of a dominant rain forest tree by a major seed predator. *Ecology*, **84**, 431–438.

Silvera, K., Skillman, J. B. & Dalling, J. W. (2003) Seed germination, seedling growth and habitat partitioning in two morphotypes of the tropical pioneer tree *Trema micrantha* in a seasonal forest in Panama. *Journal of Tropical Ecology*, **19**, 27–34.

Sinclair, A. R. E. (1995) Equilibria in plant–herbivore interactions. In *Serengeti II, Dynamics, Management, and Conservation of an Ecosystem* (ed. A. R. E. Sinclair & P. Arecese) Chicago: University of Chicago Press, pp. 91–113.

Smith, T. J., III, Chan, H. T., Melvor, C. C. & Robblee, M. B. (1989) Comparisons of seed predation in tropical tidal forests from three continents. *Ecology*, **70**, 146–151.

Stiles, F. G. (1978) Coadapted competitors: the flowering seasons of hummingbird-pollinated plants in a tropical forest. *Science*, **196**, 1177–1178.

Terborgh, J. (1992) Maintenance of diversity in tropical forests. *Biotropica*, **24**, 283–292.

Terborgh, J. & Wright, S. J. (1994) Effects of mammalian herbivores on plant recruitment in two neotropical forests. *Ecology*, **75**, 1829–1833.

Thiede, D. A. & Augspurger, C. K. (1996) Intraspecific variation in seed dispersion of *Lepidium campestre* (Brassicaceae). *American Journal of Botany*, **83**, 856–866.

Ticktin, T. (2003) Relationships between El Niño Southern Oscillation and demographic patterns in a substitute food for collared peccaries in Panama. *Biotropica*, **35**, 189–197.

Tuomisto, H., Ruokolainen, K., Poulsen, A. D. *et al.* (2002) Distribution and diversity of pteridophytes and Melastomataceae along edaphic gradients in Yansuni National Park, Ecuadorian Amazonia. *Biotropica*, **34**, 516–533.

Webb, L. J, Tracey, J. G. & Haydock, K. P. (1967) A factor toxic to seedlings of the same species associated with living roots of the non-gregarious subtropical rain forest tree *Grevillea robusta*. *Journal of Applied Ecology*, **4**, 1035–1038.

Wenny, D. G. & Levey, D. J. (1998) Directed seed dispersal by bellbirds in a tropical cloud forest. *Proceedings of the National Academy of Sciences*, **95**, 6204–6207.

Wright, S. J. (2002) Plant diversity in tropical forests: a review of mechanisms of species coexistence. *Oecologia*, **130**, 1–14.

Wright, S. J. & Duber, H. C. (2001) Poachers and forest fragmentation alter seed dispersal, seed survival, and seedling recruitment in the palm *Attalea butyraceae*, with implications for tropical tree diversity. *Biotropica*, **33**, 583–595.

CHAPTER FOURTEEN

Have the impacts of insect herbivores on the growth of tropical tree seedlings been underestimated?

FERGUS P. MASSEY
University of Sussex
MALCOLM C. PRESS
University of Sheffield
SUE E. HARTLEY
University of Sussex

Introduction

Understanding the processes underpinning high biodiversity in tropical rain forests is a major goal for ecologists. There has been a long-standing interest in the role of herbivory in the tropics, in part because it has been proposed as a mechanism contributing to the maintenance of this diversity for tree species (Janzen 1970; Connell 1971; Marquis, Chapter 13, this volume). The Janzen–Connell hypothesis predicts that herbivore damage may act in a density-, distance- or frequency-dependent manner and, if the amounts of damage are sufficient to cause changes in growth rates or mortality of seedlings (or seeds), then this could lead to the distancing of conspecifics and hence promote species diversity. In areas of high conspecific density, increased herbivory could kill or reduce the growth of more common species, allowing rarer species to persist. This hypothesis relies on the ability of herbivory to significantly affect seedling growth or mortality, and although many studies have found variation in damage levels between species (Coley 1983; Coley & Barone 1996; Blundell & Peart 1998; Howlett & Davidson 2001), few studies have found conclusive effects on seedling growth, much less on mortality (Aide & Zimmerman 1990; Webb & Peart 1999; Howlett & Davidson 2001; Horvitz & Schemske 2002; Hyatt *et al.* 2003; Pearson *et al.* 2003b). Of the studies that have demonstrated differences in growth rates or mortality between species caused by herbivory (Marquis 1984; Coley 1986; Bazzaz *et al.* 1987; Sork 1987; Osunkoya *et al.* 1993; Terborgh & Wright 1994; Sagers & Coley 1995; Whitmore & Brown 1996; Ickes *et al.* 2001; Wallin & Raffa 2001), the majority are attributable to mammalian herbivores and seed predators (Sork 1987; Osunkoya *et al.* 1993; Terborgh & Wright 1994; Ickes *et al.* 2001). Most previous studies testing the Janzen–Connell hypothesis have identified insect seed

Biotic Interactions in the Tropics: Their Role in the Maintenance of Species Diversity, ed. D. F. R. P. Burslem, M. A. Pinard and S. E. Hartley. Published by Cambridge University Press. © Cambridge University Press 2005.

predators as having significant distance-dependent effects on mortality, but few studies have identified similar trends on seedlings (see review by Hammond & Brown 1998, but see also Harms *et al.* 2000; Hyatt *et al.* 2003). Although root and shoot borers have been shown to have significant effects on performance of dipterocarp seedlings (Whitmore & Brown 1996; Bebber *et al.* 2002), there is surprisingly little evidence that defoliation by insects has similar impacts.

Can we, therefore, assume from these previous studies that insect folivores are not important in the maintenance of diversity within tropical rain forests, at least in comparison with mammals and seed predators? Such a conclusion may be premature because many of the studies that have failed to find significant impacts of insect herbivores may overlook many potentially major effects of herbivory because of difficulties associated with their measurement. For example, it is well documented that most damage to leaves occurs during their expansion phase (Coley & Aide 1991), but many studies make measurements that lack sufficient frequency at this early stage in leaf development and so are unable to detect differences in the effects of herbivory between different species. These differential impacts early in leaf development may prove crucial when trying to assess accurately the effects of herbivory on the competitive abilities of young seedlings. These studies may also disregard the impact of leaf abscission resulting from damage during the leaf-expansion phase, but accumulation rates differ markedly between species and may constitute a significant loss of resources for some species but not others (Blundell & Peart 2000). However, the relative effect of this resource loss will differ according to the ability of species and individuals to compensate.

In this chapter we consider four relatively neglected aspects of the impacts of insect herbivores on plants, based on our recent studies of dipterocarp seedlings. The Dipterocarpacae is a large family comprising about 480 species, which dominate the species-rich lowland rain forests of much of southeast Asia. Dipterocarps mast gregariously, producing recalcitrant seeds that usually result in the formation of dense seedling banks (but see Curran *et al.* 1999; Blundell & Peart 2004). Although dipterocarps are subjected to comparatively low levels of attack by insect herbivores (Becker 1983; Curran & Webb 2000), herbivory could nevertheless play an important role in determining the competitive hierarchy of the large number of closely related co-existing species over relatively small spatial scales. Specifically, we address the following themes:

(1) **Pattern of early leaf development and its effects on the degree and rates of damage**: During leaf expansion, leaves are at their most vulnerable to herbivory. Small-scale interspecific differences in development pattern, such as rates of leaf expansion, or the ability of a species to produce new leaves and compensate for herbivory, could affect the impact of damage on different species.

Table 14.1 *Growth and leaf characteristics of five dipterocarp species in order of increasing seedling shade tolerance*

Species	Growth characteristics	Timber density (kg m^{-3})^a	Leaf characteristics	Total phenolics in young (mature) leaves as % dry weight
Shorea leprosula	Fast-growing, relatively light-demanding	519	Relatively unlignified leaves	16.6 (10.2)
Shorea johorensis	Fast-growing, relatively light-demanding	525	Relatively unlignified leaves	14.3 (9.6)
Dryobalanops lanceolata	Slow-growing, relatively shade-tolerant	590	Tough waxy leaves	18.7 (10.7)
Shorea gibbosa	Relatively slow-growing, shade-tolerant	n/a	Tough mature leaves	18.3 (11.7)
Hopea nervosa	Slow-growing, shade-tolerant	629	Tough mature leaves	29.4 (15.5)

^{a}Burgess 1966
n/a = not available.

(2) **Early leaf abscission in response to herbivory**: Damage to developing leaves can cause them to abscise. We examine whether rates of abscission of damaged leaves differ among species, in terms of both frequency and stage of leaf development, and whether these differences are sufficient to affect the relative amounts of damage between species.

(3) **Delayed greening**: Many woody species in the tropics delay provisioning of leaves with photosynthetic apparatus until late in development. The functional significance of delayed greening, and of the red coloration displayed by many of these leaves, has been a topic of debate for some time (Coley & Barone 1996; Dominy *et al.* 2002). We examine the possibility that delayed provisioning of leaves serves to reduce resource loss when herbivory occurs (Kursar & Coley 1992b).

(4) **Seedling environment**: We examine the impact of environmental spatial heterogeneity (seedling position within canopy gaps and the identity of surrounding seedlings) on the amounts of damage by herbivores.

We concentrate on seedlings of five contrasting dipterocarp species, which differ in their growth rates (in both understorey and gaps) and leaf-defence characteristics, such as phenolic content and leaf toughness (Table 14.1; Massey 2002). At one extreme of the growth and defence trade-off spectrum is *Shorea leprosula*, with fast growth and leaf-turnover rates, low levels of phenolics and leaf toughness, and characterized by high levels of insect herbivore damage

(Massey 2002). At the opposite end is *Hopea nervosa*, which has a slower rate of growth, higher allocation to defences and suffers relatively low levels of damage (Table 14.1). By comparing the responses of these two extreme species together with three intermediate species to the four previously neglected aspects of herbivory outlined above, we can assess the relevance of these measures to future studies of herbivory.

Study site

All data presented here come from studies carried out in the Danum Valley Conservation Area (DVCA), Sabah, Malaysia (4° 54′ N, 117° 48′ E). The DVCA is an area of primary lowland dipterocarp forest (438 km^2) surrounded by 9730 km^2 of selectively logged forest. A largely aseasonal evergreen forest, the region has a mean annual rainfall of 2822 mm and mean annual temperature of 26.7 °C, with a mean maximum of 30.9 °C and a mean minimum of 22.5 °C (Marsh & Greer 1992). All species studied are found extensively throughout the conservation area.

Leaf development and herbivory

It has long been known that the majority of herbivore damage in the tropics occurs as leaves are expanding (Coley & Aide 1991) and previous studies have highlighted the need to include measurements on young leaves when assessing the impact of herbivores (Mooney & Gulmon 1982; Harper 1989; Coley & Barone 1996). Despite this, few studies have looked in detail at the pattern of leaf development and the accumulation of damage on young leaves, even though this is essential to understanding variation in herbivory among species with differing patterns of leaf development (but see Kursar & Coley 2003). Instead, most studies on herbivore impacts use repeated measurements made on tagged leaves, often with weeks or months between measures (Coley 1980; Aide 1993; Howlett & Davidson 2001; Pearson *et al.* 2003a). Herbivore damage is usually expressed as a rate, i.e. damage at the end of a period divided by the time, thus assuming that the rate of damage accumulation is linear. Even in studies where rates of leaf expansion have been measured on a fine timescale, the corresponding herbivore-damage pattern has not been assessed over the same timescale (Kursar & Coley 1991). If damage accumulation is assumed to be linear, comparisons of the rates of damage between species will be misleading in cases where the species being compared have very different rates of leaf production. As younger foliage may be best placed for light capture and contain higher nutrient concentrations than mature leaves (Kursar & Coley 2003), these interspecific differences in damage impact may have more significance than previously supposed. Studies looking for effects of herbivore damage need to focus work on the periods in which the damage is at its greatest and when small differences in leaf-development patterns among species could affect amounts of damage. Here we present data examining the effect of leaf-development rates on the timing

of herbivore damage on two species of dipterocarp seedlings in canopy gaps
and discuss the implications for comparing levels of herbivore damage between
species.

Leaf-expansion rates and patterns of herbivore defoliation were monitored on
20 seedlings of two dipterocarp species with contrasting growth rates and levels of herbivore damage, *Shorea leprosula* and *Dryobalanops lanceolata* (Table 14.1),
planted in six canopy gaps. On each seedling, between five and ten leaf buds were
tagged and the growth rate, herbivore defoliation and mortality of developing
leaves were monitored every two days for the first 14 days, then every five days
until fully developed (approximately 90 days). Here we are only concerned with
the rates of defoliation due to insect herbivores and therefore do not include
those leaves that were abscised during development.

For both species, the patterns of herbivore defoliation were non-linear and
closely mirrored the rates of leaf expansion (Fig. 14.1). However, for these contrasting species, both the levels and patterns of damage differed greatly. The
steady regular growth of *S. leprosula* leaf expansion was mirrored by gradually
increasing amounts of herbivore damage throughout expansion (0.24 $\pm$ 0.06%
of non-abscised leaf area removed per day, Fig. 14.1a). This eventually resulted
in a greater amount of total damage accumulated than on *D. lanceolata* (20.9%
vs. 13.2% leaf area removed for *S. leprosula* and *D. lanceolata*, respectively, 90 days
after bud burst). In contrast to *S. leprosula*, the majority of *D. lanceolata* leaves displayed a period of non-expansion followed by a period of very rapid leaf expansion (Fig. 14.1b). During these two periods *D. lanceolata* suffered very different
rates of damage. Herbivore damage was negligible during the initial period, but
later, during rapid expansion, damage rate increased to 0.38 ($\pm$ 0.03)% leaf area
removed per day, a rate which was not only very different from the earlier rate
for this species but which was also much higher than the rate for *S. leprosula*.
Therefore, although the highest rates of damage were recorded for *D. lanceolata*, *S. leprosula* demonstrated greater cumulative damage. In addition, 85% of
the total damage recorded over the first 90 days after bud burst in *D. lanceolata*
occurred as the leaves were expanding, compared with only 70% in *S. leprosula*.
This difference illustrates very clearly that rates of herbivory on young leaves
vary greatly both between and within species, and that calculations based on
linear damage accumulation are inadequate when trying to compare rates of
damage on different species.

Damage to expanding leaves, which potentially make a significant contribution to the lifetime carbon balance of the plant, may have disproportionately
large impacts on growth and fitness and must therefore be recorded accurately.
Thus, we suggest that when comparing effects of herbivory between species,
measurements of herbivore should focus on the early stages of leaf development,
where the majority of herbivory damage takes place and where there may be
large differences between species in damage impact. This is highlighted in *D.
lanceolata* where a damage rate taken over the entire expansion time would give

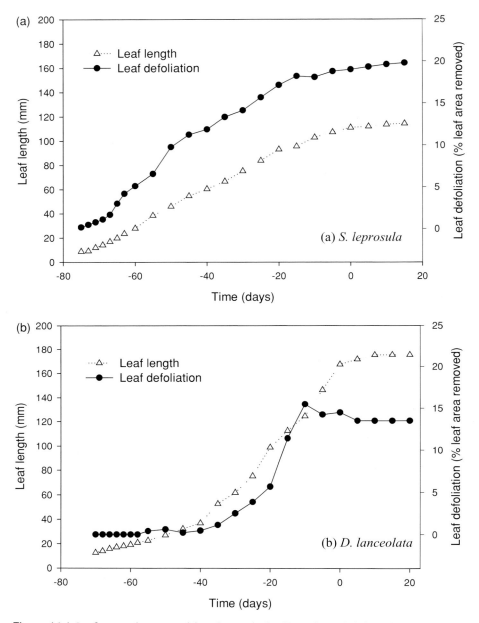

Figure 14.1 Leaf expansion as positive change in leaf length, and defoliation over 90 days following bud burst for (a) *S. leprosula* (*n* = 178) and (b) *D. lanceolata* (*n* = 156). Following convention time = 0 is used to denote full expansion. Expansion rate: *S. leprosula* (−75 to 0 days) = 1.49 ± 0.2 mm day^{-1}; *D. lanceolata* (−40 to 0 days) = 3.8 ± 0.7 mm day^{-1}. Doubling time (50 to 100 mm): *S. leprosula* = 32.37 ± 1.2 days, *D. lanceolata* = 15.63 ± 1.7 days. Plot does not include those leaves that did not expand during study.

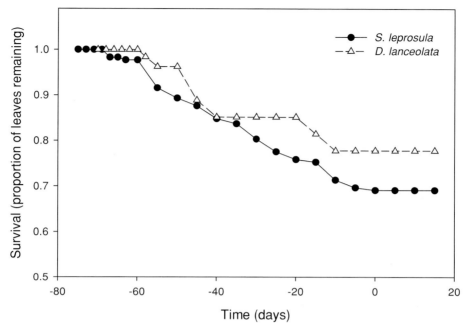

Figure 14.2 Proportion of leaves surviving during expansion period for *S. leprosula* ($n = 178$) and *D. lanceolata* ($n = 156$). Time = 0 represents full expansion of leaves.

a misleading result that is less than the rate of damage on *S. leprosula* (-70 days to full expansion: 0.15 (± 0.07)% leaf area removed per day vs. 0.24 (± 0.06)% for *S. leprosula*). The only other study that we are aware of that considers this aspect of developing leaves also suggested that rates of herbivory, although highly variable between the five species studied, were closely related to leaf-expansion rate and relate to the species' life-history strategy (Kursar & Coley 2003). The study also links variations in damage rates within species to chlorophyll content, leaf toughness and foliar nitrogen concentrations, and emphasizes that changes in expansion rates and duration can lead to large interspecific variation in damage. The timing of damage measurements is therefore critical during early expansion if these interspecific variations are to be found.

Early leaf abscission

A second issue regarding the measurement of herbivory on developing leaves arises from the fact that many damaged leaves are abscised, particularly when young. This means that if widely separated time points are used to calculate leaf area loss they may be inaccurate because they miss leaves that have already been abscised because of herbivore damage. Natural levels of abscission in response to damage may be high and, critically, may vary among species. For example, studies of leaf development for two dipterocarp species, *S. leprosula* and *D. lanceolata*, found that 31% and 21% of leaves were abscised prior to maturity (Fig. 14.2), for each species, respectively, and of those leaves abscised at least 76% and 73%, respectively, had been damaged in the five days previous to abscission (data not shown, Massey 2002).

Few studies have examined the circumstances under which herbivory will induce leaf abscission, especially on young leaves (Blundell & Peart 2000), and interspecific differences in damage-induced abscission have been largely overlooked. In a detailed study of two contrasting dipterocarp species, variation in abscission in response to damage of leaves of four age classes was measured, following the methods of Blundell and Peart (2000).

In six canopy gaps, seedlings of two species of dipterocarp were selected for study (Table 14.1): *S. leprosula* ($n = 24$) and *H. nervosa* ($n = 20$). Leaves on the planted 2-year-old seedlings were grouped into four age classes and five leaves from each age group were tagged on each seedling. The classes were:

(1) Expanding 1: leaf partially unfurled though the two halves of each leaf not yet longitudinally separated
(2) Expanding 2: fully unfurled but less than full size
(3) Expanded: full size though not yet lignified
(4) Mature: full size and lignified

Seedlings were then divided into experimental and control groups at random. On experimental plants, leaves were damaged by making two longitudinal cuts parallel to the mid-rib, removing approximately 50% of leaf tissue from leaves of all age classes. On control plants, leaves were handled in a similar manner, although no cuts were made. A census was made of all plants at 10 weeks after damage and the number of tagged leaves remaining was recorded.

Hopea nervosa showed very little abscission regardless of leaf age. In contrast, *S. leprosula* abscised nearly all damaged leaves in the youngest age class, with significant levels of abscission in age class two as well (Fig. 14.3). The consequences for assessing rates of herbivory are dramatic. Measurements of herbivore impacts based solely on the amounts of defoliation would have found no differences between these two species: 18% and 16% leaf area lost for *S. leprosula* and *H. nervosa* respectively. However, when incorporating the large difference in abscission rates, damage on *S. leprosula*, in terms of total leaf loss, was over twice that suffered by *H. nervosa*. (43.4% of leaf area compared with 21.9%; Table 14.2). It is clear that if the levels of herbivory are to be compared accurately between species, levels of abscission must be monitored.

Delayed greening

The large variation in damage-induced leaf abscission between species may reflect their different strategies over the provisioning of young leaves. Many woody species in the tropics delay provisioning of developing leaves with photosynthetic apparatus or chlorophyll, leaving them with a red or pink

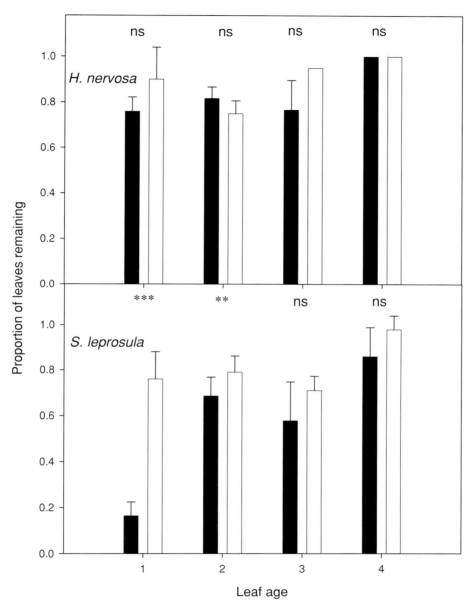

Figure 14.3 Proportion of tagged leaves retained on *H. nervosa* and *S. leprosula* seedlings. Values are means (± SE) of experimental (filled bars; 50% tissue removal) and control (open bars) leaves. Leaf ages: (1) leaf partially unfurled though two halves of leaf not yet longitudinally separated; (2) fully unfurled but less than full size; (3) expanded: full size though not yet lignified; (4) mature: full size and lignified. Significant difference between experimental and control samples using chi-squared test is indicated by *** $p < 0.001$, ** $p < 0.01$, ns not significant.

Table 14.2 *Defoliation and damage-induced leaf-abscission levels for expanding leaves and corrected levels of herbivore damage including abscised leaves for* H. nervosa *and* S. leprosula *seedlings*

	Standing damage on expanding leaves (%)	Leaf abscission prior to full expansion (%)	Total damage including abscission (% leaf area loss)
Hopea nervosa	16	7	21.9
Shorea leprosula	18	31	43.4

coloration, attributable to the accumulation of anthocyanins (Kursar & Coley 1992a; Whatley 1992; Juniper 1993). The functional significance of this phenomenon remains unclear, despite being debated for many years (Coley & Barone 1996; Dominy *et al.* 2002). Previous proposed roles have included anti-fungal defence (Coley & Aide 1989), protection against ultraviolet radiation (Gould *et al.* 1995), and anti-herbivore defence (Kursar & Coley 1992b; Juniper 1993). Although there is little evidence to suggest that anthocyanins have any significant toxic or deterrent effect on insect herbivores (Harborne 1979; Lee *et al.* 1987), delayed greening may still play a role in minimizing losses to herbivory because delaying provisioning of leaves with nitrogen-rich photosynthetic pigments and proteins during the most vulnerable stages of development could reduce the loss of nutrients to insect herbivores (Kursar & Coley 1992b).

Interspecific variation in the degree of early leaf abscission induced by herbivore damage could be linked to delayed greening. For example, species that abscise damaged leaves early in development may reduce resource loss by not provisioning those leaves. An observational study of 2-year-old seedlings of five different dipterocarp species examined the occurrence of delayed greening, the degree and timing of early leaf abscission due to herbivore damage, and the amounts of defoliation on young leaves, using a tagged set of developing leaves. The results reveal three possible strategies amongst the different species (Table 14.3). In two of the delayed-greening species (*Shorea gibbosa* and *D. lanceolata*) abscission occurred only in the earliest stages of development, when leaves were still red. Once provisioning with photosynthetic apparatus occurred, as leaves were approaching full size but not yet lignified, the levels of abscission were minimal. In contrast, the two species that lack delayed greening – *S. leprosula* and *Shorea johorensis* – exhibited abscission until late in development (Table 14.3). Both these species have relatively fast growth rates and leaf-turnover rates, and abscission could play a part in the plant accepting a certain level of leaf loss, traded against low investment in defences. These strategies appear to agree with predictions from the resource-allocation hypothesis (Coley *et al.* 1985): the fast-growing green leaf species with relatively low levels of defence and leaf investment abscise leaves quickly in favour of new growth, whilst the

Table 14.3 *Comparisons of early leaf-development characteristics for seedlings of five dipterocarp species*

Species (n = no. of seedlings)	Delayed greening	Leaf abscission prior to full expansion (% total leaves)	Defoliation during leaf expansion (% leaf removed)	Defoliation during the expansion phase as % of the total damage accumulated by mature leaves	Timescale of abscission in response to defoliation
D. lanceolata (n = 44)	yes[a]	21	6	68	Until green
S. gibbosa (n = 27)	yes[a]	17	4.5	47	Until green
S. leprosula (n = 44)	no	31	18	63	Until full expansion
S. johorensis (n = 30)	no	27	5	48	Until full expansion
H. nervosa (n = 32)	yes[b]	7	16	94	Negligible

[a] Delays greening until 3/4 full expansion
[b] Delays greening until lignification

slower-growing delayed-greening species reverse this trade-off between abcission and new growth.

A third strategy was exploited by *H. nervosa*, which rarely abscised damaged leaves even at the earliest stage (Fig. 14.3). In this case, a very high proportion of the total leaf lifetime damage occurred during expansion (96%), and delayed greening was simply a mechanism to minimize resource loss to herbivory by delaying investment in leaves (Table 14.3). This strategy may be more effective than abscission because *H. nervosa* is never a canopy emergent and so may require a greater degree of adaptation to low-light conditions than the other species. Therefore, for *H. nervosa* complete leaf loss might carry a greater cost than for more rapidly growing, less shade-tolerant species.

Seedling environment

In addition to the effects of leaf development and leaf abscission, variation in herbivore impacts, both between and within species, may be related to fine-scale variation in environmental factors. Many studies of herbivory consider only the gap and/or understorey environments, whereas there is a large degree of environmental heterogeneity within both these environments. Light (quantity and quality) is the variable that probably varies most within gaps, but important gradients in temperature and water availability have also been widely recorded (see e.g. Brown 1993). All these gradients might be expected to influence the type, abundance and activity of insect herbivores. There have been studies on the effects of gap size on herbivore-damage rates (Shure & Wilson 1993; Blundell & Peart 2001; Pearson *et al.* 2003a), but effects of climatic gradients within gaps on insect herbivory have often been ignored. Variation in seedling growth in response to difference in light levels within a gap has been considered but impacts of insect herbivores were not measured (Brown 1996).

Environmental heterogeneity within gaps is likely to affect specific insect feeding guilds differently. Classifying insects into feeding guilds – leaf chewers, leaf rollers, gall-forming and leaf miners – is a convenient way to test for changes brought about by microclimate (Root 1973; Strong *et al.* 1984).

Herbivore behaviour is likely to be influenced further by the plant species composition within gaps. Both Janzen (1970) and Connell (1971) hypothesized that seedling survival and growth in forest could be density-dependent, distance-dependent and/or frequency-dependent, with the monospecific seedling stands being more susceptible to attack by specialist-phytophagous insects than heterospecific ones (Connell *et al.* 1984). In addition, it has been proposed that plants in high densities are more 'apparent' to herbivores (Feeny 1976) and would, therefore, suffer greater amounts of damage. Neither the small-scale within-gap variation in environmental conditions, nor the effects of neighbouring seedling identity on herbivore impacts have been studied in detail in dipterocarp forests (Blundell & Peart 2001). However, both factors may be of particular interest in understanding regeneration patterns of dipterocarp seedlings. Dipterocarps form dense, monospecific seedling banks close to the maternal tree following mast-fruiting events (Whitmore 1984; Appanah 1993), with smaller numbers of seedlings establishing further away in heterospecific banks, thus forming polycultures of seedlings.

Two experiments were designed to examine the effects of both seedling position within the gap and neighbour identity on herbivore-damage levels. The first experiment studied 1-m² blocks of 25 planted *S. leprosula* seedlings at the centre and edge of five canopy gaps. Over 18 months the type of herbivore damage, as insect feeding guilds present, and amounts of defoliation were monitored. We found that both the amount and type of damage was affected by the seedling's position. For example, seedlings at the gap edge suffered greater amounts of defoliation, and had a higher density of leaf rollers, whilst those in the centre had a greater density of leaf galls (Fig. 14.4).

These results are consistent with studies comparing gap and understorey or canopy and understorey plants that have found greater amounts of damage in lower-light environments (Lowman 1985; Lowman 1992; Barone 1994; Gilbert 1995). Several processes, which are not mutually exclusive, could explain the observed pattern of damage. Firstly, the increased expansion rate of leaves in the centre of the gap could have reduced the exposure time sufficiently to offset damage (Howlett & Davidson 2001). Secondly, the abundance of herbivores,

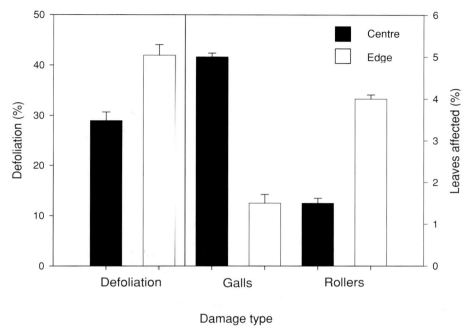

Figure 14.4 Herbivore defoliation and the percentage of leaves affected by leaf galls and leaf rollers on *S. leprosula* seedlings at the centre and the edge of canopy gaps. Seedling defoliation: paired *t*-test of block means $t = 4.31$, $p = 0.003$, $n = 9$. Distribution of feeding guilds: $\chi^2 = 130.06$, $p < 0.001$, d.f. $= 2$.

which is generally accepted to be greater in gaps than in the understorey (Benson 1978; Basset 1991), could also vary according to position within the gap. This was found to be the case here, with gall-forming insects, which are naturally protected against both predators and desiccation by the structure of the gall, preferring the centre of gaps. In contrast, leaf-rolling insects may benefit from the increased leaf production and quality in the gap, whilst still being protected from desiccation at the edge. Therefore, increases in leaf damage at the edge of the gap would be a result of differences in insect community structure within the gap. Thirdly, increases in the temperature at the centre of gaps caused by higher irradiance levels might increase desiccation of insects. Other possibilities are that seedlings at the edge of gaps were in a position within the overlap of gap-specialist herbivores and understorey specialist herbivores, and that the distribution of predators could be different across the gap (Maiorana 1981; Basset 1991).

The second experiment compared the amounts of herbivore damage on dipterocarp seedlings grown in either a monoculture or a polyculture seedling bank over 18 months. In the centre of six canopy gaps, four 1-m² blocks were planted with 25 2-year-old seedlings in a grid pattern. Two blocks contained only *S. leprosula* seedlings, while the other two blocks contained five *S. leprosula* seedlings amongst individuals of four other dipterocarp species. We found that the seedlings in a dense monospecific stand suffered greater amounts of damage than those seedlings surrounded by other species: 29% vs. 18% defoliation respectively (paired t-test of block means: $t = 8.10$, $p < 0.001$, $n = 11$). There are a number of possible explanations for the differences. Firstly, the concentrated resource in a monospecific stand might induce insects to remain on the patch for longer, with higher numbers of specialist herbivores (Root 1973). In the case of *S. leprosula*, which has relatively low defence levels, the species may also be susceptible to generalist herbivores, for which a concentrated patch of palatable species would be attractive. Secondly, locating a host is more efficient in a high-density patch of a palatable species than when the palatable seedlings are mixed amongst species with high defence levels (Tahvanainen & Root 1972; Atsatt & O'Dowd 1976; Feeny 1976; Rausher 1981). Distance to conspecifics can affect both the growth and herbivore defoliation of seedlings, and this has been attributed to higher concentrations of conspecifics acting as reservoirs for herbivores (Connell 1978; Connell *et al.* 1984; Howe *et al.* 1985; Hubbell & Foster 1986; Condit *et al.* 1994; Blundell & Peart 1998). Lastly, insects may be actively deterred by proximity to unpalatable species and have been shown to leave seedlings in a polyculture more quickly than on the same species in monoculture, even though they are on host plants (Bach 1980).

Conclusions

Impacts of insect herbivores on the performance of tree seedlings have been the focus of many studies because of their potentially important role in determining diversity in tropical rain forests. However, few studies have found strong evidence to suggest that insects significantly affect growth or mortality of tree seedlings. We suggest that this lack of evidence could be due in part to the methods used for assessing herbivory and herbivore impacts. We highlight four key issues:

(1) Variations in the rate and pattern of leaf development between species significantly affect the amounts of herbivore damage. Repeated measurements, on tagged leaves, need to be made on a timescale of a few days during the early stages of leaf development to record relative damage accurately between species. Because of the non-linear nature of damage accumulation in some species, repeated measurements weeks or months apart can lead to misleading estimates of damage rate depending upon where measurements are taken during the period of leaf expansion.

(2) Although neglected in the majority of studies, interspecific variation in amounts of insect-induced early leaf abscission results in degrees of leaf loss sufficient to change the damage-ranking between species. Rates of damage-induced leaf abscission must be included in studies to determine relative impacts of insect herbivores.

(3) We found large variation in the amount and timing of abscission between species, which could be related, in part, to whether or not the species delayed provisioning of developing leaves. Fast-growing species, which do not delay greening of developing leaves (e.g. *S. leprosula* and *S. johorensis*), follow the predictions of the resource-allocation hypothesis and accept a high level of leaf loss through abscission in favour of flushing new leaves and low defence investment. However, delayed-greening species, which have intermediate growth rates (e.g. *D. lanceolata* and *S. gibbosa*), only abscise leaves that are damaged before significant provisioning of leaves has been made.

(4) Both seedlings' position within canopy gaps and the composition of the seedling bank can affect the amount and type of herbivore damage within species. Variation in microclimate within gaps is sufficient to alter the structure of insect herbivore communities and therefore the types of damage suffered. For example, we found a greater abundance of gall-forming insect damage on seedlings at the centre of gaps and leaf-rolling insects at the edge. Changes in a seedling's susceptibility to herbivore attack due to seedling position and neighbouring seedling identity need to be considered when making both inter- and intraspecific comparisons of herbivore damage.

Competition for resources and light capture within seedling banks form important parts in determining regeneration success, and hence forest diversity. Regeneration success will be affected critically by individual seedling performance relative to other species, especially in dipterocarp forests where dense banks of seedlings can persist for years in canopy gaps and the understorey. Insect herbivores could have a crucial role in determining the condition and performance of seedlings, and we suggest that the relative impacts of herbivory on different species may have been underestimated in the past because of incorrect measurement techniques.

Acknowledgements

We thank all the staff and research assistants at Danum Valley Field Centre and K. Williamson for field assistance, and A. Blundell and D. Bebber for their comments on the manuscript. F. P. M. was supported by a NERC studentship.

References

Aide, T. M. (1993) Patterns of leaf development and herbivory in a tropical understorey community. *Ecology*, **74**, 455–466.

Aide, T. M. & Zimmerman, J. K. (1990) Patterns of insect herbivory, growth, and survivorship in juveniles of a neotropical liana. *Ecology*, **71**, 1412–1421.

Appanah, S. (1993) Mass flowering of dipterocarp forests in the aseasonal tropics. *Journal of Biosciences*, **18**, 457–474.

Atsatt, P. R. & O'Dowd, D. J. (1976) Plant defence guilds. *Science*, **193**, 24–29.

Bach, C. E. (1980) Effects of plant density and diversity on the population dynamics of a specialist herbivore: the striped cucumber beetle, *Acalymma vittate. Ecology*, **61**, 1515–1530.

Barone, J. A. (1994) Herbivores and herbivory in the canopy and understorey on Barro Colorado Island, Panama. Sarasota, Florida: Selby Botanical Gardens.

Basset, Y. (1991) Leaf production of an overstory rain-forest tree and its effects on the temporal distribution of associated insect herbivores. *Oecologia*, **88**, 211–219.

Bazzaz, F. A., Chiariello, N. R., Coley, P. D. & Pitelka, L. F. (1987) Allocating resources to reproduction and defence. *Bioscience*, **37**, 58–67.

Bebber, D., Brown, N. & Speight, M. (2002) Drought and root herbivory in understorey *Parashorea* Kurz (Dipterocarpaceae) seedlings in Borneo. *Journal of Tropical Ecology*, **18**, 795–804.

Becker, P. (1983) Effects of insect herbivory and artificial defoliation on survival of *Shorea* seedlings. In *Tropical Rain Forest: Ecology and Management* (ed. S. L. Sutton, T. C. Whitmore & A. C. Chadwick) London: Blackwell Scientific, pp. 241–252.

Benson, W. W. (1978) Resource partitioning in passion vine. *Butterflies*, **32**, 493–518.

Blundell, A. G. & Peart, D. R. (1998) Distance-dependence in herbivory and foliar condition for juvenile dipterocarp trees in the Bornean rain forest. *Oecologia*, **117**, 151–160.

(2000) High abscission rates of damaged expanding leaves: field evidence from seedlings of a Bornean rain forest tree. *American Journal of Botany*, **87**, 1693–1698.

(2001) Growth strategies of a shade-tolerant tropical tree: the interactive effects of canopy gaps and simulated herbivory. *Journal of Ecology*, **89**, 608–615.

(2004) Seedling recruitment failure following dipterocarp mast fruiting. *Journal of Tropical Ecology*, **20**, 229–231.

Brown, N. D. (1993) The implications of climate and gap microclimate for seedling growth-conditions in a Bornean lowland rain-forest. *Journal of Tropical Ecology*, **9**, 153–168.

(1996) A gradient of seedling growth from the centre of a tropical rain forest canopy gap. *Forest Ecology and Management*, **82**, 239–244.

Burgess, P. F. (1966) *Timbers of Sabah*. Forest Department Sandakan, Sabah, Malaysia.

Coley, P. D. (1980) Effects of leaf age and plant life history patterns on herbivory. *Nature*, **284**, 545–546.

(1983) Herbivory and defensive characteristics of tree species in a lowland tropical forest. *Ecological Monographs*, **53**, 209–233.

(1986) Costs and benefits of defence by tannins in a neotropical tree. *Oecologia*, **70**, 238–241.

Coley, P. D. & Aide, T. M. (1989) Red coloration of tropical young leaves – a possible antifungal defence. *Journal of Tropical Ecology*, **5**, 293–300.

(1991) Comparison of herbivory and plant diseases in temperate and tropical broadleaf forests. In *Plant–Animal Interactions: Evolutionary Ecology in Tropical and Temperate Regions* (ed. P. W. Price, T. M. Lewisohn, G. W. Fernamdes & W. W. Benson). New York: John Wiley & Sons, pp. 25–49.

Coley, P. D. & Barone, J. A. (1996) Herbivory and plant defences in tropical forests. *Annual Review of Ecology and Systematics*, **27**, 305–335.

Coley, P. D., Bryant, J. P. & Chapin, F. S. (1985) Resource availability and plant antiherbivore defence. *Science*, **230**, 895–899.

Condit, R., Hubbell, S. P. & Foster, R. B. (1994) Density-dependence in two understorey tree species in a neotropical forest. *Ecology*, **75**, 671–680.

Connell, J. H. (1971) On the role of natural enemies in preventing competitive exclusion in some marine animals and in rain forest trees. In *Dynamics of Populations* (ed. A. J. den Boer & G. Gradwell). Washington: PUDOC.

(1978) Diversity in tropical rainforests and coral reefs. *Science*, **199**, 1302–1309.

Connell, J. H., Tracey, J. G. & Webb, L. J. (1984) Compensatory recruitment, growth, and mortality as factors maintaining rain-forest tree diversity. *Ecological Monographs*, **54**, 141–164.

Curran, L. M. & Webb, C. O. (2000) Experimental tests of the spatiotemporal scale of seed predation in mast-fruiting Dipterocarpaceae. *Ecological Monographs*, **70**, 129–148.

Curran, L. M., Caniago, I., Paoli, G. D. *et al.* (1999) Impact of El Niño and logging on canopy tree recruitment in Borneo. *Science*, **286**, 2184–2188.

Dominy, N. J., Lucas, P. W., Ramsden, L. W., Riba-Hernandez, P., Stoner, K. E. & Turner, I. M. (2002) Why are young leaves red? *Oikos*, **98**, 163–176.

Feeny, P. (1976) Plant apparency and chemical defences. In *Biochemical Interactions between Plants and Insects: Recent Advances in Phytochemistry* (ed. J. W. Wallace & R. L. Mansell). London: Plenum Press, pp. 1–40.

Gilbert, G. S. (1995) Rainforest plant diseases, the canopy understorey connection. *Selbyana*, **16**, 75–77.

Gould, K. S., Kuhn, D. N., Lee, D. W. & Oberbauer, S. F. (1995) Why leaves are sometimes red. *Nature*, **378**, 241–242.

Hammond, D. S. & Brown, V. K. (1998) Disturbance, phenology and life-history characteristics factors influencing distance/density-dependent attack on tropical seeds and seedlings. In *Dynamics of Tropical Communities* (ed. D. M. Newbery, H. H. T. Prins & N. D. Brown). Oxford: Blackwell, pp. 51–78.

Harborne, J. B. (1979) Flavonoid pigments. In *Herbivores: Their Interactions with Secondary Plant Metabolites* (ed. G. A. Rosenthal & D. H.

Janzen). New York: Academic Press, pp. 619–655.

Harper, J. L. (1989) Value of a leaf. *Oecologia*, **80**, 53–58.

Harms, K. E., Wright, S. J., Calderon, O., Hernandez, A. & Herre, E. A. (2000) Pervasive density-dependent recruitment enhances seedling diversity in a tropical forest. *Nature*, **404**, 493–495.

Horvitz, C. C. & Schemske, D. W. (2002) Effects of plant size, leaf herbivory, local competition and fruit production on survival, growth and future reproduction of a neotropical herb. *Journal of Ecology*, **90**, 279–290.

Howe, H. F., Schupp, E. W. & Westley, L. C. (1985) Early consequences of seed dispersal for a neotropical tree (*Virola surinamensis*). *Ecology*, **66**, 781–791.

Howlett, B. E. & Davidson, D. W. (2001) Herbivory on planted dipterocarp seedlings in secondary logged forests and primary forests of Sabah, Malaysia. *Journal of Tropical Ecology*, **17**, 285–302.

Hubbell, S. P. & Foster, R. B. (1986) Biology, change and history and structure of tropical rain forest tree communities. In *Community Ecology* (J. M. Diamond & T. J. Case). Cambridge: Harper and Row, pp. 314–330.

Hyatt, L. A., Rosenberg, M. S., Howard, T. G. *et al.* (2003) The distance dependence prediction of the Janzen–Connell hypothesis: a meta-analysis. *Oikos*, **103**, 590–602.

Ickes, K., Dewalt, S. J. & Appanah, S. (2001) Effects of native pigs (*Sus scrofa*) on woody understorey vegetation in a Malaysian lowland rain forest. *Journal of Tropical Ecology*, **17**, 191–206.

Janzen, D. H. (1970) Herbivores and the number of tree species in tropical forests. *American Naturalist*, **104**, 501–528.

Juniper, B. E. (1993) Flamboyant flushes: a reinterpretation of non-green flush colours in leaves. In *International Dendrological Society*

Yearbook. London: International Dendrological Society, pp. 49–57.

Kursar, T. A. & Coley, P. D. (1991) Nitrogen-content and expansion rate of young leaves of rain-forest species – implications for herbivory. *Biotropica*, **23**, 141–150.

(1992a) Delayed development of the photosynthetic apparatus in tropical rain-forest species. *Functional Ecology*, **6**, 411–422.

(1992b) Delayed greening in tropical leaves – an antiherbivore defence. *Biotropica*, **24**, 256–262.

(2003) Convergence in defence syndromes of young leaves in tropical rainforests. *Biochemical, Systematics and Ecology*, **31**, 929–949.

Lee, D. W., Brammeier, S. & Smith, A. P. (1987) The selective advantages of anthocyanins in developing leaves of mango and cacao. *Biotropica*, **19**, 40–49.

Lowman, M. D. (1985) Temporal and spatial variability in insect grazing of the canopies of five Australian rainforest tree species. *Australian Journal of Ecology*, **10**, 7–24.

(1992) Herbivory in Australian rain-forests, with particular reference to the canopies of *Doryphora sassafras* (Monimiaceae). *Biotropica*, **24**, 263–272.

Maiorana, V. C. (1981) Herbivory in sun and shade. *Biological Journal of the Linnean Society*, **15**, 151–156.

Marquis, R. J. (1984) Leaf herbivores decrease fitness of a tropical plant. *Science*, **226**, 537–539.

Marsh, C. W. & Greer, A. G. (1992) Forest land-use in Sabah, Malaysia – an introduction to Danum Valley. *Philosophical Transactions of the Royal Society of London B*, **335**, 331–339.

Massey, F. P. (2002) Environmental impacts on tropical rainforest dipterocarp tree seedling regeneration. Unpublished Ph.D. thesis, University of Sheffield.

Mooney, H. A. & Gulmon, S. L. (1982) Constraints on leaf structure and function in reference to herbivory. *Bioscience*, **32**, 198–206.

Osunkoya, O. O., Ash, J. E., Graham, A. W. & Hopkins, M. S. (1993) Growth of tree seedlings in tropical rain forests of North Queensland. *Journal of Tropical Ecology*, **9**, 1–18.

Pearson, T. R. H., Burslem, D. F. R. P., Goeriz, R. E. & Dalling, J. W. (2003a) Interactions of gap size and herbivory on establishment, growth and survival of three species of neotropical pioneer trees. *Journal of Ecology*, **91**, 785–796.

(2003b) Regeneration niche partitioning in neotropical pioneers: effects of gap size, seasonal drought and herbivory on growth and survival. *Oecologia*, **137**, 456–465.

Rausher, M. D. (1981) Host plant-selection by *Battus-philenor* butterflies – the roles of predation, nutrition, and plant chemistry. *Ecological Monographs*, **51**, 1–20.

Root, R. B. (1973) Organisation of a plant–arthropod association in simple and diverse habitats: the fauna of collards (*Brasicca oleracea*). *Oecologia*, **43**, 95–124.

Sagers, C. L. & Coley, P. D. (1995) Benefits and costs of defence in a neotropical shrub. *Ecology*, **76**, 1835–1843.

Shure, D. J. & Wilson, L. A. (1993) Patch-size effects on plant phenolics in successional openings of the southern Appalachians. *Ecology*, **74**, 55–67.

Sork, V. L. (1987) Effect of predation and light on seedling establishment in *Gustavia superba*. *Ecology*, **68**, 1341–50.

Strong, D. R., Lawton, J. H. & Southwood, R. (1984) *Insects on Plants*. Oxford: Blackwell Science.

Tahvanainen, J. O. & Root, R. B. (1972) The influence of vegetation diversity of the population ecology of a specialised herbivore *Phyllotreta cruciferae* (Coleoptera: Chrysomelidae). *Oecologia*, **10**, 321–346.

Terborgh, J. & Wright, S. J. (1994) Effects of mammalian herbivores on plant recruitment in two neotropical forests. *Ecology*, **75**, 1829–1833.

Wallin, K. F. & Raffa, K. F. (2001) Effects of folivory on subcortical plant defences: can defence theories predict interguild processes? *Ecology*, **82**, 1387–1400.

Webb, C. O. & Peart, D. R. (1999) Seedling density dependence promotes coexistence of Bornean rain forest trees. *Ecology*, **80**, 2006–2017.

Whatley, J. M. (1992) Plastid development in distinctively coloured juvenile leaves. *New Phytologist*, **120**, 417–426.

Whitmore, T. C. (1984) Gap size and species richness in tropical rain forests. *Biotropica*, **16**, 239–239.

Whitmore, T. C. & Brown, N. D. (1996) Dipterocarp seedling growth in rain forest canopy gaps during six and a half years. *Philosophical Transactions of the Royal Society of London B*, **351**, 1195–1203.

CHAPTER FIFTEEN

Multi-trophic interactions and biodiversity: beetles, ants, caterpillars and plants

D. K. LETOURNEAU
University of California, Santa Cruz
L. A. DYER
Tulane University

Biodiversity as a process

Biodiversity is dynamic, with species richness and composition changing over time and space in response to ecological, evolutionary and physical processes (e.g. Todd *et al.* 2002; Prieto *et al.* 2001; Hart *et al.* 1989; Menge *et al.* 1983). The scale of changes in biodiversity can be large, such as changes in geological time from plate tectonics (e.g. Crame 2001), variable, such as expected from global climate change, or as small as those responding to localized disturbance or heterogeneity (e.g. Clark *et al.* 1982; Louton *et al.* 1996; Jansen 1997). At each of these scales, biodiversity can be conceptualized as a *process* responsive to particular biotic and abiotic factors rather than as a static attribute of a particular location. The role of biotic interactions in maintaining biodiversity in tropical ecosystems, then, can be elucidated by studies that show how biotic factors can singly or in combination, directly or indirectly, change biological diversity in those systems. Using a model system in Costa Rica we will highlight indirect trophic interactions that cause changes in biodiversity within a rainforest food web.

Whereas theoretical studies in ecology and evolution are often on the mark with respect to their appreciation of dynamic processes, their application in efforts to conserve biodiversity has been subject to shortcuts. Specifically, over the last century, many conservation efforts have focused on saving particular species at certain locations. Broader goals now target particular habitats and hotspots of endemism. The perception of biodiversity as a process, though, may be useful as well. For example, biodiversity in Yellowstone National Park is very much different with the re-introduction of wolves, even though conservation of the park as a location and habitat has not changed. Also, if eutrophication proceeds in an otherwise protected lake, biodiversity is likely to change as a

Biotic Interactions in the Tropics: Their Role in the Maintenance of Species Diversity, ed. D. F. R. P. Burslem, M. A. Pinard and S. E. Hartley. Published by Cambridge University Press. © Cambridge University Press 2005.

process responding to nutrient-inputs through the system (Carpenter & Kitchell 1993). Thus, (1) biotic interactions are a key component of biological diversity; (2) biotic interactions can be direct or indirect; and (3) there may be feedback between biodiversity and biotic interactions in terms of cause and effect.

To test for mechanisms or to measure the strength of a particular factor in driving biodiversity, for example nutrient availability or predation or structural complexity (e.g. Kohn & Leviten 1976), the effects of that factor on biodiversity change can be isolated experimentally and examined. To better understand complex processes that promote or impede biodiversity, it is necessary to determine how different factors contribute, interact and combine to influence species richness and composition in natural and managed systems (Glasser 1979; Ghazoul & McLeish 2001; Wright & Duber 2001; Brehm & Fiedler 2003; Witman & Smith 2003). At present, we know a great deal about single and combined effects that directly affect biodiversity locally and globally (e.g. Hixon & Brostoff 1996; Siemann *et al.* 1998; Schulze *et al.* 2001). In fish exclusion/inclusion experiments, Hixon and Brostoff (1996) found that high algal diversity was maintained by damselfish grazing. Siemann *et al.* (1998) found that arthropod diversity increased significantly with plant diversity in experimental grasslands with replicated plots of plants sown with different numbers of plant species. Although herbivore diversity was related to plant diversity, the number of natural enemy species explained more of the variation in herbivore diversity.

Fewer field studies are available to assess indirect effects of biotic and abiotic factors on biodiversity. Although indirect trophic effects have been a topic of recent emphasis, most of these studies have taken the form of trophic-cascades research involving biomass of different trophic levels or changes in the abundance of organisms mediated by changes in abundance on higher or lower trophic levels. In this chapter, we (1) describe the notion of trophic cascades; (2) mention early intellectual roots of the theory; (3) briefly review experimental evidence for the phenomenon in terrestrial and aquatic systems, (4) present a case of four trophic level biotic interactions influencing biodiversity in a component rain forest community; and (5) discuss emerging studies on diversity cascades. We suggest that strong interactions through a complex food web resulting in changes in biodiversity are one way to conceptualize biodiversity as a process, rather than a state or characteristic of a location or habitat. We also hope to inspire more studies on diversity cascades, in which biodiversity changes through indirect, sometimes subtle, top-down and/or bottom-up driven processes.

Trophic cascades in aquatic and terrestrial systems

For decades, ecologists have investigated the roles of top-down (predator-based regulation) and bottom-up (resource-based regulation) forces in structuring biological communities. The notion of trophic cascades involves indirect effects of top-down or bottom-up forces, such that a change in one trophic level causes a change on a non-adjacent trophic level. Indirect effects include changes in biomass, abundance, productivity (Hairston *et al.* 1960) or biotic diversity (Paine 1966). Trophic cascades theory has early roots in the classic paper by Hairston *et al.* (1960) which argues that 'the world is green' because predators regulate herbivores, allowing higher plant productivity in the community. Smith (1966) and Fretwell (1977) generalized these ideas to predict that primary producers will be resource-limited in communities that contain an odd number of functional trophic levels and consumer-limited in habitats that contain an even number of trophic levels. Complementary bottom-up hypotheses follow the basic laws of thermodynamics: biomass of herbivores and primary and secondary carnivores is dependent on total primary productivity and attenuates as energy is lost through transfer up the trophic chain (Lindeman 1942; Slobodkin 1960). Hunter and Price (1992) described bottom-up cascades as interactions that cascade up trophic webs to determine species diversity and population dynamics at higher trophic levels.

Although some of the early conceptual papers focused on community structure in terrestrial habitats, experimental manipulations of consumers in freshwater and marine systems formed the initial basis for demonstrating such cascading effects (Carpenter *et al.* 1990; Brett & Goldman 1996). In the 1990s, the number of studies increased by an order of magnitude (Persson 1999), providing more support for the role of indirect effects in structuring terrestrial communities. Top-down trophic cascades, in which plant biomass and/or tissue lost to herbivores has been modified by the regulatory action of predators, have been demonstrated in a range of terrestrial systems (e.g. Atlegrim 1989; Tscharntke 1992; Rosenheim *et al.* 1993; Marquis & Whelan 1994; Schmitz 1994; Carter & Rypstra 1995; Walker & Jones 2001). A recent study by van Bael *et al.* (2003) showed that indirect effects of bird predation defended tropical forest trees from arthropod damage in the canopy where foliage productivity is high. However, the relative importance of top-down and bottom-up trophic cascades in structuring aquatic and terrestrial communities is still debated, with conflicting conclusions derived from meta-analyses of trophic cascades (incidence and/or strength) research by Schmitz *et al.* (2000), Halaj and Wise (2001) and Shurin *et al.* (2002).

Recent studies have focused on how top-down and bottom-up cascades may work in concert to structure communities (Oksanen 1991; Schmitz 1992; Leibold 1996); vary over space, time and taxa (Power 2000; Schmitz & Sokol-Hessner 2002); act on components of complex food webs (Polis & Strong 1996; Persson 1999; Strong *et al.* 1999); and maintain heterogeneity and biodiversity (Hunter & Price 1992; Terborgh 1992; Dunne *et al.* 2002; Dyer & Stireman 2003) in complex ecosystems. The two field studies we discuss below touch on each of the aspects listed above, and concern primarily the latter question – how trophic cascades affect biotic diversity. Although various definitions are given or assumed in the literature (Hunter 2001), here we use the term diversity cascade (Dyer & Letourneau 2003), defined as an indirect effect on biodiversity that causes or is a consequence of changes on non-adjacent trophic levels. Diversity cascades can go up (resource-driven changes) or down (predator-driven changes) (Hunter & Price 1992).

Direct and indirect trophic effects on biodiversity

Changes in biodiversity can be mediated via direct trophic interactions. For example, the species diversity of predatory birds may depend, to some extent, on the availability of food resources for their prey (German & Chacon de Ulloa 1997), or arthropod predator diversity might follow herbivore diversity, which is generated by plant diversity in the habitat (Saiz et al. 2000). Predators certainly have strong population-level effects on their prey, which can manifest as indirect effects on non-prey species that cause changes in biodiversity. Such keystone predators exemplify the key role of indirect biotic interactions and biodiversity (Paine 1966; Spiller & Schoener 1990). But indirect effects on species richness and composition are more difficult to detect than direct impacts of, say, predators on their prey, or the effect of light levels on a local flora. Therefore, the importance of indirect and/or cascading effects in creating, maintaining or changing biodiversity remains in the realm of speculation, especially in complex, tropical ecosystems (e.g. Covich 1976; Rathcke & Price 1976; Thomas 1989; Sime & Brower 1998; Roldan & Simonetti, 2001 and see Chalcraft & Resetarits 2003). Some recent studies suggest that indirect trophic effects on biodiversity, however, can be large and/or common, even in complex tropical communities (e.g. Terborgh et al. 2001; Schoener & Spiller 1996; Dyer & Letourneau 2003; Letourneau et al. 2004). Indirect, trophically mediated effects on biodiversity may not be the rule in tropical systems (e.g. Ley et al. 2002). However, such biotic interactions are worth exploring because they demonstrate unpredictable consequences of discrete changes in the abundance or diversity of distant functional groups.

Trophic cascades in the *Piper* ant-shrub system in Costa Rica

We have documented a series of biotic interactions associated with *Piper cenocladum* C. DC. (Piperaceae), a neotropical rainforest understorey shrub. The *Piper* ant-plant system is highly manipulable, because a key predator species can be added or removed experimentally from the shrubs. These manipulations can be carried out under different conditions of plant-resource availability, because light and soil quality are heterogeneous in the forest understorey where these plants naturally occur. With multiple species interacting antagonistically or

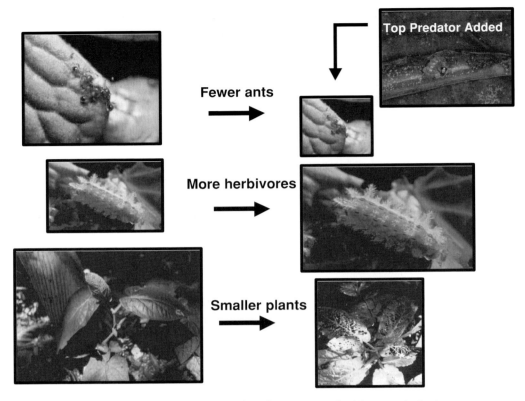

Figure 15.1 Experimental results demonstrating that, compared with control plants, shrubs with artificially introduced *Tarsobaenus letourneauae* beetle larvae had a significantly lower percentage of petioles occupied by *Pheidole bicornis* ants, a higher level of folivory and a smaller size (total leaf-area measurements were correlated to plant biomass). The size of each photograph is scaled to represent increases or decreases in the average abundance of top predators, predators, herbivores and plants.

beneficially across four trophic levels, *Piper* ant-plants support a complex arthropod community (see faunal description below).

In a previous study, we found that top predators in the system (clerid beetles) reduced the mean abundance of dominant plant-ants five-fold, and that herbivory increased three-fold, reducing the size of these tropical shrubs, on average, by half through indirect effects spanning three trophic levels (Letourneau & Dyer 1998a). The presence of a fourth trophic level predator, despite the attenuated strength of its effects, resulted in dramatically higher leaf-tissue loss than when ants were abundant and protecting the plant from susceptible herbivores (Fig. 15.1). In two simultaneous and complementary experiments run between 1994 and 1997, we also tested for effects of top predator beetles on animal and plant diversity. In these studies, we assessed the effects of top predator beetles on the diversity (H') of the endophytic fauna in *P. cenocladum*, ant species richness, herbivore richness or number of herbivore guilds, and species richness of understorey plants in the local understorey community.

Study sites

Our experimental studies on communities associated with *Piper* ant-plants (e.g. Letourneau & Dyer 1998a; 1999a, b; 2003; Letourneau *et al.* 2004) have been conducted in lowland, tropical wet forests of the La Selva Biological Station (owned and operated by the Organization for Tropical Studies). The La Selva sites, at ∼ 100 m elevation on the Caribbean side of Costa Rica, receive an average of 4200 mm annual rainfall. We have also conducted comparative studies (Letourneau & Dyer 1998b) at Corcovado National Park, Osa Peninsula, between the Pacific Ocean and the Golfo Dulce (< 100 m elevation, ∼ 3500 mm annual precipitation) at Carara Biological Reserve, on the Pacific Coast near Jaco (elevation ∼15 m, ∼ 2000 mm precipitation annually) and at Hacienda Loma Linda near the Panama border at Agua Buena, Coto Brus (at 1185 m, ∼ 3300 mm annual precipitation).

Plants

The distribution of *Piper* ant-shrubs in Costa Rica is from lowland to mid-elevation, in primary and late-successional secondary forests receiving at least 2000 mm rainfall per year. *Piper sagittifolium* C. DC., *P. obliquum* Ruiz & Pavon, and *P. fimbriulatum* C. DC. occur on the Pacific slopes of Costa Rica, and the range of *Piper cenocladum* C. DC. extends throughout the Atlantic lowlands. All but *P. sagittifolium* are closely related, as part of the *P. obliquum* complex (Burger 1971). *Piper* ant-shrubs have hollow stems when inhabited by *Pheidole bicornis* Forel ant colonies, because minor workers remove the pith. Single-celled, opalescent food bodies, rich in lipids and proteins, are produced on the adaxial side of the sheathing leaf bases (Risch & Rickson 1981). The sheaths serve as hollow petiole-like chambers that house ants. A rich and varied array of tree, liana and palm species make up the canopy over *Piper* ant-shrubs. Neighbouring understorey plants are even more diverse, including juveniles of all upper canopy species and hundreds of herbs, ferns, vines, shrubs and understorey trees (Dyer & Letourneau 1999a; Letourneau *et al.* 2004).

Herbivores

The herbivores most commonly found feeding on *Piper* ant-plants at our study sites in Costa Rica are lepidopterans, hymenopterans, orthopterans and coleopterans. Specialist Lepidoptera feeding on *P. cenocladum* foliage at La Selva include geometrids (e.g. *Eois dibapha*, *Eois apyraria*, *Epimecis* sp.) and skippers (e.g. *Quadrus cerealis*). Specialist stem-boring weevils (Coleoptera: Curculionidae: *Ambates* spp.) are common, as are leaf-feeding flea beetles (Coleoptera: Chrysomelidae: *Physimera* spp.) (L. A. Dyer, personal observation; Marquis 1991;

D. K. Letourneau, personal observations). Generalist lepidopteran folivores include an apatelodid *Tarchon* sp., a limacodid *Euclea plugma,* and a saturniid *Automeris postalbida* (L. A. Dyer, G. Gentry, H. Garcia Lopez, personal observation). Feeding damage on foliage from leafcutter ants (Hymenoptera: Formicidae: *Atta* spp.) and orthopterans (primarily Orthoptera: Tetigoniidae) is also common. Cumulative leaf area lost to arthropods is usually between 5% and 13% for *Piper* ant-shrubs.

Ant mutualists

Pheidole bicornis is a small, dimorphic species (Formicidae: Myrmicinae) that occupies all the *Piper* ant-plant species in Costa Rica (Wilson 2003). This ant both stimulates the production of food bodies by the plant and harvests these nutrient-rich, swollen epidermal cells for the brood. The ants feed primarily on food bodies produced by the plant, and carry into the plant interior soft-bodied insects, insect eggs and spores found on the leaves and flowers (Risch *et al.* 1977; Letourneau 1983; Letourneau 1998; Fischer *et al.* 2002). As omnivorous ants, the colonies kill insect eggs deposited on the plant and small larvae (2nd and 3rd trophic level), and harvest plant-produced food bodies (1st trophic level), but do not feed on photosynthetic or structural plant tissue. *Pheidole bicornis* occupies 85%–95% of *Piper* ant-plants in most forests studied in Costa Rica (Letourneau & Dyer 1998b). In addition to plant-ants, other ants, spiders and parasitoids inflict mortality on herbivores feeding on ant-plants (Chapter 17, this volume).

Top predators

Tarsobaenus letourneauae Barr = *Phyllobaenus* spp. (Coleoptera: Cleridae) occur as larvae inside the sheathing leaf bases of *P. cenocladum* where they feed on both ant brood and food bodies (Letourneau 1990; Fig. 15.1). *Dipoena* spp. spiders (Arachnida: Aranae), including *D. banksii = D. schmidti* (Araneae: Theridiidae), specialize on ants, and occur almost everywhere *Piper* ant-plants are found. They position themselves and capture ants at the entrance/exit hole on the outside of the hollow petiole chamber. Different species of congeneric clerids occur in *Piper* ant-plants on the Caribbean and Pacific sides of the Central Cordillera of Costa Rica, and several other species of predatory larvae or nymphs occur occasionally on *Piper* ant-plants, including owl fly larvae (Neuroptera: Ascalaphidae), and syrphid larvae (Diptera: Syrphidae: *Microdon* spp.).

Effect of top predators on animal biodiversity

To detect possible indirect effects of top predators on overall animal diversity in the endophytic community of *P. cenocladum* ant-plants we selected 80 naturally occurring shrubs on infertile ultisols at La Selva Biological Station and assigned three treatments in a factorial design: high-versus low-light environment,

Table 15.1 *Effects of top predator species on biodiversity parameters*

Factor	Animal diversity H'		Ant species richness (S)		Borer species richness (S)	
	$F_{1,26}$ value	P-value	χ^2-value	P-value	χ^2-value	P-value
Top predator	14.9	0.0003	14.05	0.0002	4.41	0.0395
Fertilizer	0.4	0.5		>0.05		>0.05
Light	0.04	0.9		>0.05		>0.05
Prd × Frt	0.8	0.4				
Frt × Lgt	1.4	0.2				
Prd × Lgt	0.01	0.9				
Prd × Frt × Lgt	2.0	0.2				

ANOVA (normal distribution) or chi squared analyses showing significant effects of top predator treatment on biodiversity parameters under conditions of high and low light with and without fertilizer added to individual *P. cenocladum* shrubs. Treatments are described in the text. Prd = top predator; Frt = fertilizer; Lgt = light.

fertilizer additions versus no fertilizer, and introduction of top predators versus no top predator (for details, see Dyer and Letourneau 2003). Hollow stems and petiole chambers of all shrubs were initially inhabited by *Pheidole bicornis* ant colonies. After 15 months, we dissected all 77 surviving *Piper cenocladum* shrubs and collected the entire endophytic fauna (43,188 individuals in the phyla Arthropoda, Annelida, Crustacea and Nematoda) to determine the number of invertebrate species and their relative abundance for each shrub. More than 50 species of animals were found living within the petiole chambers and stems of the harvested shrubs. Seven species of ants accounted for the majority of shrub inhabitants. *Pheidole bicornis* ants were the most abundant, ranging from 0 to over 2500 individuals per shrub. The other six ant species contributed over 1,000 individuals, usually inhabiting plants or portions of plants not occupied by *Ph. bicornis*. The remaining individuals (969) were annelids, nematodes, crustaceans and other arthropods, including collembolans, dipterans, coleopterans and at least 17 mite species.

Top predators significantly increased animal diversity (H') within the shrubs (Table 15.1). Shrubs with top predator additions had twice the animal diversity, on average, of shrubs without additions of the predatory clerid beetle (Fig. 15.2). Fifteen months of clerid beetle additions reduced the abundance of the dominant ant species, on average, to one fourth of its abundance in control plots with undisturbed colonies of *Ph. bicornis*. Ant species richness was higher when shrubs contained top predators (Fig. 15.2) because of the colonization of ants other than the dominant species (*Ph. campanae, Ph. susannae, Ph. specularis, Ph. ruida, Hypoponera* sp., *Neostruma myllorhapha*). Ant species richness was

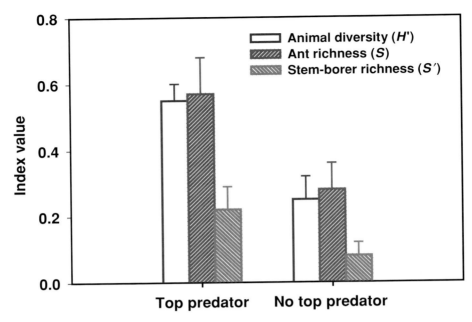

Figure 15.2 On average, compared with control plants, animal diversity (Shannon–Weiner Index), ant species richness and stem-borer species richness were significantly greater in shrubs with artificially introduced *Tarsobaenus letourneauae* beetle larvae.

significantly and positively correlated with animal species richness (Pearson's $R = 0.313$, $P = 0.0063$). The average species richness of stem borers per shrub was also greater in shrubs treated with top predators than with no predator additions (Fig. 15.2). No significant effect of fertilizer, light, or significant interaction among those factors was detected for overall animal diversity (H'), number of ant species or number of borer species.

To test for the effects of top predators on the diversity of consumers feeding (primarily externally) on *P. cenocladum* shrubs, we re-analysed data that were initially collected to test for differential effects of plant-ants on types of herbivory experienced by the shrubs. On each of the 77 harvested plants described above, we categorized and measured (e.g. leaf tissue lost to chewing insects) or counted (e.g. petiole galls) the types of feeding found on each leaf. The eight identifiable categories were lepidopteran damage, orthopteran damage, beetle damage, mid-rib miner damage, stem-borer damage, leafcutter ant damage, galls and other herbivore damage. Miner, borer and gall damage was determined to be present or absent on each shrub, and leaf-chewer categories were determined to be present if 1% or more of the shrub's foliage was damaged. The average shrub displayed damage by at least two of these eight different categories of herbivores ($\bar{X} = 2.2 \pm 0.15$ SE types). Given this rather crude assessment of the diversity of exploitation experienced by these shrubs, and the fact that some of leaves were present before the experiment began, we were surprised to find that top predators significantly affected the range of herbivore damage to shrubs (ANOVA, $F_{1,72} = 12.04$, $P = 0.0009$). Whereas shrubs with active ant colonies had on average 1.7 ± 0.2 SE types of damage, shrubs that had been inhabited by top predators had 2.7 ± 0.19 SE types of damage, on average. There was no significant effect of fertilizer or light, and no significant interactions among any of the predator or plant-resource factors.

Effect of top predators on plant biodiversity

In the understorey of tropical forests, there are many abiotic and biotic factors that maintain high alpha diversity. Most of these factors must act in some way to minimize competitive exclusion (Wright 2002) (Chapter 13, this volume). *Piper cenocladum* is a good candidate for studies of competitive interactions because it is a dominant shrub in the understorey, which facilitates its possible direct competition with other understorey plants. It also attracts a diverse array of generalist herbivores, which allows for indirect competition with plants that are attacked by these same herbivores (Strong *et al.* 1984). If the survival or reproductive success of *P. cenocladum* in the forest understorey affects other plant species through competitive or positive interactions, then effects of the predatory beetles on *Piper* shrubs (described above) potentially extend to the plant community through plant–plant interactions. Insect–plant interactions within the understorey are also possible: given the high diversity of the wet-forest understorey, other plant species may also serve as hosts to the herbivores released by the presence of top predators (Fig. 15.3). Indeed, most of the lepidopteran larvae that feed on *P. cenocladum* are generalists on plants in the Piperaceae (geometrids, limacodids and hesperiids) as well as plants in several other families (apatelodids). Thus, the effect of beetles could potentially cascade to the local plant community via species that are alternative hosts of the generalist herbivores (Dyer & Letourneau 1999a).

To examine the contribution of *P. cenocladum* to plant-diversity dynamics, changes in species richness of understorey plants were monitored over 20 months in response to direct and indirect (via herbivores) competitive interactions of this understorey shrub. Control (or 'direct competition with *Piper*') plots included naturally high densities of *P. cenocladum*, all shrubs of this species were removed from 'low-competition' plots, and herbivore-laden fragments were added to 'indirect-competition' plots. Since the experiments were under different conditions of plant-resource availability, we were able to consider variation in top-down forces as a function of bottom-up heterogeneity (Hunter & Price 1992).

In June 1994, we established 24 10-m by 10-m plots with *P. cenocladum* in 12 sites with infertile soils (ultisols) and 12 sites with fertile soils (inceptisols) at La Selva. These two soil locations were treated as two separate experiments,

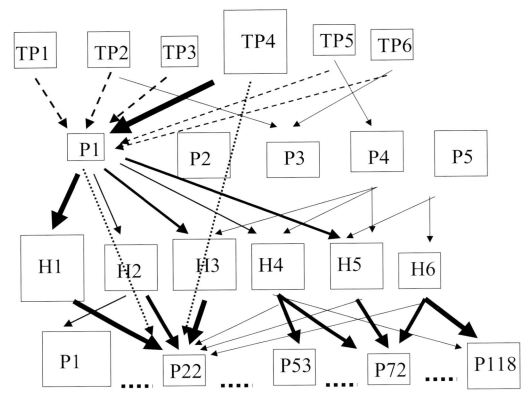

Figure 15.3 Suggested explanation for indirect effects of a single top predator on understorey plant diversity in a food web consisting of top predators (TP1 = *Microdon* syrphid fly, TP2 = ascalaphid, TP3 = *Dipoena* spider, TP4 = *Tarsobaenus* clerid beetle, TP5 = generalist ant, TP6 = generalist spider), mesopredators (P1 = *Pheidole bicornis* ants, P2 = other *Pheidole* ants, P3 = *Hypoponera* sp., P4 = *Neostruma myllorhapha*, P5 = parasitoid wasps), herbivores (H1 = *Eois* caterpillar, H2 = *Quadrus* caterpillar, H3 = *Automeris* caterpillar, H4 = *Tarchon* caterpillar, H5 = *Euclea* caterpillar, H6 = *Atta* ants) and plants (P1 to P118 are different plant species found in the understorey associated with *Piper cenocladum* (P22)), showing how some plant species can drop out of the sample because of high herbivory levels after ecological release.

since rich and poor soils were in distinct habitats (separated by about 1 km) that were not interspersed, and any attempts to analyse them together would rely on pseudoreplication. The fertile inceptisols (immature soils with poorly developed lower horizons) and the relatively infertile ultisols (highly weathered, low stock of basic cations) are the richest and poorest soils, respectively, that support natural stands of *P. cenocladum* in La Selva forests. All 24 plots contained *P. cenocladum* shrubs before any manipulations were applied. The plots were haphazardly assigned to three treatments for both experiments:

Indirect competition ('IC') plots: We selected four plots from each soil type and planted 10 small *P. cenocladum* cuttings (with three leaves per cutting) per plot. Six months after the transplants were established, we added early instars of *T. letourneauae* larvae to the plots every 2–4 months (one beetle per plant). The range of *P. cenocladum* cuttings plus shrubs in a plot was 12–16.

No competition ('NC') plots: We selected four plots from each soil type and completely removed all *P. cenocladum* shoots (range: 12–20 shrubs) from the plots by cutting the shrubs at the stem–soil interface. Roots of these plants were left in the soil.

Direct competition ('DC') plots: For both soil types, the remaining four plots were left unmanipulated. The range of *P. cenocladum* shrubs in the plots was 8–14.

We demarcated a 2-m by 4-m subplot in the centre of all plots, tagged every plant in the subplot between the heights of 20 cm and 2 m, and identified all tagged plants to either species or genus and morphospecies. Total available nitrogen and phosphorus were determined from three soil samples per plot. Soils analyses were conducted in the laboratory of R. Sanford using standard methods (e.g. Kauffman *et al.* 1993) and a mean value per plot for each nutrient was calculated.

Species richness, along with measures of herbivory and abundance of all plants in the plots, were recorded immediately before treatments were applied and subsequently at approximately 3-month intervals for 20 months. The mean levels of herbivory recorded on all plants in the subplots (at 3-month intervals) were based on visual inspection, such that herbivore damage was categorized as high (at least one leaf with greater than 50% damage) or low (all leaves with less than 50% damage). We examined the effect of nitrogen, phosphorus, light availability and herbivory on mean plant species richness and abundance, using multiple regressions. The effects of our competition manipulations on the change in species richness (i.e. number of species at the last sampling date versus the first) were tested using analysis of variance (ANOVA). Because of the intensive nature of identifying thousands of small plants to species, we were unable to achieve high sample sizes, so two components of our statistical analyses were unconventional. First, for all tests, we applied a liberal alpha of 0.1 because of low power and high probability of Type II error. Second, because of insufficient degrees of freedom, we did not use initial species richness as a covariate and final species richness as a dependent variable.

As expected, the plots in this lowland wet forest supported high alpha diversity of understorey plants, with 142 species of plants in 48 families identified in our plots. Initial plant species richness in 2-m by 4-m subplots was higher in the poor-soil experiments (17.3 $\pm$ 1.8 SE species per subplot) compared to the rich-soil sites (13 $\pm$ 0.7 SE species per subplot). Plants with high levels of herbivory

Table 15.2 *Predictors of species richness and abundance ranked by standardized parameter estimates from the species-richness model*

Independent variable	Richness parameter	Abundance parameter
Nitrogen	2.6**	2.3*
Herbivory	2.0*	−0.03
Phosphorus	−1.3	−1.5
Light	−0.80	−1.8

Tests for parameter estimates significantly different from 0 are $P < 0.1$ (*) and $P < 0.05$ (**) based on t values and 1 d.f. A significant positive parameter estimate indicates a positive effect of the independent variable on the response variable. An insignificant parameter estimate indicates no effect.

were also more frequent in the rich-soil (33% of all plants) versus poor-soil (9% of all plants) sites. We cannot attribute these differences between sites to the systematic effects of soil versus chance differences, since the experiments (i.e. soil sites) were not replicated. Results from the poor-soil sites did not reveal any associations between the response variables and herbivory, light or nutrients (species richness: $F_{4,7} = 0.7$, $P = 0.6$, $r^2 = 0.3$; abundance: $F_{4,7} = 0.2$, $P = 0.9$, $r^2 = 0.1$), nor did the competition treatments have significant effects on species richness ($F_{2,9} = 1.6$, $P = 0.3$).

For the rich-soil sites, the combination of predictor variables explained 72% of the variation in plant species richness, with soil nitrogen and herbivory being the strongest predictors of richness ($F_{4,7} = 4.4$, $P = 0.04$, $r^2 = 0.72$; Table 15.2). The predictors also explained 68% of the variation in total plant abundance, but soil nitrogen was the only variable with a significant parameter estimate ($F_{4,7} = 3.7$, $P = 0.07$, $r^2 = 0.68$; Table 15.2). Understorey plant species richness increased as nitrogen availability increased and herbivory decreased. Plant abundance also increased with nitrogen availability but did not change with herbivory. The effects of phosphorus and light on species richness and abundance contributed little to the overall regressions (Table 15.2). If herbivory does cause lower species richness, by killing seedlings and maintaining superior competitors, then the IC plots should have resulted in the lowest species richness. Overall, the plots increased in species richness in 22 months, but adding *P. cenocladum* plants with beetles acted as a classic indirect competition treatment ($F_{2,9} = 4.9$, $P = 0.04$), with species in these plots declining by 4 ± 1.5 SE per plot after 20 months. In contrast, DC and NC plots experienced increases of 11.8 ± 6.2 SE and 8.5 ± 1.6 SE species per plot. Patterns of herbivory on all of these plants support the notion that our IC plots were sources of herbivores for other species of plants, with 45% of understorey plants exhibiting high herbivory in IC plots versus 25% of understorey plants in the other plots. Herbivore damage may be a strong determinant of survival for seedlings or young plants (Letourneau *et al.* 2004) and may also affect asexual reproduction (Dyer *et al.* 2004) (Chapters 13 and 14, this volume).

The bottom-up results in the rich soils are consistent with diversity theory: increases in plant resources should cause greater diversity (Wright 2002). The fact that the poor-soil experimental plots had more species than the rich-soil plots is not necessarily inconsistent with this pattern, since the sites were not replicated and there were many other differences between these sites, including species composition, land-use history and water content (D. K. Letourneau and L. A. Dyer, unpublished data). The top-down results in rich soils suggest a couple of mechanisms. First, *P. cenocladum* shrubs may exclude competitive weedy plant species, allowing for maintenance of higher species richness. Thus, pulling *P. cenocladum* out of NC plots resulted in a smaller increase in species than seen in the DC plots (which contained ant-defended *P. cenocladum* shrubs). Second, herbivore-susceptible *P. cenocladum* fragments appeared to be important sources of seedling mortality for other species of plants, leading to an overall decline in species richness over the 22 months. Survival rates of smaller plants, between 20 cm and 60 cm, were lower on plants in the high damage category (Dyer & Letourneau 1999a). It is not surprising that a combination of top-down (herbivory) and bottom-up (soil nutrients) forces are important determinants of plant diversity, but it is impressive that these two factors can account for almost three-quarters of the variation in diversity at one site but are not associated with diversity at a nearby site with different soils and species assemblages.

Direct and indirect effects of a keystone predator

The experimental addition of top predator beetles in *P. cenocladum* ant-shrubs increased overall animal diversity of the endophytic community and reduced local understorey plant species richness. Apex predators had a direct, positive effect on ant species diversity within the shrubs. Indirect effects on herbivores included an increase in stem-borer diversity and a significantly greater number of types of herbivore damage on individual shrubs with top predator beetle additions. Further, top predator suppression of *P. bicornis* ants released herbivores that are likely to be responsible for reducing plant diversity in the local understorey plant community (Fig. 15.3). Putting the experimental results into the context of top-down 'diversity cascades' (Dyer & Letourneau 2003), the patterns of community change indirectly caused by *T. letourneauae* beetles in the IC plots provide further support for the notion that this beetle acts as a keystone predator (Paine 1966). A remarkable difference in this case is that the effects of beetle predators extended three trophic levels below the predator as opposed to effects restricted to the next lower trophic level. Changes in diversity on distant trophic levels in response to an elevated density of clerid beetles demonstrate its significance as an *indirect* keystone predator (*sensu* Schmitz 2003). It is also worth noting that the beetles acted as 'diversity-reducing' predators at the plant level (Schoener & Spiller 1996), indicating that the *P. bicornis* ants are 'diversity-enhancing' predators for understorey vegetation. Thus, our data support the hypothesis that numbers of trophic levels, particularly three versus four levels, can affect an entire community by altering populations, animal biodiversity and ultimately plant species richness.

Trophic cascades, diversity cascades and ecosystem function

Biodiversity changes over geological, evolutionary and ecological time scales as a dynamic process responding directly and indirectly to abiotic conditions and biotic interactions. Our studies on a component community involving approximately 100 species of organisms that interact with *Piper cenocladum* shrubs within a tropical rainforest reserve provide a mesocosm-level illustration of how indirect biotic interactions, interacting with local abiotic conditions, can drive changes in biodiversity. Trophic cascades research has provided important lessons about the magnitude of indirect effects of predators on plant productivity – effects that are not predictable or explainable with a restricted view of predator–prey or herbivore–plant interactions. Likewise, research on the role of plant diversity in ecosystem function has offered insights on the lack of redundancy of different plant species in terms of optimal nutrient-cycling and habitat productivity. Duffy (2002) strongly encourages the integration of theoretical and empirical approaches used by community ecologists studying trophic cascades and conservation biologists examining the role of biodiversity in ecosystem function. A small bridge between these separate research traditions is beginning to emerge as theorists consider two levels of biodiversity in natural communities: species diversity within trophic levels and trophic-level diversity within ecosystems (Paine 2002; Schmitz 2003). Schmitz (2003) and Dyer and Letourneau (2003) have shown that indirect trophic effects can also manifest as 'biodiversity cascades' and 'diversity-cascades,' respectively. Viewing biodiversity as a process may provide insights into biodiversity conservation and promote multi-trophic level concepts in ecosystem function (e.g. Estes 1995; Dirzo & Miranda 1990; but see cautionary note in Brewer *et al.* 1997). Indeed, conservation efforts based on land acquisition or preservation of charismatic species could be misguided if complex processes of change in biodiversity are not also taken into account (Micheli *et al.* 1989). The cascading changes in animal and plant diversity caused by our re-distribution of a small predatory beetle in the understorey of a rain forest are surprising, and warn of cumulative effects from small perturbations. It is possible that tropical biodiversity even in relatively intact communities could be threatened by the slow erosion of complex, tight interactions among species.

Acknowledgements

This research was supported by grants from the National Science Foundation, University Research Expeditions Program (UREP) and the University of California to D.K.L. The Organization for Tropical Studies provided logistical support, sites and facilities. G. Vega, R. Krach, C. Squassoni, R. DiGaudio, T. Haff, H. Kloeppl, J. Sorensen, S. Klas, A. Barberena, A. Lewis, A. Shelton, K. Harvey and a team of UREP volunteers assisted in the field and laboratory A. Agrawal, D. Clark, R. Marquis, T. Schoener and M. Power provided helpful suggestions on earlier drafts of the research results. M. Fitzsimmons inspired our discussion of biodiversity as a process.

References

Atlegrim, O. 1989. Exclusion of birds from bilberry stands: impact on insect larval density and damage to the bilberry. *Oecologia* **79**:136–139.

Brehm, G. & K. Fiedler. 2003. Faunal composition of geometrid moths changes with altitude in an Andean montane rain forest. *Journal of Biogeography* **30**:431–440.

Brett, M. T. & C. R. Goldman. 1996. A meta-analysis of the freshwater trophic cascade. *Proceedings of the National Academy of Sciences* **93**:7723–7726.

Brewer, S. W., M. Rejmanek, E. E. Johnstone & T. M. Caro. 1997. Top-down control in tropical forests. *Biotropica* **29**:364–367.

Burger, W. 1971. Flora Costaricensis. *Fieldiana Botany* **35**:1–227.

Carpenter, S. C., T. M. Frost, J. F. Kitchell *et al.* 1990. Patterns of primary production and herbivory in 25 North American lake ecosystems. In J. Cole, S. Findlay & G. Lovett, eds., *Comparative Analyses of Ecosystems: Patterns, Mechanisms, and Theories.* New York: Springer-Verlag, pp. 67–96.

Carpenter, S. R. & J. F. Kitchell. 1993. *The Trophic Cascade in Lakes.* New York: Cambridge University Press.

Carter, P. E. & A. L. Rypstra. 1995. Top-down effects in soybean agroecosystems: spider density affects herbivore damage. *Oikos* **72**:433–439.

Chalcraft, D. R. & W. J. Resetarits. 2003. Predator identity and ecological impacts: functional redundancy or functional diversity? *Ecology* **84**:2407–2418.

Clark, D. B., C. Guayasamin, O. Pazmino, C. Donoso & Y. P. Devillacis. 1982. The tramp ant *Wasmannia auropunctata* – autecology and effects on ant diversity and distribution on Santa Cruz Island, Galapagos. *Biotropica* **14**:196–207.

Covich, A. 1976. Recent changes in molluscan species-diversity of a large tropical lake (Lago de Peten, Guatemala). *Limnology and Oceanography* **21**:51–59.

Crame, J. A. 2001. Taxonomic diversity gradients through geological time. *Diversity & Distributions* **7**:175–189.

Dirzo, R. & A. Miranda. 1990. Contemporary neotropical defaunation and forest structure, function and diversity – a sequel to John Terborgh. *Conservation Biology* **4**:444–447.

Duffy, J. E. 2002. Biodiversity and ecosystem function: the consumer connection. *Oikos* **99**:201–219.

Dunne, J. A., R. J. Williams & N. D. Martinez. 2002. Network structure and biodiversity loss in food webs: robustness increases with connectance. *Ecology Letters* **5**:558–567.

Dyer, L. A., G. Gentry & M. A. Tobler. 2004. Fitness consequences of herbivory: impacts on asexual reproduction of tropical rainforest understorey plants. *Biotropica* **36**: 68–73.

Dyer, L. A. & D. Letourneau. 2003. Top-down and bottom-up diversity cascades in detrital vs. living food webs. *Ecology Letters* **6**:60–68.

Dyer, L. A. & D. K. Letourneau. 1999a. Relative strengths of top-down and bottom-up forces in a tropical forest community. *Oecologia* **119**:265–274.

1999b. Trophic cascades in a complex terrestrial community. *Proceedings of the National Academy of Sciences* **96**: 5072–5076.

Dyer, L. A. & J. O. Stireman. 2003. Community-wide trophic cascades and other indirect interactions in an agricultural community. *Basic and Applied Ecology* **4**:423–432.

Estes, J. A. 1995. Top-level carnivores and ecosystem effects: questions and approaches. In C. G. Jones & J. H. Lawton, eds., *Linking Species and Ecosystems*. New York: Chapman & Hall, pp. 151–158.

Fischer, R. C., A. Richter, W. Wanek & V. Mayer. 2002. Plants feed ants: food bodies of myrmecophytic *Piper* and their significance for the interaction with *Pheidole bicornis* ants. *Oecologia* **133**:186–192.

Fretwell, S. D. 1977. The regulation of plant communites by food chains exploiting them. *Perspectives in Biology and Medicine* **20**:169–185.

German, N. L. & P. Chacon de Ulloa. 1997. Diversity of understorey insects and insectivorous birds in disturbed tropical rainforest habitats. *Caldasia* **19**:507–520.

Ghazoul, J. & M. McLeish. 2001. Reproductive ecology of tropical forest trees in logged and fragmented habitats in Thailand and Costa Rica. *Plant Ecology* **153**:335–345.

Glasser, J. W. 1979. Role of predation in shaping and maintaining the structure of communities. *American Naturalist* **113**: 631–641.

Hairston, N. G., F. E. Smith & L. B. Slobodkin. 1960. Community structure, population control, and competition. *American Naturalist* **94**:421–424.

Halaj, J. & D. H. Wise. 2001. Terrestrial trophic cascades: how much do they trickle? *American Naturalist* **157**:262–281.

Hart, T. B., J. A. Hart & P. G. Murphy. 1989. Monodominant and species-rich forests of the humid tropics – causes for their co-occurrence. *American Naturalist* **133**:613–633.

Hixon, M. A. & W. N. Brostoff. 1996. Succession and herbivory: effects of differential fish grazing on Hawaiian coral-reef algae. *Ecological Monographs* **66**:67–90.

Hunter, M. D. 2001. Multiple approaches to estimating the relative importance of top-down and bottom-up forces on insect populations: experiments, life tables, and time-series analysis. *Basic and Applied Ecology* **2**:295–309.

Hunter, M. D. & P. W. Price. 1992. Playing chutes and ladders: heterogeneity and the relative roles of bottom-up and top-down forces in natural communities. *Ecology* **73**:724–732.

Jansen, A. 1997. Terrestrial invertebrate community structure as an indicator of the success of a tropical rainforest restoration project. *Restoration Ecology* **5**:115–124.

Kauffman, J. B., R. L. Sanford, D. L. Cummings, I. H. Salcedo & E. V. S. B. Sampaio. 1993. Biomass and nutrient dynamics associated with slash fires in neotropical dry forests. *Ecology* **74**:140–151.

Kohn, A. J. & P. J. Leviten. 1976. Effect of habitat complexity on population density and species richness in tropical intertidal predatory gastropod assemblages. *Oecologia* **25**:199–210.

Leibold, M. A. 1996. A graphical model of keystone predators in food webs: trophic regulation of abundance, incidence, and diversity patterns in communities. *American Naturalist* **147**:784–812.

Letourneau, D. K. 1983. Passive aggression: an alternative hypothesis for the *Piper–Pheidole* association. *Oecologia* **60**:122–126.

1990. Code of ant–plant mutualism broken by parasite. *Science* **248**:215–217.

1998. Ants, stem-borers, and fungal pathogens: experimental tests of a fitness

advantage in *Piper* ant-plants. *Ecology*
79:593–603.

Letourneau, D. K. & L. A. Dyer. 1998a.
Experimental test in lowland tropical forest
shows top-down effects through four
trophic levels. *Ecology* **79**:1678–1687.

1998b. Density patterns of *Piper* ant-plants
and associated arthropods: top predator
cascades in a terrestrial system? *Biotropica*
30:162–169.

Letourneau, D. K., L. A. Dyer & G. Vega C. 2004.
Indirect effects of a top predator on a rain
forest understorey plant community.
Ecology **85**:2144–2152.

Ley, J. A., I. A. Halliday, A. J. Tobin, R. N. Garrett
& N. A. Gribble. 2002. Ecosystem effects of
fishing closures in mangrove estuaries of
tropical Australia. *Marine Ecology: Progress
Series* **245**:223–238.

Lindeman, R. L. 1942. The trophic–dynamic
aspect of ecology. *Ecology* **23**:399–418.

Louton, J., J. Gelhaus & R. Bouchard. 1996. The
aquatic macrofauna of water-filled bamboo
(Poaceae: Bambusoideae: Guadua)
internodes in a Peruvian lowland tropical
forest. *Biotropica* **28**:228–242.

Marquis, R. J. 1991. Herbivore fauna of *Piper*
(Piperaceae) in a Costa Rican wet forest:
diversity, specificity, and impact. In P. W.
Price, T. M. Lewinsohn, G. W. Fernandes &
W. W. Benson, eds., *Plant–Animal Interactions:
Evolutionary Ecology in Tropical and Temperate
Regions*. New York: John Wiley & Sons,
pp. 179–208.

Marquis, R. J. & C. J. Whelan. 1994.
Insectivorous birds increase growth of
white oak through consumption of leaf
chewing insects. *Ecology* **75**:2007–2014.

Menge, B. A., L. R. Ashkenas & A. Matson. 1983.
Use of artificial holes in studying
community-development in cryptic marine
habitats in a tropical rocky intertidal
region. *Marine Biology* **77**:129–142.

Micheli, F., G. A. Polis, P. Dee Boersma *et al.*
1989. Human alteration of food webs.

In M. E. Soulé & K. A. Kohm, eds., *Research
Priorities for Conservation Biology*.
Washington, DC: Island Press.

Oksanen, L. 1991. Trophic levels and trophic
dynamics: a consensus emerging? *Trends in
Ecology and Evolution* **6**:58–60.

Paine, R. T. 1966. Food web complexity and
species diversity. *American Naturalist*
100:65–76.

2002. Trophic control of production in a
rocky intertidal community. *Science*
296:736–739.

Persson, L. 1999. Trophic cascades: abiding
heterogeneity and the trophic level concept
at the end of the road. *Oikos* **85**:385–397.

Polis, G. A. & D. R. Strong. 1996. Food web
complexity and community dynamics.
American Naturalist **147**:813–846.

Power, M. E. 2000. What enables trophic
cascades? Commentary on Polis *et al. Trends
in Ecology and Evolution* **15**:443–444.

Prieto, A. S., L. J. Ruiz, N. Garcia & M. Alvarez.
2001. Mollusc diversity in an *Arca zebra*
(Mollusca: Bivalvia) community, Chacopata,
Sucre, Venezuela. *Revista de Biologia Tropical*
49:591–598.

Rathcke, B. J. & P. W. Price. 1976. Anomalous
diversity of tropical ichneumonid
parasitoids: a predation hypothesis.
American Naturalist **110**:889–893.

Risch, S. J., M. McClure, J. Vandermeer & S.
Waltz. 1977. Mutualism between three
species of tropical *Piper* (Piperaceae) and
their ant inhabitants. **98**:433–444.

Risch, S. J. & F. R. Rickson. 1981. Mutualism in
which ants must be present before plants
produce food bodies. *Nature* **291**:149–
150.

Roldan, A. I. & J. A. Simonetti. 2001.
Plant–mammal interactions in tropical
Bolivian forests with different hunting
pressures. *Conservation Biology* **15**:617–623.

Rosenheim, J. A., L. R. Wilhoit & C. A. Armer.
1993. Influence of intraguild predation
among generalist insect predators on the

suppression of an herbivore population. *Oecologia* **96**:439–449.

Saiz, F., L. Yates, C. Nunez, M. Daza, M. E. Varas & C. Vivar. 2000. Biodiversity of the canopy arthropods associated to vegetation of the north of Chile, II region. *Revista Chilena de Historia Natural* **73**:671–692.

Schmitz, O. J. 1992. Exploitation in model food chains with mechanistic consumer-resource dynamics. *Theoretical Population Biology* **41**:161–181.

1994. Resource edibility and trophic exploitation in an old-field food web. *Proceedings of the National Academy of Sciences* **91**:5364–5367.

2003. Top predator control of plant biodiversity and productivity in an old-field ecosystem. *Ecology Letters* **6**:156–163.

Schmitz, O. J., P. A. Hamback & A. P. Beckerman. 2000. Trophic cascades in terrestrial systems: a review of the effects of carnivore removals on plants. *American Naturalist* **155**:141–153.

Schmitz, O. J. & L. Sokol-Hessner. 2002. Linearity in the aggregate effects of multiple predators in a food web. *Ecology Letters* **5**:168–172.

Schoener, T. W. & D. A. Spiller. 1996. Devastation of prey diversity by experimentally introduced predators in the field. *Nature* **381**:691–694.

Schulze, C. H., K. E. Linsenmair & K. Fiedler. 2001. Understorey versus canopy: patterns of vertical stratification and diversity among Lepidoptera in a Bornean rain forest. *Plant Ecology* **153**:133–152.

Shurin, J. B., E. T. Borer, E. W. Seabloom *et al.* 2002. A cross-ecosystem comparison of the strength of trophic cascades. *Ecology Letters* **5**:785–791.

Siemann, E., D. Tilman, J. Haarstad & M. Ritchie. 1998. Experimental tests of the dependence of arthropod diversity on plant diversity. *American Naturalist* **152**:738–750.

Sime, K. R. & A. V. Z. Brower. 1998. Explaining the latitudinal gradient anomaly in ichneumonid species richness: evidence from butterflies. *Journal of Animal Ecology* **67**:387–399.

Slobodkin, L. B. 1960. Ecological energy relationships at the population level. *The American Naturalist* **94**:213–236.

Smith, F. E. 1966. Population limitation in ecosystems. *Biometrics* **22**:960–975.

Spiller, D. A. & T. W. Schoener. 1990. A terrestrial field experiment showing the impact of eliminating top predators on foliage damage. *Nature* **347**:469–472.

Strong, D. R., J. H. Lawton & T. R. E. Southwood. 1984. *Insects on Plants: Community Patterns and Mechanisms*. Cambridge, Massachusetts: Harvard University Press.

Strong, D. R., A. V. Whipple, A. L. Child & B. Dennis. 1999. Model selection for a subterranean trophic cascade: root-feeding caterpillars and entomopathogenic nematodes. *Ecology* **80**:2750–2761.

Terborgh, J. 1992. Maintenance of diversity in tropical forests. *Biotropica* **24**:283–292.

Terborgh, J., L. Lopez, P. Nunez *et al.* 2001. Ecological meltdown in predator-free forest fragments. *Science* **294**:1923–1926.

Thomas, C. D. 1989. Predator–herbivore interactions and the escape of isolated plants from phytophagous insects. *Oikos* **55**:291–298.

Todd, J. A., J. B. C. Jackson, K. G. Johnson *et al.* 2002. The ecology of extinction: molluscan feeding and faunal turnover in the Caribbean Neogene. *Proceedings of the Royal Society of London B* **269**:571–577.

Tscharntke, T. 1992. Cascade effects among 4 trophic levels: bird predation on galls affects density-dependent parasitism. *Ecology* **73**:1689–1698.

van Bael, S. A., J. D. Brawn & S. K. Robinson. 2003. Birds defend trees from herbivores in a neotropical forest canopy. *Proceedings of*

the National Academy of Sciences **100**:
8304–8307.

Walker, M. & T. H. Jones. 2001. Relative roles of
top-down and bottom-up forces in
terrestrial tritrophic plant–insect
herbivore–natural enemy systems. *Oikos*
93:177–187.

Wilson, E. O. 2003. *Pheidole in the New World: a
Dominant, Hyperdiverse Ant Genus*. Cambridge,
MA: Harvard University Press.

Witman, J. D. & F. Smith. 2003. Rapid
community change at a tropical

upwelling site in the Galapagos Marine
Reserve. *Biodiversity and Conservation*
12:25–45.

Wright, S. J. 2002. Plant diversity in tropical
forests: a review of mechanisms of species
coexistence. *Oecologia* **130**:1–14.

Wright, S. J. & H. C. Duber. 2001. Poachers and
forest fragmentation alter seed dispersal,
seed survival, and seedling recruitment
in the palm *Attalea butyraceae*, with
implications for tropical tree diversity.
Biotropica **33**:583–595.

CHAPTER SIXTEEN

The trophic structure of tropical ant–plant–herbivore interactions: community consequences and coevolutionary dynamics

DOYLE McKEY
Université Montpellier 2
LAURENCE GAUME
Centre National de Recherches Scientifiques
CARINE BROUAT
Institut de Recherche pour le développement
BRUNO DI GIUSTO
Centre National de Recherches Scientifiques
LAURENCE PASCAL
Université Montpellier 2
GABRIEL DEBOUT
University of East Anglia
AMBROISE DALECKY
Institut National de Recherche Agronomique
MARTIN HEIL
Universität Duisburg-Essen

Introduction

The first part of this paper examines the consequences of an interlocking set of mutualisms, involving ants, plants, bacteria and phloem-feeding insects, for the structure and functioning of herbivore-based food webs in tropical communities. This part draws heavily from important recent work by Davidson and colleagues (Davidson 1997; Davidson *et al.* 2003) and extends their discussion of community-level implications of their findings. The second part explores how trophic interactions evolve when coevolution produces specialized symbiotic ant–plant mutualisms, and is based largely on our own work on interactions between ants and *Leonardoxa* myrmecophytes of African rainforests. The paper complements a recent general review of ant–plant protection mutualisms (Heil & McKey 2003).

Biotic Interactions in the Tropics: Their Role in the Maintenance of Species Diversity, ed. D. F. R. P. Burslem, M. A. Pinard and S. E. Hartley. Published by Cambridge University Press. © Cambridge University Press 2005.

Ant–plant–herbivore interactions and tropical food webs

How food webs function, and how trophic interactions shape communities, have long been central questions in ecology. Interactions between organisms at adjacent trophic levels – predators and prey, parasites and hosts – and competitive interactions among organisms at the same trophic level, all occupy major roles in theories to explain the great species richness and other traits of tropical forest ecosystems (Wright 2002). Following the lead of classic studies like those of Hairston *et al.* (1960) and Paine (1966), investigations of how communities function have increasingly taken into account not only these direct interactions, but also indirect interactions that extend across several trophic levels. Do natural enemies of herbivores have measurable impacts on fitness of individual plants, on relative abundance of plant species, on primary productivity or on plant species diversity? Conversely, does variation in productivity or other processes at the first trophic level, through its effects on herbivores, influence processes and patterns at higher trophic levels? The frequency of such 'trophic cascades' in different ecosystems (Schmitz *et al.* 2000; Halaj & Wise 2001; Shurin *et al.* 2002), and the respective roles of 'bottom-up' and 'top-down' forces in the web of indirect effects (Dyer & Letourneau 2003; Terborgh *et al.* 2003; van Bael *et al.* 2003), are questions widely debated today (Letourneau & Dyer, Chapter 15, this volume).

Compared with investigations of competition and predation as structuring forces, studies that address the roles of mutualisms in shaping communities are underrepresented (Wimp & Whitham 2001). A host of mutualisms form an important part of the structural glue of tropical forests. Among these are transport mutualisms, which permit, for example, the persistence of outcrossing plants at very low densities (Janzen 1971) and the escape of animal-dispersed seeds from density-dependent mortality of seeds and seedlings (Wright 2002); nutrition mutualisms, which allow plants to assimilate minerals rapidly in leached tropical soils and can affect community composition in complex ways (Kiers *et al.* 2000); and protection mutualisms, in which plants attract natural enemies of herbivores and as a result suffer less damage. This last category is particularly underrepresented in community-level studies. Protection mutualisms can by their very nature have far-reaching effects in food webs, because they involve not only direct (and often indirect) trophic exchanges between the two mutualists, but also interactions with at least one additional trophic level. Furthermore, selection acts on the mutualists to increase the strength of these interactions. Protection mutualisms may thus be particularly likely to generate effects that cascade well beyond the two mutualistic partners.

Ants are among the principal predators of arthropods in tropical forests (Novotny *et al.* 1999; Floren *et al.* 2002), and protection and nutritional mutualisms between ants and other organisms are a large part of the reason why. Ants participate in three kinds of mutualisms that are of key importance in determining their impact on the structure of ecological communities. First, many plants provide ants with direct rewards – energy-rich extrafloral nectar or pearl bodies – that encourage the presence and activity of these predators on the plant. Ants thus attracted can increase plant fitness by reducing damage caused by herbivores and/or pathogens. Second, many ants are engaged in protection mutualisms with herbivores, principally phloem-feeding hemipterans (formerly known as 'Homoptera' (von Dohlen & Moran 1995)) such as aphids, membracids, scale insects and mealybugs. The carbohydrate-rich excretions of these 'trophobionts' provide ants with an energy-rich food source. Ants also protect other myrmecophilous herbivores, most notably caterpillars of many lycaenid butterflies (Pierce *et al.* 2002). Relationships with lycaenids are important for some plants, but in terms of importance at the community level they are dwarfed by those with hemipterans. Associations between ants and their trophobionts can have widely varied effects on plants (Cushman & Addicott 1991; Oliveira & Del-Claro, this volume). In some cases, ant-tended trophobionts are probably the plant's principal herbivores and the effect on the plant is negative. In other cases, the protective benefits (e.g. against chewing insects) conferred by trophobiont-tending ants appear to outweigh the costs to the plant, resulting in a three-partner mutualism of ant, trophobiont and plant.

In all the above-mentioned kinds of mutualisms, the rewards produced for ants are carbohydrate-rich, with low contents of nitrogen and highly unbalanced amino acid compositions (Davidson & Patrell-Kim 1996). Profiting from these abundant but unbalanced food resources appears to require – or at least is greatly aided by – the involvement of ants in a third type of mutualism, with endosymbiotic microbes that help to repair such nutritional imbalances. Although little studied so far, ant–microbe mutualisms appear especially important for tropical arboricole ants, the ecological group of ants for which exudates derived directly or indirectly from plants are of greatest importance in the diet (Davidson *et al.* 2003). Ant-tended hemipterans are in turn dependent on their own microbial symbionts (Delabie 2001).

Thus ants, the most important predators of arthropods in tropical forest canopies, are largely sustained by an interlocking set of mutualisms. Without these mutualisms, food webs in tropical forest communities would probably be very different from those we know. Before developing this argument, we will briefly review what is known about the three types of mutualisms directly involving ants (see also Heil & McKey 2003).

Ant–plant protection mutualisms based on direct food rewards

Extrafloral nectary-bearing plants are diverse and abundant in many tropical ecosystems (Schupp & Feener 1991; Pemberton 1998; Rico-Gray *et al.* 1998), although their frequency varies substantially among biogeographical regions (Heil & McKey 2003), habitats (e.g. Cogni *et al.* 2003) and plant life forms (Bentley 1981; Blüthgen *et al.* 2000). A variety of functions have been proposed for extrafloral nectaries. These include nutritional benefits to plants that attract soil-nesting ants to their base, resulting in better soil quality (Wagner 1997), and the distraction of ants from visiting flowers (Wagner & Kay 2002) or tending hemipterans (Becerra & Venable 1989). However, the most frequently demonstrated benefit to plants is attraction of protective ants (recently reviewed by Heil & McKey (2003)). Herbivory induces increased rates of nectar production by existing nectaries (Heil *et al.* 2001) and can even increase the number of nectaries formed by the plant following damage (Mondor & Addicott 2003). Whereas most nectaries are fairly conspicuous structures, some are quite cryptic, and current surveys may thus underestimate their frequency. Most tropical biologists would probably be surprised, for example, to know that a plant as abundant and familiar as cassava, *Manihot esculenta* Crantz (Euphorbiaceae), secretes nectar at the base of petioles of young leaves (Bakker & Klein 1992); externally evident nectar-secreting structures are absent. The frequency of another type of direct food reward, pearl bodies – lipid- or carbohydrate-rich glandular trichomes harvested by ants (Schupp & Feener 1991) – is probably even more seriously underestimated, because unlike most nectary-bearing plants, once these ant-attractant food bodies are harvested, no evidence of any secretory structure is left behind. Only Schupp and Feener (1991) appear to have adequately surveyed the frequency of glandular trichomes, by examining plants grown in a screened growing house. Even in this study, the proportion of species whose glandular trichomes actually function as food bodies was not determined.

Within habitats, the abundance and diversity of ants at nectaries show marked variation among plant species (Hossaert-McKey *et al.* 2001) and across seasons (Rico-Gray *et al.* 1998). The importance of ants as protective agents is also likely to be highly variable among nectary-bearing plants. Ant protection appears to be a particularly important component of some plant strategies. An example is the mutualism between ants and the African wild yam *Dioscorea praehensilis* Benth. (Dioscoreaceae), in which ants provide protection during the most vulnerable stage of its unusual growth cycle (di Giusto *et al.* 2001). The plant is a perennial geophytic forest-canopy vine, which dies back to a large underground tuber at the beginning of the annual dry season. Near the end of the dry season, fuelled by tuber reserves, a single new stem repeats the climb from the ground to the forest canopy. Success depends on completing this trip as rapidly as possible, because only upon reaching the favourable light conditions of the canopy does the plant produce leaves that can restock tuber reserves and support flowering and fruiting. Rapid height growth is at such a premium that the stem does not branch until it reaches the canopy, a trip that requires at least two months. During this period, any herbivore attack of the plant's single apical meristem imposes a very great cost, because it can result in the loss of weeks of production by the entire aerial system, which lives only a single growing season. Cataphylls at the stem apex bear nectaries that attract ants to the meristem continually throughout this trip (di Giusto *et al.* 2001).

The *Dioscorea* example may be extreme (among opportunistic, non-symbiotic mutualisms) in the plant's likely dependence on ant protection. However, although the impact of ants attracted to extrafloral nectar (EFN) and other direct food rewards varies considerably, much work (most recently reviewed by Heil and McKey 2003) demonstrates that they often significantly reduce levels of attack by herbivores. Ants attracted to EFN and other direct rewards are important to the ecology of many tropical forest plants.

Mutualisms between ants and hemipteran trophobionts

Community-level comparisons of herbivory have usually focused on chewing insects (e.g. Leigh & Windsor 1982; Coley & Barone 1996), the feeding activity of which is comparatively easy to quantify. Sucking insects have been neglected, although their community effects – thanks to widespread mutualism with ants – may dwarf those of chewing insects (Davidson *et al.* 2003). The biology of phloem-sucking hemipterans includes two key features that have led to the repeated evolution of protection mutualisms with ants (Delabie 2001). First, their food contains carbohydrates and water in excess. With little metabolic cost to themselves, they can excrete large amounts of an energy-rich liquid reward that is highly prized by ants. Second, their feeding behaviour ties them to a sedentary lifestyle. Their stylets are often deeply inserted into plant tissues, and reaching a suitable sap source takes minutes or hours. This makes active escape from predators difficult. Even if phloem-sucking insects could escape, the time lost afterwards in re-inserting the stylet would lower their foraging efficiency. Phloem-feeding hemipterans have thus evolved a wide range of anti-predator defences; mutualism with tending ants figures prominently among these.

Protected by ants from many predators and parasitoids, phloem-sucking insects often reach high densities. These densities should be vastly underestimated by canopy-fogging 'knock-down' experiments, because these insects die with their stylets inserted and are not 'knocked down'. In the few studies based on direct observation in the canopy, numbers of ant-tended hemipterans rival those of ants themselves (Blüthgen *et al.* 2000; Dejean *et al.* 2000; Blüthgen & Fiedler 2002). In terms of biomass, they are thus likely to be the second most important insect group in the canopy, after ants.

For ants, association with phloem-sucking insects provides access to an abundant supply of energy-rich plant-derived resources, even on plants that do not produce direct rewards. Some hemipterans are facultatively associated with ants, whereas others are obligate myrmecophiles. Even in the latter case, associations are rarely specific, a given hemipteran often forming associations with several ant species, and vice versa (e.g. Blüthgen *et al.* 2000; Dejean *et al.* 2000). Different ant species also have different effects on the species richness, density and distribution of hemipterans (Itioka & Inoue 1999). Although plants vary in both their suitability as hosts for phloem-sucking insects (Gullan 1997; Blüthgen & Fiedler 2002) and the quality of the honeydew produced (Gullan 1997; Delabie

2001; Davidson *et al.* 2003), the host-plant ranges of many hemipteran tro-phobionts are likewise often broad (Delabie 2001; Blüthgen & Fiedler 2002). Thus, association with hemipterans makes ants not only (quasi-)herbivores, but *generalist* herbivores. Dominant ant species can thus obtain energy-rich food from a large proportion of individual plants in the community, even in species-rich tropical forests. Furthermore, while some ant-tended hemipterans are largely restricted to young growth (Blüthgen & Fiedler 2002), many others (e.g. coccids and stictococcids in African forests; D. McKey, personal observations) are capable of feeding on mature twigs. Thus, despite seasonal and other sources of varia-tion, in tropical forest ecosystems honeydew is often much more continuously available in space and time than is extrafloral nectar, production of which is usually restricted to particular developmental stages of leaves and other organs and to particular times of day. The phenology of nectar production also varies among plant species.

Ants also harvest hemipteran trophobionts for protein (Carroll & Janzen 1973; Gullan 1997; Delabie 2001). Sap-sucking hemipterans present important con-trasts with other potential sources of protein as regards their pattern of avail-ability in space and time, as we have already seen in the case of energy. Most phytophagous insects in tropical ecosystems use ephemeral resources such as young leaves, flowers and developing fruits (Coley & Barone 1996). Their den-sities thus vary markedly in time, dependent on the phenology of their host plants. Sap-feeders may thus provide ants with a more regular and predictable supply of protein than they can gain from insects depending on other resources.

Little is known in quantitative terms about ant predation on their tropho-bionts (Davidson *et al.* 2003). In those systems that have been studied, the fre-quency of predation often appears to depend on the balance between availability to ants of carbohydrates and other resources (Gullan 1997; Delabie 2001). When food of ants is supplemented with more sugar, workers prey more frequently on the aphids they tend (Offenberg 2001). For this and other reasons, the mutu-alistic benefits to trophobionts are often dependent on their density (Breton & Addicott 1992; Morales 2000), which appears often to be regulated by ants (e.g. Larsen *et al.* 2001).

Do ant–hemipteran associations have positive or negative effects on plants? In most cases, the net effects of these associations on plants in tropical ecosys-tems are unknown (Oliveira & Del-Claro, Chapter 17, this volume). Judging from temperate-zone examples (e.g. Karhu 1998; Karhu & Neuvonen 1998), these effects may be complex. In addition to the negative effects of loss of resources to sucking insects, there is also significant potential for the transmission of diseases by ant-tended hemipterans (Delabie 2001). However, little appears to be known about this aspect of the associations outside of agricultural environments. Positive effects include not only ant protection against other herbivores, mostly chewing insects, but also – in temperate-forest systems – improved physico-chemical prop-erties of soil in ant nests (Wagner 1997), which are often located at the bases of trees harbouring large populations of hemipterans. Except for epiphytes (for which ant nests appear to be of widespread importance as substrates for establishment (Longino 1986)), the latter type of benefit is likely to be less important in tropical forests, where most ants that forage in tree crowns are also tree-nesting. If trophobionts are maintained at high densities and few chewing insects are present, or if ants do not effectively protect against those chewing insects that are present (Mackay 1991), the costs of ant–hemipteran associations will outweigh their benefits. On the other hand, if the protection ants provide against other insects is greater than the cost imposed by the hemipterans they maintain, then plants, ants and hemipterans are engaged in a three-partner mutualism. It is certain that ant-tended hemipterans play a currently underestimated role in tropical forest ecosystems (Davidson *et al.* 2003). But whether, in each case, we have underestimated the cost of herbivory to plants or instead the investment of plants in protection mutualisms and indirect ('biotic') defence, remains to be established.

Nutritional mutualisms involving bacteria

The crucial role of endosymbiotic bacteria in the nutritional ecology of phloem-sucking hemipterans, including ant-tended trophobionts, is well known (Delabie 2001). Much less known, but probably also crucial to the ecology of tropical herbivore food webs (Davidson *et al.* 2003), are nutritional mutualisms between bacteria and tree-dwelling ants. Several groups of tropical arboricolous ants harbour symbiotic bacteria. At least some species of *Camponotus*, a widespread genus well represented among tropical arboricolous ants, have intracellular bacterial endosymbionts (Schröder *et al.* 1996), although to our knowledge tropical tree-dwelling *Camponotus* have not been studied. Other genera harbour bacteria in the lumen of the gut, sometimes in specialized structures. These include *Cephalotes* (Jaffé *et al.* 2001), some *Tetraponera* spp. (Billen & Buschinger 2000; van Borm *et al.* 2002), and *Cataulacus* (Caetano *et al.* 1994). All of these are groups that consume large quantities of nectar and honeydew, or whose diets also include unusual components such as epiphyllous non-vascular plants or wind-borne pollen and fungal spores. Both direct observations and isotopic data indicate that they have largely plant-derived diets, likely to be characterized by low protein content and unbalanced amino acid composition (Davidson *et al.* 2003). Although the functions of the bacterial symbionts in each case remain to be elucidated, patterns strongly suggest that they play roles in repairing these nutritional imbalances. For example, the localization of bacterial symbionts in *Tetraponera*, near complexes of Malpighian tubules, suggests that they may function in internal nitrogen-recycling (van Borm *et al.* 2002). Unravelling the functioning of ant–bacteria mutualisms is one of the key tasks facing ant ecologists over the next few years (Bourseaux-Eude & Gross 2000).

The trophic structure of opportunistic ant–plant interactions

Among this diversity of resources – direct (EFN and pearl bodies) and indirect (honeydew) energy-rich rewards, hemipteran trophobionts and other foliage-dwelling arthropods consumed as prey – what are the respective roles of each in ant diets? How important to tropical tree-dwelling ants are food sources that are not even indirectly derived from the plants they patrol, such as pollen, fungal spores and epiphylls? In short, what is the trophic structure of opportunistic ant–plant interactions, and what are the implications of this structure for understanding community processes?

First, the relative importance of these different resources appears to vary among ant groups. Dolichoderines, for example, frequently tend large aggregations of trophobionts on plants, while also using extrafloral nectar (Davidson *et al.* 2003; Davidson *et al.* 2004). In contrast, most species of formicines are 'leaf foragers' that actively and solitarily scour vegetation for diverse resources. Although many formicines also tend trophobionts, and some are specialists in that activity, they appear to figure more prominently among consumers of EFN. Formicines also account for most of the observations indicating exploitation of unusual plant-derived resources such as pith and epiphylls (lichens and perhaps other groups). Pollen and perhaps fungal spores are suspected to be important in the nutritional ecology of many pseudomyrmecines and cephalotines (Baroni Urbani & Andrade 1997).

Plant-derived food sources supply most of the energy in the diets of all these ants. Studies of the natural abundance of stable isotopes indicate that tropical tree-dwelling ants also obtain a large part of their nitrogen from plant sources (Davidson *et al.* 2003), despite the low contents of nitrogen in most of the plant-derived foods. According to the hypothesis developed by Davidson (1997; Davidson *et al.* 2003), the dependence of tropical tree-dwelling ants on energy-rich, nitrogen-poor resources derived from plants is a key feature of their ecology, with enormous consequences for the communities in which they are key predators. Davidson's analysis (Davidson 1997; Davidson *et al.* 2003) of the consequences of the energy-rich, nitrogen-poor subsidy from plants in the ecology of tropical tree-dwelling ants is a *tour de force*, and constitutes one of the most thoroughly developed terrestrial examples of an emerging field, ecological stoichiometry (Elser *et al.* 2000), that has so far been nourished principally by examples from aquatic ecosystems (e.g. Gaedke *et al.* 2002). With abundant energy but limited by nitrogen, tropical tree-dwelling ants have evolved bacterial symbioses to compensate for nutritional imbalance; strategies of defence, physiology and morphology that spare nitrogen; and predation strategies that lavishly use plant-derived energy to gain animal protein (Davidson 1997). Experimental studies are beginning to supply evidence that honeydew- and nectar-feeding ants do in fact behave as if they were limited by nitrogen (Kay 2002).

Ant–plant interactions and the structure and functioning of tropical food webs

How have these interlocking mutualisms contributed to making ants such important predators in tropical forest canopies? Davidson (1997) has argued that the abundance of nectar, pearl bodies and hemipteran honeydew has allowed tropical tree-dwelling ants to evolve energetically costly prey-foraging strategies. Fuelled by plant-derived rewards and adapted to nitrogen-poor diets, ants can reach higher densities, and maintain prey species at lower densities, than if they were solely dependent on animal prey. Plants should thereby benefit from decreased levels of herbivory. Davidson (1997) postulates in effect that top-down trophic cascades, from effects of ants on herbivores to the effects of herbivores on plants, are a dominant feature of the community ecology of tropical forests. (It should be noted that 'top-down' and 'bottom-up' effects are in this case thoroughly conflated, as the traits of predators are themselves dependent on the traits of plants.)

Such effects have been more or less well documented in the simplified interaction webs involving numerous myrmecophytes and their host-specific ants e.g. Letourneau & Dyer 1998; Letourneau & Dyer, this volume) and in several opportunistic interactions centred on EFN-bearing plants (reviewed by Heil & McKey (2003)). Do they occur at the community level? According to some current theory (Polis & Strong 1996), in highly diverse communities the effect of predators should have low penetration to lower trophic levels, because as predators reduce densities of some herbivores, other herbivores not affected by these predators take their place, so that the overall effect on plants is limited. However, there is evidence for top-down trophic cascades driven by other groups of predators in tropical forests (van Bael *et al.* 2003; Terborgh *et al.* 2003). Furthermore, as mentioned above, ants are both generalist predators and generalist herbivores; few herbivores might be unaffected by ants, and the penetration of their effects to lower trophic levels may thus remain high.

Let us assume for the moment that tree-nesting ants do generate effects that cascade from herbivores to plants, and that such effects are community-wide. This leads to the question of just how far to other community components the impact of ants may extend. As dominant predators, do they structure other guilds of natural enemies of foliage-dwelling arthropods? Almost 30 years ago, Rathcke and Price (1976) noted that parasitoids in tropical ecosystems should experience high rates of indirect predation by ants that eat parasitized hosts, and postulated that this could explain the apparently anomalous latitudinal gradient in diversity of parasitoid hymenoptera (with a peak not in the tropics, but at mid latitudes, for groups such as Ichneumonidae (Janzen 1975)). It soon became clear that tropical diversity in many parasitoid groups, as in other insects, had been underestimated, and this hypothesis was put aside. It is now being re-examined in more precise incarnations. Does indirect predation by ants shape the structure of parasitoid guilds in tropical ecosystems? Ant predation on larval insects could lead to greater representation of life-history strategies featuring rapid development in very early host stages (e.g. egg parasitoids (Gaston *et al.* 1996; Kruger & McGavin 1998)), or favour traits that otherwise reduce the probability of indirect predation. Another idea to explain parasitoid diversity gradients, the 'nasty-host hypothesis' (Gauld *et al.* 1992), could also be seen in a new light. According to this hypothesis, tropical phytophagous insects have fewer parasitoids than do temperate-zone insects because they are more often chemically defended. If ant predation is an important contributor to the selection pressure that has favoured the frequent evolution of toxicity in tropical phytophagous insects, this could be seen as another important community-level consequence of the mutualisms that sustain ants.

The impact of ants might extend even further. The steep decline of ant diversity and abundance with elevation on tropical mountains (Janzen 1973) could contribute to explaining why small mammals, which potentially compete with ants for food, increase in diversity with elevation in the tropics (Samson *et al.* 1997; Heaney 2001). Not surprisingly, some hemipteran groups rich in ant-tended species also show sharp diversity gradients with elevation (Olmstead & Wood 1990).

All these postulated effects of ants on the ecology of tropical forest communities rest, of course, on the hypothesis that tree-nesting ants, in partnership with plants and hemipteran trophobionts, collectively reduce the densities of other foliage-dwelling arthropods. To our knowledge, this hypothesis has not been tested at the community level. Doing this would require the collaboration of teams of myrmecologists and other entomologists. A large-scale manipulative study would be technically difficult, but if current hypotheses are on the right track, the results of such a study should be striking.

Evolution of trophic structure in symbiotic ant–plant mutualisms

In all tropical regions, opportunistic interactions between plants and plant-foraging ants have repeatedly given rise to tighter associations, in which ant colonies reside permanently in structures of plants. Hemipteran trophobionts are frequent third partners in such symbiotic associations. In symbiotic associations, each participant becomes a more predictable feature of the other participants' environment, creating greater opportunities for (co)evolutionary specialization (Davidson & McKey 1993).

As symbiotic ant–plant mutualisms coevolve, new selection pressures acting on resource flows can modify the trophic structure of interactions in several ways. These evolutionary changes can affect the overall rate of resource flow from plants to ants, as well as the kinds of rewards that are offered and the chemical composition of each. These two aspects, quantitative and qualitative changes, will be considered in turn.

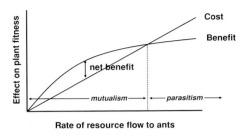

Figure 16.1 Graphical model illustrating the continuum between mutualism and parasitism in the effects of symbiotic ants on fitness of the host (adapted from Fonseca (1993)). Under certain circumstances, ants may have an interest in maintaining a rate of flow of plant resources to them that is greater than that optimal for the plant (where net benefits to the plant are maximized). See text for explanation.

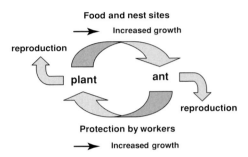

Figure 16.2 The potential for conflict between the two partners in horizontally transmitted symbiotic ant–plant mutualisms. The cycle of mutualistic benefits is provided by each partner's investment in the growth of the other. However, resources for reproduction are drawn from the same limited pool as growth, resulting in the potential for evolutionary conflicts between mutualists. Adapted from Gaume (1998).

Overall rate of resource supply

Resource flow from plant to ant should increase with coevolutionary specialization, because plants that maintain larger resident ant colonies are better protected (Rocha & Bergallo 1992), and ants that better protect the plant reap more benefits. The lineages that have been studied conform to this expectation, with specialized myrmecophytes supplying food resources to their resident ants at higher rates than do their less specialized congeners to opportunistic ants (reviewed by Heil & McKey 2003), and specialized plant-ants providing more effective protection than do opportunistic ants (Fiala *et al.* 1994; Nomura *et al.* 2000).

The cost to the plant of supporting its ants increases proportionally to investment: producing more resources entails greater costs. However, benefits to the plant plateau. At some point, supporting more ants confers little or no increased protection (Fonseca 1993). At very high rates of resource supply, costs may outweigh benefits (Fig. 16.1). Somewhere below this point, there is a rate of resource supply at which net benefit is maximized. If the interests of the plant and its ant colony were entirely congruent, this rate optimal for the plant would also be optimal for the ants. However, because ant–plant symbioses are horizontally transmitted (plant seeds and ant foundresses disperse independently), neither individual partner has an interest in the other's reproduction. Because resources allocated to growth (which increases mutualistic benefits) and reproduction of the tree or colony come from the same limited pool, this can lead to a conflict of evolutionary interests (Fig. 16.2). Selection on each partner should favour adaptations that increase its own benefits. These may come at the expense of partner reproduction. *Allomerus* plant-ants, for example, protect *Cordia nodosa* host-plants but castrate floral buds; resources thereby diverted from repro-duction are channelled into increased growth, benefiting the ants (Yu & Pierce 1998). In turn, plants can modify the distribution of inflorescences (Raine *et al.* 2002), or of domatia (Izzo & Vasconcelos 2002), to limit contacts between ants and flowers and thereby presumably reduce such parasitism. The outcome of such battles for control over resource flow may be a coevolutionary tug-of-war with constantly shifting position. Alternatively, one of the partners may evolve an 'unbeatable strategy' that stabilizes the system – at least until it is beaten by the next surprising counter-adaptation.

Qualitative changes

As seen in a preceding section, food webs in opportunistic, non-symbiotic ant–plant interactions can be complex, with energy coming from nectar, honeydew and sometimes other plant-derived resources, and protein (and other nutrients) being supplied to some extent by these same resources, supplemented by bodies of hemipteran trophobionts and of insects captured on plants. Are food webs in specialized, symbiotic ant–plant mutualisms of comparable complexity?

The evolutionary conflicts developed above can shape not only the rate at which plants supply resources to ants, but also the kinds of resources offered. This is because some kinds of resources are more easily controlled by ants and others by the plant (Gaume *et al.* 1998; Itino *et al.* 2001). The clearest example is the contrast between rewards produced directly by the plant (EFN, pearl bodies) and those produced indirectly (hemipteran honeydew). Plants control their pro-duction of direct rewards, responding to variation in both resource availability (Linsenmair *et al.* 2001) and the likelihood of herbivore attack (Heil *et al.* 2001), and sometimes to the presence of the mutualist ant (Risch & Rickson 1981). Parasitic manipulation of plant production of direct rewards (Letourneau 1990) appears to be rare. In contrast, ants can more easily control rates at which plant resources are channelled to them in the form of honeydew (and hemipteran bodies), by regulating the number of trophobionts (among other mechanisms). While severe overexploitation of the plant by trophobiont-tending ants would lead to greatly reduced growth of the ant colony as well as of the plant (Fonseca 1993), traits leading to overexploitation that simply reduces plant reproduction would not be counter-selected. Also, because there is always a non-zero proba-bility that the plant, or the ant colony, will die from causes independent of the efficacy of mutualistic exchange (e.g. physical disturbances), even exploitation levels that lead to modest reduction of host-plant growth could be favoured, if they increase the likelihood that the ants can reproduce before such an event occurs. If trophobiont-tending ants do exploit their hosts at levels greater than optimal for the latter, selection could favour plants that evolve mechanisms to reduce the rate of resource flow through trophobionts and increase the depen-dence of their symbiotic ants on direct rewards, whose production is more easily controlled by the plant.

Is there any evidence for such a scenario? Several observations suggest it. First, myrmecophyte symbioses often seem to originate from three-partner ant–plant–hemipteran mutualisms. The high frequency of such tripartite systems in myrmecophytes is striking; systems including indirect rewards for ants are much more numerous than those in which only direct rewards are produced (Davidson & McKey 1993; Oliveira & Del-Claro, this volume). One plausible interpretation of this finding is that ant association with trophobionts is often necessary, because the latter assure a regular and predictable, if not always rich, supply of protein that is usually not provided by direct rewards, at least those produced by unspecialized plants. This could have been a particularly crucial advantage at the outset of symbiosis, when predatory ants first became permanent residents of a single individual plant. Unable to buffer temporal variation in the abundance of phytophagous insects by hunting on many different plants, predators would have been dependent on hemipteran-supplied protein. Alternatively, these interactions may have been parasitic from the start, with ants tending hemipterans in stems while foraging little on plant surfaces. However, if this were the case, it would be difficult to understand why selection favoured the retention – and elaboration – of plant structures for housing ants.

Second, some of the most specialized myrmecophyte systems do not include trophobionts. In these, plants have often evolved direct rewards much richer in nitrogen than those produced by other myrmecophytes (e.g. Heil *et al.* 1998; Fischer *et al.* 2002). Did these symbioses originate via other pathways, not involving trophobionts, or have trophobionts been lost in these systems and replaced by direct food rewards? If the latter is the case, is there any evidence that evolution was driven by the conflicts described above?

The *Leonardoxa* case

Leonardoxa, a genus of small to medium-sized trees found in rain forests from northern Gabon to south-eastern Nigeria, offers an opportunity to examine these hypotheses, because very closely related taxa present dramatic contrasts in both myrmecophytic specialization and the structure of ant–plant–herbivore food webs. As currently circumscribed, the genus consists of a single polytypic species, *L. africana*, comprising of four vicariant taxa. These have been described as subspecies (McKey 2000), but they are clearly on different adaptive trajectories and are perhaps best considered as species *in statu nascendi*. One of these taxa, *L. a. gracilicaulis*, found in northern Gabon and southern Cameroon, is not a myrmecophyte. Like plants of many related genera, it has foliar nectaries that attract a diversity of opportunistic ants. Exclusion experiments show that the protective effect of these ants is at best weak (Gaume 1998). The three other taxa are all myrmecophytes bearing swollen twigs that function as ant domatia. Subspecies *rumpiensis* is an intriguing tree of submontane forests in the West Cameroonian Dorsal. Thus far little studied, it is inhabited by a diversity of poorly known ant species. The two other taxa are lowland-forest myrmecophytes, each associated with a different, closely related formicine ant. Subspecies *letouzeyi* is found north of the West Cameroonian Dorsal, in the Cross River area straddling Cameroon and extreme south-eastern Nigeria. It is associated with the ant *Aphomomyrmex afer*. Subspecies *africana* occurs south of the Dorsal, and is restricted to a narrow band of coastal forest southward to near the border between Cameroon and Equatorial Guinea. It is associated with the ant *Petalomyrmex phylax*. Both of these formicine genera are monotypic, and the two ants are sister species among extant taxa (Chenuil & McKey 1996). *Aphomomyrmex* shares numerous traits with related genera such as *Cladomyrma* and *Myrmelachista*, and is not host-specific, being known from at least one other host (*Vitex*, in the Lamiaceae (formerly placed in Verbenaceae)). In contrast, *Petalomyrmex* has evolved several morphological specializations, all of which appear to be adaptations to its sole host, *L. a. africana*, which is the most highly specialized myrmecophyte in the *L. africana* complex. The specializations of *Petalomyrmex* include most notably the strongly dorsiventrally flattened alates, matched to the slit-shaped prostoma (and *Petalomyrmex* entrance holes dug at the prostoma) of its host (Brouat *et al.* 2001).

Another difference between the two ants is that *Aphomomyrmex*, like most related genera, tends hemipteran trophobionts in the domatia of its hosts, whereas *Petalomyrmex*, alone among studied members of the tribe Myrmelachistini in this respect, never tends trophobionts. Mapping of characters on phylogenies of ants (Chenuil & McKey 1996) and plants (Brouat *et al.* 2004) indicates that as the system specialized, the lineage leading to *Petalomyrmex* lost an ancestral association with hemipteran trophobionts, while the host of this ant, *L. a. africana*, evolved larger and more numerous foliar nectaries and presumably higher rates of supply of direct rewards to its mutualist ant.

What selective pressures could have driven these evolutionary changes in the trophic structure of the symbiosis? Three arguments suggest that they resulted from evolutionary conflicts between mutualists and are due to adaptations of the plant, which has thereby gained a greater degree of control of resource flow. First, there is no clear reason why ants should cease to tend hemipteran trophobionts when plants increase the production of direct rewards, if by tending they can gain additional resources. (However, hemipteran-tending might be strongly counter-selected in resident ants if hemipteran-transmitted pathogens reduce the growth or survival probability of the ants' sole individual host.) Second, the absence of hemipterans in ant associations with *L. a. africana* appears to be due to a trait of the plant. The inner walls of the stem domatia of this plant are covered by a layer of sclerenchyma (Brouat 2000) that may constitute a physical barrier to phloem-sucking insects. Third, field studies of the tripartite mutualisms involving *L. a. letouzeyi*, *Aphomomyrmex*, and its hemipteran trophobionts provide circumstantial evidence of the kinds of conflicts between ants and plants that are postulated to have favoured plant adaptations permitting increased control.

In a population of *L. a. letouzeyi* studied in Korup National Park, *Aphomomyrmex* was observed to tend two different hemipterans, with the coccid *Houardia abdita* being the sole or dominant trophobiont in about half the trees (= *Aphomomyrmex* colonies) studied and the pseudococcid *Paraputo anomala* the sole trophobiont in the other trees (Gaume *et al.* 1998). Among colonies overall, benefits to the tree of ant occupancy, as estimated by the amount of chewing-herbivore damage accumulated by leaves, increased with worker density in relation to number of leaves. However, comparing the two groups, this relationship was much stronger in trees whose ant colonies tended *Paraputo*. These results suggest that increasing the amount of resources supplied to ants resulted in increased protection, but that the strength of the relationship depended strongly on the identity of the third partner (Gaume *et al.* 1998). Not only did *Aphomomyrmex* confer greater benefits to the plant when it tended *Paraputo*, but the cost incurred by the plant may also be lower with this partner, if trophobiont biomass is correlated with the cost they impose. For comparable worker densities (number of workers in relation to domatia volume), dry biomass of associated trophobionts was between two and three times as great in trees whose colonies tended *Houardia* as in those with *Paraputo*. *Houardia* were dense and found in every domatium; *Paraputo* were sparse and mostly restricted to domatia in younger twigs.

Ant colonies tending *Houardia* thus appeared to impose greater costs and confer fewer benefits to their host plants than did those tending *Paraputo*. Their lower level of mutualism appears to pay off; for comparable colony size, production of alates was higher in colonies tending *Houardia* (Gaume & McKey 2002). Although the reasons for the differences in costs and benefits between colonies associated with different trophobionts must still be elucidated by experimental studies, observations suggest the following explanation (Gaume *et al.* 1998). Colonies tending *Houardia* depend on foliar nectaries for energy, exploiting the trophobiont for protein and using this abundant resource to increase the production of sexual brood. In contrast, colonies tending *Paraputo* exploit the honeydew they produce and are not dependent on foliar nectar. However, they apparently either do not consume *Paraputo* bodies or, if they do, gain insufficient protein from the low-density populations of this trophobiont. Limited by protein, they more actively patrol young leaves to hunt insect prey.

Plants that evolve adaptations favouring *Paraputo* over *Houardia* should have a selective advantage. Loss of nectaries would favour *Paraputo*, and this could explain why some individual *L. a. letouzeyi* lack foliar nectaries (McKey 2000). However, the plant has only limited control; *Houardia* remains in the system, and ant colonies that tend it require nectaries to reach the high density required for protective effect. Selection in opposing directions on nectary number could explain why this trait shows much greater among-individual variation in *L. a. letouzeyi* than in any of the other *Leonardoxa*, including the non-myrmecophyte *L. a. gracilicaulis* (McKey 2000).

Another possible plant response would be to evolve traits that exclude trophobionts altogether. By increasing the severity of nitrogen limitation, this could lead ants to intensify patrolling to capture insect prey, as we propose occurs when *Aphomomyrmex* tends *Paraputo*. However, temporal variation in herbivore abundance, coupled to the plant's phenology, may make complete dependence on such prey impossible for ants resident in a single host. Traits excluding trophobionts would thus usually be advantageous only if accompanied by the provision of direct rewards that satisfy all ant needs, including that for nitrogen. (This could be accomplished by an increase in the quality of direct rewards, by the reduction of ant requirements for nitrogen and other resources, or by some combination of the two.) Once direct rewards supply sufficient protein as well as energy, a situation results in which ants are likely to have more to gain by protecting the plant as a direct (and relatively constant and reliable) protein source than by using it to grow indirect protein sources (phytophagous insects), the abundance of which is highly variable in space and time. Control of resource flow by the plant thus has the effect of bringing the ants' interests closer to its own.

Another possible selective pressure favouring simplification of food webs in specialized ant–plant systems is the increased energetic efficiency that would result from elimination of a trophic level between plant and ant (Gaume *et al.* 1998; Itino *et al.* 2001). It is unclear how much the energetic efficiency of resource transfer to ants would be enhanced by eliminating hemipterans. This would depend in part on the extent of metabolic transformation of phloem sap before its excretion as honeydew.

Figure 16.3 summarizes our hypotheses about the evolution of trophic structure during specialization of ant–plant mutualism in *Leonardoxa*. Ants associated with the myrmecophilic *L. a. gracilicaulis* (Fig. 16.3a) eat foliar nectar, honeydew, hemipteran trophobionts and prey, in proportions that probably vary among species. *Aphomomyrmex*, symbiotic in *L. a. letouzeyi*, uses all of these resources, and probably epiphylls as well (D. McKey and L. Gaume, unpublished observations), like some other formicine ants (Davidson *et al.* 2003). When they tend *Houardia* (Fig. 16.3b), *Aphomomyrmex* obtain relatively more of their carbon from nectar and more of their protein from trophobionts; when they tend *Paraputo* (Fig. 16.3c), these ants obtain more carbon from honeydew and more nitrogen from chewing-insect prey. In both cases, ants use the plant to 'ranch' animal protein, either wild or domesticated. In contrast, our observations indicate that trophic structure in the highly specialized *L. a. africana* / *Petalomyrmex* symbiosis (Fig. 16.3d) is very simple, with ants obtaining both carbon and nitrogen mostly directly from the host. First, there is no indication that *Petalomyrmex* interacts in any way with epiphylls, which are much less commonly observed on *L. a. africana* than on *L. a. letouzeyi*. The latter grows in shadier, more humid forests; it also has tougher, perhaps longer-lived, leaves. Second, trophobionts are absent from

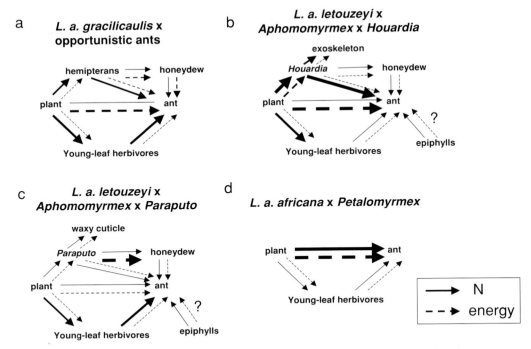

Figure 16.3 Summary of hypotheses about the evolution of trophic structure during the specialization of interactions between *Leonardoxa africana* and ant associates. a, The non-myrmecophytic, basal *L. a.* subsp. *gracilicaulis*, associated with a diversity of opportunistic ants. b, The 'transitional' myrmecophyte *L. a.* subsp. *letouzeyi*, associated with *Aphomomyrmex afer*, when colonies tend the coccid *Houardia abdita*. c, The same plant and ant combination, when colonies tend the pseudococcid *Paraputo anomala*. d, The highly specialized myrmecophyte *L. a.* subsp. *africana* associated with *Petalomyrmex phylax*, which does not tend trophobionts. Width of arrows indicates hypothesized relative magnitude of flows of nitrogen and energy in these systems.

the *L. a. africana/Petalomyrmex* system. Third, patrolling by *Petalomyrmex* workers is sufficiently intense that their main protective effect is probably to deter visits (and oviposition) by adult insects and to remove eggs and small immature stages. While *Petalomyrmex* workers do eat small caterpillars experimentally placed on the plant and dismember larger insects, transporting the pieces into domatia (L. Gaume, G. Debout and D. McKey, unpublished observations), the amount of protein they gain from herbivores they find under natural circumstances is probably limited.

According to these hypotheses, the adaptive function of patrolling young leaves has changed during evolutionary specialization of the system (Fig. 16.4). Opportunistic ants on *L. a. gracilicaulis* visit young leaves to hunt prey and to harvest foliar nectar. While nectaries are active only on mature leaves of *L. a.*

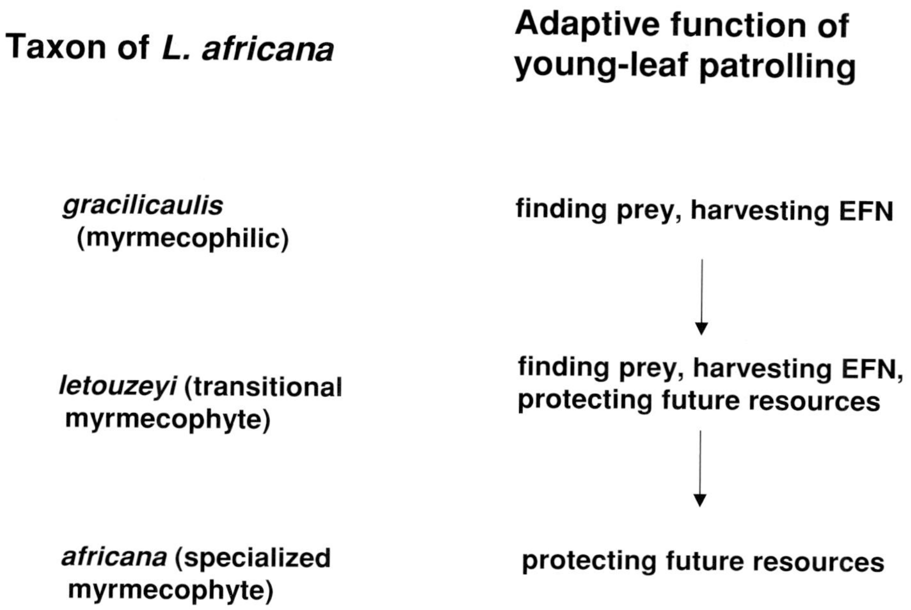

Taxon of *L. africana*	**Adaptive function of young-leaf patrolling**
gracilicaulis (myrmecophilic)	finding prey, harvesting EFN
	↓
letouzeyi (transitional myrmecophyte)	finding prey, harvesting EFN, protecting future resources
	↓
africana (specialized myrmecophyte)	protecting future resources

Figure 16.4 Summary of hypotheses about changes in the adaptive function of ant patrolling of young leaves during the evolutionary specialization of mutualism between *Leonardoxa* and its ant associates.

africana, young leaves also secrete nectar in *L. a. gracilicaulis* (Gaume 1998), as in most myrmecophilic plants. In the transitional myrmecophyte *L. a. letouzeyi*, *Aphomomyrmex* patrols young leaves to hunt prey (especially when associated with *Paraputo*, according to our hypotheses), but also to protect future resources and harvest nectar. In the highly specialized *L. a. africana*, *Petalomyrmex* appears to encounter little food on young leaves and patrols them to protect a future resource.

Preliminary data on natural abundance of nitrogen isotopes (A. Cason and D. McKey, unpublished) also indicate that *Petalomyrmex* obtains its nitrogen directly from plants. In general, the heavier isotope of nitrogen, ^{15}N, increases by 3–4‰ with each increasing trophic level (Scrimgeour *et al.* 1995). Values of ^{15}N for *Petalomyrmex* workers ranged from 1.2 to 5.6 ($n = 13$ colonies), values within the range of those obtained for ants that appear to derive most of their nitrogen from plant exudates (Davidson *et al.* 2003). Furthermore, ^{15}N values for *Petalomyrmex* colonies were correlated with those for their individual hosts and enriched on average by 2.9‰ relative to the plant (A. Cason and D. McKey, unpublished), indicating that plant and ant are separated by only a single trophic level (Scrimgeour *et al.* 1995). However, the degree of enrichment expected depends on a number of factors, including whether animals use plant amino acids or break down plant proteins (Davidson *et al.* 2003), and this is not known for

Petalomyrmex. The involvement of bacterial symbionts (abundant in hindguts of both *Aphomomyrmex* and *Petalomyrmex*; C. Rouland, personal communication) might also affect enrichment (Davidson *et al.* 2003). Firm conclusions about the relative roles of plant- and insect-derived nitrogen in the diet of *Petalomyrmex* thus await further work.

Forest understorey as a habitat for plant-ants

In forest understorey, tree-dwelling ants are likely to be even more nitrogen-limited than those inhabiting other strata. Lower plant growth rates, well-defended long-lived leaves and other plant parts, and pulsed production of young leaves all contribute to variable and often low densities of phytophagous insects, probably including phloem-sucking insects. Tied to a single host in a prey-poor environment, understorey plant-ants may thus be particularly dependent on plant-derived protein and thus particularly adapted to a low and imbalanced nitrogen supply. Coevolution with an understorey ant-plant may thus have sharpened adaptations of ants such as *Petalomyrmex phylax* to scarcity of nitrogen. Nitrogen content (dry mass) of bodies of its workers is $10.8 \pm 0.5\%$ ($n = 13$ colonies; A. Cason and D. McKey, unpublished), within the range of mean values for other exudate-feeders and lower than values for legionary and other predatory ants (Davidson & Patrell-Kim 1996). If *Petalomyrmex* obtains most of its nitrogen directly from its host plant, as our observations suggest, it does indeed have a nitrogen-poor diet, for foliar nectar of *L. a. africana*, the only resource *Petalomyrmex* has been observed to harvest from its host, contains only $0.73 \pm 0.45\%$ nitrogen ($n = 4$ trees; A. Cason and D. McKey, unpublished). Using the standard conversion factor for 'crude protein', this represents about 4.6% dry mass protein. The abundant bacterial gut symbionts of *Petalomyrmex* (C. Rouland, personal communication), completely unstudied, seem likely to play significant roles in this ant's nutritional ecology, perhaps by synthesizing essential amino acids or recycling nitrogenous wastes, as suggested for symbioses between bacteria and other tropical tree-dwelling ants (van Borm *et al.* 2002; Davidson *et al.* 2003; Gil *et al.* 2003).

The ecological diversity of ant–plant–herbivore interactions

This brief, partial discussion of the particular exigencies of forest-understorey environments opens the door to a vast and still poorly explored domain – the comparative biology of ant–plant mutualisms. The role of gradients of resource availability (e.g. between canopy and understorey, between pioneers and plants of mature forest) in structuring ant–plant interactions (Davidson *et al.* 1991; Davidson & McKey 1993) is but one theme. The diversity of nutritional ecologies encountered among tropical tree-dwelling ants is another important theme, in which broad-scale ecological and phylogenetic patterns are just now becoming clear (Davidson *et al.* 2003). A third central theme, so far based almost entirely on observations and correlations (Gaume *et al.* 1998; Davidson *et al.* 2003), is the ecological diversity of the hemipteran trophobionts tended by ants. A fourth theme, barely addressed so far, is the ecological diversity of other phytophagous insects with which ants interact. Ecologically important traits of herbivores such as size, development time, phenology, and chemical composition should vary among host plants in predictable ways, affecting the traits required of effective ant mutualists (Meunier *et al.* 1999), but patterns in this diversity among ant-plant systems have scarcely been examined. The task is now to combine all these elements in highly integrative studies that consider each of these participants in multitrophic interactions as an actor that evolves, not simply a part of the environment of one or another focal partner. Ant–plant–herbivore interactions can then become an even richer set of biological models for testing questions of general importance in ecology and evolutionary biology (Heil & McKey 2003).

Perspectives

Finally, our reading suggests several more questions about the trophic structure of ant–plant–herbivore interactions that could be rewarding to explore.

(1)   How important are non-hemipteran insects as food sources for specialist plant-ants? Why do some specialist plant-ants simply discard insects found on the plant (e.g. Janzen 1966), whereas others consume them as prey? Do some specialist plant-ants use the host plant to 'ranch' these insects, just as they ranch phloem-sucking hemipterans? We have argued that herbivores other than hemipteran trophobionts would often be unreliable protein sources for host-restricted ants, owing to their intermittent and unpredictable availability, mirroring host-plant phenology. It is interesting that the case most suggestive of an important role of insects captured on the plant in the nutrition of the resident ant colony – *Azteca* plant-ants on *Cecropia* (Sagers *et al.* 2000) – involves a host plant with relatively continuous production of young leaves.

(2)   How important is resource flow from ant to plant? In an increasing number of 'protection mutualisms', ants appear also to confer nutritional benefits on plants (e.g. Sagers *et al.* 2000; Alvarez *et al.* 2001; Belin-Depoux & Bastien 2002), just as in many ant–epiphyte 'nutritional mutualisms'. Are adaptations to enhance such flows most widespread in forest-understorey ant-plants, as Janzen (1974) long ago suggested? Traits such as the canaliculate sclerenchyma lining the inner wall of domatia in *Leonardoxa a. africana* (Brouat 2000) suggest the possibility of exchange between the resident *Petalomyrmex* colony and its host, although what might be exchanged, and in which direction materials might flow, are completely unknown. The best information is on *Maieta guianensis*, in which leaf pouches bear protuberances whose likely role is to absorb nutrients from ant wastes. Experiments

have demonstrated this function for nitrogen (Solano & Dejean 2004). The only other manipulative study of understorey plants of which we are aware – of *Piper*, a plant that appears to lack such specialized absorptive structures – showed that ant-mediated nitrogen uptake by plants occurs but may not be quantitatively important (Fischer *et al.* 2003).

(3)   Although attention has focused on nitrogen/carbohydrate ratios, nitrogen is probably not the only nutrient in short supply in the unbalanced dietary regimes of tropical tree-dwelling ants (and of their hemipteran trophobionts). Stoichiometric analysis of ant–plant–herbivore systems should be extended to include other crucial elements. Given the status of phosphorus as a limiting nutrient for tropical-forest plants, especially on old, weathered soils (Vitousek & Farrington 1997) – and thus its likely scarcity in plant secretions – this element would appear to be a suitable candidate. Nitrogen/phosphorus ratios may affect growth rates of individuals, demography of populations, and thereby community processes (Elser *et al.* 2000). Do these ratios vary in the food of different groups of tropical tree-dwelling ants? In the round of studies that is sure to be stimulated by the work of Davidson *et al.* (2003), it would be interesting to include both elements right from the start.

(4)   With rising levels of atmospheric CO_2, will nitrogen be even more diluted in extrafloral nectar and honeydew? Will tropical tree-dwelling ants be subjected to even more severe nitrogen limitation?

(5)   The ability of introduced ants to form mutualisms with plants (Fleet & Young 2000) and with hemipterans – native or introduced (Helms & Vinson 2002) – makes ants often formidable biological invaders (Lach 2003; O'Dowd *et al.* 2003). By reducing the densities of both herbivores and of beneficial insects that visit nectaries, invasive ants may destroy many kinds of interactions.

(6)   The second part of this paper has focused on trophic rewards produced by plants for ants. How does the plant regulate the ratio of rate of supply of food resources with that of other resources, such as nesting space (Fonseca 1999)? If nesting space is more strongly limiting than food, ants might channel a greater proportion of plant-derived resources into reproduction (and dispersal) relative to colony growth. If food is more strongly limiting than nest space, this could favour parasitic manipulation of the rate at which the plant supplies food. Integrative studies must consider morphogenetic, physiological and evolutionary interactions between the two kinds of rewards.

(7)   Some ant-garden plants appear to constitute a partial exception to the statement that symbiotic ant–plant mutualisms are horizontally transmitted. *Anthorrhiza* myrmecophytes within the territory of a single *Dolichoderus* sp. colony are close relatives, owing to dispersal and planting of seeds by the

ants (Maeyama & Matsumoto 2000). In such systems, ants have an interest in the reproduction of their hosts. Does this have consequences for resource flows within these systems?

(8)   As for *Leonardoxa* and its ants, increasingly detailed phylogenetic information is accumulating for the species-rich ant–*Macaranga* associations (Davies *et al.* 2001; Blattner *et al.* 2001; Feldhaar *et al.* 2003; Vogel *et al.* 2003), and studies address a broad range of biological questions (e.g. Itioka *et al.* 2000; Nomura *et al.* 2000; Itino & Itioka 2001; and others reviewed by Heil and McKey (2003)). However, these radiations include only a fraction of the ecological diversity represented in ant–plant symbioses. Studies of ant and plant radiations that present ecological contrasts with these systems are now necessary to explore the diversity of trophic structures, and their evolution, in these complex biotic interactions.

Acknowledgments

We thank David Burslem and the Organizing Committee for the invitation to participate in the Special Symposium in Aberdeen, and Sue Hartley for editorial help. Nadir Alvarez and Martine Hossaert-McKey provided useful comments on various drafts of the manuscript. Comments of two reviewers, Dinah Davidson and Paulo Oliveira, greatly improved the final version. Our studies of ant–plant–herbivore interactions have been funded by the French government, through grants from the CNRS and the Institut Français de la Biodiversité; and by the National Geographic Society's Committee for Research and Exploration. We thank the Ministry of Research and Higher Education of the Republic of Cameroon for permission to carry out research in Cameroon. Many colleagues helped with fieldwork. Alain Ngomi, Luc and Marie Moliko, and Jean-Jacques Ndoung Tom are especially thanked for their assistance and hospitality in the field.

References

Alvarez, G., Armbrecht, I., Jimenez, E., Armbrecht, H. & Ulloa-Chacon, P. 2001. Ant–plant association in two *Tococa* species from a primary rain forest of Colombian Choco (Hymenoptera: Formicidae). *Sociobiology* **38**: 585–602.

Bakker, F. M. & Klein, M. E. 1992. Transtrophic interactions in cassava. *Experimental and Applied Acarology* **14**: 293–311.

Baroni Urbani, C. & de Andrade, M. L. 1997. Pollen eating, storing and spitting by ants. *Naturwissenschaften* **84**: 256–258.

Becerra, J. X. & Venable, D. L. 1989. Extrafloral nectaries: a defense against ant–Homoptera mutualism? *Oikos* **55**: 276–280.

Belin-Depoux, M. & Bastien, D. 2002. Regard sur la myrmécophilie en Guyane française. Les dispositifs d'absorption de *Maieta guianensis* et la triple association *Philodendron*-fourmis-Aleurodes. *Acta Botanica Gallica* **149**: 299–318.

Bentley, B. L. 1981. Ants, extrafloral nectaries, and the vine life-form: an interaction. *Tropical Ecology* **22**: 127–133.

Billen, J. & Buschinger, A. 2000. Morphology and ultrastructure of a specialized bacterial pouch in the digestive tract of *Tetraponera* ants (Formicidae, Pseudomyrmecinae). *Arthropod Structure and Development* **29**: 259–266.

Blattner, F. R., Weising, K., Bänfer, G., Maschwitz, U. & Fiala, B. 2001. Molecular analysis of phylogenetic relationships among myrmecophytic *Macaranga* species. *Molecular Phylogenetics and Evolution* **19**: 331–344.

Blüthgen, N. & Fiedler, K. 2002. Interactions between weaver ants *Oecophylla smaragdina*, homopterans, trees and lianas in an Australian rain forest canopy. *Journal of Animal Ecology* **71**: 793–801.

Blüthgen, N., Verhaagh, M., Goitia, W., Jaffé, K., Morawetz, W. & Barthlott, W. 2000. How plants shape the ant community in the Amazonian rainforest canopy: the key role of extrafloral nectaries and homopteran honeydew. *Oecologia* **125**: 229–240.

Bourseaux-Eude, C. & Gross, R. 2000. New insights into symbiotic associations between ants and bacteria. *Research in Microbiology* **151**: 513–519.

Breton, L. M. & Addicott, J. F. 1992. Density-dependent mutualism in an aphid–ant interaction. *Ecology* **73**: 2175–2180.

Brouat, C. 2000. Origine et evolution des mutualismes plantes-fourmis: le cas de *Leonardoxa*, myrmécophyte à domaties caulinaires. Unpublished thesis, Université Montpellier II, France.

Brouat, C., Garcia, N., Andary, C. & McKey, D. 2001. Plant lock and ant key: pairwise coevolution of an exclusion filter in an ant–plant mutualism. *Proceedings of the Royal Society of London B* **268**: 2131–2141.

Brouat, C., McKey, D. & Douzery, E. 2004. Differentiation and gene flow in a geographic mosaic of plants coevolving with ants: phylogeny of the *Leonardoxa africana* complex (Leguminosae: Caesalpinioideae) using AFLP markers. *Molecular Ecology* **13**: 1157–1171.

Caetano, F. H., Jaffé, K. & Crewe, R. W. 1994. The digestive tract of *Cataulacus* ants: presence of microorganisms in the ileum. In *Les Insectes Sociaux* (ed. A. Lenoir, G. Arnold & M. Lepage). Paris: Université Paris Nord, p. 391.

Carroll, C. R. & Janzen, D. H. 1973. Ecology of foraging by ants. *Annual Review of Ecology and Systematics* **4**: 231–257.

Chenuil, A. & McKey, D. 1996. Molecular phylogenetic study of a myrmecophyte symbiosis: did *Leonardoxa*-ant associations diversify via cospeciation? *Molecular Phylogenetics and Evolution* **6**: 270–286.

Cogni, R., Freitas, A. V. L. & Oliveira, P. S. 2003. Interhabitat differences in ant activity on plant foliage: ants at extrafloral nectaries of *Hibiscus pernambucensis* in sandy and mangrove forests. *Entomologia Experimentalis et Applicata* **107**: 125–131.

Coley, P. D. & Barone, J. A. 1996. Herbivory and plant defenses in tropical forests. *Annual Review of Ecology and Systematics* **27**: 305–335.

Cushman, J. H. & Addicott, J. F. 1991. Conditional interactions in ant–plant–herbivore mutualisms. In *Ant–Plant Interactions* (ed. C. R. Huxley & D. F. Cutler). Oxford: Oxford University Press, pp. 92–103.

Davidson, D. W. 1997. The role of resource imbalances in the evolutionary ecology of tropical arboreal ants. *Biological Journal of the Linnean Society* **61**: 153–181.

Davidson, D. W. & McKey, D. 1993. The evolutionary ecology of symbiotic ant–plant relationships. *Journal of Hymenoptera Research* **1**: 13–83.

Davidson, D. W. & Patrell-Kim, L. 1996. Tropical arboreal ants: why so abundant? In *Neotropical Biodiversity and Conservation* (ed. A. C. Gibson), Mildred E. Mathias Botanical Garden Publication No. 1. Los Angeles: University of California Press, pp. 127–140.

Davidson, D. W., Foster, R. B., Snelling, R. R. & Lozada, P. W. 1991. Variable composition of some tropical ant–plant symbioses. In *Plant–Animal Interactions: Evolutionary Ecology in Tropical and Temperate Regions* (ed. P. W. Price, T. M. Lewinsohn, G. W. Fernandes & W. W. Benson). New York: John Wiley & Sons, pp. 145–162.

Davidson, D. W., Cook, S. C., Snelling, R. R. & Chua, T. H. 2003. Explaining the abundance of ants in lowland tropical rainforest canopies. *Science* **300**: 969–972.

Davidson, D. W., Cook, S. C. & Snelling, R. R. 2004. Liquid feeding performances of ants (Formicidae): ecological and evolutionary implications. *Oecologia* **139**: 255–266.

Davies, S. J., Lum, S. K. Y., Chan, R. & Wang, L. K. 2001. Evolution of myrmecophytism in western Malesian *Macaranga* (Euphorbiaceae). *Evolution* **55**: 1542–1559.

Dejean, A., McKey, D., Gibernau, M. & Belin, M. 2000. The arboreal ant mosaic in a Cameroonian rainforest (Hymenoptera: Formicidae). *Sociobiology* **35**: 403–423.

Delabie, J. H. C. 2001. Trophobiosis between Formicidae and Hemiptera (Sternorrhyncha and Auchenorrhyncha): an overview. *Neotropical Entomology* **30**: 501–516.

di Giusto, B., Anstett, M. C., Dounias, E. & McKey, D. 2001. Variation in the effectiveness of biotic defence: the case of an opportunistic ant–plant protection mutualism. *Oecologia* **129**: 367–375.

Dyer, L. A. & Letourneau, D. 2003. Top-down and bottom-up diversity cascades in detrital vs. living food webs. *Ecology Letters* **6**: 60–68.

Elser, J. J., Sterner, R. W., Gorokhova, E. *et al.* 2000. Biological stoichiometry from genes to ecosystems. *Ecology Letters* **3**: 540–550.

Feldhaar, H., Fiala, B., Gadau, J., Mohamed, M. & Maschwitz, U. 2003. Molecular phylogeny of *Crematogaster* subgenus *Decacrema* ants (Hymenoptera: Formicidae) and the colonization of *Macaranga* (Euphorbiaceae)

trees. *Molecular Phylogenetics and Evolution* **27**: 441–452.

Fiala, B., Grunsky, H., Maschwitz, U. & Linsenmair, K. E. 1994. Diversity of ant–plant interactions: protective efficacy in *Macaranga* species with different degrees of ant association. *Oecologia* **97**: 186–192.

Fischer, R. C., Richter, A., Wanek, W. & Mayer, V. 2002. Plants feed ants: food bodies of myrmecophytic *Piper* and their significance for the interaction with *Pheidole bicornis* ants. *Oecologia* **133**: 186–192.

Fischer, R. C., Wanek, W., Richter, A. & Mayer, V. 2003. Do ants feed plants? A ^{15}N labelling study of nitrogen fluxes from ants to plants in the mutualism of *Pheidole* and *Piper*. *Journal of Ecology* **91**: 126–134.

Fleet, R. R. & Young, B. L. 2000. Facultative mutualism between imported fire ants (*Solenopsis invicta*) and a legume (*Senna occidentalis*). *Southwestern Naturalist* **45**: 289–298.

Floren, A., Biun, A. & Linsenmair, K. E. 2002. Arboreal ants as key predators in tropical lowland rainforest trees. *Oecologia* **131**: 137–144.

Fonseca, C. R. 1993. Nesting space limits colony size of the plant-ant *Pseudomyrmex concolor*. *Oikos* **67**: 473–482.

1999. Amazonian ant-plant interactions and the nesting space limitation hypothesis. *Journal of Tropical Ecology* **15**: 807–825.

Gaedke, U., Hochstadter, S. & Straile, D. 2002. Interplay between energy limitation and nutritional deficiency: empirical data and food web models. *Ecological Monographs* **72**: 251–270.

Gaston, K. J., Gauld, I. D. & Hanson, P. 1996. The size and composition of the hymenopteran fauna of Costa Rica. *Journal of Biogeography* **23**: 105–113.

Gauld, I. D., Gaston, K. J. & Janzen, D. H. 1992. Plant allelochemicals, tritrophic interactions and the anomalous diversity of tropical parasitoids: the nasty host hypothesis. *Oikos* **65**: 353–357.

Gaume, L. 1998. Mutualisme, parasitisme, et évolution des symbioses plantes-fourmis: le cas de *Leonardoxa* (légumineuse) et de ses fourmis associées. Unpublished thesis, Ecole Nationale Supérieure Agronomique de Montpellier.

Gaume, L. & McKey, D. 2002. How identity of the homopteran trophobiont affects sex allocation in a symbiotic plant-ant: the proximate role of food. *Behavioral Ecology and Sociobiology* **51**: 197–205.

Gaume, L., McKey, D. & Terrin, S. 1998. Ant–plant–homopteran mutualism: how the third partner affects the interaction between a plant-specialist ant and its myrmecophyte host. *Proceedings of the Royal Society of London B* **265**: 569–575.

Gil, R., Silva, F. J., Zientz, E. *et al.* 2003. The genome sequence of *Blochmannia floridanus*: comparative analysis of reduced genomes. *Proceedings of the National Academy of Sciences* **100**: 9388–9393.

Gullan, P. J. 1997. Relationships with ants. In *Soft Scale Insects: Their Biology, Natural Enemies and Control* (ed. Y. Ben-Dov & C. J. Hodgson). Amsterdam: Elsevier, pp. 351–373.

Hairston, N. G., Sr, Smith, F. E. & Slobodkin, L. B. 1960. Community structure, population control, and competition. *American Naturalist* **44**: 421–425.

Halaj, J. & Wise, D. H. 2001. Terrestrial trophic cascades: how much do they trickle? *American Naturalist* **157**: 262–281.

Heaney, L. R. 2001. Small mammal diversity along elevational gradients in the Philippines: an assessment of patterns and hypotheses. *Global Ecology and Biogeography* **10**: 15–39.

Heil, M. & McKey, D. 2003. Protective ant–plant interactions as model systems in ecological and evolutionary research. *Annual Review of Ecology, Systematics and Evolution* **34**: 425–553.

Heil, M., Fiala, B., Kaiser, W. & Linsenmair, K. E. 1998. Chemical contents of *Macaranga* food bodies: adaptations to their role in ant attraction and nutrition. *Functional Ecology* **12**: 117–122.

Heil, M., Koch, T., Hilpert, A., Fiala, B., Boland, W. & Linsenmair, K. E. 2001. Extrafloral nectar production of the ant-associated plant, *Macaranga tanarius*, is an induced, indirect, defensive response elicited by jasmonic acid. *Proceedings of the National Academy of Sciences* **98**: 1083–1088.

Helms, K. R. & Vinson, S. B. 2002. Widespread association of the invasive ant *Solenopsis invicta* with an invasive mealybug. *Ecology* **83**: 2425–2438.

Hossaert-McKey, M., Orivel, J., Labeyrie, E., Pascal, L., Delabie, J. H. C. & Dejean, A. 2001. Differential associations with ants of three co-occurring extrafloral nectary-bearing plants. *Ecoscience* **8**: 325–335.

Itino, T. & Itioka, T. 2001. Interspecific variation and ontogenetic change in antiherbivore defense in myrmecophytic *Macaranga* species. *Ecological Research* **16**: 765–774.

Itino, T., Itioka, T., Hatada, A. & Hamid, A. A. 2001. Effects of food rewards offered by ant-plant *Macaranga* on the colony size of ants. *Ecological Research* **16**: 775–786.

Itioka, A. & Inoue, T. 1999. The alternation of mutualistic ant species affects the population growth of their trophobiont mealybug. *Ecography* **22**: 169–177.

Itioka, T., Nomura, M., Inui, Y., Itino, T. & Inoue, T. 2000. Difference in intensity of ant defense among three species of *Macaranga* myrmecophytes in a southeast Asian dipterocarp forest. *Biotropica* **32**: 318–326.

Izzo, T. J. & Vasconcelos, H. L. 2002. Cheating the cheater: domatia loss minimizes the effects of ant castration in an Amazonian ant-plant. *Oecologia* **133**: 200–205.

Jaffé, K., Caetano, F. H., Sánchez, P. *et al.* 2001. Sensitivity of colonies and individuals of *Cephalotes* ants to antibiotics imply a feeding symbiosis with gut microorganisms. *Canadian Journal of Zoology* **79**: 1120–1124.

Janzen, D. H. 1966. Coevolution of mutualism between ants and acacias in Central America. *Evolution* **20**: 249–275.

1971. Euglossine bees as long-distance pollinators of tropical plants. *Science* **171**: 203–205.

1973. Sweep samples of tropical foliage insects: effects of seasons, vegetation types, elevation, time of day, and insularity. *Ecology* **54**: 687–708.

1974. Tropical blackwater rivers, animals, and mast fruiting by the Dipterocarpaceae. *Biotropica* **6**: 69–103.

1975. The peak in North-American ichneumonid species richness lies between 38 degrees and 42 degrees N. *Ecology* **62**: 532–537.

Karhu, K. J. 1998. Effects of ant exclusion during outbreaks of a defoliator and a sap-sucker on birch. *Ecological Entomology* **23**: 185–194.

Karhu, K. J. & Neuvonen, S. 1998. Wood ants and a geometrid defoliator of birch: predation outweighs beneficial effects through the host plant. *Oecologia* **113**: 509–516.

Kay, A. 2002. Applying optimal foraging theory to assess nutrient availability ratios for ants. *Ecology* **83**: 1935–1944.

Kiers, E. T., Lovelock, C. E., Krueger, E. L. & Herre, E. A. 2000. Differential effects of tropical arbuscular mycorrhizal fungal inocula on root colonization and tree seedling growth: implications for tropical forest diversity. *Ecology Letters* **3**: 106–113.

Kruger, O. & McGavin, G. C. 1998. The influence of ants on the guild structure of *Acacia* insect communities in Mkomazi Game Reserve, north-east Tanzania. *African Journal of Ecology* **36**: 213–220.

Lach, L. 2003. Invasive ants: unwanted partners in ant–plant interactions? *Annals of the Missouri Botanical Garden* **90**: 91–108.

Larsen, K. J., Staehle, L. M. & Dotseth, E. J. 2001. Tending ants (Hymenoptera: Formicidae) regulate *Dalbulus quinquenotatus* (Homoptera: Cicadellidae) population dynamics. *Environmental Entomology* **30**: 757–762.

Leigh, E. G., Jr & Windsor, D. M. 1982. Forest production and regulation of primary consumers on Barro Colorado Island. In *The Ecology of a Tropical Forest. Seasonal Rhythms and Long-term Changes* (ed. E. G. Leigh, Jr, A. S. Rand & D. M. Windsor). Washington DC: Smithsonian Institution Press, pp. 111–122.

Letourneau, D. K. 1990. Code of ant–plant mutualism broken by parasite. *Science* **248**: 215–217.

Letourneau, D. & Dyer, L. A. 1998. Experimental test in lowland tropical forest shows top-down effects through four trophic levels. *Ecology* **79**: 1678–1687.

Linsenmair, K. E., Heil, M., Kaiser, W. M., Fiala, B., Koch, T. & Boland, W. 2001. Adaptations to biotic and abiotic stress: *Macaranga* ant-plants optimize investment in biotic defence. *Journal of Experimental Botany* **52**: 2057–2065.

Longino, J. T. 1986. Ants provide substrate for epiphytes. *Selbyana* **9**: 100–103.

MacKay, D. A. 1991. The effects of ants on herbivory and herbivore numbers on foliage of the mallee eucalypt, *Eucalyptus incrassata* Labill. *Australian Journal of Ecology* **16**: 471–483.

Maeyama, T. & Matsumoto, T. 2000. Genetic relationship of myrmecophyte (*Anthorrhiza caerulea*) individuals within and among territories of the arboreal ant (*Dolichoderus* sp.) detected using random amplified polymorphic DNA markers. *Australian Journal of Ecology* **25**: 273–282.

McKey, D. 2000. *Leonardoxa africana* (Leguminosae: Caesalpinioideae): a complex of mostly allopatric subspecies. *Adansonia* **22**: 71–109.

Meunier, L., Dalecky, A., Berticat, C., Gaume, L. & McKey, D. 1999. Worker size variation and the evolution of ant–plant mutualisms: comparative morphometrics of workers of two closely related plant-ants, *Petalomyrmex phylax* and *Aphomyrmex afer* (Formicinae). *Insectes Sociaux* **46**: 171–178.

Mondor, E. B. & Addicott, J. F. 2003. Conspicuous extra-floral nectaries are inducible in *Vicia faba*. *Ecology Letters* **6**: 495–497.

Morales, M. A. 2000. Mechanisms and density dependence of benefit in an ant-membracid mutualism. *Ecology* **81**: 482–489.

Nomura, M., Itioka, T. & Itino, T. 2000. Variations in abiotic defense within myrmecophytic and non-myrmecophytic species of *Macaranga* in a Bornean dipterocarp forest. *Ecological Research* **15**: 1–11.

Novotny, V., Basset, Y., Auga, J. *et al.* 1999. Predation risk for herbivorous insects on tropical vegetation: a search for enemy-free space and time. *Australian Journal of Ecology* **24**: 477–483.

O'Dowd, D. J., Green, P. T. & Lake, P. S. 2003. Invasional 'meltdown' on an oceanic island. *Ecology Letters* **6**: 812–817.

Offenberg, J. 2001. Balancing between mutualism and exploitation: the symbiotic interaction between *Lasius* ants and aphids. *Behavioral Ecology and Sociobiology* **49**: 304–310.

Olmstead, K. L. & Wood, T. K. 1990. Altitudinal patterns in species richness of neotropical treehoppers (Homoptera, Membracidae): the role of ants. *Proceedings of the Entomological Society of Washington* **92**: 552–560.

Paine, R. T. 1966. Food web complexity and species diversity. *American Naturalist* **100**: 65–76.

Pemberton, R. W. 1998. The occurrence and abundance of plants with extrafloral nectaries, the basis for antiherbivore defensive mutualisms, along a latitudinal gradient in east Asia. *Journal of Biogeography* **25**: 661–668.

Pierce, N. E., Braby, M. F., Heath, A. *et al.* 2002. The ecology and evolution of ant association in the Lycaenidae (Lepidoptera). *Annual Review of Entomology* **47**: 733–771.

Polis, G. A. & Strong, D. R. 1996. Food web complexity and community dynamics. *American Naturalist* **147**: 813–846.

Rathcke, B. J. & Price, P. W. 1976. Anomalous diversity of tropical ichneumonid parasitoids: a predation hypothesis. *American Naturalist* **110**: 889–893.

Raine, N. E., Willmer, P. & Stone, G. N. 2002. Spatial structuring and floral avoidance behavior prevent ant–pollinator conflict in a Mexican ant-acacia. *Ecology* **83**: 3086–3096.

Rico-Gray, V., Garcia-Franco, J. G., Palacios-Rios, M., Diaz-Castelazo, C., Parra-Tabla, V. & Navarro, J. A. 1998. Geographical and seasonal variation in the richness of ant–plant interactions in Mexico. *Biotropica* **30**: 190–200.

Risch, S. J. & Rickson, F. 1981. Mutualism in which ants must be present before plants produce food bodies. *Nature* **291**: 149–150.

Rocha, C. F. D. & Bergallo, H. G. 1992. Bigger ant colonies reduce herbivory and herbivore residence time on leaves of an ant-plant: *Azteca muelleri* vs. *Coelomera ruficornis* on *Cecropia pachystachya*. *Oecologia* **91**: 249–252.

Sagers, C. L., Ginger, S. M. & Evans, R. D. 2000. Carbon and nitrogen isotopes trace nutrient exchange in an ant–plant mutualism. *Oecologia* **123**: 582–586.

Samson, D. A, Rickart, E. A. & Gonzales, P. C. 1997. Ant diversity and abundance along an elevational gradient in the Philippines. *Biotropica* **29**: 349–363.

Schröder, D., Deppisch, H., Obermayer, M. *et al.* 1996. Intracellular endosymbiotic bacteria of *Camponotus* species (carpenter ants): systematics, evolution and ultrastructural characterization. *Molecular Microbiology* **21**: 479–489.

Schmitz, O. J., Hamback, P. A. & Beckerman, A. P. 2000. Trophic cascades in terrestrial systems: a review of the effects of carnivore removals on plants. *American Naturalist* **155**: 141–153.

Schupp, E. W. & Feener, D. H. 1991. Phylogeny, life form, and habitat dependence of ant-defended plants in a Panamanian forest. In *Ant–Plant Interactions* (ed. C. R. Huxley & D. F. Cutler). Oxford: Oxford University Press, pp. 175–197.

Scrimgeour, C. M., Gordon, S. C., Handley, L. L. & Woodford, J. A. T. 1995. Trophic levels and anomalous ^{15}N of insects on raspberry (*Rubus idaeus* L.). *Isotopes in Environmental and Health Studies* **31**: 107–115.

Shurin, J. B., Borer, E. T., Seabloom, E. W. *et al.* 2002. A cross-ecosystem comparison of the strength of trophic cascades. *Ecology Letters* **5**: 785–791.

Solano, P. J. & Dejean, A. 2004. Ant-fed plants: comparison between three geophytic myrmecophytes. *Biological Journal of the Linnean Society* **83**: 433–439.

Terborgh, J., López, L., Nuñez, P. *et al.* (2003) Ecological meltdown in predator-free forest fragments. *Science* **294**: 1923–1926.

van Bael, S. A., Brawn, J. D. & Robinson, S. K. 2003. Birds defend trees from herbivores in a neotropical forest canopy. *Proceedings of the National Academy of Sciences* **100**: 8304–8307.

van Borm, S., Buschinger, A., Boomsma, J. J. & Billen, J. 2002. *Tetraponera* ants have gut symbionts related to nitrogen-fixing root-nodule bacteria. *Proceedings of the Royal Society of London B* **269**: 2023–2027.

Vitousek, P. M. & Farrington, H. 1997. Nutrient limitation and soil development: experimental test of a biogeochemical theory. *Biogeochemistry* **37**: 63–75.

Vogel, M., Bänfer, G., Moog, U. & Weising, K. 2003. Development and characterization of chloroplast microsatellite markers in *Macaranga* (Euphorbiaceae). *Genome* **46**: 845–857.

von Dohlen, C. & Moran, N. A. 1995. Molecular phylogeny of the Homoptera: a paraphyletic taxon. *Journal of Molecular Evolution* **41**: 211–223.

Wagner, D. 1997. The influence of ant nests on *Acacia* seed production, herbivory and soil nutrients. *Journal of Ecology* **85**: 83–93.

Wagner, D. & Kay, A. 2002. Do extrafloral nectaries distract ants from visiting flowers? An experimental test of an overlooked hypothesis. *Evolutionary Ecology Research* **4**: 293–305.

Wimp, G. M. & Whitham, T. G. 2001. Biodiversity consequences of predation and host plant hybridization on an aphid–ant mutualism. *Ecology* **82**: 440–452.

Wright, S. J. 2002. Plant diversity in tropical forests: a review of mechanisms of species coexistence. *Oecologia* **130**: 1–14.

Yu, D. W. & Pierce, N. E. 1998. A castration parasite of an ant–plant mutualism. *Proceedings of the Royal Society of London B* **265**: 375–382.

CHAPTER SEVENTEEN

Multitrophic interactions in a neotropical savanna: ant–hemipteran systems, associated insect herbivores and a host plant

PAULO S. OLIVEIRA
Universidade Estadual de Campinas
KLEBER DEL-CLARO
Universidade Federal de Uberlândia

Introduction

In many habitats ants form a major part of the arthropod fauna found on vegetation, and recent studies have shown that the abundance and diversity of ant–plant associations is particularly remarkable in the tropical region. For instance, one-third of the plant species in a Panamanian forest (Schupp & Feener 1991) and over 20% of the woody species in a Brazilian savanna (Oliveira & Oliveira-Filho 1991) were found to produce ant rewards. Furthermore, 312 ant–plant interactions were recorded in one Mexican coastal site (Rico-Gray 1993), and the ant–plant community in an Amazonian rainforest comprised 377 plants per ha (Fonseca & Ganade 1996). In the tropics many ant species use plant surfaces as a foraging substrate to search for both live and dead animal prey, as well as for different types of plant-derived food products (Carroll & Janzen 1973). Ant activity on foliage can be promoted by the occurrence of predictable and immediately renewable food sources, including extrafloral nectar, honeydew from phloem-feeding hemipterans, and secretions from lepidopteran larvae (see Way 1963; Bentley 1977; Buckley 1987; Koptur 1992; Pierce *et al.* 2002). In fact, plant- and insect-derived liquid foods appear to provide a large amount of the energy supply of foliage-dwelling ants (Tobin 1994; Davidson *et al.* 2003). Although food resources located on foliage are probably more often found and exploited by arboreal species, ground-nesting ants frequently extend their foraging areas onto the plant substrate as well (Rico-Gray 1993; Blüthgen *et al.* 2000; Davidson *et al.* 2003). Intense ant activity on vegetation has resulted in a multitude of ant–plant–herbivore interactions, ranging from facultative to obligate associations (reviewed by Beattie 1985; Davidson & McKey 1993; Bronstein 1998). From the

Biotic Interactions in the Tropics: Their Role in the Maintenance of Species Diversity, ed. D. F. R. P. Burslem, M. A. Pinard and S. E. Hartley. Published by Cambridge University Press. © Cambridge University Press 2005.

plant's standpoint, the outcomes of many of these interactions are largely mediated by how ant behavioural patterns can affect herbivore performance on a given host plant (Chapters 15 and 16, this volume).

Ant patrolling activity on leaves may affect insect herbivores in different ways, and this in turn may result in positive, negative or neutral consequences for plants (Bronstein 1994a, b; Beattie & Hughes 2002). The distribution of ants within the plant crown and their behaviour towards herbivores may depend on the nature of the food source being exploited by the foragers (e.g. McKey 1984; Völkl 1992; Oliveira 1997, this volume). For instance, liquid food is typically supplied on foliage in the form of extrafloral nectar and insect honeydew, and general aggression exhibited by ants at these two sources ('ownership behaviour', see Way 1963) can have consequences that are markedly variable, both for the herbivore and for the plant (Oliveira *et al.* 2002; Beattie & Hughes 2002, and citations therein). By showing aggressive behaviour towards herbivores on extrafloral nectary-bearing plants, visiting ants can positively affect plant fitness by decreasing herbivore damage to vegetative and reproductive plant parts (reviewed by Koptur 1992). On the other hand, by protecting sap-feeding hemipterans from their natural enemies, honeydew-gathering ants can negatively affect plant fitness by increasing hemipteran damage to the host plant (Rico-Gray & Thien 1989). However, the outcomes of such plant–ant–herbivore systems may not always be so straightforward; whereas some studies have demonstrated that ant visitation to extrafloral nectaries may afford no apparent benefit to plants, others have shown that ant–Hemiptera interactions may be beneficial to the host plant (see reviews in Beattie 1985; Koptur 1992; Beattie & Hughes 2002).

Three decades ago, Carroll and Janzen (1973) first suggested that honeydew-producing hemipterans could function as insect analogues of extrafloral nectaries because of tending ants' deterrence of other herbivores associated with the plant. In short, ant–Hemiptera interaction could positively affect plant fitness if the benefits of ant-derived protection from herbivory outweighed the losses incurred by hemipteran feeding (Carroll & Janzen 1973; Janzen 1979). In theory, in order for the ant–hemipteran association to affect plant fitness positively, the ant-tended partner should not be the primary herbivore, and the ants should effectively deter non-hemipteran herbivores (Messina 1981; see also Horvitz & Schemske 1984 on ant-tended lepidopteran larvae).

A number of factors (e.g. time, habitat type, identity, abundance and behaviour of species partners, and severity of herbivore damage) may influence the final results of plant–ant–herbivore associations, and only by considering the variation of associated costs and benefits inherent in these systems can we understand their complex dynamics and the range of possible outcomes (Thompson 1988; Cushman 1991; Bronstein 1994a, b; Gaume *et al.* 1998). Although mutualism is defined as an interaction between two species that is beneficial to both (Boucher *et al.* 1982), some mutualisms can only be understood in the context of the

Figure 17.1 Schematic outline of the interaction system involving ants and *Guayaquila*
xiphias treehoppers on shrubs of *Didymopanax vinosum* in the cerrado savanna of Brazil.
(A) *G. xiphias* prefers to aggregate at the apex of the single reproductive branch. Tending
ants collect honeydew from treehoppers day and night, and have a positive impact on
treehopper survival. Ant presence reduces the abundance of the principal natural
enemies of *G. xiphias*, **(B)** salticid spiders, **(C)** predatory syrphid flies and **(D)** parasitoid

community, and by assessing the influence of other species and other trophic levels on the pairwise relationship (reviewed by Bronstein & Barbosa 2002) (this volume).

In this chapter we report on our research on ant–plant–herbivore interactions in the cerrado savanna of Brazil, with special emphasis on ant–Hemiptera interactions and their effects on associated insect herbivores. We first present the natural history of the study system, present experimental data supporting ant-derived benefits to hemipterans, and describe the relevant behavioural aspects involving the participant species. In the second part we examine the ways through which ant–hemipteran associations can affect damage from different types of associated herbivores, illustrate the intricacy of the effects and infer the consequences of such multitrophic systems for the host plant. Finally, we emphasize the relevance of multitrophic systems for community ecology and conservation of biodiversity.

Ant–Hemiptera interactions in the cerrado savanna
The treehopper Guayaquila xiphias *and its ant attendants*

In the cerrado of Mogi-Guaçu (SE Brazil; 22° 18′ S, 47° 10′ W), aggregations of *Guayaquila xiphias* (Fabr.) (Membracidae) commonly infest shrubs of *Didymopanax vinosum* March. (Araliaceae), where they are tended by a diverse assemblage of honeydew-gathering ants. The vegetation consists of a dense scrub of shrubs and trees, known as cerrado *sensu stricto* (Oliveira-Filho & Ratter 2002). Treehoppers occur on *D. vinosum* throughout the year, and levels of infestation on a plant range from one female with her egg mass to over 200 individuals in a single aggregation (mean ± SD = 18.8 ± 23.6; no. of aggregations N = 222). Although females of *G. xiphias* show parental care by guarding the egg mass and young nymphs, in the presence of ants they may abandon the first brood to ants and produce an additional clutch. Nymphs complete their development about 3 weeks after hatching, and newly emerged adults disperse from natal aggregations. The treehoppers are normally located on stems and feed near growing meristems, but in the flowering season (March to September) they tend to aggregate close to the inflorescence at the apex of the single reproductive branch (Fig. 17.1). Across diurnal and nocturnal censuses, a total of 21 ant species have been recorded collecting honeydew from *Guayaquila* aggregations, the most frequent ones being *Camponotus rufipes* Fabr., *C. crassus* Mayr, *C. renggeri* Emery

---

wasps. **(E)** Untended or poorly tended treehopper aggregations and brood-guarding females attract ants by flicking accumulated honeydew onto lower leaves and beneath the host plant. Upon discovering scattered honeydew droplets on the ground, alerted ants eventually find the treehoppers as they climb onto the plant, and begin tending activities at the newly discovered food source. See text for further details.

(Formicinae) and *Ectatomma edentatum* Roger (Ponerinae). Daily turnover of ant species at a given treehopper aggregation occurs frequently (an account of the whole ant assemblage is given by Del-Claro & Oliveira 1999). The relevance of hemipteran honeydew as an energy (and perhaps water) supply for cerrado ants is such that some species (*C. rufipes* and *E. edentatum*) tend *Guayaquila* on a round-the-clock basis, and *C. rufipes* may even build satellite nests of dry grass to house groups of treehoppers. Indeed, even after having discovered an alternative sugar source on the host plant, honeydew-gathering ants do not desert *Guayaquila* aggregations and keep tending levels unchanged (Del-Claro & Oliveira 1993).

Treehoppers are attacked by three main types of natural enemies on shrubs of *D. vinosum* (see Fig. 17.1). A total of 15 species of salticid spiders may prey on nymphs and adults. Predatory larvae of *Ocyptamus arx* (Fluke) (Diptera: Syrphidae) suck empty the entire body contents of the treehoppers, and occasionally feed on egg masses if no nymph or adult treehopper is present on the plant. Treehopper egg masses are parasitized by *Gonatocerus* wasps (Myrmaridae).

How tending ants benefit Guayaquila xiphias

A series of controlled ant-exclusion experiments performed in the cerrado enabled us to assess the nature of the benefits afforded by tending ants to *G. xiphias*, and identify the variable outcomes of the interaction (Del-Claro & Oliveira 2000). In March 1992 and 1993 we tagged 44 *D. vinosum* shrubs (1–2 m tall). Each plant had one incipient aggregation (*G. xiphias* female with her brood), and was randomly assigned to a control (ants present; $N = 22$ plants) or treatment group (ants excluded; $N = 22$ plants). Ants were prevented from climbing onto treated plants by applying sticky Tanglefoot resin to the trunk base, and pruning grass bridges within 0.5 m. Treehoppers and their natural enemies were censused every 2–3 days on control and treatment plants during 16 days (08.00 to 16.00 h). The occurrence of a second egg mass laid by the resident treehopper female was also recorded in each plant class. In both years of experimental manipulations the initial size of treehopper aggregations did not differ between control and treatment plants.

Increased ant density near *Guayaquila* aggregations markedly affected the spatial distribution and foraging behaviour of spiders, syrphid flies and parasitoid wasps on the host plant. Spiders were not only warded off from the vicinity of the treehoppers by tending ants, but also jumped off the plants after attacks by large ants such as *C. rufipes* and *E. edentatum*. The two ant species were also seen attacking adult syrphid flies near treehopper aggregations. Aggression by tending ants was strong enough to keep parasitoid wasps away from brood-guarding females, and significantly decreased their success in approaching the egg mass. Whereas on ant-excluded plants the wasps were more frequent near (< 5 cm) hemipteran egg masses, on ant-visited plants they were seen mostly on leaves (Fig. 17.2).

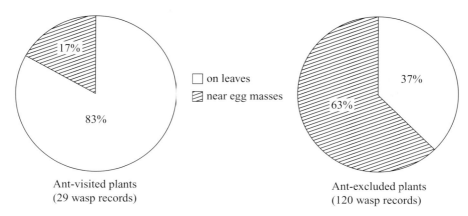

Figure 17.2 Distribution of parasitoid wasps (*Gonatocerus* sp.) on shrubs hosting *Guayaquila xiphias* treehoppers in the presence or absence (*N* = 22 plants in each group) of tending ants. The spatial distribution of wasps (on leaves versus near egg masses) is significantly affected by ant attendance to brood-guarding females (χ^2 = 17.46, *P* = 0.001, d.f. = 1). Modified from Del-Claro & Oliveira (2000).

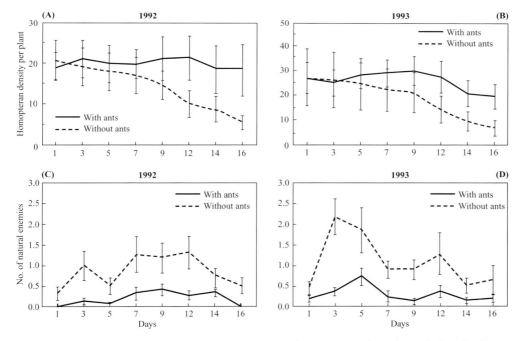

Figure 17.3 Density of *Guayaquila xiphias* treehoppers per plant through time in the presence or absence of tending ants (*N* = 22 plants in each group), in (**A**) March 1992 and (**B**) 1993. Ant tending had a positive effect on treehopper survival in 1992 (treatment × time: *F* = 4.33, d.f. = 7, *P* = 0.0001), but not in 1993 (*F* = 1.11, d.f. = 7, *P* = 0.35). Ant presence decreased significantly the abundance of natural enemies of *Guayaquila* on plants, in (**C**) 1992 (treatment: *F* = 11.54, d.f. = 1, *P* = 0.0015) and (**D**) 1993 (*F* = 11.51, d.f. = 1, *P* = 0.0015). Values are means ± 1 SE. Repeated-measures ANOVA performed on square-root transformed data. Modified from Del-Claro & Oliveira (2000).

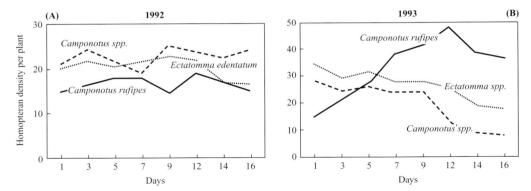

Figure 17.4 Density of *Guayaquila xiphias* treehoppers per plant through time as a function of the species identity of ant partners. (**A**) 1992: Group 1. *Camponotus rufipes* (N = 6); Group 2. *C. crassus, C. renggeri*, and *C. sp.* (N = 7); Group 3. *Ectatomma edentatum* (N = 9). (**B**) 1993: Group 1. *C. rufipes* (N = 6); Group 2. *C. crassus, C. aff. blandus*, and *C. renggeri* (N = 7); Group 3. *E. edentatum* and *E. planidens* (N = 9). While in 1992 all ants had a positive impact on *Guayaquila* (treatment × time: F = 0.5, d.f. = 14, P = 0.93), in 1993 ant-derived benefits to homopteran survival differed among the species (treatment × time: F = 5.16, d.f. = 14, P = 0.0001). Values are means. Repeated-measures ANOVA performed on square-root transformed data. Modified from Del-Claro & Oliveira (2000).

The ant-exclusion experiments unequivocally demonstrated that tending ants have a positive impact on treehopper survival, and decrease the abundance of the natural enemies of *Guayaquila* on the host plant (Fig. 17.3). Two years of experimental manipulations, however, have shown that ant-derived effects on hemipteran survival can vary both with time and with the species of tending ant (Figs. 17.3, 17.4). Whereas in 1992 *Camponotus* and *Ectatomma* species were equally beneficial to *Guayaquila*, in 1993 only *C. rufipes* had a positive effect on treehopper survival (Fig. 17.4). The experiments also revealed that ant-tending can positively affect treehopper fecundity, because brood-guarding females transfer parental care to ants and lay an additional clutch more often than untended females (91% vs. 54% of the cases; N = 22 females in each group; P = 0.018, χ^2 = 5.61, d.f. = 1).

Enhancement of interaction through treehopper behaviour

Accumulated honeydew can be flicked by untended phloem-feeding hemipterans with the hind legs or caudae, or by contraction of the rectum or entire abdomen (Hölldobler & Wilson 1990). Although the occurrence of scattered honeydew droplets beneath untended or poorly tended hemipterans has long been noted (e.g. Way 1954; Douglas & Sudd 1980), only more recently has the flicking of honeydew been investigated experimentally and its consequences for ant–Hemiptera interaction evaluated in greater detail (Del-Claro & Oliveira 1996). Aggregations of *Guayaquila xiphias* are never seen unattended by ants in the cerrado, and constancy of ant-tending is due largely to the treehopper's capacity to provide cues that promote contact with potential ant partners on the host plant. Upon oviposition, while sitting on egg masses, or while guarding newly eclosed nymphs, females of *G. xiphias* commonly flick accumulated honeydew onto lower leaves and onto the ground beneath the host plant (Fig. 17.1). We first speculated that honeydew flicking by the female could function to attract ants at early stages of brood development (Del-Claro & Oliveira 1993). We noted, however, that developing nymphs in poorly tended aggregations also show the same behaviour. In view of the important benefits afforded by ant-tending to both female and brood (see above), promotion of early contact with ants would presumably be highly advantageous for *Guayaquila*. Indeed, field observations of ants (*Camponotus*, *Cephalotes* and *Ectatomma*) at flicked honeydew strongly supported the hypothesis of ant attraction. Upon discovering the droplets on the ground, the alerted scout ant typically searches around the immediate vicinity and eventually climbs onto the host plant (Fig. 17.1). While ascending the plant the forager further licks additional flicked honeydew scattered on lower foliage (Fig. 17.1), and eventually encounters the female and her brood near the apical meristem. The ant then collects some secretion from the nymphs and returns directly to the nest nearby, where additional workers are recruited to exploit the newly discovered food source. The ant colony then establishes a long-term fidelity to the path leading to the treehoppers, and tending activities endure as the nymphs develop. In a series of field experiments we were able to demonstrate that flicked honeydew induces ground-dwelling ants to climb onto the host plant and begin tending activities. Pieces of honeydew-soaked filter paper placed beneath treehopper-free plants induced significantly more ground-dwelling ants to climb onto the plant than did control water-soaked papers (Del-Claro & Oliveira 1996).

Ant–*Guayaquila* interactions, associated herbivores, and host plant
Damage by associated herbivores to Didymopanax vinosum
Shrubs of *D. vinosum* are infested by four principal non-hemipteran insect herbivores: (1) *Liothrips didymopanicis* Del-Claro & Mound (Thysanoptera: Phlaeothripidae): thrips consume the apical leaf primordia and young leaves. Feeding damage alters host-plant architecture by causing folding of leaves, and by leading to growth of lateral shoots due to death of the apical meristem (Fig. 17.5A, B). Severe damage by thrips may cause death of the host plant (Del-Claro & Mound 1996). While moving between folded leaves and shoot tips, the thrips may become vulnerable to tending ants, and predation by *Camponotus rufipes* and *Cephalotes pusillus* (Klug) was observed twice on plants with *Guayaquila* treehoppers. (2) *Caralauca olive* Jesmar (Coleoptera: Chrysomelidae): adults mate on the host plant and feed mainly on mature adult leaves. Chewing activity

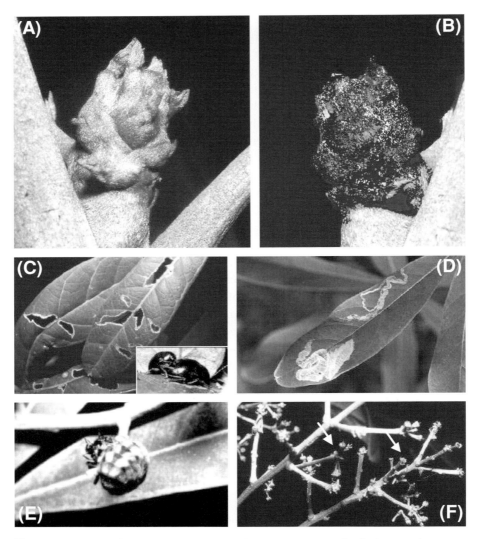

Figure 17.5 Types of herbivore damage to *Didymopanax vinosum* shrubs in cerrado savanna. **(A)** An intact meristem and **(B)** an apical meristem entirely destroyed by *Liothrips didymopanicis* thrips. **(C)** Leaves with characteristic chewing marks made by *Caralauca olive* beetles (inset), and **(D)** mines made by developing lepidopteran larvae. **(E)** A larva of *Panthiades polibetes* resting on a floral bud while tended by *Camponotus*. **(F)** Inflorescence of *D. vinosum* presenting signs of floral damage (arrows) by *P. polibetes* feeding activity.

by the beetles produces characteristic marks on leaf blades (Fig. 17.5C). (3) Leaf-miners (Lepidoptera, undetermined family): mining/feeding activity by developing larvae leaves easily detectable tunnels within the leaf blade (Fig. 17.5D). (4) *Panthiades polibetes* Cramer (Lepidoptera: Lycaenidae): larvae feed on floral buds (diameter 2 mm). The cryptic caterpillars rest on floral buds and are tended

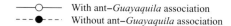

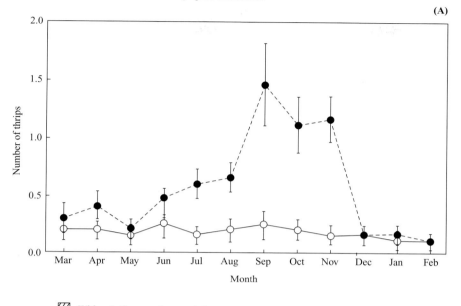

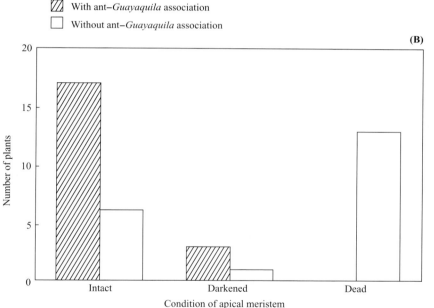

Figure 17.6 (A) Infestation levels by the thrips *Liothrips didymopanicis* on *Didymopanax vinosum* shrubs through time, in the presence (control) or absence (treatment) of an ant-*Guayaquila xiphias* association on the plant ($N = 20$ plants in each group). Thrips abundance is negatively affected by ant–treehopper interactions ($F = 19.33$, d.f. $= 1$, $P = 0.0001$). Values are means ± 1 SE. Repeated-measures ANOVA performed on square-root transformed data. **(B)** After 12 months of experimental exclusion, damage to the apical meristem by thrips was significantly greater on plants without ants and *G. xiphias* than on plants with the association ($\chi^2 = 19.26$, d.f. $= 2$, $P = 0.0001$, $N = 20$ plants in each group).

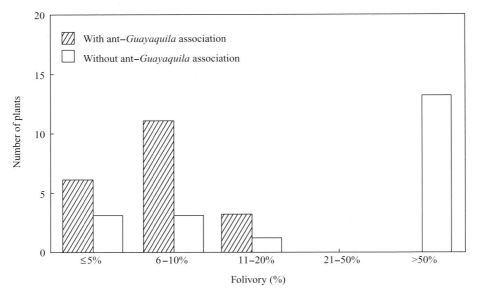

Figure 17.7 Levels of folivory (pooled for *Caralauca olive* beetles and leaf-mining lepidopteran larvae) on *Didymopanax vinosum* shrubs, in the presence (control) or absence (treatment) of an ant–*Guayaquila xiphias* association on the plant (N = 20 plants in each group). After 12 months of experimental exclusion, percentage of leaf surface damaged was significantly higher on plants without ants and *G. xiphias* than on plants with the association (χ^2 = 19.27, d.f. = 4, P = 0.0007).

from the earliest instars by at least seven ant species (*Camponotus* spp., *Cephalotes clypeatus* (Fabr.) and *Ectatomma edentatum*) that feed on larval secretions (Fig. 17.5E). Worker ants from a single colony may simultaneously tend and collect liquid from both *Panthiades* and *Guayaquila* if these co-occur on a given shrub of *D. vinosum*. The inflorescences of plants infested by *Panthiades* may have several buds destroyed by larval feeding (Fig. 17.5F).

How ant–Guayaquila *interactions affect associated herbivores*

The effect of ants tending *Guayaquila* on associated herbivores was evaluated in monthly censuses on shrubs of *D. vinosum* from March 1992 to February 1993. Plants were tagged at the beginning of the dry season (March), when

Figure 17.8 Monthly infestation pattern by larvae of the myrmecophilous *Panthiades polibetes* butterflies on experimental *Didymopanax vinosum* shrubs, in the presence (control) or absence (treatment) of an ant–*Guayaquila xiphias* association on the plant. In all months the butterflies infested in preference the plants hosting ant-tended treehoppers. March (χ^2 = 21.56, d.f. = 1, P = 0.001), April (χ^2 = 7.15, d.f. = 1, P = 0.007), May (χ^2 = 9.29, d.f. = 1, P = 0.002).

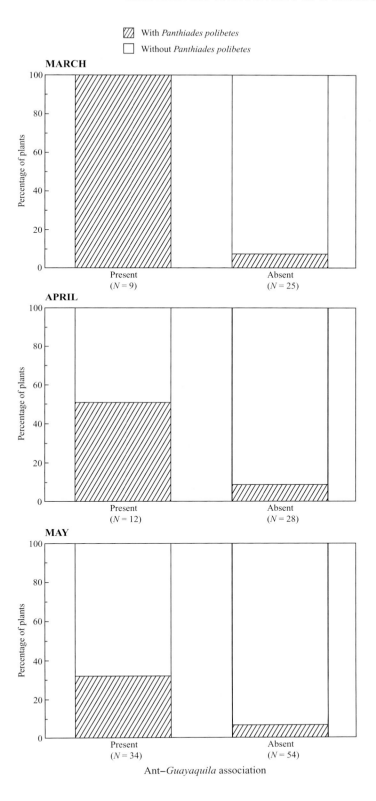

Ant–*Guayaquila* association

growing meristems are intact (Fig. 17.5A), leaf and bud expansion begin, and herbivore activity is very low (Del-Claro & Mound 1996; Del-Claro & Oliveira 1999). Shrubs were approximately the same height (1–2 m tall), had similar numbers of leaves, and were *c.* 4 m from each other. Infestation levels and/or damage by associated herbivores were measured on experimentally manipulated plants, with and without ant–*Guayaquila* associations (hereafter 'with AG' and 'without AG', respectively). After dispersal of newly emerged adult treehoppers from natal aggregations, with-AG plants were repeatedly re-infested by gravid *Guayaquila* females from March 1992 to February 1993. In the period, sequential occupation of such plants by ant–*Guayaquila* associations lasted 9.4 ± 2.4 months (mean ± SD). The assemblage of tending ants on tagged with-AG plants included *Camponotus crassus*, *C. renggeri*, *C. rufipes*, *Cephalotes pusillus* and *Ectatomma edentatum*. To avoid establishment of ant–hemipteran association on without-AG treatment plants, occasional scout ants were excluded by applying sticky Tanglefoot resin to the trunk base. Upon landing on such ant-excluded plants, potentially colonizing *Guayaquila* females would normally feed on plant sap and flick accumulated honeydew (see above) for nearly 1 hour before abandoning the plant.

Infestation by thrips was monitored monthly during 1 year by counting the number of adults on the plants ($N = 20$ in each experimental group). In March 1992 all plants had intact growing meristems. After 1 year the condition of the apical meristem relative to damage by thrips was re-evaluated as intact, darkened or dead (see Fig. 17.5A, B). Damage by beetles and leaf-mining caterpillars was pooled under 'folivory', which was estimated visually as the percentage leaf area eaten and/or covered by mines (Fig. 17.5C, D). In March 1992 all plants had 5% folivory or less. After 1 year, levels of folivory were re-evaluated under five categories (5%, 6–10%, 11–20%, 21–50%, > 50%). Folivore damage was evaluated on the same plants used to investigate abundance and damage by thrips ($N = 20$ in each experimental group).

The effect of ant–*Guayaquila* interaction on infestation of *D. vinosum* shrubs by *Panthiades polibetes* (Fig. 17.5E, F) was investigated by monitoring with-AG and without-AG plants (different from the ones described above) from March to May 1992. At the beginning of each month, we tagged groups of plants in these two categories bearing young inflorescences (bud diameter 2 mm) and free from *Panthiades*. The numbers of plants in with-AG and without-AG groups each month were, respectively, 9 and 25 (March), 12 and 28 (April), and 34 and 54 (May). Tagged plants were free from attack by thrips (see above), lateral vegetative shoot tips were not present; and the stem's apical meristem originated the inflorescence axis (Fig. 17.1). With-AG plants were visited by *Camponotus crassus*, *C.* aff. *blandus* Fr. Smith, *C. renggeri*, *C. sericeiventris* Guerin, *C. abdominalis* Fabr. and *Ectatomma edentatum*. A band of Tanglefoot was applied to the base of each without-AG plant. Tagged plants of both groups were re-checked for *P. polibetes* infestation on the last day of each month, and were not re-used

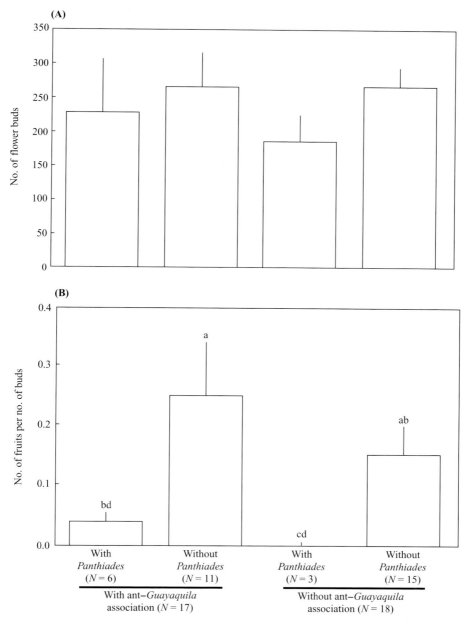

Figure 17.9 Production of flower buds and fruit set in experimental *Didymopanax vinosum* shrubs, in the presence (control) or absence (treatment) of an ant–*Guayaquila xiphias* association on the plant, and as a function of infestation by myrmecophilous bud-consuming *Panthiades polibetes* caterpillars. **(A)** Production of floral buds did not differ among treatments (Kruskal–Wallis test, $H = 1.83$, d.f. $= 3$, $P = 0.609$). **(B)** Fruit set differed significantly among the four groups ($H = 10.32$, d.f. $= 3$, $P = 0.016$); treatments designated by the same letter above bars are not significantly different. Values are means ± 1 SE.

in subsequent months. Results are presented as the percentage of with-AG and without-AG plants infested by *P. polibetes* in each month. The effect of bud-destroying *P. polibetes* on the reproductive output of experimental plants was evaluated by dividing the number of fruits by the number of buds produced by each shrub of *D. vinosum*.

Results show that occurrence of ant–*Guayaquila* interactions can have variable effects on the abundance and/or damage caused by associated herbivores of *D. vinosum*. Damage by thrips to the apical meristem, and by chewing and mining insects to leaves, were both significantly lower on with-AG plants than on without-AG plants (Figs. 17.6, 17.7). The lycaenid *P. polibetes*, on the other hand, preferentially infested plants with ant-*Guayaquila* interactions (Fig. 17.8). Presence of *P. polibetes* larvae on a plant led to lower fruit production, and a higher proportion of with-AG plants had lycaenid larvae than did without-AG plants. On with-AG plants bud consumption by ant-tended larvae of *P. polibetes* reduced the reproductive output of *D. vinosum* by 84% (Fig. 17.9).

Discussion

The association between *Guayaquila xiphias* and tending ants on shrubs of *Didymopanax vinosum* in the cerrado savanna can be summarized as follows. Hemipteran honeydew is an important promoter of ant activity on foliage, and the treehoppers are also able to attract potential tending ants by flicking accumulated honeydew beneath the host plant. Ownership behaviour and aggression by ants in the vicinity of the treehoppers keep parasitoid wasps away from brood-guarding *Guayaquila* females. Presence of tending ants decreases the abundance of predators and parasitoids on the host plant, and has a positive impact on hemipteran survival over time. Moreover, ant tending confers a direct reproductive benefit to *G. xiphias* females, which can transfer parental care to ants and lay an additional clutch. Two years of field experiments revealed that the outcomes of ant–*G. xiphias* associations are dynamic in nature, and that benefits from ant tending may vary with time and/or with the species of ant partner. Similar findings have been reported for other ant–Hemiptera systems (e.g. McEvoy 1979; Bristow 1983; 1984; Cushman & Whitham 1989; Cushman & Addicott 1991; Buckley & Gullan 1991; Völkl & Kroupa 1997). Additional aspects of the ecology and natural history of ant–*G. xiphias* systems, in the context of related research in the area, can be found in our earlier work (Del-Claro & Oliveira 1993; 1996; 1999; 2000).

New insights in the study of the evolutionary ecology of ant–Hemiptera and ant–Lepidoptera systems, including reviews of the most important works, can be found in the recent literature (e.g. Morales 2000; Billick *et al.* 2001; Beattie & Hughes 2002; Bronstein & Barbosa 2002; Oliveira *et al.* 2002; Pierce *et al.* 2002; Billick & Tonkel 2003). In the remainder of this discussion we will focus on the interactions generated by the presence of ant–Hemiptera associations on plants, with emphasis on the impacts of tending ants on associated herbivores, and on the consequences of such effects for the host plant.

Ant behaviour on foliage, and herbivore deterrence

Ant behaviour required to deter herbivores from feeding or ovipositing on plants is similar to that needed to protect ant-tended insects from their natural enemies, and in both cases ownership behaviour by the ants is sufficient to expel intruders from the ants' immediate foraging area (Way 1963; Bentley 1977). Indeed, several ant species associated with extrafloral nectaries or honeydew-producing insects probably confer protection on the plants and the insects through similar behaviours (DeVries 1991; Koptur 1992). For instance, the aggressive *Camponotus rufipes*, an abundant ant on cerrado foliage (Oliveira & Brandão 1991), is very effective at both removing potential herbivores from plants with extrafloral nectaries (Oliveira *et al.* 1987), and protecting *Guayaquila* treehoppers from their natural enemies (Del-Claro & Oliveira 2000). Ant aggressiveness, however, is not an essential trait for herbivore deterrence on foliage. For instance, 'timid' minute *Petalomyrmex* ants efficiently protect *Leonardoxa* trees from chewing and sucking herbivores (Gaume *et al.* 1997), and 'passive' *Pheidole* may confer protection on ant-inhabited *Piper* saplings by removing eggs of insect folivores (Letourneau 1983).

Although not fortuitous, ant-derived protection against herbivores on plants hosting ant–hemipteran associations can be regarded as a by-product of the increased alertness and general aggression shown by ants near the hemipterans. It is also possible that honeydew-gathering ants are protein-limited, leading to increased searching for insect prey and thereby to increased plant protection.

Effects on host plants by honeydew-gathering ants

Since the hypothesis was first proposed by Carroll and Janzen (1973), several studies have supported the idea that honeydew-producing hemipterans can protect plants by attracting tending ants that deter other herbivores. Room (1972) showed that *Crematogaster* ants tending hemipterans on a mistletoe species protect the plant from other herbivores and allow increased shoot growth of the mistletoe. *Formica* ants associated with *Publilia* treehoppers attack leaf-chewing beetles and reduce defoliation to goldenrod (*Solidago*), resulting in increased seed production and growth by stems bearing membracids (Messina 1981). Another type of benefit conferred to a scale's host plant by tending ants was reported by Bach (1991), who demonstrated that sanitation by honeydew-gathering ants reduces leaf death and abscission caused by fungal infection on accumulated honeydew (see also Queiroz & Oliveira 2001).

Herbivore deterrence by tending ants has also been recorded on plants that regularly house ant colonies, and that produce food rewards for ants. Stout (1979) reported that *Myrmelachista* ants tending mealybugs inside *Ocotea* trees can remove insect eggs from young stems and leaves, and suggested that ant inhabitants could protect the plant from herbivores. *Crematogaster* ants inhabiting *Macaranga* trees tend scales inside the stems, and not only remove herbivores from leaves but also prune foreign plants that come in contact with their host plant (Fiala *et al.* 1989). Finally, benefits to a plant may be mediated by the type of honeydew-producing hemipteran tended by ant inhabitants. Gaume *et al.* (1998) have shown that net benefits against herbivory conferred by *Aphomomyrmex* ants on myrmecophytic *Leonardoxa* vary with the type of sap-sucking hemipteran (coccids or pseudococcids) tended by resident ants inside the tree's hollow twigs.

Negative effects to host plants caused by ants tending honeydew-producing insects have also been documented by several authors both in temperate and tropical habitats, and indicate that other secondary interactions may complicate potential ant-derived benefits to the plant. For example, Fritz (1983) has demonstrated that although *Formica* ants tending *Vanduzea* treehoppers reduced adult density and oviposition by leaf-mining beetles on black locust, they also indirectly protected beetle larvae by excluding their main hemipteran predator. As a result of the ants' opposite effects on adult and immature beetles, Fritz (1983) found no apparent benefit or harm to black locust in having ants tending *Vanduzea*. In the system studied by Buckley (1983) in Australia, *Iridomyrmex* ants feeding on honeydew from *Sextius* treehoppers also collected extrafloral nectar from the host plant (*Acacia*). Because ants were more attracted to treehoppers than to nectaries, ant-derived protection against other herbivores was disrupted and resulted in an overall negative effect of the ant–Hemiptera interaction on plant growth and seed set (Buckley 1983; see also DeVries & Baker 1989). Similar results were obtained by Rico-Gray and Thien (1989) in Mexican sand-dune communities, where honeydew-producing mealybugs shift ant attention away from the extrafloral nectaries of *Schomburgkia* orchids, resulting in increased damage to plant reproductive organs and reduced fruit set. In the same type of habitat, however, the impact of ant–aphid interactions on *Paullinia* seed set varies among years from negative to neutral (Rico-Gray & Castro 1996). How patterns of ant attendance at extrafloral nectaries versus honeydew-producing hemipterans on a given plant can affect the herbivore-deterrent effects of ants is still debatable, and probably depends on the plant and insect species involved (see Becerra & Venable 1989; Fiala 1990; Rashbrook *et al.* 1992; Del-Claro & Oliveira 1993; Gaume *et al.* 1997; 1998).

In perhaps the most detailed study to date of the interactions involving a plant with extrafloral nectaries (*Calathea*), visiting ants, and ant-tended lepidopteran larvae (*Eurybia*), Horvitz and Schemske (1984) emphasized the variable and interdependent character of the components of such complex insect–plant systems, and the variable magnitude of positive and negative effects on plants. Their experiments show that in the absence of ants, damage to reproductive tissues by *Eurybia* caterpillars lowered *Calathea* seed production by 66%, far more than the 33% reduction in the presence of tending ants. Given that *Eurybia* is a specialist herbivore capable of frequently infesting *Calathea*, the ant–*Eurybia* association seems advantageous for the plant even though seed set is greatest without the caterpillars. Species-specific variation in ant effects on seed production and among-site differences in ant communities further increase the unpredictability of the outcomes in such a complex interaction system (Horvitz & Schemske 1984).

The multispecies system around ant–Guayaquila associations

The results presented here on the system involving ant-*Guayaquila* associations on *Didymopanax* shrubs revealed that effects of tending ants on other species of herbivores ranged from negative to positive. Three types of herbivores (thrips, chewing beetles and leaf-mining caterpillars) were negatively affected by ants, and their damage was reduced on plants hosting ant–*Guayaquila* associations. The bud-destroying lycaenid *Panthiades polibetes*, on the other hand, not only preferentially infested plants with ants and *Guayaquila* but also shifted ant attention partly away from treehoppers towards liquid-rewarding caterpillars. This multitrophic interaction system is unique in its complexity because it involves four types of herbivores, each exploiting plant tissue in a distinct mode, and all being affected by a focal ant–Hemiptera association. As opposed to typical ant–plant systems in which ant-derived protective effects on plants are mediated by plant traits inducing ownership behaviour by ants (e.g. extrafloral nectaries, food bodies, ant domatia; reviewed by Beattie and Hughes 2002), the multispecies interactive system around ant–*Guayaquila* associations is similar to those treated by other studies in that the analyses of pairwise interactions cannot predict the overall impact on the plant from all species involved (see Price *et al.* 1980; Thompson 1988; Cushman 1991; Bronstein 1998; and citations therein).

The full range of effects inherent in multispecies mutualistic systems, in which other species and trophic levels affect the nature and outcome of pairwise interactions, has recently been addressed by Bronstein and Barbosa (2002), who looked at examples from numerous studies. We have shown that ants tending *Guayaquila xiphias* aggregations can have contrasting impacts on damage caused by other herbivores, with varying consequences for *Didymopanax vinosum* shrubs. To understand the net interactive effects of each component in such a complex multitrophic system, one should consider the factors acting both together and separately (e.g. Horvitz & Schemske 1984; Gaume *et al.* 1998). Although an evaluation of the net effects of each component in our system was not possible because of constraints for establishing different treatment groups of appropriate size, a range of potential interactive effects could be identified in the studied system. Some of these are indicated below.

By reducing damage to leaves and shoot tips of *D. vinosum*, tending ants may indirectly benefit the *Guayaquila* by preserving plant quality (see Messina 1981;

Fritz 1982; Bach 1991). Indeed, increased thrips activity on plants free from ant-tended *Guayaquila* increased the probability of mortality of the plant's principal apical meristem. Such damage can kill the host plant (Del-Claro & Mound 1996). Since severe damage by thrips impedes inflorescence development in *D. vinosum*, deterrence by tending ants is likely to be beneficial for *Guayaquila* treehoppers because they normally feed at the apex of the single reproductive branch (Fig. 17.1; see Del-Claro & Oliveira 1999).

The infestation pattern of the myrmecophilous *Panthiades polibetes* on *D. vinosum* shrubs indicates that presence of ant–*Guayaquila* interaction can act as a strong stimulus inducing oviposition by the butterfly, resulting in decreased fruit set. Because myrmecophilous lycaenids may use ants as cues for oviposition, the presence of ant-tended treehoppers may further increase the stimulus by increasing ant density on foliage (see Atsatt 1981a; Pierce & Elgar 1985; a negative effect of ant presence on butterfly oviposition has been shown by Freitas and Oliveira 1996). By efficiently attracting tending ants onto the host plant through honeydew flicking, *Guayaquila* treehoppers create a potentially enemy-free space (Atsatt 1981b) for colonization by competing myrmecophilous *Panthiades*. Therefore the presence of ant foragers promotes species coexistence of ant-tended insects on the host plant. Whether ants are a limited resource on cerrado foliage, and whether competition for their services can adversely affect coexisting *Guayaquila* and *Panthiades*, are questions that remain to be addressed (see Addicott 1978; Cushman & Addicott 1989).

One might also expect a range of reciprocal interactive effects among the species of associated herbivores of *D. vinosum*. For example, early attack by thrips to apical meristem and young leaves (altering plant architecture, and leaf quality) should directly affect subsequent use of the host plant by both chewing and mining insects, and vice versa. Similarly, because ant-tended larvae of *Panthiades* depend on floral resources for growth, host-plant use by the butterfly should also be negatively associated with previous damage by thrips to the apical meristem and to young leaves.

To summarize, the variable character of the numerous components of the complex multitrophic system described here requires sampling at different sites and times for a more realistic evaluation of the impact of the interactions on the host plant. The combination of negative and positive effects among participant species, and their interdependence, should probably result in temporal and spatial variation in the overall impact of the interactions on plants (see Horvitz & Schemske 1984; Thompson 1988; 1994).

At this point, we briefly comment on the uniqueness of the cerrado for research on interaction systems involving ants. At least two factors account for the prevalence of ants on cerrado foliage, and make ant–plant–herbivore interactions especially pervasive in this biome. First, many arboreal ant species use hollowed-out stems as nest sites, and this per se causes intense ant-patrolling activity on leaves (Morais & Benson 1988). Second, the wide occurrence of predictable food rewards in the form of extrafloral nectaries and insect-derived secretions further promotes ant foraging on cerrado foliage. In the study area in Mogi-Guaçu for example, species with extrafloral nectaries represented 20% of the woody flora (21 of 104 species) surveyed by Oliveira and Leitão-Filho (1987), whereas ant-tended treehoppers occurred on 30% of the 93 plant species examined by Lopes (1995). We are just beginning to understand the role of natural enemies, ants in particular, in structuring herbivore communities and controlling herbivory levels in the cerrado. A range of potential interactive effects of herbivory and abiotic factors on plant fitness and plant population dynamics further provides excellent opportunities for future experimental work (Marquis *et al.* 2002). Thus current available information clearly suggests that cerrado vegetation is unique in providing promising scenarios for research on ant–plant–herbivore interactions (see Oliveira *et al.* 2002; and included references).

Final considerations

Terrestrial communities based on living plants involve at least three interacting trophic levels: plants, herbivores and natural enemies of herbivores (Price *et al.* 1980). In recent decades it has become increasingly apparent that consideration of the third trophic level is essential for understanding not only insect–plant interactions but also whole communities (Thompson 1994). Reports on a variety of interaction systems, including herbivory (Lawton and McNeill 1979; Price *et al.* 1980), mutualism (Gilbert 1980), seed predation and seed dispersal (Heithaus *et al.* 1980), pollination (Horvitz & Schemske 1988) and competition (Price *et al.* 1986), have all emphasized the multispecific character of the interactions, the relevance of the effects from some third species or trophic level, and the need for analysing plant–animal interactions within a broader community context (Chapter 15). Despite the inherent complexity of multitrophic interactions, one current goal in community ecology is to identify across different trophic levels which populations are limited by resource availability (bottom-up forces) and which are limited by consumption by higher trophic levels (top-down forces), and how trophic controls vary at both local and global spatial scales (see Thompson 2002; Dyer & Coley 2002; and citations therein).

Programs for conservation of biodiversity have been concerned primarily with the maintenance of species diversity and ecosystem functions, and the preservation of genetic variation within populations. However, as priorities have moved towards a more landscape-level view of biodiversity, the so-called 'interaction biodiversity' has received greater attention (Thompson 1997; this volume). The numerous studies reported in this chapter have shown in various ways that interspecific interactions can shape the organization of communities, and therefore act as links between species and ecosystems. Biodiversity should be viewed and evaluated also in ways that embrace the extreme richness inherent to plant–animal interactions, including the species' ecological roles, the kinds of interactions and their outcomes, trophic web structure, selection pressures, habitat heterogeneity and geographical variation (Price 2002). Conservation of 'interaction biodiversity' (Thompson 1997) should therefore be regarded as an integral part of strategies to maintain viable conserved communities.

Acknowledgements

We thank David Burslem and the Organizing Committee for the invitation to participate in the Special Symposium of the British Ecological Society in Aberdeen. The final version of the manuscript was considerably improved by the comments of two anonymous referees, and by the editing of Susan Hartley. H. P. Dutra, K. S. Brown, A. V. L. Freitas, R. Cogni, V. Rico-Gray and H. L. Vasconcelos provided helpful suggestions on early versions of the manuscript; J. R. Trigo and A. X. Linhares helped with the statistical analyses. Our studies in the cerrado were supported by Brazilian Federal (CNPq, CAPES) and State (FAPEMIG) agencies, and by research grants from the Universidade Estadual de Campinas (FAEP). We thank the Instituto de Botânica de São Paulo for permission to work in its cerrado reserve.

References

Addicott, J. F. (1978) Competition for mutualists: aphids and ants. *Canadian Journal of Zoology*, **56**, 2093–2096.

Atsatt, P. R. (1981a) Ant-dependent food plant selection by the mistletoe butterfly *Ogyris amaryllis* (Lycaenidae). *Oecologia*, **48**, 60–63.

  (1981b) Lycaenid butterflies and ants: selection for enemy-free space. *American Naturalist*, **118**, 638–654.

Bach, C. E. (1991) Direct and indirect interactions between ants (*Pheidole megacephala*), scales (*Coccus viridis*) and plants (*Pluchea indica*). *Oecologia* **87**, 233–239.

Beattie, A. J. (1985) *The Evolutionary Ecology of Ant–Plant Mutalisms*. Cambridge: Cambridge University Press.

Beattie, A. J. & Hughes, L. (2002) Ant–plant interactions. In *Plant–Animal Interactions: An Evolutionary Approach* (ed. C. M. Herrera & O. Pellmyr). Oxford: Blackwell Science, pp. 211–235.

Becerra, J. X. & Venable, D. L. (1989) Extrafloral nectaries: a defense against ant–Homoptera mutualism? *Oikos*, **55**, 276–280.

Bentley, B. L. (1977) Extrafloral nectaries and protection by pugnacious bodyguards. *Annual Review of Ecology and Systematics*, **8**, 407–428.

Billick, I. & Tonkel, K. (2003) The relative importance of spatial vs. temporal variability in generating a conditional mutualism. *Ecology*, **84**, 289–295.

Billick, I., Weidmann, M. & Reithel, J. (2001) The importance of ant-tending to maternal care in the membracid species, *Publilia modesta*. *Behavioral Ecology and Sociobiology*, **51**, 41–46.

Blüthgen, N., Verhaagh, M., Goitía, W., Jaffé, K., Morawetz, W. & Barthlott, W. (2000) How plants shape the ant community in the Amazonian rainforest canopy: the key role of extrafloral nectaries and homopteran honeydew. *Oecologia*, **125**, 229–240.

Boucher, D. H., James, S. & Keeler, K. H. (1982) The ecology of mutualism. *Annual Review of Ecology and Systematics*, **13**, 315–347.

Bristow, C. M. (1983) Treehoppers transfer parental care to ants: a new benefit of mutualism. *Science*, **220**, 532–533.

(1984) Differential benefits from ant attendance to two species of Homoptera on New York iron weed. *Journal of Animal Ecology*, **53**, 775–826.

Bronstein, J. L. (1994a) Conditional outcomes in mutualistic interactions. *Trends in Ecology and Evolution*, **9**, 214–217.

(1994b) Our current understanding of mutualism. *Quarterly Review of Biology*, **69**, 31–51.

(1998) The contribution of ant-plant protection studies to our understanding of mutualism. *Biotropica*, **30**, 150–161.

Bronstein, J. L. & Barbosa, P. (2002) Multitrophic/multispecies mutualistic interactions: the role of non-mutualists in shaping and mediating mutualisms. In *Multitrophic Level Interactions* (ed. T. Tscharntke & B. A. Hawkins). Cambridge: Cambridge University Press, pp. 44–66.

Buckley, R. C. (1983) Interaction between ants and membracid bugs decreases growth and seed set of host plant bearing extrafloral nectaries. *Oecologia*, **58**, 132–136.

(1987) Interactions involving plants, homoptera, and ants. *Annual Review of Ecology and Systematics*, **18**, 11–138.

Buckley, R. C. & Gullan, P. (1991) More aggressive ant species (Hymenoptera: Formicidae) provide better protection for soft scales and mealybugs (Homoptera: Coccidae, Pseudococcidae). *Biotropica*, **23**, 282–286.

Carroll, C. R. & Janzen, D. H. (1973) Ecology of foraging by ants. *Annual Review of Ecology and Systematics*, **4**, 231–257.

Cushman, J. H. (1991) Host-plant mediation of insect mutualisms: variable outcomes in herbivore–ant interactions. *Oikos*, **61**, 138–144.

Cushman, J. H. & Addicott, J. F. (1989) Intra- and interspecific competition for mutualists: ants as a limited and limiting resource for aphids. *Oecologia*, **79**, 315–321.

(1991) Conditional interactions in ant–plant–herbivore mutualisms. In *Ant–Plant Interactions* (ed. C. R. Huxley & D. F. Cutler). Oxford: Oxford University Press, pp. 92–103.

Cushman, J. H. & Whitham, T. G. (1989) Conditional mutualism in membracid–ant association: temporal, age-specific, and density-dependent effects. *Ecology*, **70**, 1040–1047.

Davidson, D. W., Cook, S. C., Snelling, R. R. & Chua, T. H. (2003) Explaining the abundance of ants in lowland tropical rainforest canopies. *Science*, **300**, 969–972.

Davidson, D. W. & McKey, D. (1993) The evolutionary ecology of symbiotic ant–plant relationships. *Journal of Hymenoptera Research*, **2**, 13–83.

Del-Claro, K. & Mound, L. A. (1996) Phenology and description of a new species of *Liothrips* (Thysanoptera: Phlaeothripidae) from *Didymopanax* (Araliaceae) in Brazilian cerrado. *Revista de Biologia Tropical*, **44**, 193–197.

Del-Claro, K. & Oliveira, P. S. (1993) Ant-homoptera interaction: do alternative sugar sources distract tending ants? *Oikos*, **68**, 202–206.

(1996) Honeydew flicking by treehoppers provides cues to potential tending ants. *Animal Behaviour*, **51**, 1071–1075.

(1999) Ant-Homoptera interactions in neotropical savanna: the honeydew-producing treehopper *Guayaquila xiphias* (Membracidae) and its associated ant fauna on *Didymopanax vinosum* (Araliaceae). *Biotropica*, **31**, 135–144.

(2000) Conditional outcomes in a neotropical treehopper–ant association: temporal and species-specific effects. *Oecologia*, **124**, 156–165.

DeVries, P. J. (1991) Mutualism between *Thisbe irenea* butterflies and ants, and the role of ant ecology in the evolution of larval–ant associations. *Biological Journal of the Linnean Society*, **43**, 179–195.

DeVries, P. J. & Baker, I. (1989) Butterfly exploitation of a plant-ant mutualism: adding insult to herbivory. *Journal of the New York Entomological Society*, **97**, 332–340.

Douglas, J. M. & Sudd, J. H. (1980) Behavioral coordination between an aphid and the ant that tends it: an ethological analysis. *Animal Behaviour*, **28**, 1127–1139.

Dyer, L. A. & Coley, P. D. (2002) Tritrophic interactions in tropical versus temperate communities. In *Multitrophic Level Interactions* (ed. T. Tscharntke & B. A. Hawkins). Cambridge: Cambridge University Press, pp. 67–88.

Fiala, B. (1990) Extrafloral nectaries versus ant-Homoptera mutualism: a comment on Becerra and Venable. *Oikos*, **59**, 281–282.

Fiala, B., Maschwitz, U., Pong, T. Y. & Helbig, A. J. (1989) Studies of a South East Asian ant–plant association: protection of *Macaranga* trees by *Crematogaster borneensis*. *Oecologia*, **79**, 463–470.

Fonseca, C. R. & Ganade, G. (1996) Asymmetries, compartments and null interactions in an Amazonian ant–plant community. *Journal of Animal Ecology*, **65**, 339–347.

Freitas, A. V. L. & Oliveira, P. S. (1996) Ants as selective agents on herbivore biology: effects on the behaviour of a non-myrmecophilous butterfly. *Journal of Animal Ecology*, **65**, 205–210.

Fritz, R. S. (1982) An ant-tended treehopper mutualism: effects of *Formica subsericea* on survival of *Vanduzea arquata*. *Ecological Entomology*, **7**, 267–276.

(1983) Ant protection of a host plant's defoliator: consequence of an

ant-membracid mutualism. *Ecology*, **64**, 789–797.

Gaume, L., McKey, D. & Anstett, M.-C. (1997) Benefits conferred by 'timid' ants: active anti-herbivore protection of the rainforest tree *Leonardoxa africana* by the minute ant *Petalomyrmex phylax*. *Oecologia*, **112**, 209–216.

Gaume, L., McKey, D. & Terrin, S. (1998) Ant–plant–homopteran mutualism: how the third partner affects the interaction between a plant-specialist ant and its myrmecophyte host. *Proceedings of the Royal Society of London B*, **265**, 569–575.

Gilbert, L. E. (1980) Food web organization and the conservation of neotropical diversity. In *Conservation Biology: An Evolutionary-Ecological Perspective* (ed. M. E. Soulé & B. A. Wilcox). Sunderland: Sinauer, pp. 11–33.

Heithaus, E. R., Culver, D. C. & Beattie, A. J. (1980) Models of some ant–plant mutualisms. *American Naturalist* **116**, 347–361.

Hölldobler, B. & Wilson, E. O. (1990) *The Ants*. Cambridge, MA: Harvard University Press.

Horvitz, C. C. & Schemske, D. W. (1984) Effects of ants and ant-tended herbivore on seed production of a neotropical herb. *Ecology*, **65**, 1369–1378.

(1988) A test of the pollinator limitation hypothesis for a neotropical herb. *Ecology*, **69**, 200–206.

Janzen, D. H. (1979) New horizons in the biology of plant defenses. In *Herbivores: Their Interactions with Secondary Plant Metabolites* (ed. G. A. Rosenthal & D. H. Janzen). New York: Academic Press, pp. 331–350.

Koptur, S. (1992) Extrafloral nectary-mediated interactions between insects and plants. In *Insect–Plant Interactions*. Vol. 4 (ed. E. Bernays). Boca Raton: CRC Press, pp. 81–129.

Lawton, J. H. & McNeill, S. (1979) Between the devil and the deep blue sea: on the problem of being a herbivore. *Symposium of the British Ecological Society* **20**, 223–244.

Letourneau, D. K. (1983) Passive aggression: an alternative hypothesis for the *Piper–Pheidole* association. *Oecologia*, **60**, 122–126.

Lopes, B. C. (1995) Treehoppers (Homoptera: Membracidae) in the southeast Brazil: use of host plants. *Revista Brasileira de Zoologia* **12**, 595–608.

Marquis, R. J., Morais, H. C. & Diniz, I. R. (2002) Interactions among cerrado plants and their herbivores: unique or typical? In *The Cerrados of Brazil: Ecology and Natural History of a Neotropical Savanna* (ed. P. S. Oliveira & R. J. Marquis). New York: Columbia University Press, pp. 306–328.

McEvoy, P. B. (1979) Advantages and disadvantages to group living in treehoppers (Homoptera: Membracidae). *Miscellaneous Publications of the Entomological Society of America*, **11**, 1–13.

McKey, D. (1984) Interaction of the ant-plant *Leonardoxa africana* (Caesalpiniaceae) with its obligate inhabitants in a rainforest in Cameroon. *Biotropica*, **16**, 81–99.

Messina, F. J. (1981) Plant protection as a consequence of ant–membracid mutualism: interactions on Goldenrod (*Solidago* sp.). *Ecology*, **62**, 1433–1440.

Morais, H. C. & Benson, W. W. (1988) Recolonização de vegetação de cerrado após queimada, por formigas arborícolas. *Revista Brasileira de Biologia*, **48**, 459–466.

Morales, M. A. (2000) Mechanisms and density dependence of benefit in an ant–membracid mutualism. *Ecology*, **81**, 482–489.

Oliveira, P. S. (1997) The ecological function of extrafloral nectaries: herbivore deterrence by visiting ants and reproductive output in *Caryocar brasiliense* (Caryocaraceae). *Functional Ecology*, **11**, 323–330.

Oliveira, P. S. & Brandão, C. R. F. (1991) The ant community associated with extrafloral nectaries in Brazilian cerrados. In *Ant–Plant Interactions* (ed. D. F. Cutler & C. R. Huxley). Oxford: Oxford University Press, pp. 198–212.

Oliveira, P. S., Freitas, A. V. L. & Del-Claro, K. (2002) Ant foraging on plant foliage: contrasting effects on the behavioral ecology of insect herbivores. *The Cerrados of Brazil: Ecology and Natural History of a Neotropical Savanna* (ed. P. S. Oliveira & R. J. Marquis). New York: Columbia University Press, pp. 287–305.

Oliveira, P. S. & Leitão-Filho, H. F. (1987) Extrafloral nectaries: their taxonomic distribution and abundance in the woody flora of cerrado vegetation in southeast Brazil. *Biotropica*, **19**, 140–148.

Oliveira, P. S. & Oliveira-Filho, A. T. (1991) Distribution of extrafloral nectaries in the woody flora of tropical communities in Western Brazil. In *Plant–Animal Interactions: Evolutionary Ecology in Tropical and Temperate Regions* (ed. P. W. Price, T. M. Lewinsohn, G. W. Fernandes & W. W. Benson). New York: John Wiley & Sons, pp. 163–175.

Oliveira, P. S., Silva, A. F. da & Martins, A. B. (1987) Ant foraging on extrafloral nectaries of *Qualea grandiflora* (Vochysiaceae) in cerrado vegetation: ants as potential antiherbivore agents. *Oecologia*, **74**, 228–230.

Oliveira-Filho, A. T. & Ratter, J. A. (2002) Vegetation physiognomies and woody flora of the cerrado biome. In *The Cerrados of Brazil: Ecology and Natural History of a Neotropical Savanna* (ed. P. S. Oliveira & R. J. Marquis). New York: Columbia University Press, pp. 91–120.

Pierce, N. E., Braby, M. F., Heath, A. *et al.* (2002) The ecology and evolution of ant association in the Lycaenidae (Lepidoptera). *Annual Review of Ecology and Systematics*, **47**, 733–771.

Pierce, N. E. & Elgar, M. A. (1985) The influence of ants on host plant selection by *Jalmenus evagora*, a myrmecophilous lycaenid butterfly. *Behavioral Ecology and Sociobiology*, **16**, 209–222.

Price, P. W. (2002) Species interactions and the evolution of biodiversity. In *Plant–Animal Interactions: An Evolutionary Approach* (ed. C. M. Herrera & O. Pellmyr). Oxford: Blackwell Science, pp. 3–25.

Price, P. W., Bouton, C. E., Gross, P., McPheron, B. A., Thompson, J. N. & Weis, A. E. (1980) Interactions among three trophic levels: influence of plants on interactions between insect herbivores and natural enemies. *Annual Review of Ecology and Systematics*, **11**, 41–65.

Price, P. W., Westoby, M., Rice, B. *et al.* (1986) Parasite mediation in ecological interactions. *Annual Review of Ecology and Systematics*, **17**, 487–505.

Queiroz, J. M. & Oliveira, P. S. (2001) Tending-ants protect honeydew-producing whiteflies (Homoptera: Aleyrodidae). *Environmental Entomology*, **30**, 295–297.

Rashbrook, V. K., Compton, S. G. & Lawton, J. H. (1992) Ant–herbivore interactions: reasons for the absence of benefits to a fern with foliar nectaries. *Ecology*, **73**, 2167–2174.

Rico-Gray, V. (1993) Use of plant-derived food resources by ants in the dry tropical lowlands of coastal Veracruz, Mexico. *Biotropica*, **25**, 301–315.

Rico-Gray, V. & Castro, G. (1996) Effect of an ant–aphid interaction on the reproductive fitness of *Paullinia fuscecens* (Sapindaceae). *Southwestern Naturalist*, **41**, 434–440.

Rico-Gray, V. & Thien, L. B. (1989) Ant–mealybug interaction decreases reproductive fitness of *Schomburgkia tibicinis* (Orchidaceae) in Mexico. *Journal of Tropical Ecology*, **5**, 109–112.

Room, P. M. (1972) The fauna of the mistletoe *Tapinanthus bangwensis* growing on cocoa in Ghana: relationships between fauna and mistletoe. *Journal of Animal Ecology*, **41**, 611–621.

Schupp, E. W. & Feener, D. H. (1991) Phylogeny, lifeform, and habitat dependence of ant-defended plants in a Panamanian forest. In *Ant–Plant Interactions* (ed. C. R. Huxley & D. F. Cutler). Oxford: Oxford University Press, pp. 175–197.

Stout, J. (1979) An association of an ant, a mealy bug and an understorey tree from a Costa Rican rain forest. *Biotropica*, **11**, 309–311.

Thompson, J. N. (1988) Variation in interspecific interactions. *Annual Review of Ecology and Systematics*, **19**, 65–87.

(1994) *The Coevolutionary Process*. Chicago: University of Chicago Press.

(1997) Conserving interaction biodiversity. In *The Ecological Basis of Conservation: Heterogeneity, Ecosystems, and Biodiversity* (ed. S. T. A. Pickett, R. S. Ostfeld, M. Shachak & G. E. Likens). New York: Chapman & Hall, pp. 285–293.

(2002) Plant–animal interactions: future directions. In *Plant–Animal Interactions: An Evolutionary Approach* (ed. C. M. Herrera & O. Pellmyr). Oxford: Blackwell Science, pp. 236–247.

Tobin, J. E. (1994) Ants as primary consumers: diet and abundance in the Formicidae. In *Nourishment and Evolution in Insect Societies* (ed. J. H. Hunt & C. A. Nalepa). Oxford: Westview Press, pp. 279–308.

Völkl, W. (1992) Aphids and their parasitoids: who actually benefits from ant-attendance? *Journal of Animal Ecology*, **61**, 273–281.

Völkl, W. & Kroupa, A. S. (1997) Effects of adult mortality risks on parasitoid foraging tactics. *Animal Behaviour*, **54**, 349–359.

Way, M. J. (1954) Studies of the association of the ant *Oecophylla longinoda* and the scale insect *Saissetia zanzibarensis*. *Bulletin of Entomological Research*, **45**, 113–134.

(1963) Mutualism between ants and honeydew-producing homoptera. *Annual Review of Entomology*, **8**, 307–344.

PART IV

Biotic interactions in human-dominated landscapes

CHAPTER EIGHTEEN

The alteration of biotic interactions in fragmented tropical forests

WILLIAM F. LAURANCE

Smithsonian Tropical Research Institute and National Institute for Amazonian Research (INPA)

Introduction

Tropical rainforests are renowned for their ecological complexity and the seeming ubiquity of coevolved relationships among species (Janzen 1969; Gilbert 1980). Unfortunately, these forests are being destroyed and fragmented at alarming rates, to the extent that many tropical protected areas are becoming virtual islands in a sea of heavily degraded land (Laurance & Bierregaard 1997; DeFries *et al.* 2002). The future of tropical biodiversity will be largely determined by the extent to which natural ecological processes and communities can be maintained in isolated fragments of forest.

Here I summarize available information on the alteration of biotic interactions, such as predation and key symbioses like pollination and seed dispersal, in fragmented tropical forests. This review is necessarily preliminary, given the great diversity of species and ecological interactions in the tropics and the fact that the alteration of biotic linkages is among the most poorly understood consequences of habitat fragmentation.

Initial impacts of fragmentation

Forest fragmentation leads to the reduction and isolation of remnant forest patches. Although each fragmented landscape is unique, a common pattern is that most forest fragments in human-dominated landscapes are small, ranging in size from a hectare or less to a few hundred hectares (Gascon *et al.* 2000; Cochrane & Laurance 2002). The biota of such small forest patches are expected to be vulnerable to edge effects and many other negative consequences of fragmentation (e.g. Laurance *et al.* 1998, 2002).

As forest-clearing proceeds, some species that are initially present in a landscape may be completely extirpated by habitat loss, and others may persist in tiny numbers in fragments, especially if they are rare or patchily distributed (termed the sample effect; Wilcox 1980). Forest-dependent species will be demographically and genetically isolated to varying degrees in fragments, depending on

Biotic Interactions in the Tropics: Their Role in the Maintenance of Species Diversity, ed. D. F. R. P. Burslem, M. A. Pinard and S. E. Hartley. Published by Cambridge University Press. © Cambridge University Press 2005.

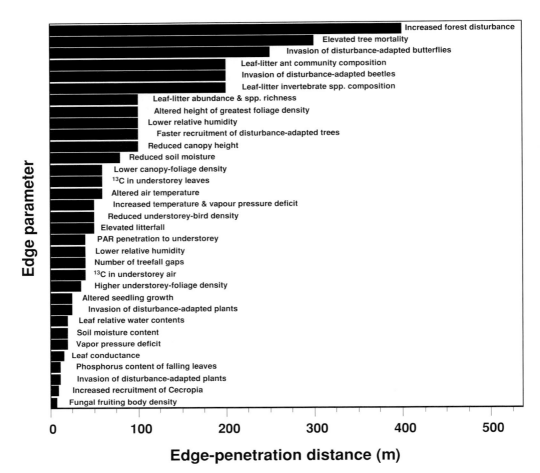

Figure 18.1 Penetration distances of different edge effects into Amazonian forest remnants (after Laurance *et al.* 2002). The distances shown here are typical; landscape factors such as edge age and the type of modified vegetation adjoining fragments can also have a significant influence on the intensity and penetration of edge effects.

their intrinsic characteristics (such as their mobility and tolerance of degraded habitats; Diamond *et al.* 1987; Laurance 1991) as well as attributes of the particular landscape (such as the degree to which the fragment is isolated from other forests and the types of degraded lands surrounding the fragment). When isolated, small populations (< 100 individuals) are vulnerable to local extinction from random demographic and genetic effects, frequently exacerbated by environmental variations and local catastrophes (Leigh 1981; Shafer 1981; Lande 1988).

In addition to physical isolation, an important driver of ecological change in forest fragments is edge effects (Fig. 18.1) – diverse physical and biotic alterations associated with the abrupt, artificial margins of habitat fragments (Janzen 1986;

Lovejoy *et al.* 1986; Murcia 1995; Laurance *et al.* 2002). Fragmented tropical forests often have sharply elevated tree mortality, a proliferation of disturbance-adapted trees and vines (Viana *et al.* 1997; Laurance *et al.* 1998), and hotter, drier conditions near forest boundaries (Kapos 1989). Most edge effects penetrate anywhere from a few metres to a few hundred metres into fragments, although some phenomena (such as surface fires and certain invading species) can penetrate as far as 2–10 km into forests (Laurance 2000). As a result of such changes, some forest-dependent species decline near fragment margins whereas disturbance-adapted and generalist species increase, sometimes dramatically (Lovejoy *et al.* 1986; Restrepo *et al.* 1999; Laurance *et al.* 2002).

Fragmented forests are not merely reduced and isolated; they also face additional threats from hunting, logging, surface fires, pollution, and climatic and atmospheric change (e.g. Peres 2001; Weathers *et al.* 2001; Cochrane & Laurance 2002; Milner-Gulland *et al.* 2003; Laurance 2004). The impacts of such ancillary threats on fragmented populations are poorly understood, in part because they vary greatly in intensity among different landscapes and regions, complicating efforts to develop simple generalizations. However, because hunting is nearly ubiquitous and can have important impacts on tropical vertebrates and their ecological roles in fragmented forests (Cullen *et al.* 2000; Peres 2001), its effects will be considered briefly here.

Many species are highly vulnerable in fragmented and hunted tropical landscapes, including large predators and herbivores, many understorey birds, some primates, a variety of insects and various plant species (Lovejoy *et al.* 1986; Didham *et al.* 1998; Restrepo *et al.* 1999; Laurance *et al.* 2002). Clearly, the diverse changes in species composition, species abundances, forest dynamics and physical environment alter some – and possibly many – biotic interactions in tropical forest fragments. As discussed below, what is far more poorly understood is the degree to which these altered interactions accelerate local species extinctions or disrupt key ecological processes in fragmented forests, especially over longer time scales ranging from centuries to millennia (e.g. Kellman *et al.* 1996; Corlett & Turner 1997).

Pollination

Most tropical plants are pollinated by insects, birds or bats, and rely on pollinators to maintain genetic diversity through outcrossing (Bawa 1990; Ricklets & Renner 1994; Momose *et al.* 1998). Recent analyses confirm that inbreeding depression in small, isolated populations of plants can reduce seed set, germination, survival and resistance to stress; and that populations with reduced genetic variation can experience reduced growth and increased extinction rates (cf. Keller and Waller 2002 and references therein).

Three factors largely determine the vulnerability of a plant species to the loss of its pollinators in fragmented forests. The first is the degree to which the plant is reliant on pollinators for reproduction. Many tropical plants are obligate

outbreeders, being either dioecious or self-incompatible, or possess mixed mating systems that are highly outcrossed through various mechanisms that select
for non-self pollen or outcrossed seed (Bawa & Opler 1975; Renner & Ricklefs
1995; Kenta *et al.* 2002). Many species are also rare, with effective populations
often encompassing hundreds to many thousands of hectares (Murawski &
Hamrick 1991; Chase *et al.* 1996; Stacy *et al.* 1996; Nason *et al.* 1998), and they frequently have clumped distributions, and thus can be vulnerable to population
isolation and an erosion of genetic diversity. Hence, for most tropical plants,
animal pollinators are critical.

The second factor is the degree to which the plant depends on specific pollinator species. Plants that rely on just one or a few pollinator species may be
more vulnerable to habitat disturbance than are those can utilize a suit of different pollinators (but see Aizen *et al.* 2002). Only a relatively small number
of plant species, such as orchids and figs, have obligate relationships with a
single bee or wasp pollinator (Bawa 1990). A restricted suite of pollinators is
also common on faunally depauperate islands, where a number of plants rely
on specific birds (Cox 1983) or flying foxes (Cox *et al.* 1991; Rainey *et al.* 1995)
for pollination and seed dispersal. The majority of other plant species, however, have a higher degree of redundancy in their pollinators (e.g. Horvitz &
Schemske 1990). Notably, most tropical trees rely on insects and are pollination
generalists (Ashton 1969; Bawa & Opler 1975; Stacy *et al.* 1996; Momose *et al.*
1998).

The third factor is the impact of habitat fragmentation on the pollinators. At
present, only a few generalizations can be made about the responses to fragmentation of pollinators, as species respond to habitat fragmentation in individual-
istic ways depending on their tolerance of edge and degraded habitats, population density, trophic position and other factors (Terborgh 1974; Laurance 1991;
Didham *et al.* 1998). Among Amazonian butterflies, for example, a number of
forest-interior species (especially in the Ithomiinae) decline in forest fragments,
whereas light-loving generalist species increase (Brown & Hutchings 1997). Of 12
euglossine-bee species studied by Powell & Powell (1987), four declined in abundance in recently isolated Amazonian forest fragments but other euglossines
increased in the modified habitats surrounding the fragments (see also Becker
et al. 1991). Larger pollinators, such as many hummingbirds (Aizen & Feinsinger
1994a; Stouffer & Bierregaard 1995; Aldrich & Hamrick 1998) and nectar- and
fruit-eating bats (Kalko 1998; Law & Lean 1999; Sampaio 2000), appear relatively
resilient to habitat fragmentation and can often traverse the degraded lands
surrounding fragments (Steffan-Dewenter & Tscharntke 1999). When subjected
to both fragmentation and hunting, however, flying foxes have often declined
precipitously, especially on oceanic islands (Cox *et al.* 1991; Rainey *et al.* 1995).
Even when pollinators do not decline markedly in fragmented forests, changes
in the distribution of resources can alter pollinator foraging behaviour with

respect to flower fidelity and foraging distance, which affects the quality and quantity of pollen arriving at the plant (Aizen & Feinsinger 1994a; Aldrich & Hamrick 1998; Ghazoul & McLeish 2001).

To date, only a few studies have provided unequivocal evidence of reduced plant reproduction in response to fragmentation-induced changes in pollinators. One of the most important studies is that of Aizen and Feinsinger (1994a), who worked not in tropical rainforest but in dry subtropical forest in the Argentinian Chaco. Among 16 plant species they studied (most of which were mainly bee-pollinated), 13 had reduced seed production in forest fragments. Moreover, among three self-compatible species, plants in fragments appeared to receive most pollen from themselves or from closely related neighbours, suggesting that pollinators were unable to transport pollen between different fragments (Aizen & Feinsinger 1994a).

Other evidence of negative fragmentation effects is more anecdotal. In the central Amazon, Dick (2001a) observed an absence of small beetles and stingless bees on mass-flowering trees in cattle pastures, and suggested that plants that are pollinated exclusively by small beetles, such as those in the nutmeg family (Myristicaceae), could be especially vulnerable to fragmentation. In Costa Rican forest fragments, Ghazoul and McLeish (2001) found no decrease in the abundance of *Trigona* bee pollinators for the widespread tree *Anacardium excelsum*, but observed that pollination success and seed production was positively correlated with fragment size. They proposed that *Trigona* bees rarely move among fragments, reducing genetic variability, fertilization and seed set among trees in small fragments. Aldrich and Hamrick (1998) found that reproduction of the bat-dispersed tree *Symphonia globulifera* in forest fragments in Costa Rica was dominated by a few large, isolated trees in pastures that had high rates of selfing, suggesting that fragmentation might eventually cause a genetic bottleneck in the population. Finally, Murren (2002, 2003) found that epiphytic orchids (*Catasetum viridiflavum*) on small man-made islands in Lake Gatun, Panama had lower seed set than did plants in nearby mainland areas, suggesting that pollination by euglossine bees was reduced on islands, although there was no evidence of reduced gene flow among island populations.

A collapse of plant–pollinator mutualisms has also been observed on oceanic islands, which are somewhat analogous to habitat fragments in being isolated and having relatively depauperate biotas. In Hawaii, the loss or severe decline of native sphingid-moth pollinators has been implicated in the very low fruit set of endemic *Brighamia* (Campanulaceae). Moreover, Hawaiian plants like *Stenogyne kanehoana* (Lamiaceae) have not set seeds for many years because their honeycreeper pollinators have been extirpated (Renner 1998). The decimation of flying foxes on many Pacific islands has also led to reduced reproduction in endemic plants that rely on these large bats for pollination (Cox *et al.* 1991; Rainey *et al.* 1995).

In other cases, however, fragmentation has had little effect on plant reproduction because native pollinators are unaffected or recover quickly from population declines. Two species of hummingbird-pollinated plants in the Argentinian Chaco actually had stable or increased reproduction, evidently because hummingbirds preferred the flower-rich fragments and readily moved among them (Aizen & Feinsinger 1994a). After being decimated by a hurricane, populations of fig wasps recovered quickly (within 5–16 months) because the wasps were able to disperse at least 60 km from other forests that were unaffected by the hurricane (Bronstein & Hossaert-McKey 1995). Harrison (2003) inferred that some fig-wasp species in an isolated forest fragment (4500 ha) in Borneo had arrived by dispersing from forests at least 30 km away, although wasps that pollinated dioecious figs were poorly represented and appeared to be much more dispersal-limited than monoecious-fig wasp species. Nason and Hamrick (1997) found that small populations of the neotropical tree *Spondias mombin* in forest fragments actually had higher rates of pollen immigration than did trees in intact forests, suggesting that fragmentation did not result in genetic isolation of populations. Indeed, studies of genetic markers suggest that small fragments and isolated trees in pastures can act as important stepping stones for many pollinators (Boshier *et al.* 1995; Chase *et al.* 1996; Stacy *et al.* 1996; Dick 2001b; White *et al.* 2002; Dick *et al.* 2003; Murren 2003). Finally, Murcia (1996) found no effect of fragment size or edge effects on pollination of 16 plant species in Colombian cloud forests.

In some fragmented landscapes, plant reproduction and gene flow have been maintained despite severe declines of natural pollinators, when exotic animal species take over the pollination duties. Isolated canopy trees (*Dinizia excelsa*) in Amazonian pastures were genetically 'rescued' by abundant Africanized honeybees (*Apis mellifera scutellata*), which more than compensated for the decline of native insect pollinators and contributed to an increase in the production of outcrossed seeds in pasture trees, relative to those in intact forest (Dick 2001a, 2001b; Dick *et al.* 2003). Similarly, in the Argentinian Chaco, the proportion of flower visits by Africanized honeybees increased in forest fragments relative to intact forest (Aizen & Feinsinger 1994b). In Hawaii, a number of endemic plant species that formerly relied on honeycreepers for pollination are now being pollinated by introduced bird species (Renner 1998). Although exotic pollinators help to maintain plant reproduction in some cases, it is possible that they may displace native pollinators and thereby contribute to the breakdown of natural plant–pollinator interactions.

In summary, habitat fragmentation and related disturbances alter many plant–pollinator interactions, but the effects of these changes on plant reproduction vary considerably among different studies and landscapes. In some cases, ecological redundancy of pollinators and exotic pollinators limit the impact of declining native pollinators, but in many other cases plant fecundity is clearly being reduced. In general, pollen movement among the most isolated fragments is more likely to be mediated by large pollinators, such as birds, bats and large bees, than by small insects (Steffan-Dewenter & Tscharntke 1999). Surprisingly, self-compatible and self-incompatible plants may show little overall difference in their vulnerable to fragmentation, despite the latter being more reliant on animal pollinators (Aizen *et al.* 2002). However, canopy trees, which are typically pollination generalists, could be less vulnerable to fragmentation than are understorey plants, which rely on fewer pollinator species (Aizen & Feinsinger 1994a; Momose *et al.* 1998; Renner 1998). Plant–pollinator relationships are often asymmetrical, with the pollinator being more specialized on the host plant than vice versa (as in some solitary bees; Renner 1998), and thus host-specific pollinators could be vulnerable in fragmented forests if their host plants decline.

Seed dispersal, seed predation and herbivory

Although some tropical plants rely on wind, gravity or water to disperse their propagules, the bulk of species have animal-dispersed fruits or seeds. Birds and bats are the main dispersers for smaller-seeded plants, whereas terrestrial and arboreal mammals disperse larger-seeded species (Levey *et al.* 1994). A distinction can be drawn between primary dispersers (such as monkeys) that remove fruit or seeds directly from tree crowns, and secondary dispersers (such as terrestrial mammals) that remove fallen fruit and seeds (Wright 2003).

Many seed-dispersing animals are also seed predators, although different species vary greatly in the degree to which they consume seeds; for example, peccaries and white-tailed deer are almost exclusively predators of large seeds, whereas tapirs act primarily as large-seed dispersers and agoutis are major seed predators but aid plant recruitment by caching and burying seeds (Brewer *et al.* 1997). Insects (notably bruchid beetles) are usually seed predators (Janzen 1970). Seed dispersal can provide a critical escape from severe density-dependent seed predation, seedling herbivory and pathogen attack near parent plants (Janzen 1970; Connell 1971), and there is a strong indication that some tree species would be unable to recruit without mammalian dispersers (Augspurger 1984; Chapman & Chapman 1995).

How does forest fragmentation affect dispersal and predation of plant propagules? At least on small man-made islands, these processes appear to be profoundly disrupted. On tiny (< 1 ha) islands created in the late 1980s in Venezuela, predators of vertebrates have disappeared and some of their prey, including rodents, howler monkeys, iguanas and leaf-cutter ants, have increased 10- to 100-fold in density. As a consequence, seedlings and saplings of canopy trees are severely reduced, providing evidence of a trophic cascade that is ramifying throughout the animal and plant communities (Terborgh *et al.* 2001). On equally small (< 2 ha) but older (c. 80-year-old) islands in Lake Gatun, Panama, most mammals have disappeared but rats (*Proechimys semispinosus*) have increased sharply in abundance, and predation on large seeds is sharply elevated relative to mainland forest (Asquith *et al.* 1997). On Lake Gatun islands that are too small to support any mammals, tree communities are dominated by a few large-seeded species (Putz *et al.* 1990; Leigh *et al.* 1993) that may out-compete smaller-seeded trees when key predators of large seeds, such as agouti and paca, are absent (Terborgh 1992).

It is unclear, however, whether such trends are typical of 'real' forest fragments, which are surrounded by a matrix of modified land and subjected to varying levels of hunting pressure. Fragmented forests in Los Tuxtlas, Mexico, have experienced sharply reduced grazing pressure and increased plant regeneration, leading to dense monospecific carpets of seedlings on the forest floor (Dirzo & Miranda 1991). As discussed by Asquith *et al.* (1997), this may arise from defaunation of larger (> 1 kg) predatory and herbivorous mammals by hunters and, perhaps more importantly, from suppression of small-rodent populations by abundant vipers (*Bothrops* spp.). The net effect is that plant recruitment at Los Tuxtlas is very different from that observed on small islands.

Plant recruitment is often altered in fragmented forests. In Costa Rican fragments, Guariguata *et al.* (2002) found higher seed predation by small rodents, and reduced dispersal, for one canopy tree species (*Dipterix panamensis*) but not for a second species (*Carapa guianensis*). Tabarelli *et al.* (1999) found that small (< 15 ha) forest fragments of Brazilian Atlantic forest isolated for over 50 years had more abiotically dispersed trees and shrubs, and fewer animal-dispersed trees, than did larger (370–7900 ha) fragments, probably because of large-scale declines of frugivores in small fragments. In Tanzania, a number of tree species in small (< 9 ha) forest fragments show reduced recruitment, apparently because of sharp declines in frugivorous birds and primates in the fragments (Cordeiro & Howe 2001, 2003).

In at least two studies, fragmentation reduced plant recruitment by promoting an influx of seed predators from surrounding degraded habitats. In a national park in Indonesian Borneo, large influxes of vertebrate seed predators from adjoining degraded lands led to a dramatic collapse in recruitment of dipterocarp canopy trees during a major masting event (Curran *et al.* 1999). Similar patterns were observed at Pasoh Forest Reserve in Peninsular Malaysia, where pigs (*Sus scrofa*) achieved abnormally high population densities by feeding on oil palms surrounding the reserve and then severely damaged the understorey vegetation (Ickes *et al.* 2001) and dipterocarp recruitment within the reserve. Moreover, when the surrounding oil palms were felled in 2001, pig populations crashed and dipterocarp recruitment in the reserve recovered following a general flowering event the following year (S. J. Wright, personal communication).

Some evidence suggests that large-seeded plants may be more vulnerable than small-seeded species in fragmented landscapes because the large vertebrates on which they rely for dispersal, such as ungulates and primates, are often sharply reduced in abundance by the combined effects of hunting and fragmentation (Fuentes 2000; Peres 2001). In central Panama, hunting and fragmentation had negative effects on long-distance recruitment of the large-seeded palm *Attalea butyraceae* (Wright & Duber 2001). Seed predation and dispersal of seeds away from presumptive parent trees were reduced on small islands and in areas subjected to heavy hunting pressure, relative to lightly hunted and intact-forest sites. Seedling densities were highest around parent trees on islands and in heavily poached sites, suggesting that human disruption of mammal communities may eventually reduce local tree diversity by disrupting natural density-dependent seed predation (Wright & Duber 2001). Small-seeded plants can also be vulnerable if their key dispersers respond negatively to fragmentation, as occurs in some species dispersed by birds (e.g. Cordeiro & Howe 2001, 2003) or by dung beetles (e.g. Chapman *et al.* 2003).

Forest fragmentation can also affect rates of leaf herbivory, although this phenomenon has been relatively little studied in the tropics. As discussed above, Terborgh *et al.* (2001) found intense herbivory on seedlings and saplings on small man-made islands in Venezuela, following population explosions of leaf-cutting ants and other herbivores on the islands. Forest fragments in the Amazon and Brazilian Atlantic region also show atypically high densities of leaf-cutting ants, especially near fragment margins, which could increase mortality and turnover of young trees (Vasconcelos & Cherrett 1995; Laurance 2003). Herbivorous insects generally increase near the margins of Amazonian forest fragments, possibly in response to increased leaf flush there (Lovejoy *et al.* 1986; Didham 1997). Herbivory from mammals (mainly rodents) on seedlings and saplings was evidently greater on small islands than on large islands and mainland areas in Panama (Asquith *et al.* 1997). With the exception of Benitez-Malvido *et al.* (1999), who found little difference in insect damage to leaves between fragmented and intact Amazonian forests, most studies have found increased herbivory in forest fragments. The decline of many predators in fragments (see below), and the concomitant increases in herbivorous vertebrates and insects, may partially explain this pattern.

It must be emphasized that the effects of habitat fragmentation on plant recruitment and survival are often context specific, depending on the nature of the fragmented landscape, its constituent fauna and the relative impact of hunting on predatory and herbivorous mammals. As proposed by Wright (2003), the indirect effects of hunting on plant communities could depend on whether large carnivores are being hunted (such as for the fur trade) or whether hunting is principally for food, in which case agoutis, deer and other herbivorous mammals are mainly taken. In the former case, abundances of larger-bodied seed predators and herbivores may actually increase, whereas in the latter case they are likely to decline.

Predation

Large predators such as jaguars, tigers and some raptors are more vulnerable to local extinction in fragmented forests than are species at lower trophic levels, because predators require large home ranges and are frequently persecuted by humans. The disappearance of large predators may have important ecological impacts in fragmented landscapes, especially if top-down regulation of prey species is common (Terborgh 1992). The phenomenon of 'mesopredator release' was coined to describe population explosions of medium-sized omnivores (e.g. agoutis, coati mundis, opossums, raccoons, coyotes) in fragmented habitats where large predators have disappeared (Soulé *et al.* 1988; Terborgh 1992; Crooks & Soulé 1999).

Mesopredator release has been documented in fragmented temperate habitats (Soulé *et al.* 1988; Crooks & Soulé 1999) but its importance in the tropics is uncertain. The near extirpation of jaguars, pumas and harpy eagles on Barro Colorado Island, Panama might have led to unusually high densities of medium-sized mammals such as agoutis, coati mundis, sloths and howler monkeys (Glanz 1982; Terborgh 1992), which could account for elevated nest predation on the island (Loiselle & Hoppes 1983; Sieving 1992). It is unclear, however, whether medium-sized mammals are actually hyper-abundant on Barro Colorado Island or merely appear to be so, as they are habituated to humans and therefore conspicuous to researchers on the island (Wright *et al.* 1994). In tropical Australia, atypically high abundances of native rodents in fragmented rainforests may result from the loss of larger rainforest-dependent predators, although edge-related forest disturbances could also favour rodents (Laurance 1994, 1997).

Another potential impact of forest fragmentation is that generalist omnivores, seed predators and brood parasites may flood into fragments from surrounding degraded lands (Curran *et al.* 1999; Ickes *et al.* 2001), causing negative impacts on nesting birds, other small vertebrates and plant communities, especially near forest edges. This phenomenon has been demonstrated in many fragmented temperate and boreal forests (e.g. Gates & Gysel 1978; Wilcove 1985; Paton 1994), but relatively few such studies have been conducted in the tropics. Experimental studies provided no evidence of increased nest or seed predation near forest edges in central Amazonia (Meyer 1999; Bruna 2002) and tropical Australia (Laurance *et al.* 1993; Laurance & Grant 1994; Harrington *et al.* 1997), although Lindell (2000) found increased predation on quail eggs near forest edges in Costa Rica.

It is clear, nevertheless, that important changes in predator and insectivore communities do occur in fragmented forests. In the central Amazon, predation on understorey invertebrates may be reduced because insectivorous bats, understorey birds and army ants all decline sharply (Laurance *et al.* 2002). In tropical Australia, larger rainforest-dependent predators, such as spotted-tailed quolls (Fig. 18.2), rufous owls and larger pythons, decline sharply in forest fragments and are replaced by smaller, generalist predators like barn owls, red-bellied

Figure 18.2 The decline in fragmented landscapes of forest-dependent predators, such as this spotted-tailed quoll (*Dasyurus maculatus*) from tropical Australia, may lead to population explosions of omnivorous mammals that in turn prey on nesting birds, invertebrates, seeds and other rainforest species (photo by S. E. Williams).

blacksnakes and white-tailed rats (Laurance 1997). As discussed above, at least on small man-made islands, striking trophic cascades have been observed following the disappearance of large predators (Terborgh *et al.* 1997, 2001). Such wholesale changes in predator communities must alter predation pressure on different groups of vertebrates and invertebrates.

Other ecological interactions

Fragmentation could potentially alter a great diversity of ecological interactions. For example, army ants require extensive areas for foraging and disappear from small forest fragments in the Amazon and East Africa; in turn, obligate ant-following birds, which prey on insects that flee the marauding ants, also frequently disappear (Harper 1989; Bierregaard *et al.* 1992; Peters & Kramer 2003). The disappearance of collared peccaries (*Tayassu pecari*) from Amazonian forest fragments leads to declines of several frog species that require the muddy pools created by wallowing peccaries, for reproduction (Zimmerman & Bierregaard 1986). The decline of dung beetles in fragmented forests could have large effects on the phoretic mites that rely on these beetles for dispersal (Klein 1989). Altered levels of insect herbivory in fragments could potentially affect the vulnerability of rainforest plants to infection by fungal pathogens (Benitez-Malvido *et al.* 1999).

It is apparent that these and a great many other ecological interactions are being altered by habitat fragmentation. For some phenomena, such as population explosions of leaf-cutting ants along forest-fragment margins (Vasconcelos & Cherrett 1995; Laurance 2003) that could negatively affect plant recruitment, we have some inkling of their potential impacts on forest ecosystems. For many others – such as the potential effects of fragmentation on disease and parasite dynamics, and on ecological interactions that occur belowground or in the forest canopy – very little is known. In the future, researchers should continue to document the remarkably varied impacts of habitat fragmentation on biotic interactions, but we must also attempt to determine the effects of these alterations on species survival and key ecological processes, and their generality across a range of fragmented landscapes.

Acknowledgements

I thank Rhett Harrison, Christopher Dick, Joseph Wright, Helene Muller-Landau, Heraldo Vasconcelos and three anonymous referees for helpful comments on drafts of the manuscript.

References

Aizen, M. A., Ashworth, L. & Galetto, L. (2002) Reproductive success in fragmented habitats: do compatibility systems and pollination specialization matter? *Journal of Vegetation Science* **13**, 885–892.

Aizen, M. A. & Feinsinger, P. (1994a) Forest fragmentation, pollination, and plant reproduction in a Chaco dry forest. *Ecology* **75**, 330–351.

(1994b) Habitat fragmentation, native insect pollinators, and feral honey bees in Argentine 'Chaco Serrano'. *Ecological Applications* **4**, 378–392.

Aldrich, P. R. & Hamrick, J. L. (1998) Reproductive dominance of pasture trees in a fragmented tropical forest mosaic. *Science* **281**, 103–105.

Ashton, P. S. (1969) Speciation among tropical forest trees: some deductions in the light of recent evidence. *Biological Journal of the Linnean Society* **1**, 155–196.

Asquith, N. M., Wright, S. J. & Clauss, M. J. (1997) Does mammal community composition control recruitment in neotropical forests? Evidence from Panama. *Ecology* **78**, 941–946.

Augspurger, C. K. (1984) Seedling survival of tropical tree species: interactions of dispersal distance, light gaps, and pathogens. *Ecology* **65**, 1705–1712.

Bawa, K. S. (1990) Plant–pollinator interactions in tropical rain forests. *Annual Review of Ecology and Systematics* **21**, 399–422.

Bawa, K. S. & Opler, P. A. (1975) Dioecism in tropical forest trees. *Evolution* **29**, 167–179.

Becker, P., Moure, J. B. & Peralta, F. (1991) More about euglossine bees in Amazonian forest fragments. *Biotropica* **23**, 586–591.

Benitez-Malvido, J., Garcia-Guzman, G. & Kossman-Ferraz, I. (1999) Leaf-fungal incidence and herbivory on tree seedlings in tropical rainforest fragments: an experimental study. *Biological Conservation* **91**, 143–150.

Bierregaard, R. O., Lovejoy, T. E., Kapos, V., dos Santos, A. & Hutchings, R. (1992) The biological dynamics of tropical rainforest fragments. *BioScience* **42**, 859–866.

Boshier, D. H., Chase, M. R. & Bawa, K. S. (1995) Population genetics of *Cordia alliodora* (Boraginaceae), a neotropical tree. III. Gene flow, neighborhood, and population structure. *American Journal of Botany* **82**, 484–490.

Brewer, S. W., Rejmanek, M., Johnstone, E. & Caro, T. (1997) Top-down control in tropical forests. *Biotropica* **29**, 364–367.

Bronstein, J. L. & Hossaert-McKey, M. (1995) Hurricane Andrew and a Florida fig pollination mutualism: resilience of an obligate interaction. *Biotropica* **27**, 373–381.

Brown, K. S. Jr & Hutchings, R. W. (1997) Disturbance, fragmentation, and the dynamics of diversity in Amazonian forest butterflies. In *Tropical Forest Remnants: Ecology, Management, and Conservation of Fragmented Communities* (ed. W. F. Laurance & R. O. Bierregaard). Chicago: University of Chicago Press, pp. 91–110.

Bruna, E. M. (2002) Effects of forest fragmentation on *Heliconia acuminata* seedling recruitment in central Amazonia. *Oecologia* **132**, 235–243.

Chapman, C. A. & Chapman, L. J. (1995) Survival without dispersers: seedling recruitment under parents. *Conservation Biology* **9**, 675–678.

Chapman, C. A., Chapman, L. J., Vulinec, K., Zanne, A. & Lawes, M. J. (2003) Fragmentation and alteration of seed dispersal processes: an initial evaluation of dung beetles, seed fate, and seedling diversity. *Biotropica* **35**, 382–393.

Chase, M. R. Moller, C., Kesseli, R. & Bawa, K. S. (1996) Distant gene flow in tropical trees. *Nature* **383**, 398–399.

Cochrane, M. A. & Laurance, W. F. (2002) Fire as a large-scale edge effect in Amazonian forests. *Journal of Tropical Ecology* **18**, 311–325.

Connell, J. H. (1971) On the role of natural enemies in preventing competitive exclusion in some marine animals and in rain forest trees. In *Dynamics of Populations* (ed. P. den Boer & G. Gradwell). Wageningen: Centre for Agricultural Publishing and Documentation, pp. 298–313.

Cordeiro, N. J. & Howe, H. F. (2001) Low recruitment of trees dispersed by animals in African forest fragments. *Conservation Biology* **15**, 1733–1741.

(2003) Forest fragmentation severs mutualism between seed dispersers and an endemic African tree. *Proceedings of the National Academy of Sciences* **100**, 14052–14056.

Corlett, R. T. & Turner, I. M. (1997) Long-term survival in tropical forest remnants in Singapore and Hong Kong. In *Tropical Forest Remnants: Ecology, Management, and Conservation of Fragmented Communities* (ed. W. F. Laurance & R. O. Bierregaard). Chicago: University of Chicago Press, pp. 333–346.

Cox, P. A. (1983) Extinction of the Hawaiian avifauna resulted in a change of pollinators for the ieie, *Freycinetia arborea*. *Oikos* **41**, 195–199.

Cox, P. A., Elmquist, T., Pierson, E. D. & Rainey, W. E. (1991) Flying foxes as strong interactions in South Pacific island ecosystems: a conservation hypothesis. *Conservation Biology* **5**, 448–454.

Crooks, K. R. & Soulé, M. E. (1999) Mesopredator release and avifaunal extinctions in a fragmented system. *Nature* **400**, 563–566.

Cullen, L., Bodmer, R. E. & Padua, C. V. (2000) Effects of hunting in habitat fragments of the Atlantic forest, Brazil. *Biological Conservation* **95**, 49–56.

Curran, L. M., Caniago, I., Paoli, G. *et al.* (1999) Impact of El Niño and logging on canopy tree recruitment in Borneo. *Science* **286**, 2184–2188.

DeFries, R. S., Houghton, R. A., Hansen, M. C., Field, C. B., Skole, D. & Townsend, J. (2002) Carbon emissions from tropical deforestation and regrowth based on satellite observations for the 1980s and 1990s. *Proceedings of the National Academy of Sciences* **99**, 14256–14261.

Diamond, J. M., Bishop, K. D. & van Balen, S. (1987) Bird survival in an isolated Javan island or mirror? *Conservation Biology* **1**, 132–142.

Dick, C. W. (2001a) Habitat change, African honeybees and fecundity in the Amazonian tree *Dinizia excelsa* (Fabaceae). In *Lessons from Amazonia: The Ecology and Conservation of a Fragmented Forest* (ed. R. O. Bierregaard, C. Gascon, T. Lovejoy & R. Mesquita). New Haven: Yale University Press, pp. 146–157.

(2001b) Genetic rescue of remnant tropical trees by an alien pollinator. *Proceedings of the Royal Society of London B* **268**, 2391–2396.

Dick, C. W., Etchelecu, G. & Austerlitz, F. (2003) Pollen dispersal of tropical trees (*Dinizia excelsa*: Fabaceae) by native insects and African honeybees in pristine and fragmented Amazonian rainforest. *Molecular Ecology* **12**, 753–764.

Didham, R. K. (1997) The influence of edge effects and forest fragmentation on leaf-litter invertebrates in central Amazonia. In *Tropical Forest Remnants: Ecology, Management, and Conservation of Fragmented Communities* (ed. W. F. Laurance & R. O. Bierregaard). Chicago: University of Chicago Press, pp. 55–70.

Didham, R. K., Lawton, J. H., Hammond, P. M. & Eggleton, P. (1998) Trophic structure stability and extinction dynamics of beetles (Coleoptera) in tropical forest fragments. *Philosophical Transactions of the Royal Society of London B* **353**, 437–451.

Dirzo, R. & Miranda, A. (1991) Altered patterns of herbivory and diversity in the forest understory: a case study of the possible consequences of contemporary defaunation. In *Plant–Animal Interactions: Evolutionary Ecology in Tropical and Temperate Regions* (ed. P. Price, T. Lewinsohn, G. Fernandes & W. Benson). New York: John Wiley & Sons, pp. 273–287.

Fuentes, M. (2000) Frugivory, seed dispersal and plant community ecology. *Trends in Ecology and Evolution* **15**, 487–488.

Gascon, C., Williamson, G. B. & Fonseca, G. A. B. (2000) Receding edges and vanishing reserves. *Science* **288**, 1356–1358.

Gates, J. E. & Gysel, L. W. (1978) Avian nest dispersion and fledging success in field-forest ecotones. *Ecology* **59**, 871–883.

Ghazoul, J. & McLeish, M. (2001) Reproductive ecology of tropical forest trees in logged and fragmented habitats in Thailand and Costa Rica. *Plant Ecology* **153**, 335–345.

Gilbert, L. E. (1980) Food web organization and the conservation of neotropical diversity. In *Conservation Biology: An Evolutionary–Ecological Perspective* (ed. M. E. Soulé & B. A. Wilcox). Sunderland, MA: Sinauer Associates, pp. 11–33.

Glanz, W. E. (1982) The terrestrial mammal fauna of Barro Colorado Island: censuses and long-term changes. In *The Ecology of a Tropical Forest: Seasonal Rhythms and Long-term Changes* (ed. E. G. Leigh, A. S. Rand & D. Windsor). Washington DC: Smithsonian Institution Press, pp. 455–468.

Guariguata, M. R., Arias-Le Claire, H. & Jones, G. (2002) Tree seed fate in a logged and fragmented forest landscape, northeastern Costa Rica. *Biotropica* **34**, 405–415.

Harper, L. H. (1989) The persistence of ant-following birds in small Amazonian forest fragments. *Acta Amazônica* **19**, 249–263.

Harrington, G. N., Irvine, A. K., Crome, F. H. J. & Moore, L. A. (1997) Regeneration of large-seeded trees in Australian rainforest fragments: a study of higher-order interactions. In *Tropical Forest Remnants: Ecology, Management, and Conservation of Fragmented Communities* (ed. W. F. Laurance & R. O. Bierregaard). Chicago: University of Chicago Press.

Harrison, R. D. (2003) Fig wasp dispersal and the stability of a keystone plant resource in Borneo. *Proceedings of the Royal Society of London B* **270**, 576–579.

Horvitz, C. C. & Schemske, D. W. (1990) Spatiotemporal variation in insect mutualists of a neotropical herb. *Ecology* **71**, 1085–1097.

Ickes, K., DeWalt, S. J. & Appanah, S. (2001) Effects of native pigs (*Sus scrofa*) on the understory vegetation in a Malaysian lowland rain forest: an exclosure study. *Journal of Tropical Ecology* **17**, 191–206.

Janzen, D. H. (1969) Allelopathy by myrmecophytes: the ant *Azteca* as an allelopathic agent of *Cecropia*. *Ecology* **50**, 147–153.

(1970) Herbivores and the number of tree species in tropical forests. *American Naturalist* **104**, 501–528.

(1986) The eternal external threat. In *Conservation Biology: An Evolutionary-Ecological Perspective* (ed. M. E. Soulé & B. A. Wilcox). Sunderland, MA: Sinauer Associates, pp. 286–303.

Kalko, E. K. V. (1998) Organization and diversity of tropical bat communities through space and time. *Zoology: Analysis of Complex Systems* **101**, 281–297.

Kapos, V. (1989) Effects of isolation on the water status of forest patches in the Brazilian Amazon. *Journal of Tropical Ecology* **5**, 173–185.

Keller, F. K. & Waller, D. M. (2002) Inbreeding effects in wild populations. *Trends in Ecology and Evolution* **17**, 230–241.

Kellman, M., Tackaberry, R. & Meave, J. (1996) The consequences of prolonged fragmentation: lessons from tropical gallery forests. In *Forest Patches in Tropical Landscapes* (ed. J. Schelhas & R. Greenberg). Washington, DC: Island Press, pp. 37–59.

Kenta, T., Shimuzu, K., Nagagawa, M., Okada, K., Hamid, A. & Nakashizuka, T. (2002) Multiple factors contribute to outcrossing in a tropical emergent *Dipterocarpus tempehes*, including a new pollen tube guidance mechanism for self-incompatability. *American Journal of Botany* **89**, 60–66.

Klein, B. C. (1989) Effects of forest fragmentation on dung and carrion beetle communities in central Amazonia. *Ecology* **70**, 1715–1725.

Lande, R. (1988) Genetics and demography in biological conservation. *Science* **241**, 1455–1460.

Laurance, W. F. (1991) Ecological correlates of extinction proneness in Australian tropical rain forest mammals. *Conservation Biology* **5**, 79–89.

(1994) Rainforest fragmentation and the structure of small mammal communities in tropical Queensland. *Biological Conservation* **69**, 23–32.

(1997) Responses of mammals to rainforest fragmentation in tropical Queensland: a review and synthesis. *Wildlife Research* **24**, 603–612.

(2000) Do edge effects occur over large spatial scales? *Trends in Ecology and Evolution* **15**, 134–135.

(2003) Rainforests at risk from altered ant ecology. News feature, *BioMedNet* (www.bmn.com), 17 March.

(2004) Forest–climate interactions in fragmented tropical landscapes. *Philosophical Transactions of the Royal Society of London B* **359**, 345–352.

Laurance, W. F. & Bierregaard, R. O., eds. (1997) *Tropical Forest Remnants: Ecology, Management, and Conservation of Fragmented Communities.* Chicago: University of Chicago Press.

Laurance, W. F., Ferreira, L. V., Rankin-de Merona, J. & Laurance, S. G. (1998) Rain forest fragmentation and the dynamics of Amazonian tree communities. *Ecology* **79**, 2032–2040.

Laurance, W. F., Garesche, J. & Payne, C. W. (1993) Avian nest predation in modified and natural habitats in tropical Queensland: an experimental study. *Wildlife Research* **20**, 711–723.

Laurance, W. F. & Grant, J. D. (1994) Photographic identification of ground-nest predators in Australian tropical rainforest. *Wildlife Research* **21**, 241–248.

Laurance, W. F., Lovejoy, T. E., Vasconcelos, H. L. *et al.* (2002) Ecosystem decay of Amazonian forest fragments: a 22-year investigation. *Conservation Biology* **16**, 605–618.

Law, B. S. & Lean, M. (1999) Common blossom bats (*Syconycteris australis*) as pollinators in fragmented Australian tropical rainforest. *Biological Conservation* **91**, 201–212.

Leigh, E. G. (1981) The average lifetime of a population in a varying environment. *Journal of Theoretical Biology* **90**, 213–239.

Leigh, E. G., Wright, S. J., Herre, E. A. & Putz, F. E. (1993) The decline of tree diversity on newly isolated tropical islands: a test of a null hypothesis and some implications. *Evolutionary Ecology* **7**, 76–102.

Levey, D. J., Moermond, T. C. & Denslow, J. S. (1994) Frugivory: an overview. In *La Selva: Ecology and Natural History of a Neotropical Rain Forest* (ed. L. McDade, K. S. Bawa, H. Hespenheide & G. Hartshorn). Chicago: University of Chicago Press, pp. 282–294.

Lindell, C. (2000) Egg type influences predation rates in artificial nest experiment. *Journal of Field Ornithology* **71**, 16–21.

Loiselle, B. A. & Hoppes, W. G. (1983) Nest predation in insular and mainland lowland rain forest in Panama. *Condor* **85**, 93–95.

Lovejoy, T. E., Bierregaard, R. O., Rylands, A. *et al.* (1986) Edge and other effects of isolation on Amazon forest fragments. In *Conservation Biology: The Science of Scarcity and Diversity* (ed. M. E. Soulé). Sunderland, MA: Sinauer, pp. 257–285.

Meyer, J. (1999) Efeitos da fragmentação de floresta sobre predação em ovos de aves e remoção de sementes. Unpublished M.Sc. thesis. Manaus, Brazil: National Institute for Amazonian Research (INPA).

Milner-Gulland, E. J., Bennett, E. L. & Society for Conservation Biology 2002 Annual Meeting Wild Meat Group (2003) Wild meat: the bigger picture. *Trends in Ecology and Evolution* **18**, 351–357.

Momose, K., Yunoto, T., Nagamitsu, T. *et al.* (1998) Pollination biology in a lowland dipterocarp forest in Sarawak, Malaysia. I. Characteristics of the plant-pollination community in a lowland dipterocarp forest. *American Journal of Botany* **85**, 1477–1501.

Murawski, D. A. & Hamrick, J. L. (1991) The effect of the density of flowering individuals on the mating systems of nine tropical tree species. *Journal of Heredity* **67**, 167–174.

Murcia, C. (1995) Edge effects in fragmented forests: implications for conservation. *Trends in Ecology and Evolution* **10**, 58–62.

(1996) Forest fragmentation and the pollination of neotropical plants. In *Forest Patches in Tropical Landscapes* (ed. J. Schelhas & R. Greenberg). Washington DC: Island Press, pp. 19–36.

Murren, C. J. (2002) Effects of habitat fragmentation on pollination: pollinators, pollinia viability and reproductive success. *Journal of Ecology* **90**, 100–107.

(2003) Spatial and demographic population genetic structure in *Catasetum viridiflavum* across a human-dominated habitat. *Journal of Evolutionary Biology* **13**, 333–340.

Nason, J. D. & Hamrick, J. L. (1997) Reproductive and genetic consequences of forest fragmentation: two case studies of neotropical canopy trees. *Journal of Heredity* **88**, 264–276.

Nason, J. D., Herre, E. A. & Hamrick, J. L. (1998) The breeding structure of a tropical keystone plant resource. *Nature* **391**, 685–687.

Paton, P. W. C. (1994) The effect of edge on avian nest success: how strong is the evidence? *Conservation Biology* **8**, 17–26.

Peres, C. A. (2001) Synergistic effects of subsistence hunting and habitat fragmentation on Amazonian forest vertebrates. *Conservation Biology* **15**, 1490–1505.

Peters, M. & Kramer, M. (2003) Vulnerability of driver ants and ant-following birds in fragmented forests in Kenya. Published abstract, *Annual Meeting of the Association for Tropical Biology and Conservation*, Aberdeen, Scotland.

Powell, A. H. & Powell, G. V. V. (1987) Population dynamics of male euglossine bees in Amazonian forest fragments. *Biotropica* **19**, 176–179.

Putz, F. E., Leigh, E. G. & Wright, S. J. (1990) The arboreal vegetation on 70-year-old islands in the Panama Canal. *Garden* **14**, 18–23.

Rainey, W. E., Pierson, E. D., Elmqvist, T. & Cox, P. A. (1995) The role of flying foxes (Pteropodidae) in oceanic island ecosystems of the Pacific. In *Ecology, Evolution and Behavior of Bats* (ed. P. A. Racey & S. H. Swift). London: Symposia of the Zoological Society of London, vol. **67**, pp. 47–62.

Renner, S. S. (1998) Effects of habitat fragmentation on plant pollinator interactions in the tropics. In *Dynamics of Tropical Communities* (ed. D. M. Newberry, H. Prins & N. D. Brown). London: 37th Symposium of the British Ecological Society, pp. 339–358.

Renner, S. S. & Ricklefs, R. E. (1995) Dioecy and its correlates in the flowering plants. *American Journal of Botany* **82**, 596–606.

Restrepo, C., Gomez, N. & Heredia, S. (1999) Anthropogenic edges, treefall gaps, and fruit–frugivore interactions in a neotropical montane forest. *Ecology* **80**, 668–685.

Ricklefs, R. E. & Renner, S. S. (1994) Species richness within families of flowering plants. *Evolution* **48**, 1619–1636.

Sampaio, E. (2000) The effects of fragmentation on structure and diversity of bat communities in a central Amazonian tropical rain forest. Unpublished Ph.D. thesis, University of Tuebingen.

Shafer, M. L. (1981) Minimum population sizes for species conservation. *BioScience* **31**, 131–134.

Sieving, K. E. (1992) Nest predation and differential insular extinction among selected forest birds of central Panama. *Ecology* **73**, 2310–2328.

Soulé, M. E., Bolger, D. T., Alberts, A., Wright, J., Sorice, M. & Hill, S. (1988) Reconstructed dynamics of rapid extinctions of chaparral-requiring birds in urban habitat islands. *Conservation Biology* **2**, 75–92.

Stacy, E. A., Hamrick, J. L., Nason, J. D., Hubbell, S. P., Foster, R. B. & Condit, R. (1996) Pollen dispersal in low-density populations of three neotropical tree species. *American Naturalist* **148**, 275–298.

Steffan-Dewenter, I. & Tscharntke, T. (1999) Effects of habitat isolation on pollinator communities and seed set. *Oecologia* **121**, 432–440.

Stouffer, P. C. & Bierregaard, R. O. (1995) Effects of forest fragmentation on understory hummingbirds in Amazonian Brazil. *Conservation Biology* **9**, 1085–1094.

Tabarelli, M., Mantovani, W. & Peres, C. A. (1999) Effects of habitat fragmentation on plant guild structure in the montane Atlantic forest of southeastern Brazil. *Biological Conservation* **91**, 119–128.

Terborgh, J. (1974) Faunal equilibria and the design of wildlife preserves. In *Tropical Ecological Systems* (ed. F. B. Golley &

E. Medina). New York: Springer-Verlag, pp. 369–380.

(1992) Maintenance of diversity in tropical forests. *Biotropica* **24**, 283–292.

Terborgh, J., Lopez, L., Nuñez V. *et al.* (2001) Ecological meltdown in predator-free forest fragments. *Science* **294**, 1923–1926.

Terborgh, J., Lopez, L., Tello, J., Yu, D. & Bruni, A. (1997) Transitory states in relaxing ecosystems of tropical land-bridge islands. In *Tropical Forest Remnants: Ecology, Management, and Conservation of Fragmented Communities* (ed. W. F. Laurance & R. O. Bierregaard). Chicago: University of Chicago Press, pp. 256–274.

Vasconcelos, H. L. & Cherrett, J. M. (1995) Changes in leaf-cutting ant populations (Formicidae: Attini) after the clearing of mature forest in Brazilian Amazonia. *Studies on Neotropical Fauna and Environment* **30**, 107–113.

Viana, V. M., Tabanez, A. & Batista, J. (1997) Dynamics and restoration of forest fragments in the Brazilian Atlantic moist forest. In *Tropical Forest Remnants: Ecology, Management, and Conservation of Fragmented Communities* (ed. W. F. Laurance & R. O. Bierregaard). Chicago: University of Chicago Press, pp. 351–365.

Weathers, K. C., Cadenasso, M. L. & Pickett, S. T. A. (2001) Forest edges as nutrient and pollutant concentrators: potential synergisms between fragmentation, forest canopies, and the atmosphere. *Conservation Biology* **15**, 1506–1514.

White, G. M., Boshier, D. & Powell, W. (2002) Increased pollen flow counteracts fragmentation in a tropical dry forest: an example from *Swietenia humilis* Zuccarini. *Proceedings of the National Academy of Sciences* **99**, 2038–2042.

Wilcove, D. S. (1985) Nest predation in forest tracts and the decline of migratory songbirds. *Ecology* **66**, 1211–1214.

Wilcox, B. A. (1980) Insular ecology and conservation. In *Conservation Biology: An Evolutionary–Ecological Perspective* (ed. M. E. Soulé & B. A. Wilcox). Sunderland, MA: Sinauer Associates, pp. 95–117.

Wright, S. J. (2003) The myriad consequences of hunting for vertebrates and plants in tropical forests. *Perspectives in Plant Ecology, Evolution and Systematics* **6**, 73–86.

Wright, S. J. & Duber, H. C. (2001) Poachers and forest fragmentation alter seed dispersal, seed survival, and seedling recruitment in the palm *Attalea butyraceae*, with implications for tropical tree diversity. *Biotropica* **33**, 583–595.

Wright, S. J., Gompper, M. E. & DeLeon, B. (1994) Are large predators keystone species in neotropical forests? The evidence from Barro Colorado Island. *Oikos* **71**, 279–294.

Zimmerman, B. L. & Bierregaard, R. O. (1986) Relevance of the equilibrium theory of island biogeography and species–area relations to conservation with a case from Amazonia. *Journal of Biogeography* **13**, 133–143.

CHAPTER NINETEEN

Effects of natural enemies on tropical woody-plant invasions

SAARA J. DeWALT

Rice University

Plant invasions pose a current and increasing threat to species diversity and composition of forests worldwide. Biological invasions are natural ecological processes and the movement of plants across geographic barriers has always occurred (Sauer 1988), but humans have greatly accelerated the rate of introductions and moved plants across barriers that probably would not have been spanned naturally. Many of these plant introductions, whether deliberate or accidental, have had negative effects in the areas of introduction. Plant invasions have led to native species loss, altered ecosystem-level processes and caused enormous economic and environmental damage in various ecosystems, including tropical forests (Vitousek *et al.* 1987; Gordon 1998; Parker *et al.* 1999; Mack *et al.* 2000). Remote tropical islands and fragmented landscapes are particularly vulnerable to invasion by non-native plants (Laurance *et al.* 2002; Denslow 2003), but continental species-rich tropical rainforests also are invasible (Usher 1991; Rejmánek 1996a), particularly following natural or human disturbance (Whitmore 1991). Because of such negative impacts, species invasions are seen as one of the primary agents of global change (Vitousek *et al.* 1996), and tropical forests are unlikely to be immune (Fine 2002).

This chapter explores how biotic interactions, particularly herbivory and pathogen attack, may affect the abundance and distribution of invasive woody species in tropical rainforests. Specifically, I examine whether herbivores and fungal pathogens (natural enemies) are important determinants of species' abundance and distribution in their native ranges and whether the absence or reduced impact of these natural enemies may explain why certain introduced woody plants are successful invaders in tropical rainforests. I focus on insect herbivory, because of the greater degree of host specialization shown by insects than by vertebrates, and most studies I review consider insect rather than vertebrate herbivory. I then examine whether, and in which cases, classical biological control programs may be successful at substantial or complete control of invasive

Biotic Interactions in the Tropics: Their Role in the Maintenance of Species Diversity, ed. D. F. R. P. Burslem, M. A. Pinard and S. E. Hartley. Published by Cambridge University Press. © Cambridge University Press 2005.

species of tropical forests. Classical biological control, i.e. introduction of natural enemies from the native range, seeks to increase enemy loads to curb plant population growth rates.

Few studies have tested hypotheses of invasion relating to biotic interactions in tropical forests. Therefore, in this chapter I present many examples from temperate species of research related to the effects of natural-enemy loads on plant abundance and distribution. The implicit assumption is that the effects of natural enemies should be as important on tropical plants as on temperate plants.

Invasive tropical woody plant species

Worldwide, 653 woody species are considered invasive in the tropics and subtropics, and most are reported to invade highly disturbed forests (Binggeli *et al.* 1998). Of these, Binggeli *et al.* (1998) counted 37 species as invading tropical or subtropical 'natural' forest (as opposed to disturbed forest or plantation). Similarly, from searching the literature, Rejmánek (1996a) identified 42 woody species as being invasive in at least one area of old-growth tropical forest. Many other woody invaders have been found in disturbed, open areas in the tropics, but these 37 to 42 species are of particular interest because they directly threaten to disrupt ecological processes in otherwise intact forest stands. Half of the species Rejmánek (1996a) listed are invasive only on islands while the other half are invasive in continental forests. Gaps and fragmentation are likely to increase their susceptibility to invasion (invasibility), but at least 13 of the continental forest invaders are not restricted to gaps or forest fragments and are known to invade forest understorey (Rejmánek 1996a). For example, *Psidium guajava* (Myrtaceae) grows in closed forest in the Bundongo forest of western Uganda (Sheil *et al.* 2000); *Clidemia hirta* (Melastomataceae) invades forest understorey of Peninsular Malaysia, East Africa, Borneo and various islands in the Pacific and Indian Oceans (Wester & Wood 1977; Rejmánek 1996a); and *Miconia calvescens* proliferates in lowland forest of Tahiti and Hawaii (Gagné *et al.* 1992; Meyer 1996).

Of the 13 woody tropical forest invaders not restricted to gaps, at least seven are pioneers restricted to *open* habitats in their native ranges that invade *closed* tropical forests where they have been introduced (Table 19.1). They include species native to the New World and Old World tropics. This phenomenon may occur for more species, but information is scant for habitat distributions of many of these species in their native ranges.

In addition to changes in habitat distribution, other biological and ecological characteristics often differ between conspecifics in their native and introduced ranges. Some species are more abundant (Fowler *et al.* 1996; Williamson & Fitter 1996), have faster growth rates (Blossey & Nötzold 1995; Fowler *et al.* 1996), grow

Table 19.1 *Non-native invasive woody species of old-growth (OG) tropical forests that are found primarily in open areas in their native range*

Species (Family)	Native range	Habitat in native range	Introduced range	Habitat in introduced range	Reference or personal communication
Clidemia hirta (Melastomataceae)	Central and South America	Open, disturbed areas	Hawaii, Asia Africa	Open areas, OG forest	Wester & Wood (1977)
Cryptostegia grandiflora (Asclepiadaceae)	Madagascar	Open, disturbed areas	Northern Australia, Puerto Rico	Open areas, OG forest	L. Civeyrel pers. comm.
Dioscorea alata (Dioscoreaceae)	Asia	Gaps, forest edges	Tanzania	OG forest	R. Ganesan pers. comm.
Duranta erecta (Verbenaceae)	Central and South America	Open, dry, coastal areas	Tanzania, Uganda, Madagascar	OG forest	P. Acevedo pers. comm.
Litsea glutinosa (Lauraceae)	Asia	Gaps, forest edges	Mauritius	OG forest	R. Ganesan pers. comm.
Miconia calvescens (Melastomataceae)	Central and South America	Gaps, steep slopes	Tahiti, Hawaii	Open areas, OG forest	Binggeli et al. (1998)
Thunbergia grandiflora (Acanthaceae)	India	Gaps, forest edges	Singapore, Australia	OG forest	R. Ganesan pers. comm.

Invasive range is from Rejmánek (1996a). References or personal communications refer to the source of information for the habitat in which the species occurs in the native range.

to greater stature (Pritchard 1960; Blossey & Nötzold 1995), produce larger seeds (Buckley *et al.* 2003) or have higher seed production (Noble 1989; Paynter *et al.* 1996) where they have been introduced relative to where they are native. These differences suggest a positive change in environmental conditions or increased vigour of genotypes in the introduced range.

Importance of biotic interactions to invasion success

Biotic interactions affecting establishment of non-native species include some of the classic forms of ecological interactions: mutualism, predation and competition. Establishment and proliferation might occur only when organisms are present to pollinate non-native plants' flowers, disperse their seeds and aid in nutrient uptake (e.g. mycorrhizal associations, Richardson *et al.* 2000). Seed predation, herbivory and pathogen attack on non-native species should be equal to or less than those of native species to allow inclusion in the area of introduction (Keane & Crawley 2002). Finally, proliferation may be thwarted by competitors, native or non-native, already present (Tilman 1997). In several experimental studies, communities with greater species diversity have been found to be less susceptible to invasion (Tilman 1997; Naeem *et al.* 2000; Prieur-Richard *et al.* 2000), but others have found the opposite relationship (Levine & d'Antonio 1999; Smith & Knapp 1999; Stohlgren *et al.* 1999). Indeed, various environmental factors such as resource availability and disturbance may co-vary positively with overall species richness, leading to positive associations between non-native and native species diversity or abundance. Davis *et al.* (2000) suggested that resource levels may be the most important factor in establishment and spread of invasive species, but only given adequate dispersal and an absence of natural enemies. Rejmánek (1996a) found a poor correspondence between invasibility and diversity of tropical sites, suggesting that high tropical forest diversity is not a substantial barrier to invasion (Fine 2002). In general, the relative competitive ability of non-native species in tropical forests is likely to depend on myriad factors, including life-history characteristics of the invading species themselves, amount of resources and species richness within the recipient community, disturbance regime and the degree to which mutualisms and predation occur (Lonsdale 1999).

Hypotheses accounting for plant invasion

Several hypotheses have been proposed to account for the paradox that certain species are more common, vigorous and widely distributed among habitats in their introduced range than their native range. These hypotheses are not mutually exclusive and include enemy release, genetic shifts, evolution of increased competitive ability, increased resource availability and vacant niche.

(1) The enemy-release hypothesis (ERH) posits that plants in their introduced range experience a release from the herbivores and pathogens that limit them in their native range, which leads to a demographic release including higher survival, growth or reproduction (Elton 1958; Crawley 1987; Keane & Crawley 2002).

(2) The genetic-shift hypothesis states that conspecifics differ between their native and introduced ranges because of genetic differences in patterns of biomass allocation, photosynthetic traits or growth. These genetic differences may arise from the particulars of the number and identity of the original introductions or may result from natural selection after introduction (DeWalt *et al.* 2004a).

(3) The 'evolution of increased competitive ability' hypothesis (EICA) is a more specific version of the genetic-shift hypothesis and posits that genetic differences are due to a release from natural enemies and subsequent selection for faster-growing and less-well-defended plants (Blossey & Nötzold 1995; Blossey & Kamil 1996).

(4) The increased-resource-availability hypothesis states that resources are less limiting in the area of introduction than in the native range and that non-native species are able to capture these resources more efficiently than native species (Denslow 2003).

(5) The vacant-niche hypothesis postulates that non-native plants fill niches unoccupied by native species (see criticisms of the idea of a vacant niche in Herbold and Moyle 1986; Fryer 1991). This hypothesis predicts that abundance may be higher in the introduced than the native range because of reduced competition for resources. This hypothesis has also been framed in the light of functional diversity: communities will be more susceptible to invasion by species of unrepresented functional groups (Symstad 2000). Lack of functional diversity or presence of vacant niches may explain much of the success of introduced plants, particularly on remote tropical islands (Rejmánek 1996a; Fine 2002).

These hypotheses all propose a limitation to abundance or habitat distribution in the area of origin and then a change, shift or release from this limitation in the area of introduction. The two most frequently cited of these hypotheses, ERH and EICA, posit an important role for herbivores and pathogens in plant invasions. Therefore, in the remainder of this chapter, I focus on evidence for these two hypotheses.

The primary difference between ERH and EICA is that the changes in plant abundance, size or habitat distribution resulting from reduced herbivore and pathogen loads in the area of introduction involve phenotypic plastic responses on an ecological time scale (within the lifetime of an individual plant) under the ERH, but involve genetic responses on an evolutionary time scale (over generations) under the EICA hypothesis. The success of newly introduced species may be attributed to plastic phenotypic responses to a lack of herbivores or pathogens. Older introductions, on the other hand, may have displayed plastic responses at first but have since diverged genetically in patterns of biomass allocation, physiology and growth rates because of selection for greater vigour in the absence of natural enemies. Thus, there may be a two-step process by which invasive species become more abundant and vigorous in their introduced range: an initial phenotypic response to lack of natural enemies followed by a genetic change in resource allocation that optimizes performance in the absence of natural enemies.

Enemy-release hypothesis

Two predictions follow from the enemy-release hypothesis. The first prediction is that herbivores and fungal pathogens typically limit plant abundance through depression of plant growth, survival or reproduction. Many temperate and tropical studies have found that herbivores do decrease growth, survival or reproduction (e.g. Schierenbeck *et al.* 1994; Louda & Potvin 1995; Sagers & Coley 1995; Louda & Rodman 1996; Root 1996) and that pathogens cause tree-seedling death (Augspurger 1984; Augspurger & Kelly 1984; Stanosz 1994; Packer & Clay 2000). Herbivores and pathogens also differentially affect the relative abundance of some plants in adjacent habitats, such as open and shaded areas (Harper 1969; Louda *et al.* 1987; Louda & Rodman 1996). However, other studies have found less support for the hypothesis that herbivores or pathogens limit plant populations (Simberloff *et al.* 1978; McNaughton 1983; Whitham *et al.* 1991). The effects of herbivores and pathogens may differ among plant species and sites, or may depend on other biotic or abiotic factors, such as light or nutrient availability (Whitham *et al.* 1991; Hammond & Brown 1998). Thus, we cannot always assume that this first prediction holds for all species and communities.

The second prediction of the ERH is that natural enemies are less abundant or have less impact on plant populations in the introduced than in the native range and that this reduction in enemy load leads to demographic release. Natural enemies may include both above- and below-ground herbivores (vertebrate or invertebrate) that feed on any plant part (e.g. leaves, seeds, roots) and pathogens (viral or fungal). There are several ways that natural enemies could be absent or scarce in areas of introduction. First, generalist herbivores may be less common in the area of introduction, as may be the case on islands (Carlquist 1974) or areas of low plant species richness (Siemann *et al.* 1998; Prieur-Richard *et al.* 2002). Second, specialist enemies from the native range may be absent from the introduced range if they are not introduced with the plant. Third, native specialists may not switch hosts, and generalists may not include the non-native plant in their diet (Keane & Crawley 2002; Siemann & Rogers 2003c).

In essence, the enemy-release hypothesis has elements akin to the Janzen–Connell hypothesis, which proposes that plants from seeds dispersed far from maternal trees should have greater survivorship than ones falling closer because of the spatial escape from species-specific pathogens or herbivores that are more numerous near maternal trees (Janzen 1970; Connell 1971). Thus, species introductions are an extreme case of long-distance dispersal (Muller-Landau *et al.* 2003) and introduced plants would be predicted to become more numerous and closely spaced because of the escape from natural enemies (Reinhart *et al.* 2003).

Most tests of the ERH have taken one of two directions: (1) *interspecific* tests that compare native and non-native plants growing in the same community in terms of their herbivore or pathogen loads or the effects of these natural enemies on growth and survival (Blaney & Kotanen 2001a; Agrawal & Kotanen 2003; Siemann & Rogers 2003a), and (2) *intraspecific* tests that compare these variables between conspecifics in their native vs. introduced ranges (Memmott *et al.* 2000; Wolfe 2002; Beckstead & Parker 2003; Mitchell & Power 2003; Reinhart *et al.* 2003; DeWalt *et al.* 2004b). Interspecific tests address only the second prediction of the ERH, that natural enemies have limited effects on plants in their introduced range.

The success of biocontrol agents sometimes has been used to support the idea that natural-enemy release accounts for plant invasions, but there are several problems with this approach. As Keane and Crawley (2002) point out, biocontrol agents may themselves be released from their natural enemies (predators and parasitoids), which would lead to greater control of the plant species in the introduced range. In addition, the success of biocontrol agents does not show that these agents affect the demography of plants in their native range, the first prediction of the ERH. I therefore do not review any biocontrol cases in this discussion of the effects of natural enemies on plant invasions.

Despite the fact that interspecific studies do not provide a complete test of the ERH, they can provide important information about whether non-native invasive species are superior competitors to native species because of a reduction in natural-enemy loads. Interspecific studies have found mixed support for the prediction that natural enemies have less effect on non-native than native species. Agrawal and Kotanen (2003) grew 30 taxonomically paired plant species from old-field habitats (i.e. abandoned agricultural fields) in a common garden in Toronto, Canada. In contrast to predictions of the hypothesis, they found that most of the 15 non-native species suffered levels of leaf herbivory equal to or even greater than the 15 congeneric or confamilial native plants. Effects of herbivory on demographic parameters were not examined. In other studies of old-field and meadow vegetation, post-dispersal seed predation by granivores and fungal pathogens was similar for native and non-native plants (Blaney & Kotanen 2001a, b).

Siemann and Rogers (2003a) excluded above-ground fungal pathogens and insect herbivores using pesticides from seedlings of the invasive Chinese tallow tree (*Sapium sebiferum*) and the native hackberry tree (*Celtis laevigata*) for three years in three habitats: mesic forest, floodplain forest and coastal prairie sites in east Texas, United States. Consistent with the second prediction of the ERH hypothesis, in all habitats the non-native *S. sebiferum* seedlings in the control treatment (water only) suffered less damage by insect herbivores than the native species in the control treatment. Contrary to predictions, however, the relative benefit of insect herbivore exclusion on survivorship and growth was greater on *S. sebiferum* seedlings than on the native tree seedlings. They hypothesized that the competitive advantage of *S. sebiferum* over native vegetation results from lower allocation of resources to defence and greater allocation to growth and reproduction (support for the EICA hypothesis, see below). However, outbreak of specialist herbivores may control the native species' populations on a supra-annual time scale (Siemann & Rogers 2003a). Thus, even three-year studies may be too short to capture these important periodic episodes. Suppression of above-ground fungal pathogens had no effect on seedling survival or growth of either species, suggesting that fungal pathogens are not as important as insect herbi-vores in structuring the forest community in this area.

In contrast to the previous studies, an interspecific study that excluded insect and mammalian herbivores from one native and one introduced *Lonicera* vine species in open fields in South Carolina found that native species benefited more from enemy exclusion than the introduced species (Schierenbeck *et al.* 1994). Less leaf area was lost to insects and mammals for the introduced *L. japonica* than the native *L. sempervirens*. In addition, when damaged, *L. japonica* was able to com-pensate for leaf area lost to herbivory, whereas the native showed progressively less compensation in total biomass with increasing herbivory (Schierenbeck *et al.* 1994). Thus, the escape from herbivores at least partially explains the pro-liferation of the non-native *L. japonica* in the United States.

Biogeographical observational and experimental studies using the intraspe-cific test approach generally have found support for the ERH. Examining records for the number of species of viral and fungal pathogens on 473 species natural-ized in the United States from Europe, Mitchell and Power (2003) found that 84% fewer fungi and 24% fewer viruses infect plants in the United States compared with Europe. Interestingly, a similar phenomenon was found for the number of parasites on vertebrate and invertebrate animals in their native and introduced ranges (Torchin *et al.* 2003).

In addition to these meta-analyses, some field and greenhouse studies have found that natural-enemy abundance or damage is greater on plants in their native ranges. For example, both generalist and specialist herbivores caused more damage on the temperate perennial plant *Silene latifolia* in its native range of Europe than its introduced range of the United States (Wolfe 2002). Similar results were found for *Cytisus scoparius*, a shrub native to Europe that has been introduced and become invasive in New Zealand, Australia and the United States. A pair of studies found that the number of phytophagous species found on *C. scoparius* was lower in California (Bossard & Rejmánek 1994) than in Europe (Waloff & Richards 1977). In addition, specialist herbivores were more abundant on *Cytisus scoparius* in the native range than in New Zealand or Australia (Memmott *et al.* 2000). Interestingly, the number of generalist herbivores was similar between the two ranges. In another intraspecific biogeographical comparison, incidence of seed-eating insect larvae in flowerheads was greater on various Asteraceae species in their native range in Britain than in New Zealand, where they have been introduced (Fenner & Lee 2001).

Evidence for demographic release resulting from lower natural-enemy loads in the introduced range has been found in most but not all studies. Waloff and Richards (1977) found that adult survival and growth of *C. scoparius* were limited in the native range of Europe by insect herbivores. In a study conducted 20 years later, Bossard and Rejmánek (1994) found that insect herbivores do not affect growth of *C. scoparius* in the introduced range of California. Thus, there seems to be support that enemy release has occurred for *C. scoparius* in several areas where it has been introduced (Waloff & Richards 1977; Bossard & Rejmánek 1994; Memmott *et al.* 2000). In soil-sterilization experiments conducted in native and introduced ranges of *Prunus serotina*, Reinhart *et al.* (2003) found that the soil microbial community apparently negatively affects seedling growth and survival of black cherry in its native range in the United States, but positively affects these demographic parameters in the introduced range of north-western Europe. Their study suggested that *P. serotina* in Europe has escaped its soil pathogens but benefits from mycorrhizal associations, leading to the positive effects of soil microbes found in the introduced range. Klironomos (2002) hypothesized that introduced plants may accumulate species-specific soil-borne pathogens at a slower rate than native species, which would account for their ability to exist at high densities.

In contrast to these studies, escape from soil-borne natural enemies does not seem to explain the invasion of the coastal dune grass, *Ammophila arenaria*, into California coastal dunes. Sterilization of soils from both the native and introduced ranges had significant positive effects on seed germination and seedling growth in greenhouse studies conducted in the native (van der Putten & Peters 1997) and introduced ranges (Beckstead & Parker 2003). These results do not preclude the possibility that above-ground, rather than below-ground, herbivores or pathogens are significantly reduced in the introduced range (Beckstead & Parker 2003).

Test of the ERH with a tropical invader

The prediction that release from natural-enemy regulation in areas of introduction accounts in part for observed changes in plant abundance and habitat distribution was also tested experimentally for the invasive neotropical shrub

Clidemia hirta (Melastomataceae; DeWalt *et al.* 2004b). *Clidemia hirta* is an example of a tropical species in Table 19.1 that does not occur in forest understorey in its native range, but invades both high- and low-light environments in its introduced range. *Clidemia hirta* was planted into understorey and open habitats where it is native (Costa Rica) and where it has been introduced (Hawaii), and pesticides were applied to examine the effects of fungal pathogen and insect herbivore exclusion. Understorey light levels did not differ significantly between the Hawaiian and Costa Rican understorey sites (DeWalt *et al.* 2004b). Approximately one year after transplant, standing percentage leaf area missing on plants in the control (water only) treatment was five times greater on plants in Costa Rica than in Hawaii and did not differ between habitats. Plants in Costa Rica were attacked by leaf-feeding lepidoptera and weevils, gall-forming cecidomyiid flies, fungal pathogens and stem borers. In Hawaii, few plants had damage that could be attributed to insects. Those that were damaged were affected by non-native insects. A few plants in open sites were affected by *Liothrips urichi* (Phlaeothripidae), a thrips released as a biocontrol agent in the 1950s, and *Amorbia emigratella* (Tortricidae), a generalist introduced moth.

The results supported a demographic release from natural enemies in the introduced range. In understorey sites in Costa Rica, *C. hirta* survivorship over 13 months was 12% greater if sprayed with insecticide, 19% greater with fungicide, and 41% greater with both insecticide and fungicide compared with control plants sprayed only with water (Fig. 19.1). In contrast, suppression of natural enemies did not affect survival in open sites in Costa Rica or in either habitat in Hawaii, where survival was almost 100%. Fungicide application increased relative growth rates of plants that survived to the end of the experiment in both habitats of Costa Rica but not in Hawaii. These results suggest that fungal pathogens limit growth but not survivorship of *C. hirta* where it is native. For *C. hirta*, its absence from forest understorey in its native range probably results in part from the strong pressures of natural enemies. Its invasion into Hawaiian forests, and potentially on other islands and continental areas of Asia and Africa, may be aided by a release from these herbivores and pathogens (DeWalt *et al.* 2004b). However, escape from above-ground natural enemies does not explain the demographic release in high-light environments in Hawaii.

Light-dependent escape from natural enemies

One of the more interesting results of the study on *C. hirta* was that natural enemies seemed to influence its effective shade tolerance (DeWalt *et al.* 2004b). Physiologically, *C. hirta* appears relatively shade-tolerant (able to persist in low-light conditions); it occurs in understorey in Hawaii and elsewhere in its introduced range and had high survival and positive relative growth rates under

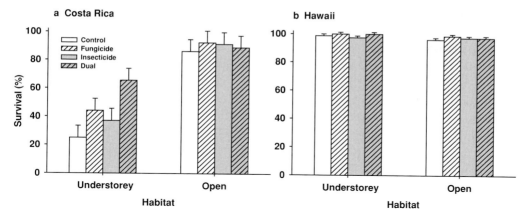

Figure 19.1 Survival of *Clidemia hirta* plants to harvest in understorey and open habitats in (a) Costa Rica and (b) Hawaii in four treatments in which natural enemies were excluded. Control plants were sprayed with water; fungicide-treated plants were sprayed with the systemic fungicide benomyl; insecticide-treated plants were sprayed with the systemic chloronicotinyl insecticide called imidacloprid and the contact pyrethroid insecticide cyfluthrin; and plants in the dual treatment were sprayed with benomyl and both insecticides. Survivorship was higher in Costa Rican understorey sites when plants were sprayed either with fungicide or insecticide. No effect of either type of pesticide was found in the open or in either habitat of Hawaii. Least-square means + 1 SE are shown. Reproduced from DeWalt *et al.* (2004b) with permission.

low-light levels in greenhouse studies (Baruch *et al.* 2000; DeWalt *et al.* 2004a). In the presence of natural enemies, such as in its native range, it is *effectively* shade-intolerant and does not occur in forest understorey.

The effect of natural enemies could be habitat-dependent if the likelihood of attack differs among habitats or if the consequences of attack differ among habitats. Hammond and Brown (1998) found that Janzen–Connell mortality processes were light-dependent; seedlings of the tropical tree *Carapa guianensis* in Guyana had higher survivorship farther from parent trees in the understorey, but distance from parent tree did not predict mortality rates for seedlings in gaps. They could not distinguish, however, whether seedlings in gaps vs. understorey differed in their susceptibility to attack or in the ability to compensate after attack. Few studies have distinguished between the two possibilities.

No systematic variation in herbivore attack on *C. hirta* was found between open and understorey sites in Costa Rica, so differences in the likelihood of herbivore or pathogen attack did not seem to be the causal mechanism for exclusion of *C. hirta* from shaded areas in its native range. Instead, it seems that the low growth and survival of plants in understorey areas when exposed to natural levels of herbivory and pathogen attack resulted from physiological intolerance of the plants to loss of leaf area and photosynthate in shaded conditions (DeWalt *et al.*

2004b). Without such attack, *C. hirta* seems able to grow and reproduce in forest understorey, such as in Hawaii.

Other studies have also found that herbivores and pathogens may significantly affect shade tolerance. Rankings of shade tolerance in several temperate tree species in an old-growth beech–maple forest differed between plots with and without white-tailed deer (Z. T. Long & W. P. Carson, unpublished data). In addition, it was proposed that the relative shade intolerance of *Prunus sebifera* (black cherry) in north-eastern US forests may be due to soil pathogens (such as *Pythium* spp.), which thrive in the moist understorey, rather than to physiological requirements for high light (Packer & Clay 2003). Indeed, fungal pathogens generally are found in areas with higher soil moisture and humidity, such as those in deep shade (Augspurger 1983). Thus, the growth–mortality trade-off shown for tropical plants grown in open vs. shaded areas (e.g. Kitajima 1994) may be relaxed when species are moved away from accumulated loads of pathogens and herbivores. Resources usually allocated to defence may be allocated for growth and reproduction without a concomitant reduction in probability of survival.

The degree of enemy release is likely to differ among areas of introduction. For example, enemy release may not be substantial in areas with native congeners of the invading species. Generalist pathogens and herbivores might have a greater probability of including a non-native species in their diet if it is chemically and morphologically similar to species already in the diet. Host-switching of specialists from native species to non-native congeners might also be more prevalent than from native species to unrelated non-native species. Indeed, there is some evidence that introduced plants with congeneric native species are less likely to proliferate than those without native congeners (Darwin's naturalization hypothesis; Darwin 1859; Mack 1996; Rejmánek 1996b, 1999). Other comparisons of native vs. alien genera on Hawaii and New Zealand have failed to find such a pattern, however (Daehler 2001; Duncan & Williams 2002). In fact, more non-native plants in New Zealand and Hawaii had native congeners than did not have native congeners. In these cases, invasion success was attributed to non-native species having life-history traits similar to their native congeners that were favoured in the area of introduction (Daehler 2001; Duncan & Williams 2002).

If tropical plant distributions are more strongly affected by herbivores and pathogens than temperate plant distributions, then enemy release may be a more general mechanism of invasion for tropical than for temperate plants. The latitudinal gradient in herbivory and pathogen attack reported by Coley and Barone (1996) would support this hypothesis. Based on 42 studies, Coley and Barone (1996) found that average rates of leaf-area loss to herbivores and pathogens were greater on tropical trees (11.1% per year) than on temperate broad-leaved trees (7.1%). It is not known whether this difference in rates of herbivory and pathogen attack leads to differences in degree of natural-enemy regulation of plant communities (Coley & Barone 1996). The Janzen–Connell effect of pathogens and herbivores increasing the distance from parents to seedlings has been shown for both temperate and tropical plants (Augspurger 1984; Gilbert & DeSteven 1996; Packer & Clay 2000), but density-dependent mortality (presumably due to host-specific predators or pathogens) seems no more prevalent among tropical trees than temperate trees (HilleRisLambers *et al.* 2002). Thus, it seems that natural enemies should be significant agents of selection in both temperate and tropical forest ecosystems.

The lack of support for the ERH in some studies, particularly the interspecific studies, may result from the short duration of the experiments and the lumping of generalist and specialist herbivores into one category. Specialist herbivores, i.e. those that are limited to hosts of the same genus or family, likely exert stronger selective pressure on plant populations than generalist herbivores. For example, in an insect exclusion study, Carson and Root (2000) found that the outbreak of a specialist chrysomelid beetle dramatically altered the trajectory of succession in old fields, whereas chronic low levels of herbivory by generalists did little to alter community dynamics. Only by studying the long-term effects (10-yr) of suppressing insects using insecticide were they able to demonstrate the importance of insect herbivory (Carson & Root 2000). Studies that found similar levels of herbivory on native and introduced plants (e.g. Agrawal & Kotanen 2003) may have run for too short a time (in this case less than one year) to see the effects of these periodic episodes of intense herbivory.

In general, more long-term biogeographical experiments that exclude natural enemies (i.e. pesticides, exclosures, soil sterilization) are needed to test the relative effects of natural enemies on invasive plants in their native and introduced ranges. These intraspecific tests are the most powerful tests of the ERH because they allow both predictions to be tested, i.e. that natural enemies limit plant populations in their native range, and that they are absent (or nearly absent) in the introduced range and this absence leads to a demographic release. Thus far, few of the intraspecific biogeographical studies conducted have manipulated natural-enemy loads and, to my knowledge, only that of DeWalt *et al.* (2004b) has studied a tropical plant.

It would be particularly informative to conduct these types of biogeographical experiments with the other woody invaders limited to high-light environments in their native ranges that invade old-growth tropical forests in their introduced ranges (Table 19.1). Such experiments would help determine whether enemy release explains changes in their relative shade tolerance between native and introduced ranges, as was found for *C. hirta* (DeWalt *et al.* 2004b). These types of studies will help determine whether or not enemy release is a general mechanism of invasion and will address the role of biotic interactions in tropical forests in general.

Evolution-of-increased-competitive-ability hypothesis

Enemy release may ultimately lead to genetic shifts in resource allocation away from defence and toward greater growth and reproduction (Blossey & Nötzold 1995). Such genetic shifts can be detected with common-garden experiments using native and introduced genotypes.

Tests of the EICA and the more general genetic-shift hypotheses have produced mixed results. Species for which there is evidence of genetically determined greater growth rates, size or fecundity for introduced compared with native genotypes include the temperate plants *Lythrum salicaria* (Blossey & Nötzold 1995; Blossey & Kamil 1996), *Sapium sebiferum* (Siemann & Rogers 2001) and *Eschscholzia californica* (Leger & Rice 2003). The role of natural enemies in driving these differences can only be inferred in most studies. For example, Siemann and Rogers (2001) found in a 14-yr common-garden experiment that *Sapium sebiferum* genotypes from Taiwanese populations (part of the native range but not necessarily source of introduced genotypes) had higher tannin concentrations and grew more slowly than non-native Texas genotypes, which had lower tannin concentrations. The difference in levels of secondary compounds suggests that introduced genotypes have different patterns of resource allocation, potentially driven by lowered herbivore loads in Texas than China. In a 17-yr common garden in Hawaii with the same genotypes, the concentration of tannins was again higher for Taiwanese than Texas genotypes (Siemann & Rogers 2003b). In contrast to the results of the Texas common garden, however, the Taiwanese genotypes outperformed the Texas genotypes in terms of basal area. In addition, Texas genotypes suffered greater levels of herbivory, presumably from the introduced Asian generalist herbivore *Adoretus sinicus* (Chinese rose beetle; Siemann & Rogers 2003b). Experiments on this tree species show the strongest evidence of any species yet studied for evolution of increased competitive ability in areas of the introduced range where herbivores are absent or at low densities. Nevertheless, alternate explanations, such as use of non-source populations to represent native genotypes, cannot be dismissed.

In contrast to the previous example, little evidence was found for greater growth rates or reproduction of introduced compared with native genotypes of *Hypericum perforatum* (Vilà *et al.* 2003), four temperate biennial species (Willis *et al.* 2000) or *Clidemia hirta* (DeWalt *et al.* 2004a). Genetic differences between native American and introduced European populations of *Solidago canadensis* were found in a common garden study conducted in Europe, but generally the differences were in the opposite direction to that predicted by the EICA hypothesis (van Kleunen & Schmid 2003). Variation among populations within continents was high, limiting the power to detect differences between continents (van Kleunen & Schmid 2003).

Ability to detect post-introduction evolution or genetic shifts of introduced genotypes may be limited because of inadequate sampling of the native range and uncertainty of the geographic location of original source populations. Eco-types within the native range may differ in growth rates and patterns of alloca-tion and defence. In addition, herbivores and pathogen loads may differ across species' ranges (e.g. with latitude as discussed above). If native but non-source ecotypes are collected and used in common-garden studies, they may differ from introduced genotypes because of this variation within the native range and not because of post-introduction evolution.

Indeed, differences in growth form and biomass allocation patterns were found between *C. hirta* sampled from Costa Rica (native range) and Hawaii (intro-duced range) in a common-garden experiment, but there was no compelling pat-tern that suggested that Hawaiian genotypes grow to greater sizes than Costa Rican genotypes or have genetically determined allocation patterns that would account for their greater abundance or expanded habitat distribution in Hawaii than in Costa Rica (DeWalt *et al.* 2004a). Instead, differences in growth form and allocation can probably be attributed to the fact that Costa Rican and Hawaiian genotypes are fairly genetically different (DeWalt & Hamrick 2004). Specifically, Costa Rican and Hawaiian populations had a low genetic identity (Nei's $I = 0.62$) as determined using allozymes (DeWalt & Hamrick 2004). The source of *C. hirta* on Hawaii is not known, but source populations may be from South America rather than Central America given the genetic distance between Costa Rican and Hawaiian populations.

In a similar example, significant differences in morphology and physiology were found among plants grown from seeds of *Bromus tectorum* collected from native European, invasive North American and introduced but non-invasive New Zealand populations, but these differences were not attributable to post-introduction evolution. Instead, different native populations were probably the source of the North American and New Zealand populations, and these native populations have different genetically determined morphology and physiology (C. L. Kinter & R. N. Mack, unpublished data). Thus, if source populations are not known, it is difficult to test whether genetic shifts, particularly in response to natural-enemy loads, actually have occurred.

Introduction of natural enemies for use as biocontrol

The high environmental and economic costs of non-native invasive plant species are increasingly apparent (Parker *et al.* 1999; Pimentel *et al.* 2000), but the meth-ods most effective for control of these species are often far from obvious (McEvoy & Coombs 1999). If tropical plants are limited by natural enemies in their native range and become more abundant and invade forest understorey in their intro-duced range because of a lack of natural enemies, then biocontrol may be effective. Indeed, biocontrol programs were deemed successful against *Senecio jacobaea* in the northwestern United States (McEvoy *et al.* 1991), *Opuntia ficus-indica* in South Africa (Zimmermann & Moran 1982), and *Hypericum perforatum* in

California (Huffaker & Kennett 1959). However, biocontrol introductions have not provided complete or substantial control of invasive plants in many other cases (Debach & Rosen 1991; McFadyen 1998). In addition, biocontrol always entails a risk of damage to non-target organisms (Howarth 1991) and, therefore, may not be appropriate when the possibility of host-switching is high. Nevertheless, biocontrol may be warranted if the other options are to do nothing or to use mechanical and chemical control that may be more ecologically damaging.

Biocontrol agents often control species only in part of their introduced range or only in particular habitats. For example, the *Liothrips urichi* introduced against *Clidemia hirta* in Fiji and Hawaii only lay eggs and larvae only feed on plants in open habitats (Reimer & Beardsley 1989). This species therefore does not affect *C. hirta* growing in forest understorey. The invasive weed *H. perforatum* is apparently restricted to shaded areas in California because of significant consumption of leaves by a biocontrol agent in open but not shaded areas (Huffaker & Kennett 1959).

Biological control programs for tropical forest invaders may be more effective and less risky when the ecology and genetics of the weed species have been studied in native and introduced ranges. For example, introduction of non-native biocontrol agents to reduce the abundance of an invasive plant species may not be warranted if natural enemies do not limit the species in its native range. From tests of the enemy-release hypothesis, we can determine which species' population dynamics could be checked by introduction of herbivores or pathogens. It should be noted, however, that biocontrol agents may be released from their own parasites and parasitoids (their natural enemies), which may make them even more effective.

Knowing the original source population(s) would also narrow the geographic range to be searched for biocontrol agents. There is evidence that natural enemies taken from source populations may provide more effective control than those taken from other parts of the species' native range (McFadyen 1998; Gaskin & Schaal 2002). Different biotypes of *Dactylopius opuntiae*, a biological control insect for cactus weeds (*Opuntia* spp.) in South Africa, are specific to different *Opuntia* species (Hoffmann *et al.* 2002). Establishment and reproduction of leafy spurge flea beetle (*Aphthona* spp.) differs between genetically diverse leafy spurge genotypes (*Euphorbia esula*) in its introduced range of North America (Lym & Carlson 2002). Some fungal pathogens also are not only species-specific but also genotype-specific (McFadyen 1998). Detailed records of the introduction history generally are lacking, but molecular techniques can help identify source populations if extensive sampling is conducted in the native range (e.g. Amsellem *et al.* 2000; Novak & Mack 2001; Schaal *et al.* 2003). Identification of the source population would also allow common-garden studies to determine whether post-introduction shifts in vigour have occurred. Loss of defence chemicals due to an absence of herbivores in the introduced range might make introduced species even more vulnerable to introduction of biocontrol agents (Siemann & Rogers 2003b).

Identification of the herbivores and pathogens that cause the most impact on population growth rates of non-native invaders in shaded conditions in their native ranges may indicate which types of biocontrol agents will have the greatest effects in the introduced range. Should biocontrol agents be sought that attack specific tissues such as reproductive parts, cause death of emerging seedlings (e.g. damping-off fungi), attack plants of a particular size, or cause partial defoliation or complete defoliation? Limiting the number of introductions of biological control agents reduces the risk of non-target effects and may provide sufficient control. Indeed, a single agent was considered responsible in over half of 28 biocontrol projects that introduced multiple agents and were deemed 'successful' (Denoth *et al.* 2002).

Modelling the effects of biocontrol agents

Stage-structured matrix-projection models may be useful tools for estimating which among several potential agents is likely to cause the greatest decline in population growth rates of target species (McEvoy & Coombs 1999). These matrix models use probabilities of survival, growth, and retrogression (vital rates) as well as the fecundity of each life-history stage over time periods generally of one year to estimate the asymptotic population growth rate under particular environmental circumstances (Caswell 2001). Perhaps the most important aspect of matrix-model analysis with respect to invasive-species management is elasticity analysis, which projects the relative contribution of each matrix element (probabilities of stasis or growth of each stage, and fecundity of adult stages) to the population growth rate under current conditions (de Kroon *et al.* 1986). In theory, biocontrol agents that affect stages or transitions with the largest elasticities will cause the largest decline in the population growth rate (Shea & Kelly 1998; McEvoy & Coombs 1999).

I used matrix-projection models to investigate the impacts of hypothetical biocontrol agents on the population dynamics of *Clidemia hirta*, an invasive tropical shrub, in two rainforests on the island of Hawaii (DeWalt 2003). Stage-structured matrix models were parameterized with field data collected over three years from 2906 plants in a recently invaded forest with an open overstorey (Laupahoehoe) and 600 plants in a less recently invaded forest with a closed canopy (Waiakea). Asymptotic population growth rates (the dominant eigenvalues of the matrices, λ) were greater than one for both populations in all years, demonstrating that both populations were growing.

Elasticity analyses suggested that several life-history transitions would have to be targeted by biological control agents to cause substantial declines in the growth rates of the two Hawaiian forest populations (DeWalt 2003). Parker (2000) found similar results for expanding populations of the non-native invasive shrub

Cytisus scoparius in the northwestern United States. In contrast, the elasticity matrices of many tropical plants studied to date generally are dominated by one vital rate, such as stasis (staying in the same stage) of large reproductive adults (Pinard 1993; Alvarez-Buylla 1994; Horvitz & Schemske 1995). The elasticity matrix also was dominated by few transitions for *Carduus nutans* (Asteraceae), an invasive thistle in New Zealand (Shea & Kelly 1998).

Simulated reductions in recruitment (fecundity), survival of particular stages, or survival across vegetative stages suggested that large declines in population growth rate can only be achieved by reducing survival of *C. hirta* across all vegetative stages (DeWalt 2003). Projection analyses suggested that continuous reductions of *C. hirta* survival by 12% at the Waiakea population and 64% at the Laupahoehoe population would cause the *C. hirta* populations eventually to go locally extinct. These reductions in survival are theoretically possible in Hawaii given that, in another study (DeWalt *et al.* 2004b), fungal pathogens and insect herbivores caused high levels of mortality for the experimental *C. hirta* planted into forest understorey in its native range in Costa Rica. The Lepidoptera, weevils, cecidomyiid flies, fungal pathogens and stem borers found on *C. hirta* in that experiment should therefore be explored as potential biological control agents for *C. hirta* in Hawaii because they substantially reduced survivorship and growth. Further field studies are needed to determine which of these natural enemies had the strongest effects on demographic rates, whether they attack multiple life-history stages, and whether introduction of one agent would sufficiently control *C. hirta* in Hawaiian forests.

Conclusions

We lack too much information to be able to make generalizations about the effects of natural enemies on invasive species in forest ecosystems, particularly tropical forests. In this review, I have used numerous examples from the literature on temperate plants with the assumption that the mechanisms of invasion do not differ substantially between temperate and tropical forests, so that results of tests of hypotheses of invasion will shed light on the invasion of any species. However, much more work is needed on tropical invasive plants and the environments they invade to be able to begin to assess whether there are important differences in the effects of natural enemies on plant invasions in temperate and tropical forests. One fundamental question is whether herbivores and pathogens are more important determinants of abundance, dispersion and species richness in tropical than temperate forests (Chapters 6 and 13, this volume). There is a clear need to examine the hypotheses of enemy-release and evolution of increased competitive ability for tropical forest invaders, and to determine whether shade tolerance is greatly increased in the absence of natural enemies. Experimental exclusion studies in native and introduced ranges, identification and use of source populations in common-garden studies, and demographic modelling will help to shed light on the ambiguities that seem plague the understanding of the role of herbivores and fungal pathogens in invasive species proliferation. Understanding the role that natural enemies play in plant invasions of tropical forests may also help to elucidate the role of biotic interactions in structuring tropical forest communities.

Acknowledgements

The chapter was refined by comments from J. Denslow, J. Dalling, P. Fine and K. Ickes, and two anonymous reviewers. Financial support for fieldwork described in this chapter was provided by the Organization for Tropical Studies, United States Department of Agriculture Forest Service, the National Science Foundation, Sigma Xi Grants-in-Aid-of-Research, and Louisiana State University.

References

Agrawal, A. A. & Kotanen, P. M. (2003) Herbivores and the success of exotic plants: a phylogenetically controlled experiment. *Ecology Letters*, **6**, 712–715.

Alvarez-Buylla, E. R. (1994) Density dependence and patch dynamics in tropical rain forests: matrix models and applications to a tree species. *American Naturalist*, **143**, 155–191.

Amsellem, L., Noyer, J. L., Le Bourgeois, T. & Hossaert-McKey, M. (2000) Comparison of genetic diversity of the invasive weed *Rubus alceifolius* Poir. (Rosaceae) in its native range and in areas of introduction, using amplified fragment length polymorphism (AFLP) markers. *Molecular Ecology*, **9**, 443–455.

Augspurger, C. K. (1983) Seed dispersal of the tropical tree *Platypodium elegans* and the escape of its seedlings from fungal pathogens. *Journal of Ecology*, **71**, 759–772.

 (1984) Seedling survival of tropical tree species: interactions of dispersal distance, light-gaps, and pathogens. *Ecology*, **65**, 1705–1712.

Augspurger, C. K. & Kelly, C. K. (1984) Pathogen mortality of tropical tree seedlings: experimental studies of effects of dispersal distance, seedling density, and light conditions. *Oecologia*, **61**, 211–217.

Baruch, Z., Pattison, R. R. & Goldstein, G. (2000) Responses to light and water availability of four invasive Melastomataceae in the Hawaiian Islands. *International Journal of Plant Sciences*, **161**, 107–118.

Beckstead, J. & Parker, I. M. (2003) Invasiveness of *Ammophila arenaria*: release from soil-borne pathogens? *Ecology*, **84**, 2824–2831.

Binggeli, P., Hall, J. B. & Healey, J. R. (1998) *A review of invasive woody plants in the tropics*. School of Agricultural and Forest Sciences Publication Number 13. Bangor: University of Wales.

Blaney, C. S. & Kotanen, P. M. (2001a) Effects of fungal pathogens on seeds of native and exotic plants: a test using congeneric pairs. *Journal of Applied Ecology*, **38**, 1104–1113.

 (2001b) Post-dispersal losses to seed predators: an experimental comparison of native and exotic old field plants. *Canadian Journal of Botany*, **79**, 284–292.

Blossey, B. & Kamil, J. (1996) What determines the increased competitive ability of invasive non-indigenous plants? In *IX International Symposium on Biological Control* (ed. V. C. Moran & J. H. Hoffmann), Stellenbosch, South Africa: University of Cape Town, pp. 3–9.

Blossey, B. & Nötzold, R. (1995) Evolution of increased competitive ability in invasive nonindigenous plants: a hypothesis. *Journal of Ecology*, **83**, 887–889.

Bossard, C. C. & Rejmánek, M. (1994) Herbivory, growth, seed production, and resprouting of an exotic invasive shrub. *Biological Conservation*, **67**, 193–200.

Buckley, Y. M., Downey, P., Fowler, S. V. *et al.* (2003) Are invasives bigger? A global study of seed size variation in two invasive shrubs. *Ecology*, **84**, 1434–1440.

Carlquist, S. (1974) *Island Biology*. New York: Columbia University Press.

Carson, W. P. & Root, R. B. (2000) Herbivory and plant species coexistence: community regulation by an outbreaking phytophagous insect. *Ecological Monographs*, **70**, 73–99.

Caswell, H. (2001) *Matrix Population Models: Construction, Analysis, and Interpretation*, 2nd edn. Sunderland, MA: Sinauer.

Coley, P. D. & Barone, J. A. (1996) Herbivory and plant defenses in tropical forests. *Annual Review of Ecology and Systematics*, **27**, 305–335.

Connell, J. H. (1971) On the role of natural enemies in preventing competitive exclusion in some marine animals and in rain forest trees. In *Advanced Study Institute on Dynamics of Numbers in Populations* (ed. P. J. Den Boer & G. Gradwell). Wageningen: Centre for Agricultural Publishing and Documentation, pp. 298–312.

Crawley, M. J. (1987) What makes a community invasible? In *Colonization, Succession, and Stability* (ed. A. J. Gray, M. J. Crawley & P. J. Edwards). Oxford: Blackwell, pp. 429–453.

Daehler, C. C. (2001) Darwin's naturalization hypothesis revisited. *American Naturalist*, **158**, 324–330.

Darwin, C. (1859) *The Origin of Species*. Oxford: Oxford University Press.

Davis, M. A., Grime, J. P. & Thompson, K. (2000) Fluctuating resources in plant communities: a general theory of invasibility. *Journal of Ecology*, **88**, 528–534.

de Kroon, H., Plaisier, A., van Groenendael, J. & Caswell, H. (1986) Elasticity: the relative contributions of demographic parameters to population growth rate. *Ecology*, **67**, 1427–1431.

Debach, P. & Rosen, D. (1991) *Biological Control by Natural Enemies*. Cambridge: Cambridge University Press.

Denoth, M., Frid, L. & Myers, J. H. (2002) Multiple agents in biological control: improving the odds? *Biological Control*, **24**, 20–30.

Denslow, J. S. (2003) Weeds in paradise: thoughts on the invasibility of tropical islands. *Annals of the Missouri Botanical Garden*, **90**, 119–127.

DeWalt, S. J. (2003) The invasive tropical shrub *Clidemia hirta* (Melastomataceae) in its native and introduced ranges: tests of hypotheses of invasion. Unpublished Ph.D. dissertation, Louisiana State University.

DeWalt, S. J., Denslow, J. S. & Hamrick, J. L. (2004a) Biomass allocation, growth, and photosynthesis of genotypes from the native and introduced ranges of the tropical shrub *Clidemia hirta*. *Oecologia*, **138**, 521–531.

DeWalt, S. J., Denslow, J. S. & Ickes, K. (2004b) Natural-enemy release facilitates habitat expansion of the invasive tropical shrub *Clidemia hirta*. *Ecology*, **85**, 471–483.

DeWalt, S. J. & Hamrick, J. L. (2004) Genetic variation of introduced Hawaiian and native Costa Rican populations of an invasive tropical shrub, *Clidemia hirta* (Melastomataceae). *American Journal of Botany*, **91**, 1155–1163.

Duncan, R. P. & Williams, P. A. (2002) Darwin's naturalization hypothesis challenged. *Nature*, **417**, 608–609.

Elton, C. S. (1958) *The Ecology of Invasions of Animals and Plants*. London: Methuen.

Fenner, M. & Lee, W. G. (2001) Lack of pre-dispersal seed predators in introduced

Asteraceae in New Zealand. *New Zealand Journal of Ecology*, **25**, 95–99.

Fine, P. V. A. (2002) The invasibility of tropical forests by exotic plants. *Journal of Tropical Ecology*, **18**, 687–705.

Fowler, S. V., Harman, H. M., Memmott, J. et al. (1996) Comparing the population dynamics of broom, *Cytisus scoparius*, as a native plant in the United Kingdom and France and as an invasive alien weed in Australia and New Zealand. In *IX International Symposium on Biological Control* (ed. V. C. Moran & J. H. Hoffmann) Stellenbosch: University of Cape Town, pp. 19–26.

Fryer, G. (1991) Biological invasions into tropical nature reserves. In *Ecology of Biological Invasions in the Tropics* (ed. P. S. Ramakrishnan). New Delhi: International Scientific Publications, pp. 87–101.

Gagné, B. H., Loope, L. L., Medeiros, A. C. & Anderson, S. J. (1992) *Miconia calvescens*: a threat to native forests of the Hawaiian Islands. *Pacific Science*, **46**, 390–391.

Gaskin, J. F. & Schaal, B. A. (2002) Hybrid *Tamarix* widespread in US invasion and undetected in native Asian range. *Proceedings of the National Academy of Sciences*, **99**, 11256–11259.

Gilbert, G. S. & DeSteven, D. (1996) A canker disease of seedlings and saplings of *Tetragastris panamensis* (Burseraceae) caused by *Botryosphaeria dothidea* in a lowland tropical forest. *Plant Disease*, **80**, 684–687.

Gordon, D. (1998) Effects of invasive, non-indigenous plant species on ecosystem processes: lessons from Florida. *Ecological Applications*, **8**, 975–989.

Hammond, D. S. & Brown, V. K. (1998) Disturbance, phenology and life-history characteristics: factors influencing distance/density-dependent attack on tropical seeds and seedlings. In *Dynamics of Tropical Communities* (ed. D. M. Newbery, H. H. T. Prins & N. D. Brown). Oxford: Blackwell, pp. 51–78.

Harper, J. L. (1969) The role of predation in vegetational diversity. *Brookhaven Symposia in Biology*, **22**, 48–62.

Herbold, B. & Moyle, P. B. (1986) Introduced species and vacant niches. *American Naturalist*, **128**, 751–760.

HilleRisLambers, J., Clark, J. S. & Beckage, B. (2002) Density-dependent mortality and the latitudinal gradient in species diversity. *Nature*, **417**, 732–735.

Hoffmann, J. H., Impson, F. A. C. & Volchansky, C. R. (2002) Biological control of cactus weeds: implications of hybridization between control agent biotypes. *Journal of Applied Ecology*, **39**, 900–908.

Horvitz, C. C. & Schemske, D. W. (1995) Spatiotemporal variation in demographic transitions of a tropical understorey herb: projection matrix analysis. *Ecological Monographs*, **65**, 155–192.

Howarth, F. G. (1991) Environmental impacts of classical biological control. *Annual Review of Entomology*, **36**, 485–509.

Huffaker, C. B. & Kennett, C. E. (1959) A ten-year study of vegetational changes associated with biological control of Klamath weed. *Journal of Range Management*, **12**, 69–82.

Janzen, D. H. (1970) Herbivores and the number of tree species in tropical forests. *American Naturalist*, **104**, 501–528.

Keane, R. M. & Crawley, M. J. (2002) Exotic plant invasions and the enemy release hypothesis. *Trends in Ecology and Evolution*, **17**, 164–170.

Kitajima, K. (1994) Relative importance of photosynthetic traits and allocation patterns as correlates of seedling shade tolerance of 13 tropical trees. *Oecologia*, **98**, 419–428.

Klironomos, J. N. (2002) Feedback with soil biota contributes to plant rarity and invasiveness in communities. *Nature*, **417**, 67–70.

Laurance, W. F., Lovejoy, T. E., Vasconcelos, H. L. et al. (2002) Ecosystem decay of Amazonian forest fragments: a 22-year investigation. *Conservation Biology*, **16**, 605–618.

Leger, E. A. & Rice, K. J. (2003) Invasive California poppies (*Eschscholzia californica* Cham.) grow larger than native individuals under reduced competition. *Ecology Letters*, **6**, 257–264.

Levine, J. M. & d'Antonio, C. M. (1999) Elton revisited: a review of evidence linking diversity and invasibility. *Oikos*, **87**, 15–26.

Lonsdale, W. M. (1999) Global patterns of plant invasions and the concept of invasibility. *Ecology*, **80**, 1522–1536.

Louda, S. M., Dixon, P. M. & Huntly, N. J. (1987) Herbivory in sun versus shade at a natural meadow-woodland ecotone in the Rocky Mountains. *Vegetatio*, **72**, 141–149.

Louda, S. M. & Potvin, M. A. (1995) Effect of inflorescence-feeding insects in the demography and lifetime fitness of a native plant. *Ecology*, **76**, 229–245.

Louda, S. M. & Rodman, J. E. (1996) Insect herbivory as a major factor in the shade distribution of a native crucifer (*Cardamine cordifolia* A. Gray, bittercress). *Journal of Ecology*, **84**, 229–237.

Lym, R. G. & Carlson, R. B. (2002) Effect of leafy spurge (*Euphorbia esula*) genotype on feeding damage and reproduction of *Aphthona* spp.: implications for biological weed control. *Biological Control*, **23**, 127–133.

Mack, R. N. (1996) Biotic barriers to plant naturalization. In *IX International Symposium on Biological Control* (ed. V. C. Moran & J. H. Hoffmann) Stellenbosch: University of Cape Town, pp. 39–46.

Mack, R. N., Simberloff, D., Lonsdale, W. M., Evans, H., Clout, M. & Bazzaz, F. A. (2000) Biotic invasions: causes, epidemiology, global consequences, and control. *Ecological Applications*, **10**, 689–710.

McEvoy, P. B. & Coombs, E. M. (1999) Biological control of plant invaders: regional patterns, field experiments, and structured population models. *Ecological Applications*, **9**, 387–401.

McEvoy, P. B., Cox, C. & Coombs, E. (1991) Successful biological control of ragwort, *Senecio jacobaea*, by introduced insects in Oregon. *Ecological Applications*, **1**, 430–442.

McFadyen, R. E. C. (1998) Biological control of weeds. *Annual Review of Entomology*, **43**, 369–393.

McNaughton, S. J. (1983) Compensatory plant growth as a response to herbivory. *Oikos*, **40**, 329–336.

Memmott, J., Fowler, S. V., Paynter, Q., Sheppard, A. W. & Syrett, P. (2000) The invertebrate fauna on broom, *Cytisus scoparius*, in two native and two exotic habitats. *Acta Oecologica*, **21**, 213–222.

Meyer, J.-Y. (1996) Status of *Miconia calvescens* (Melastomataceae), a dominant invasive tree in the Society Islands (French Polynesia). *Pacific Science*, **50**, 66–76.

Mitchell, C. E. & Power, A. G. (2003) Release of invasive plants from fungal and viral pathogens. *Nature*, **421**, 625–627.

Muller-Landau, H. C., Levin, S. A. & Keymer, J. E. (2003) Theoretical perspectives on evolution of long-distance dispersal and the example of specialized pests. *Ecology*, **84**, 1957–1967.

Naeem, S., Knops, J. M. H., Tilman, D., Howe, K. M., Kennedy, T. & Gale, S. (2000) Plant diversity increases resistance to invasion in the absence of covarying extrinsic factors. *Oikos*, **91**, 97–108.

Noble, I. R. (1989). Attributes of invaders and the invading process: terrestrial and vascular plants. In *Biological Invasions: A Global Perspective* (ed. J. A. Drake, F. DiCastri, R. H. Groves, *et al.*) New York: Wiley, pp. 301–313.

Novak, S. J. & Mack, R. N. (2001) Tracing plant introduction and spread: genetic evidence from *Bromus tectorum* (Cheatgrass). *Bioscience*, **51**, 114–122.

Packer, A. & Clay, K. (2000) Soil pathogens and spatial patterns of seedling mortality in a temperate tree. *Nature*, **404**, 278–281.

(2003) Soil pathogens and *Prunus serotina* seedling and sapling growth near conspecific trees. *Ecology*, **84**, 108–119.

Parker, I. M. (2000) Invasion dynamics of *Cytisus scoparius*: a matrix model approach. *Ecological Applications*, **10**, 726–743.

Parker, I. M., Simberloff, D., Lonsdale, W. M. *et al.* (1999) Toward a framework for understanding the ecological effects of invaders. *Biological Invasions*, **1**, 3–19.

Paynter, Q., Fowler, S. V., Hinz, H. L. *et al.* (1996) Are seed-feeding insects of use for the biological control of broom? In *IX International Symposium on Biological Control* (ed. V. C. Moran & J. H. Hoffmann) Stellenbosch: University of Cape Town, pp. 495–501.

Pimentel, D., Lach, L., Zuniga, R. & Morrison, D. (2000) Environmental and economic costs of nonindigenous species in the United States. *BioScience*, **50**, 53–65.

Pinard, M. (1993) Impacts of stem harvesting on populations of *Iriartea deltoidea* (Palmae) in an extractive reserve in Acre, Brazil. *Biotropica*, **25**, 2–14.

Prieur-Richard, A.-H., Lavorel, S., Grigulis, K. & Dos Santos, A. (2000) Plant community diversity and invasibility by exotics: invasion of Mediterranean old fields by *Conzya bonariensis* and *Conzya canadensis*. *Ecology Letters*, **3**, 412–422.

Prieur-Richard, A.-H., Lavorel, S., Linhart, Y. B. & Dos Santos, A. (2002) Plant diversity, herbivory and resistance of a plant community to invasion in Mediterranean annual communities. *Oecologia*, **130**, 96–104.

Pritchard, T. (1960). Race formation in weedy species with special reference to *Euphorbia cyparissias* L. and *Hypericum perforatum* L. In *The Biology of Weeds: A Symposium of the British Ecological Society, Oxford, 2–4 April 1959* (ed. J. L. Harper). Oxford: Blackwell, pp. 61–66.

Reimer, N. J. & Beardsley, J. W. Jr (1989) Effectiveness of *Liothrips urichi* (Thysanoptera: Phlaeothripidae) introduced for biological control of *Clidemia hirta* in Hawaii. *Environmental Entomology*, **18**, 1141–1146.

Reinhart, K. O., Packer, A., van der Putten, W. H. & Clay, K. (2003) Plant-soil biota interactions and spatial distribution of black cherry in its native and invasive ranges. *Ecology Letters*, **6**, 1046–1050.

Rejmánek, M. (1996a) Species richness and resistance to invasions. In *Biodiversity and Ecosystem Processes in Tropical Forests* (ed. G. Orians, R. Dirzo & J. H. Cushman), Vol. 122. New York: Springer, pp. 153–172.

(1996b) A theory of seed plant invasiveness: the first sketch. *Biological Conservation*, **78**, 171–181.

(1999) Invasive plant species and invasible ecosystems. In *Invasive Species and Biodiversity Management* (ed. O. T. Sandlund, P. J. Schei & A. Viken). Dordrecht: Kluwer, pp. 79–102.

Richardson, D. M., Allsopp, N., d'Antonio, C. M., Milton, S. J. & Rejmánek, M. (2000) Plant invasions – the role of mutualisms. *Biological Reviews*, **75**, 65–93.

Root, R. B. (1996) Herbivore pressure on goldenrods (*Solidago altissima*): its variation and cumulative effects. *Ecology*, **77**, 1074–1087.

Sagers, C. L. & Coley, P. D. (1995) Benefits and costs of defense in a neotropical shrub. *Ecology*, **76**, 1835–1843.

Sauer, J. D. (1988) *Plant Migration: The Dynamics of Geographic Patterning in Seed Plant Species.* Berkeley, CA: University of California Press.

Schaal, B. A., Gaskin, J. F. & Caicedo, A. L. (2003) Phylogeography, haplotype trees, and invasive plant species. *Journal of Heredity*, **94**, 197–204.

Schierenbeck, K. A., Mack, R. N. & Sharitz, R. R. (1994) Effects of herbivory on growth and biomass allocation in native and introduced species of *Lonicera*. *Ecology*, **75**, 1661–1672.

Shea, K. & Kelly, D. (1998) Estimating biocontrol agent impact with matrix models: *Carduus nutans* in New Zealand. *Ecological Applications*, **8**, 824–832.

Sheil, D., Jennings, S. & Savill, P. (2000) Long-term permanent plot observations of vegetation dynamics in Bundongo, a Ugandan rain forest. *Journal of Tropical Ecology*, **16**, 765–800.

Siemann, E. & Rogers, W. E. (2001) Genetic differences in growth of an invasive tree species. *Ecology Letters*, **4**, 514–518.

(2003a) Herbivory, disease, recruitment limitation, and success of alien and native tree species. *Ecology*, **84**, 1489–1505.

(2003b) Increased competitive ability of an invasive tree may be limited by an invasive beetle. *Ecological Applications*, **13**, 1503–1507.

(2003c) Reduced resistance of invasive varieties of the alien tree *Sapium sebiferum* to a generalist herbivore. *Oecologia*, **135**, 451–457.

Siemann, E., Tilman, D., Haarstad, J. & Ritchie, M. (1998) Experimental tests of the dependence of arthropod diversity on plant diversity. *American Naturalist*, **152**, 738–750.

Simberloff, D., Brown, B. J. & Lowrie, S. (1978) Isopod and insect root borers may benefit Florida mangroves. *Science*, **201**, 630–632.

Smith, M. D. & Knapp, A. K. (1999) Exotic plant species in a C4-dominated grassland: invasibility, disturbance, and community structure. *Oecologia*, **120**, 605–612.

Stanosz, G. R. (1994) Benomyl and acephate applications increase survival of sugar maple seedlings during their first growing season in northern Pennsylvania. *Canadian Journal of Forest Research*, **24**, 1107–1111.

Stohlgren, T. J., Binkley, D., Chong, G. W. *et al.* (1999) Exotic plant species invade hot spots of native plant diversity. *Ecological Monographs*, **69**, 25–46.

Symstad A. J. (2000) A test of the effects of functional group richness and composition on grassland invasibility. *Ecology* **81**, 99–109.

Tilman, D. (1997) Community invasibility, recruitment limitation, and grassland biodiversity. *Ecology*, **78**, 81–92.

Torchin, M. E., Lafferty, K. D., Dobson, A. P., McKenzie, V. J. & Kuris, A. M. (2003) Introduced species and their missing parasites. *Nature*, **412**, 628–630.

Usher, M. B. (1991) Biological invasions into tropical nature reserves. In *Ecology of Biological Invasions in the Tropics* (ed. P. S. Ramakrishnan). New Delhi: International Scientific Publications, pp. 21–34.

van der Putten, W. H. & Peters, B. A. M. (1997) How soil-borne pathogens may affect plant competition. *Ecology*, **78**, 1785–1795.

van Kleunen, M. & Schmid, B. (2003) No evidence for an evolutionary increased competitive ability in an invasive plant. *Ecology*, **84**, 2816–2823.

Vilà, M., Gómez, A. & Maron, J. L. (2003) Are alien plants more competitive than their native conspecifics? A test using *Hypericum perforatum* L. *Oecologia*, **137**, 211–215.

Vitousek, P. M., D'Antonio, C. M., Loope, L. L. & Westbrooks, R. (1996) Biological invasions as global environmental change. *American Scientist*, **84**, 468–478.

Vitousek, P. M., Walker, L. R., Whiteaker, L. D., Mueller-Dombois, D. & Matson, P. A. (1987) Biological invasion by *Myrica faya* alters ecosystem development in Hawaii. *Science*, **238**, 802–804.

Waloff, N. & Richards, O. W. (1977) The effect of insect fauna on growth, mortality and natality of broom, *Sarothamnus scoparius*. *Journal of Applied Ecology*, **14**, 787–798.

Wester, L. L. & Wood, H. B. (1977) Koster's curse (*Clidemia hirta*), a weed pest in Hawaiian forests. *Environmental Conservation*, **4**, 35–41.

Whitham, T. G., Maschinski, J., Larson, K. C. & Paige, K. N. (1991) Plant responses to herbivory: the continuum from negative to positive and underlying physiological mechanisms. In *Plant–Animal Interactions:*

Evolutionary Ecology in Tropical and Temperate Regions (ed. P. W. Price, T. M. Lewinsohn, G. W. Fernandes & W. W. Benson). New York, NY: Wiley, pp. 227–256.

Whitmore, T. C. (1991) Invasive woody plants in perhumid tropical climates. In *Ecology of Biological Invasions in the Tropics* (ed. P. S. Ramakrishnan). New Delhi: International Scientific Publications, pp. 35–40.

Williamson, M. & Fitter, A. (1996) The characters of successful invaders. *Biological Conservation*, **78**, 163–170.

Willis, A. J., Memmott, J. & Forrester, R. I. (2000) Is there evidence for the post-invasion evolution of increased size among invasive plant species? *Ecology Letters*, **3**, 275–283.

Wolfe, L. M. (2002) Why alien invaders succeed: support for the escape-from-enemy hypothesis. *American Naturalist*, **160**, 705–711.

Zimmermann, H. G. & Moran, V. C. (1982) Ecology and management of cactus weeds in South Africa. *South African Journal of Science*, **78**, 314–320.

CHAPTER TWENTY

New mix of alien and native species coexists in Puerto Rico's landscapes

ARIEL E. LUGO AND THOMAS J. BRANDEIS

USDA Forest Service

'There is no controversy among scientists that nonindigenous species cause extinctions of native species.' . . . 'The increase in nonindigenous species-induced rates of extinction of native species on both local and global scales is a fact.'

D. M. Lodge and K. Shrader-Frechette 2003, p. 34 and 36

'The evidence so far points to the conclusion that invaders often cause extinction on oceanic islands and in lakes but rarely in the sea or in large land masses.'

G. J. Vermeij 1996, p. 6

Introduction

The advent of the Homogeocene (Putz 1997), Homogecene (McKinney & Lockwood 1999; Lockwood & McKinney 2001) or Homogocene (Lodge & Shrader-Frechette 2003) – the era of human domination of the world – is both a challenge and an opportunity to test the ingenuity of humans. Will we be able to establish a new and sustainable balance with the rest of the world's biota? To do so requires active management of biodiversity based on understanding the function and dynamics of ecosystems. Appropriately, the approach to the study of the biota is undergoing a shift from a taxonomic, distributional and evolutionary focus, to a paradigm that considers biodiversity and ecosystem function (Naeem 2002). This new approach is holistic and quantitative, and helpful in understanding the role of biodiversity in the Homogeocene (Lugo 1995, 2002a).

Much of the current literature on tropical biodiversity focuses on the negative effects of non-indigenous or alien species, particularly invasive ones. The general belief is that invasive alien species are a major cause of native-species extinction, as well as a large burden on the economy of many countries (Allendorf & Lundquist 2003). A special issue of *Biological Conservation* (Carey *et al.* 1996), a special section in *Conservation Biology* (Allendorf & Lundquist 2003), the Ecological Society of America's *Issues in Ecology* (Mack *et al.* 2000a) and a technical report in *Ecological Applications* (Mack *et al.* 2000b) all underscore the role of invading

Biotic Interactions in the Tropics: Their Role in the Maintenance of Species Diversity, ed. D. F. R. P. Burslem, M. A. Pinard and S. E. Hartley. Published by Cambridge University Press. © Cambridge University Press 2005.

alien species in causing extinction of native species. However, the evidence to sustain the claims usually does not support sweeping generalizations such as the first one quoted above. Usually, such statements emerge from simulation models of global change (Walker & Steffen 1997; Sala *et al.* 2001), examples from particular locations (Hobbs & Mooney 1997) or unsubstantiated generalizations and extrapolations (Myers 1986, Alonso *et al.* 2001).

Dependence on global models to anticipate massive extinction of endemic species introduces confusion to the literature because references to these studies ignore their assumptions, the complex patterns of response of the models, and the dearth of empirical evidence to substantiate the predictions. As an example, Sala *et al.* (2001) modelled through the twenty-first century and, out of countless potential drivers of species extinctions, they ranked the potential effects of five of them: land use, climate, nitrogen deposition, biotic exchange and atmospheric CO_2. Biotic exchange, i.e. the effects of alien species on the native biota, ranked fourth overall as a driver of extinctions in the model simulations. For the tropics (Dirzo 2001) and other locations, the effect of biotic exchange on species extinction was negligible. Change in land use was the main driver of species extinctions. Many research studies support that conclusion and consistently show that the level of species extinctions is generally low except for particular locations such as lakes, rivers and some islands (below). In spite of the above, references to Sala's work, for example in Lodge and Shrader-Frechette (2003), ignore the context or limitations of the research.

Comprehensive large-scale analyses invariably fail to demonstrate that alien species are the root cause of massive native-species extinction. For example, Case (1996) studied the extinction and invasion of bird species in insular and mainland sites, including the Pacific Islands, and found direct correlation between rates of extinction of native species and rates of alien-species invasions. However, the two events responded to different causal forces. Anthropogenic disturbances including habitat conversion and degradation, and creation of new types of habitats, caused the extinction of native bird species well before the invasion of alien species. The aliens took advantage of empty niches in new or altered habitats (Grime 2002). Similarly, Davis (2003) points out that about 4000 plant species introduced into North America north of Mexico during the past 400 years naturalized. They now constitute nearly 20 per cent of the continent's vascular plant species without any evidence that a single native species has gone extinct from these introductions. Wohlgemuth *et al.* (2002) document the same trend for Europe, and Pitman *et al.* (2002) did so for the neotropical flora, concluding that habitat loss was the main cause for the extinctions they documented. Davis (2003) concluded that compared with intertrophic interactions and habitat loss, competition from alien species is not likely to be a common cause of extinctions of long-term resident species at global, metacommunity and even most community levels. However, alien species can affect native species through pathogens, change in disturbance regime or predation (Cowie & Robinson 2003; Davis 2003).

The debate on the extinction of species lacks information collected at the proper scales of time and space, leading invariably to confusion and a debate on values. For example, increasing evidence shows that species richness increases in most regions of the world owing to species introductions and the formation of new habitats by human activity (Case 1996; Rich & Woodruff 1996; Weber 1997; Grime 2002; Wohlgemuth et al. 2002; Davis 2003). The arrival of new species at these sites creates opportunities for the development of new ecological and evolutionary relationships (Abbott 1992; Vermeij 1996; Mooney & Cleland 2001; Davis 2003). However, many dismiss this information as unimportant because of the social value attributed to indigenous species, and the belief that alien species disproportionably affect them. For example, Lodge and Shrader-Frechette (2003) wrote:

A focus on total species diversity at the local scale, including nonindigenous species, ignores the basis of the fear of the Homogocene that is shared by many in society, not just environmentalists: the high value placed on the uniqueness of regional biota. (p. 34)

Value-laden arguments, if unsupported by sound scientific evidence, lead to positions based on the 'nasty necessity of eradication' of species (Temple 1990) with a 'shoot first, ask questions later' approach (Allendorf & Lundquist 2003; Simberloff 2003). However, Zavaleta et al. (2001) discuss how such eradication approaches can also result in unexpected changes to other ecosystem components, including accidental adverse effects on native ecosystems. Moreover, some analyses suggest that homogenization of the biota is not synonymous with low diversity. Different regions of the world will be more similar to each other than they are now, but they will also be more diverse (Davis 2003).

Clearly, the issues associated with the spread of alien species are complex but we lack sufficient information to move the debate from a value-laden one to a scientifically based ecological discussion. Moreover, the focus of scientific attention is likely to influence how managers and policy makers design programs to address biodiversity issues. If the information they receive leads them to an eradication approach, such an approach will prevail. In Hawaii, agencies at all levels of government are funding an eradication program against *Eleutherodactylus portoricensis*, a frog introduced from Puerto Rico (Leone 2003). This expensive effort is an example of how a legitimate concern for conserving biodiversity might doom the effort. A poor scientific base for the control actions leads to actions that are expensive and ineffective, and the danger of causing a backlash in the political system that funds such activities (Lugo 1999).

The role that habitat conditions play in species establishment and turnover deserves increased scientific attention because it provides an alternative management approach for dealing with biodiversity issues. A relevant hypothesis to test is that which states that change in habitats, rather than the globalization of the world's biota, is a major cause of changes in global biodiversity. The management alternatives based on a habitat or ecosystem approach are dramatically different from those based on eradication. For example, one could address the growth of aquatic weeds on reservoirs through continuous removal of plants or by dealing with water quality and hydrology. The information demand and corrective measures for each approach are different.

Typically missing from information used to assess tropical biodiversity are assessments of the disturbance regime (natural and anthropogenic), data about current and past land use, and large-scale quantitative assessment of the distribution and long-term trends in the abundance of species. Without this kind of information, it is difficult to assess correctly the ecological role of biodiversity and avoid a generalized war on groups of species based on their time of arrival at particular geographic locations. Reliable field data are required for supporting informed decisions about the changes in biodiversity taking place in the tropics, and the role of alien species in particular. Repetitive forest inventories represent one mechanism for gathering objective and long-term, large-scale quantitative information about the relationship between biodiversity and anthropogenic and natural forces of change. Rich and Woodruff (1996) provide an example of this application of large-scale inventories for the temperate zone.

We report the preliminary results of an island-wide forest inventory in Puerto Rico currently in progress, focusing on species composition and their importance value in different forest types. We also analyse 22-year trends of species composition based on three island-wide forest inventories. For the most recent inventory (2002), we analyse the species composition by life zone (*sensu* Holdridge 1967), land-cover type, level of stand maturity and tree-species representation in three size classes. Our focus is on two measures of biodiversity at a large-scale level: the ranked importance value of species, and species composition in terms of their classification as alien or native, including endemic. Our objective is to shed light on the working hypothesis that alien tree species predominate in human-disturbed environments and that they are forming new forest types (*sensu* Lugo & Helmer 2004) by mixing with native species, including endemic species. We discuss our results in terms of their implications for current debates about the ecological roles of invasive alien species and the approaches needed for their management.

The significance of Puerto Rico

In many of the tropical areas, the increasing rate of deforestation and expanding human influences over the biota and landscape exacerbate the sense of urgency for the conservation of biodiversity. The focus is on anthropogenic disturbances (such as land-cover change), and introduction of new fire regimes and alien species to sites previously devoid of these influences. Because of the magnitude

Table 20.1 *List of peculiarities of Puerto Rico that allow it to be an excellent location to study the potential future states of continental tropical forests*

- An old biota with over 40 million years of evolutionary history (Graham 1996).
- Human habitation goes back for millennia.
- Located in the trade wind and hurricane region. It is anticipated that global climate-change phenomena will be detected early in this region in changing frequency and intensity of climatic events.
- A region with a diverse natural disturbance regime (hurricanes, volcanoes, earthquakes).
- A population density of > 400 people/km^2.
- Territory was almost completely deforested (Birdsey and Weaver 1982) and has recovered to > 40 per cent forest cover (Helmer *et al.* 2002).
- A location with one of the highest rates of land-use transformation in the twentieth century (Rudel *et al.* 2000).
- Landscape has evolved from highly fragmented to less so, and from almost 100 per cent forested to almost 100 percent agricultural, and then to about 15 per cent urban (Lugo 2002b).
- Excellent scientific knowledge of the biota (Figueroa Colón 1996a).
- Excellent documentation of environmental changes.
- Excellent infrastructure for accessing forests and for their scientific study.
- The economy is mostly dependent on outside subsidies for its food, energy and goods (Scatena *et al.* 2002). This means that human activity is so intense that the island cannot support its basic needs without significant outside help. For example, almost 100 per cent of the energy and food consumed in the island is imported.

of some of these anthropogenic effects, we often fail to consider the capacity of the biota to react to, and recover from, natural and anthropogenic disturbances (Lugo *et al.* 2002).

The significance of a country like Puerto Rico for studying these issues is that conditions in the island reflect a high intensity of human activity (Table 20.1) while the island is on the recovery phase of a full deforestation event. The deforestation event lasted from the time most of the island forests were converted for agricultural use during the nineteenth century, until large-scale abandonment of agricultural fields in the mid twentieth century. Since the 1960s, forest area in Puerto Rico has steadily increased to the current 45 per cent forest cover (Helmer *et al.* 2002).

Puerto Rico is an ideal location to study and anticipate the Homogeocene (Lugo 2004a). Having experienced the cycle of deforestation and reforestation following land-cover and land-use change allows scientists to evaluate ecological phenomena, such as the invasion of alien species, from a different perspective than that visualized in locations where the process of invasion is in its early stages. There are areas in Puerto Rico where the natural recovery of vegetation after the deforestation event is in its 80th year, long enough to see changes in

Table 20.2 *Number of plots, and forest area sampled (ha) by forest type*

Forest type	Number of plots	Area sampled
Dry forest	21	1.47
Young karst forest	54	3.78
Mature karst forest	7	0.49
Reverted moist forest	6	0.42
Young moist forest	36	2.52
Mature moist forest	7	0.49
Reverted wet forest	2	0.14
Young wet forest	13	0.91
Mature wet forest	7	0.49
Lower montane wet forest	3	0.21
Upper montane wet forest	5	0.35
Shade coffee forest	16	1.12
Urban forest	6	0.42
Total number of plots	**183**	**12.81**

All forest types are subtropical *sensu* Holdridge (1967). The methods describe the definition of each forest type.

species composition of whole forests. Moreover, the forests of the island have experienced low levels of species extinctions (Lugo 1988; Figueroa Colón 1996b).

Another advantage of Puerto Rico as a case study is the abundance of ecological studies at spatial scales that vary from traditional community-level studies to repetitive island-wide forest inventories. We review these data, and report results from long-term forest inventories. Because of the importance of these data to our discussion, we summarize in Appendix 20.1 the methodology used in the island-wide forest inventories and the indices we use to evaluate forest structure and composition. In addition, we summarize in Appendix 20.2 results from the latest inventory according to forest type (Table 20.2). We use these data to describe the pattern of species distributions in relation to human activity.

Dominance and species composition in space and time

Three island-wide forest inventories in Puerto Rico allowed us the opportunity of assessing spatial and temporal patterns of species dominance and composition on a large scale. Results (Appendix 20.2) reveal the following patterns among secondary forest stands throughout Puerto Rico:

- High importance values (>10 per cent) for the top-ranked tree species (Figs. 20.1 and 20.2, pp. 502 and 505).
- The highest importance value of the top-ranked tree species corresponded to young karst forest and urban forest, followed by shade-grown coffee, and

moist-forest stands, and finally by dry, wet and montane forests (Figs. 20.1 and 20.2).

- A gradient in the percentage of alien species from high percentages in urban, young karst and moist forests to low and/or almost no presence in dry, wet and montane forests (Table 20.3, p. 501).
- A gradient in the percentage of endemic species in the opposite direction to the gradient in alien species, i.e. high percentage in dry, wet and montane forests, and low percentage or absence in young karst, moist and urban forest (Table 20.3, p. 501).
- The species–area curves for the disturbed areas are steep and reach a higher number of species at the 1-ha sample than undisturbed and mature native forests (Fig. 20.3, p. 506).
- Over a period of 22 years, species composition and the importance values of top-ranked species changed in forest stands (Fig. 20.4, p. 508).
- The rapid spread of *Spathodea campanulata*, a species introduced to the island over a century ago, reflects the recent availability of abandoned lands for colonization. This species is shade-intolerant, cannot grow under close canopy and rapidly loses dominance when the canopy closes after about 40 years of growth (Aide *et al.* 2000).
- Although alien species continue to dominate several types of forests, the number of native species in seedling and sapling classes was higher than their presence in the tree size class (Table 20.3, p. 501).
- There is no indication that the dominant alien species will decline in importance in the near future because they are regenerating and maintaining their importance values (Table 20.5, p. 507) even in mature stands where we found fourteen alien tree species (Table 20.6, p. 508).
- Some native species grew only in mature stands. *Dendropanax arboreus* seems to prefer older abandoned coffee shade. *Manilkara bidentata*, a species that indicates fairly undisturbed forest stands, was rarely found: as a minor overstorey species in mature wet forest, and once as a seedling in mature karst forest.
- At least sixteen endemic tree species thrive in these new forest stands, and we found thirty-one endemic tree species in the 2002 inventory (Table 20.4, p. 503). The endemic *Cordia borinquensis* was present and regenerating in moist and wet mature stands only. The presence of endemic species in all size classes suggests that the high dominance of invasive alien species does not inhibit the establishment and growth of these endemic tree species.

These results confirm and expand the earlier analysis of Lugo and Helmer (2004) who concluded that the establishment of alien species and their mixing with native species resulted in the formation of new forest types. These forests are new because the species composition is different from that of the native forests described for the island.

Observations on individual species and forest types

- *Spathodea campanulata* continues to play a dominant role in the island's secondary forests, and will continue to do so well into the future because continuous change in land cover creates suitable habitat for the species.

- Dry forests, although degraded, do not have an alien species that fills the role of *S. campanulata* in moist and wet forests. Native species such as *Bursera simaruba*, *Andira inermis* and *Leucaena leucocephala* make up much of the dry forest. Apparently, the harsh natural conditions of dry forests prevent the dominance of alien species, but not their presence, as aliens contributed to 13.6 per cent of the dry forest tree species (Table 20.3). A similar situation occurred at the other extreme of the moisture gradient in the montane wet and rain forests. Montane forest, perhaps because of its unsuitability for other land uses and hence historical lack of disturbance, shows little colonization by alien species.

- In shade-grown coffee, the poor regeneration of the traditional shade coffee trees (*Inga* and *Erythrina* spp.) may allow *S. campanulata* and *Guarea guidonia* to dominate stands that were formerly managed as shade coffee once those stands are abandoned. *Spathodea campanulata* and *G. guidonia* dominate not only the tree size class of abandoned shade coffee, but also the sapling and seedling size classes. Therefore, these species will continue to be the major components in this forest type for a long time. However, abandoned shade coffee still holds an impressive number of native species in the sapling and seedling stages (Table 20.3), and we can expect that even if a small percentage of these trees reach the overstorey, forest diversity will increase over time.

- Alien species such as *S. campanulata* and *Albizia procera* dominated the urban forest. Chinea (2002) found that *A. procera* was more common on bulldozed abandoned agricultural land, and hypothesized that the species had a competitive advantage on these greatly disturbed sites. Under these conditions, *A. procera* showed high importance value, as did the dominant species in young karst. The gradient in importance values from high in young karst and urban, followed by intermediate values in shade coffee and moist forest, and low values in mature karst, dry, wet and montane forests (Table 20.3) suggests that the gradient of anthropogenic disturbance is a factor in the degree of dominance of the high-ranked species in these forests.

- Although alien species play prominent roles in karst, moist and wet forests, numerous native tree species, including endemic species, grow in the sapling and seedling classes, indicating that they have colonized secondary forests and are regenerating. The steep slope of the species–area curve and their higher species count than that of native undisturbed forest stands further support the notion that alien species enrich these emerging new forests and that native and endemic species are not inhibited or prevented by their presence in these forest stands.

Driving forces of change

The results of our large-scale analysis and those of Chinea and Helmer (2003) are similar to many small-scale studies in Puerto Rico (García Montiel & Scatena 1994; Zimmerman *et al.* 1995; Molina Colón 1998; Foster *et al.* 1999; Chinea 2002; Thompson *et al.* 2002). These studies establish that past land uses affect the species composition and alien-species invasion of sites. In addition, the type of land use and its intensity also have an effect. For example, different species occur and different levels of forest structure develop to the present if the past land use (at the time of the 1982 survey) was:

- Abandoned pastures (Aide *et al.* 1995; 1996; 2000).
- Active and abandoned coffee plantations (Birdsey & Weaver 1982; Franco *et al.* 1997; Rivera & Aide 1998; Rudel *et al.* 2000; Marcano Vega *et al.* 2002).
- Abandoned sugar cane, tobacco and other crop lands (Thomlinson *et al.* 1996; Rudel *et al.* 2000; Álvarez Ruiz 2002; Chinea 2002).
- Abandoned subsistence agriculture (Molina Colón 1998).
- Abandoned human habitation (Molina Colón 1998).
- Abandoned roads, even those with pavement (Heyne 1999).

Bulldozing also affects both the species richness and proportion of aliens in recovering forests (Chinea 2002). Roadside forests show gradients of species composition and alien species invasion in proportion to distance to pavement (Lugo & Gucinski 2000). In short, the preponderance of evidence from studies in Puerto Rico is that the land-use history of sites is a principal factor to consider when interpreting vegetation structure and composition.

Anthropogenic disturbances interact with natural disturbances and edaphic, climatic and geomorphologic gradients to create a new disturbance regime. The changes in species composition caused by these forces remain visible for decades and probably centuries (Thompson *et al.* 2002, Dupouey *et al.* 2002). The disturbance regime of most tropical forests continues to evolve because of increased human activity, global change, and their interaction with the natural disturbances that occur everywhere. For this reason, the structure and species composition of forests affected by these new disturbance regimes will also change. What might appear an out of control invasion by tree species, such as the monoculture forests of *Spathodea* in the lowlands of Puerto Rico, is in reality a response to an abandoned degraded agricultural site that within decades will be a diverse forest of a new mix of alien and native species (Lugo 2004b).

For the reasons given, management and conservation recommendations benefit from information based on whole ecosystem analysis. To better understand the biology of the biota it is best to consider the context of the conditions under which it exists. Future work on invasive species will require an understanding of the physical conditions at the site, including natural disturbances, and a historical perspective on past land covers and land uses. Such studies would be more effective in guiding policy and conservation actions.

Generality and implications of these results

The phenomena we describe are not exclusive to Puerto Rico. Foster *et al.* (2002) and Lugo and Helmer (2004), review similar phenomena in the eastern United States, and Dupouey *et al.* (2002) do so for Europe where agricultural use left a legacy of environmental change that has lasted for millennia. Turner *et al.* (1995) found a substitution of native species by alien species over a 100-year period in an isolated fragment of lowland tropical rain forest in Singapore. The historical reality is that wherever humans dominate a landscape, plant and animal species that are useful to, or dependent on, people follow and thrive in the conditions created by humans (Crosby 1986). Many of these species naturalize, and form new species assemblages without harm to local species and without harm to ecological processes (Davis 2003). These invasions have important ecologic, economic and social benefits that require our attention. A dramatic tropical example of the close relationship between human economic activity and the composition and dynamics of the landscape is that of the Maya, who actively introduced species, changed the species composition of forests and managed the landscape to suit their economy and life styles (Gómez-Pompa & Brainbridge 1995). Forests so designed and managed by people created confusion among modern ecologists who thought that they were dealing with pristine natural ecosystems (Rico Gray *et al.* 1985).

Because it appears inevitable that the species composition of landscapes dominated by human activity will change, and given the likely invasion of those landscapes by alien species, it behoves us to learn about this phenomenon and develop policies and tools to cope with change. The following are some implications derived from the Puerto Rico case study:

- Given sufficient time, invasive alien species modify the conditions of degraded sites such that native species, including endemic species, can grow in sites previously unavailable to them.
- Diverse forests develop after abandonment of deforested and degraded sites. These forests contain new combinations of tree species that include alien, native and endemic species.
- Even if widespread invasive species such as *S. campanulata* dominate many types of forests and persist for decades, their presence and dominance is not incompatible with the development of a diverse forest community.
- Continuous forest inventory data is a useful tool for objective evaluation of forest communities over large geographic areas and over a long time.

New ecosystems emerging under intense anthropogenic disturbances offer an opportunity for conservation of biota in the Homogeocene. These systems are dynamic and subject to active management to achieve particular management objectives. The long-term and large-scale nature of this study allows a different perspective on the effects of invasive alien tree species. It appears that for trees in the Caribbean, a management strategy of eradication is not necessary if the goal is to sustain forest cover, high species richness and regeneration of endemic species.

Acknowledgements

This study is in collaboration with the USDA Forest Service Inventory and Analysis Program, Southern Research Station, the Puerto Rico Conservation Foundation, the USDA State and Private Forestry Program, the Urban Forest Research Unit at the USDA Forest Service Northeastern Research Station, Citizens for the Karst, and the University of Puerto Rico (UPR). It is part of the USDA Forest Service contribution to the National Science Foundation Long-Term Ecological Research Program at the Luquillo Experimental Forest (Grant BSR-8811902 to the Institute for Tropical Ecosystem Studies of the UPR and the International Institute of Tropical Forestry, USDA Forest Service); and the NASA-IRA Grant NAG8-1709, under UPR subcontract 00-CO-11120105-011. We thank H. Nunci, G. Reyes and M. Alayón for their help in the production of the manuscript. We also thank E. Helmer, R. Ostertag and E. Medina for reviewing the manuscript.

References

Abbott, R. J. 1992. Plant invasions, interspecific hybridization and evolution of new plant taxa. *Trends in Ecology and Evolution* 7:401–405.

Aide, T. M., J. K. Zimmerman, L. Herrera, M. Rosario & M. Serrano. 1995. Forest recovery in abandoned tropical pastures in Puerto Rico. *Forest Ecology and Management* 77:77–86.

Aide, T. M., J. K. Zimmerman, M. Rosario & H. Marcano. 1996. Forest recovery in abandoned cattle pastures along an elevational gradient in northeastern Puerto Rico. *Biotropica* 28:537–548.

Aide, T. M., J. K. Zimmerman, J. B. Pascarella, L. Rivera & H. Marcano Vega. 2000. Forest regeneration in a chronosequence of tropical abandoned pastures: implications for restoration. *Restoration Ecology* 8:328–338.

Allendorf, F. W. & L. L. Lundquist. 2003. Introduction: population biology, evolution, and control of invasive species. *Conservation Biology* **17**: 24–30.

Alonso, A., F. Dallmeier, E. Granek & P. Raven. 2001. Biodiversity: connecting with the tapestry of life. *Smithsonian Institution/Monitoring and Assessment of Biodiversity Program, and President's Committee of Advisors on Science and Technology.* Washington, DC: Smithsonian Institution.

Álvarez Ruiz, M. 2002. Effects of human activities on stand structure and composition, and genetic diversity of *Dacryodes excelsa* Vahl (tabonuco) forests in Puerto Rico. Unpublished Ph.D. dissertation, University of Puerto Rico.

Birdsey, R. A. & P. L. Weaver. 1982. The forest resources of Puerto Rico. *Resource Bulletin SO-85*, Southern Forest Experiment Station. New Orleans, LA: USDA Forest Service.

Brandeis, T. J. 2003. Puerto Rico's forest inventory. *Journal of Forestry* January/February: 8–13.

Carey, J. R., P. Moyle, M. Rejmánek & G. Vermeij, eds. 1996. Invasion biology. *Biological Conservation* **78**:1–214.

Case, T. J. 1996. Global patterns in the establishment and distribution of exotic birds. *Biological Conservation* **78**:69–96.

Chinea, J. D. 2002. Tropical forest succession on abandoned farms in the Humacao municipality of eastern Puerto Rico. *Forest Ecology and Management* **167**:195–207.

Chinea, J. D. & E. H. Helmer. 2003. Diversity and composition of tropical secondary forests recovering from large-scale clearing: results from the 1990 inventory in Puerto Rico. *Forest Ecology and Management* **180**:227–240.

Cowie, R. H. & A. C. Robinson. 2003. The decline of native Pacific island faunas: changes in status of the land snails of Samoa through the 20th century. *Biological Conservation* **10**:55–65.

Crosby, A. W. 1986. *Ecological Imperialism: the Biological Expansion of Europe, 900–1900.* Cambridge, UK: Cambridge University Press.

Davis, M. A. 2003. Biotic globalization: does competition from introduced species threaten biodiversity? *BioScience* **53**:481–489.

Dirzo, R. 2001. Tropical forests. In F. S. Chapin III, O. E. Sala & E. Huber-Sannwald, eds. *Global Biodiversity in a Changing Environment: Scenarios for the 21st Century.* New York: Springer Verlag, pp. 251–276.

Dupouey, J. L., E. Dambrine, J. D. Laffite & C. Moares. 2002. Irreversible impact of past land use on forest soils and biodiversity. *Ecology* **83**:2978–2984.

Figueroa Colón, J., ed. 1996a. The scientific survey of Puerto Rico and the Virgin Islands: an eighty-year reassessment of the island's natural history. *Annals of the New York Academy of Sciences* **776**:1–272.

Figueroa Colón, J. 1996b. Phytogeographic trends, centers of high species richness and endemism, and the question of extinctions in the native flora of Puerto Rico. *Annals of the New York Academy of Sciences* **776**:89–102.

Foster, D. R., M. Fluet & E. R. Boose. 1999. Human or natural disturbance: landscape-scale dynamics of the tropical forests of Puerto Rico. *Ecological Applications* 9:555–572.

Foster, D. R., G. Motzkin & D. Orwig, eds. 2002. Insights from historical geography to ecology and conservation: lessons from the New England landscape. *Journal of Biogeography* **29**:1269–1590.

Franco, P. A., P. L. Weaver & S. Eggen-McIntosh. 1997. Forest resources of Puerto Rico, 1990. *Resource Bulletin SRS-22*, Southern Research Station. Asheville, NC: USDA Forest Service.

García Montiel, D. & F. N. Scatena. 1994. The effect of human activity on the structure and composition of a tropical forest in Puerto Rico. *Forest Ecology and Management* **63**:57–78.

Gómez-Pompa, A. & D. A. Bainbridge. 1995. Tropical forestry as if people mattered. In A. E. Lugo & C. Lowe, eds. *Tropical Forests: Management and Ecology.* New York: Springer Verlag, pp. 397–422.

Graham, A. 1996. Paleobotany of Puerto Rico. From Arthur Hollic's (1928) scientific survey paper to the present. *Annals of the New York Academy of Sciences* **776**:103–114.

Grime, J. P. 2002. Declining plant diversity: empty niches or functional shifts? *Journal of Vegetation Science* **13**:457–460.

Helmer, E. H., O. Ramos, T. del M. López, M. Quiñones & W. Diaz. 2002. Mapping forest type and land cover of Puerto Rico, a component of the Caribbean biodiversity hot spot. *Caribbean Journal of Science* **38**:165–183.

Heyne, C. M. 1999. Soil and vegetation recovery on abandoned paved roads in a humid tropical rain forest, Puerto Rico. Unpublished Masters thesis, University of Nevada, Las Vegas.

Hobbs, R. J. & H. A. Mooney. 1997. Broadening the extinction debate: population deletions and additions in California and western Australia. *Conservation Biology* **12**:271–283.

Holdridge, L. R. 1967. *Life Zone Ecology*. San José, Costa Rica: Tropical Science Center.

Leone, D. 2003. Citric acid effective against noisy frogs. *Honolulu Star Bulletin*. Tuesday, June 24, p. A1.

Little, E. L., R. O Woodbury & F. H. Wadsworth. 1974. *Trees of Puerto Rico and the Virgin Islands*, vol. 2. USDA Forest Service Agriculture Handbook 449. Washington, DC: USDA.

Lockwood, J. L. & M. L. McKinney, eds. 2001. *Biotic Homogenization*. New York: Kluwer Academic/Plenum Publishers.

Lodge, D. M. & K. Shrader-Frechetter. 2003. Nonindigenous species: ecological explanation, environmental ethics, and public policy. *Conservation Biology* **17**:31–37.

Lugo, A. E. 1988. Estimating reductions in the diversity of tropical forest species. In E. O. Wilson & F. M. Peter, eds. *Biodiversity*. Washington, DC: National Academy Press, pp. 58–70.

1995. Management of tropical biodiversity. *Ecological Applications* **5**:956–961.

1999. Will concern for biodiversity spell doom to tropical forest management? *The Science of the Total Environment* **240**:123–131.

2002a. El manejo de la biodiversidad en el siglo XXI. *Interciencia* **26**:484–490.

2002b. Can we manage tropical landscapes? An answer from the Caribbean perspective. *Landscape Ecology* **17**:601–615.

2004a. The Homogeocene in Puerto Rico. In D. J. Zarin, J. Alavalapati, F. E. Putz & M. Schmink, eds. *Working Forest in the American Tropics: Conservation through Sustainable Management?* New York: Colombia University Press, pp. 366–375.

2004b The outcome of alien tree invasions in Puerto Rico. *Frontiers in Ecology and the Environment* **2**:265–273.

Lugo, A. E. & H. Gucinski. 2000. Function, effects, and management of forest roads. *Forest Ecology and Management* **133**:249–262.

Lugo, A. E. & E. Helmer. 2004. Puerto Rico's new forests. *Forest Ecology and Management* **190**:145–161.

Lugo, A. E., F. N. Scatena, W. Silver, S. Molina Colón & P. G. Murphy. 2002. Resilience of tropical wet and dry forests in Puerto Rico. In L. H. Gunderson & L. Pritchard Jr, eds. *Resilience and the Behavior of Large-scale Systems*. Washington, DC: Island Press, pp. 195–225.

Mack, R. N., D. Simberloff, W. M. Lonsdale, H. Evans, M. Clout & F. A. Bazzaz. 2000a. Biotic invasions: causes, epidemiology, global consequences, and control. *Issues in Ecology* **5**:1–20.

2000b. Biotic invasions: causes, epidemiology, global consequences, and control. *Ecological Applications* **10**:689–710.

Marcano Vega, H., T. M. Aide & D. Báez. 2002. Forest regeneration in abandoned coffee plantations and pastures in the Cordillera Central of Puerto Rico. *Plant Ecology* **161**:75–87.

McKinney, M. L. & J. L. Lockwood. 1999. Biotic homogenization: a few winners replacing many losers in the next mass extinction. *Trends in Ecology and Evolution* **14**:450–452.

Molina, S. & S. Alemañy. 1997. *Species Codes for the Trees of Puerto Rico and the US Virgin Islands*. Asheville, NC: USDA Forest Service, Southern Forest Experiment Station.

Molina Colón, S. 1998. Long-term recovery of a Caribbean dry forest after abandonment of different land-uses in Guánica, Puerto Rico. Unpublished Ph.D. dissertation, University of Puerto Rico.

Mooney, H. A. & E. E. Cleland. 2001. The evolutionary impacts of invasive species. *Proceedings of the National Academy of Sciences* **98**:5446–5451.

Myers, N. 1986. Tackling mass extinction of species: a great creative challenge. *The Horace M. Albright Lectureship in Conservation*. Berkeley, CA: University of California College of Natural Resources.

Naeem, S. 2002. Ecosystem consequences of biodiversity loss: the evolution of a paradigm. *Ecology* 83:1537–1552.

Pitman, N. C. A., P. M. Jorgensen, R. S. R. Williams, S. Leon-Yanez & R. Valencia. 2002. Extinction-rate estimates for a modern neotropical flora. *Conservation Biology* 16:1427–1431.

Putz, F. E. 1997. Florida's forests in the year 2020 and deeper into the Homogeocene. *Journal of the Public Interest Environmental Conference* 1:91–97.

Rich, T. C. G. & E. R. Woodruff. 1996. Changes in the vascular plant floras of England and Scotland between 1930–1960 and 1987–1988: the BSBI monitoring scheme. *Biological Conservation* 75:217–229.

Rico Gray, V., Gómez-Pompa, A. & Chan, C. 1985. Las selvas manejadas por los mayas de Yohaltun, Campeche, México. *Biotica* 10:321–327.

Rivera, L. W. & T. M. Aide. 1998. Forest recovery in the karst region of Puerto Rico. *Forest Ecology and Management* 108: 63–75.

Rudel, T. K., M. Pérez Lugo & H. Zichal. 2000. When fields revert to forest: development and spontaneous reforestation in post-war Puerto Rico. *Professional Geographer* 52:186–397.

Sala, O. E., F. S. Chapin III & E. Huber-Sannwald. 2001. Potential biodiversity change: global patterns and biome comparisons. In F. S. Chapin III, O. E. Sala & E. Huber-Sannwald, eds. *Global Biodiversity in a Changing Environment: Scenarios for the 21st Century.* New York: Springer Verlag, pp. 351–367.

Scatena, F. N., S. J. Doherty, H. T. Odum & P. Kharecha. 2002. An EMERGY evaluation of Puerto Rico and the Luquillo Experimental Forest. *General Technical Report IITF-GTR-9.* Río Piedras, PR: USDA Forest Service, International Institute of Tropical Forestry.

Simberloff, D. 2003. How much information on population biology is needed to manage introduced species? *Conservation Biology* 17:83–92.

Temple, S. A. 1990. The nasty necessity: eradicating exotics. *Conservation Biology* 4:113–115.

Thomlinson, J. R., M. I. Serrano, T. del M. López, T. M. Aide & J. K. Zimmerman. 1996. Land-use dynamics in a post-agricultural Puerto Rican landscape (1936–1988). *Biotropica* 28:525–536.

Thompson, J., N. Brokaw, J. K. Zimmerman *et al.* 2002. Land use history, environment, and tree composition in a tropical forest. *Ecological Applications* 12:1344–1363.

Turner, I. M., K. S. Chua, J. S. Y. Ong, B. C. Soong & H. T. W. Tan. 1995. A century of plant species loss from an isolated fragment of lowland tropical rain forest. *Conservation Biology* 10:1129–1244.

Vermeij, G. J. 1996. An agenda for invasion biology. *Biological Conservation* 78:3–9.

Walker, B. & W. Steffen. 1997. An overview of the implications of global change for natural and managed terrestrial ecosystems. *Conservation Ecology* 1(2):2. URL:http://www.consecol.org/vol1/iss2/art2

Weber, E. F. 1997. The alien flora of Europe: a taxonomic and biogeographic review. *Journal of Vegetation Science* 8:565–572.

Wohlgemuth, T., M. Burgi, C. Scheidegger & M. Schultz. 2002. Dominance reduction of species through disturbance-a proposed management principle for central European forests. *Forest Ecology and Management* 166:1–15.

Zavaleta, E. S., R. J. Hobbs & H. A. Mooney. 2001. Viewing invasive species removal in a whole-ecosystem context. *Trends in Ecology and Evolution* 16:454–459.

Zimmerman, J. K., T. M. Aide, M. Rosario, M. Serrano & L. Herrera. 1995. Effects of land management and a recent hurricane on forest structure and composition in the Luquillo Experimental Forest, Puerto Rico. *Forest Ecology and Management* 77:65–76.

Appendix 20.1 Methodology used in island-wide forest inventories in Puerto Rico

We report data from a 2002 island-wide forest inventory of Puerto Rico. The previous tree inventories corresponded to 1980 (Birdsey & Weaver 1982) and 1990 (Franco *et al.* 1997), and were limited to moist and wet secondary forest and shade coffee. For the 2002 inventory (Brandeis 2003), we used a geographical information system (GIS) to place a standard grid of 2400-ha hexagons over the entire island and installed an inventory plot at the centre of each hexagon (367 sampling points) to sample all forest types. In areas of special interest, the sampling grid was 'intensified' by using the GIS to generate smaller, equal-sized hexagons within each of the base hexagons. The degree of intensification depended on the size of the inventoried area and the expected variation of its forests.

Using previous studies as a guide, we chose a sampling grid that was three times more intense than the standard forest-inventory base grid for inventorying the forests of the northern karst belt. We used a geology map and GIS to select an additional eighty-five sampling points, which fell on calcareous parent material. We used a similar approach to add sampling points within forest types of limited extent such as upper montane wet forest, and moist and wet forests growing on ultramaphic parent material. We intensified the base hexagonal twelve times, and used vegetation and geology maps in the GIS to extract additional sampling points. The inventory of urban forests in the San Juan Bay Estuary's watershed was also a 12 × base-grid intensification, which produced 108 systematic sampling points.

Field crews visited each sampling point and installed a forest inventory plot if the vegetation there met the USDA Forest Service's Forest Inventory and Analysis program's definition of forest: at least 10 per cent tree canopy coverage, and a minimum area of 0.4 ha, or at least 37 m wide if a strip. The reported number of plots by forest type and area sampled totalled 183 plots and 12.81 ha (Table 20.2).

Sampling plots were actually a cluster of four, 7.32-m radius subplots where total height and diameter at breast height (DBH) were measured on all trees with a DBH $\geq$ 12.5 cm. We also measured total height and DBH on all saplings (2.5 cm $\leq$ DBH $\leq$ 12.4 cm), and all seedlings (height $\leq$ 30 cm and DBH $\leq$ 2.4 cm). We identified and tallied all saplings and seedlings within 2.07-m radius micro-plots nested within each subplot. The total sampled area of each plot was 0.07 ha, but because there were 36.5 m between subplot centres, each plot is more representative of a larger area than is a single plot of the same area. Species designation as native, endemic or alien followed Molina and Alemañy (1997), and Little *et al.* (1974).

Forest-type categorization

We categorized inventoried forest stands using a combination of field-crew assessment of the forest type, land-use history and GIS coverage of life zone and geology. Karst forests were located on limestone geology in the moist-forest life zone. The other moist, wet and rain forests were located on volcanic geology with the exception of six stands growing on ultramaphic substrates. These were included with the moist- and wet-forest life zones. Dry-forest stands were located on limestone and alluvial substrates.

We used two methods to analyse temporal change in species composition. First, working only with stands surveyed in 2002, we subjectively derived stand ages (young, mature) from a combination of current and historic inventory data for moist, wet and rain forest. The mature versus young stand categories rely heavily on the field crew's assessment of the stand. Crews generally used the density of large overstorey trees, in addition to their experience in the local forests, as an indicator of stand maturity. When we classify a stand as mature, we do not imply that the stand is a primary forest, nor an undisturbed forest. Mature stands include late secondary forests and forest stands with little evidence of past disturbance. All forest stands suffered some kind of land-cover change in the past and thus they are secondary forests of different ages. The small number of stands categorized as mature is noteworthy (Table 20.2, p. 489). The lack of mature forest plots may reflect some bias in the methods used to assign that category to stands, but we believe it accurately reflects the scarcity of stands in this age class. We did not attempt to separate dry, lower montane wet or upper montane wet-forest stands inventoried in 2002 into young and mature subcategories, and stands that were shade-grown coffee were categorized slightly differently.

We categorized shade-coffee stands by estimating time since abandonment using past inventory data and the field crew's assessment of land-use history. The oldest category of abandoned shade-coffee stands were abandoned coffee stands in 1980 that had not lost forest cover to the present. Old abandoned shade-coffee stands were those abandoned after the 1990 inventory but recorded as actively managed in 1980. Stands abandoned after the 1990 inventory were designated young abandoned. If there was any doubt as to the history of the shade-coffee stand, the default chosen was the middle value: old abandoned. As a result of the changes in sampling design, we excluded in 2002 some stands of shade coffee included in the 1980 and 1990 inventories. In the 2002 forest inventory, there were only two recently abandoned shade-coffee stands and only three that we categorized as abandoned prior to 1980.

The second way we analysed temporal change was to group all stands of similar forest type and land-use history together, regardless of maturity, and compare these broader categories across inventory years. We grouped moist and wet forest into one broader category of secondary forest. We also examined shade-coffee stands as a single category, i.e. without estimating the time since abandonment. Using these broader categories, we examined changes in species composition in the data from 1980, 1990 and 2002 inventories. We excluded urban, dry and montane forest types from this analysis.

By urban forest, we mean forest stands within the San Juan Metropolitan Area, specifically in the watershed of Río Piedras, which discharges into the San Juan Bay estuary. These forests were in the moist-forest life zone, but we exclude them from summary data by life zone and do not examine temporal change in these stands because this was the first inventory of this forest type. However, they were not mature stands. Reverted forest stands are those that experienced a land-cover change between the 1990 and the 2002 inventories. These stands were returned to an earlier stage of succession during that time interval.

Importance values

Importance value, expressed as the average of relative density and relative basal area, is the metric used to gauge a tree species' presence and relative importance in each forest type. We calculated importance values for each species for trees and saplings. Seedling relative frequency was calculated, not a true importance value. Seedling relative frequency for the more commonly found species may be truncated. According to national inventory protocols, field crews count up to six individuals of a seedling species per micro-plot, and then stop. A value of six may actually represent more seedlings of that species, but there is no way to know exactly how many were in the micro-plot. We assume a species is regenerating at a site if it shows importance values in the tree and sapling size classes, as well as in a relative frequency for the seedlings.

For comparison of results from the previous forest inventories, we limited the analysis to those forest types inventoried each time and normalized the data to account for different sampling area. This limited the analysis to moist and wet secondary forests and shade coffee. The comparison involved the top-ranked forty-eight species in each inventory. We express all comparisons as percentage of the parameter (total importance value and/or number of species) in each inventory.

Species–area curves

We also constructed species–area curves using 2002 data for five of the forest types we studied and one for all forests combined. We limited the analysis to a one-hectare sample size because we have comparative information for undisturbed and mature native forests.

Appendix 20.2 Results of island-wide forest inventories by forest type
Dry, moist and wet forests

The importance value of the top-ranked tree species in moist secondary forests was higher, and the slope of the ranked importance-value curve was steeper, for the top four ranked species than for the top four of secondary forests in dry and wet life zones (Fig. 20.1). *Spathodea campanulata*, an alien species, had the highest importance value in moist forest. Alien tree species constituted 6.3 to

Table 20.3 *Percentage of endemic and alien tree species, contribution of aliens to importance value, and number of species by size class in various forest types in Puerto Rico*

Forest type	Percentage of total number of species		Percentage of IV	Number of species			
	Endemic species	Alien species	Alien species	Seedlings	Saplings	Trees	Total
Urban forest	0	24.2	65.6	NA[a]	20	19	33
Young karst forest	6.3	22.5	43.6	137	82	70	160
Young moist forest	5.0	20.0	49.1	106	61	67	140
Reverted moist forest	0.3	22.6	52.5	20	13	15	31
Reverted wet forest	5.6	22.2	54.0	14	3	6	18
Young inactive shade coffee	0	27.3	22.9	14	12	9	22
Old inactive shade coffee	6.2	21.0	38.3	55	30	34	81
Oldest inactive shade coffee	0	19.0	18.4	30	13	20	42
Young wet forest	6.9	16.7	25.6	54	26	38	72
Mature karst forest	6.1	8.5	12.5	58	34	36	82
Dry forest	5.7	13.6	7.8	65	56	30	88
Mature moist forest	10.0	6.3	41.4	58	28	38	80
Mature wet forest	4.1	13.5	25.6	54	24	43	74
Lower montane wet forest	26.7	0	0	22	16	17	30
Upper montane wet forest	15.5	5.2	2.6	44	22	27	58

[a] Data are not available.

The listing of forest types is in decreasing level of anthropogenic disturbance. Data are for the year 2002. Importance value is IV.

22.6 per cent of the species pool in moist secondary forests but accounted for 41.4 to 52.5 per cent of the importance value (Table 20.3). In wet forests, the share of the importance value of alien species is also higher than their representation in the species pool. The slope of the ranked importance value curves for the top four species in dry and wet forests was relatively flat (Fig. 20.1).

In dry forests, alien tree species made up 13.6 percent of the species pool but their share of the importance value was smaller (Table 20.3). *Bursera simaruba* and *Andira inermis*, both native, were the most important dry-forest species, and both species show signs of regeneration. *Leucaena leucocephala*, however, dominates the sapling and seedling classes. Of the eighty-eight species found in dry forest, seventy-five were native. We found dry-forest endemic species *Psidium insulanum, Thouinia striata, Rondeletia inermis*, and *Tabebuia haemantha*, along with indications that the last three species are regenerating (Table 20.4). *Prosopis pallida, Pithecellobium dulce, Albizia lebbeck, Tamarindus indica, S. campanulata, Melicoccus bijugatus, Acacia farnesiana, Persea americana, Sterculia apetala, Crossopetalum rhacoma, Parkinsonia aculeata* and *Swietenia macrophylla* were the alien species found

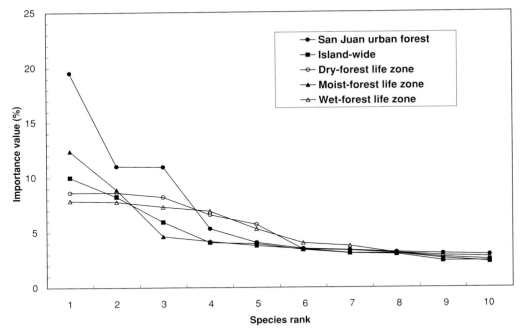

Figure 20.1 Ranked importance-value curves for secondary forests in three forest life zones, the urban forest, and an island-wide composite of secondary forests. All data are for 2002.

in dry forest, in their order of importance as an overstorey species. Most also showed sapling and seedling regeneration.

Lower and upper montane wet forests

What is most striking about the lower montane wet forests is the lack of alien species found (Table 20.3). In the upper montane wet forest, aliens make up a small proportion of the tree flora. *Syzygium jambos* and *S. campanulata* were minor components. We found *Spathodea jambos* regeneration, but not that of *S. campanulata*. Endemic tree species found in the montane forest include *Micropholis garcini-ifolia*, *Henriettea squamulosum*, *Cordia borinquensis*, *Croton poecilanthus*, *Hirtella rugosa*, *Byrsonima wadsworthii*, *Psychotria maleolens*, *Eugenia borinquensis*, *Ocotea moschata*, *Tetrazygia urbanii* and *Miconia pycnoneura*. These species made up 15.5 to 26.7 per cent of the tree flora, the highest values in the study (Table 20.3). Almost all of these endemic species also grew as seedlings (Table 20.4).

Karst forests

Spathodea campanulata dominates young secondary moist forest in the karst region. There were thirty-six alien species found in young karst stands (Table 20.3). The highest proportion of aliens occurred in the young karst forest.

Table 20.4 *Importance value of trees and saplings and relative density of seedlings of endemic species in various forest types throughout Puerto Rico*

Stand age/state	Species	Trees	Saplings	Seedlings
Shade-coffee forest				
	Thespesia grandiflora	1.37	0.51	
	Thouinia striata	1.38		1.38
	Wallenia pendula			1.38
	Antirhea obtusifolia	0.34		
	Cyathea portoricensis		0.51	
	Eupatorium portoricense			1.38
Dry forest				
	Psidium insulanum	3.99		
	Rondeletia inermis		0.21	0.56
	Tabebuia haemantha			0.19
	Thouinia striata		2.23	3.36
	Thouinia portoricensis		0.44	0.56
Moist forest				
Reversion	*Tabebuia haemantha*			4.82
Young	*Cordia borinquensis*	0.43	0.60	0.44
Young	*Gesneria pedunculosa*			0.30
Young	*Neea buxifolia*	0.22		
Young	*Tabebuia haemantha*	0.55	2.85	1.63
Young	*Thespesia grandiflora*	0.14		
Young	*Thouinia striata*		1.31	
Young	*Thouinia portoricensis*			0.07
Mature	*Byrsonima wadsworthii*	0.87		
Mature	*Cordia borinquensis*	0.70	2.20	0.68
Mature	*Miconia pachyphylla*		0.83	
Mature	*Sapium laurocerasus*	1.66		
Mature	*Tabebuia haemantha*			1.37
Mature	*Thespesia grandiflora*	0.89		
Mature	*Thouinia striata*	0.41		0.68
Mature	*Xylosma schwaneckeana*	0.86		
Karst forest				
Young	*Brunfelsia densifolia*	0.34		
Young	*Byrsonima wadsworthii*			0.05
Young	*Eupatorium portoricense*		0.11	
Young	*Gesneria pedunculosa*			2.22
Young	*Neea buxifolia*	0.27		
Young	*Psychotria maricaensis*		0.16	
Young	*Sapium laurocerasus*	0.41	0.40	0.05

(cont.)

Table 20.4 (*cont.*)

Stand age/state	Species	Trees	Saplings	Seedlings
Young	*Thespesia grandiflora*	0.13	0.12	
Young	*Thouinia striata*	2.34	3.00	2.26
Young	*Xylosma schwaneckeana*		0.10	
Mature	*Cordia borinquensis*	0.63	1.17	1.37
Mature	*Gesneria pedunculosa*			0.69
Mature	*Neea buxifolia*			3.44
Mature	*Rondeletia inermis*		0.84	0.69
Mature	*Thouinia striata*	1.58	1.73	0.34
Wet/rain forest				
Reversion	*Heterotrichum cymosum*			19.35
Young	*Henriettea squamulosum*		1.57	
Young	*Heterotrichum cymosum*			0.17
Young	*Hirtella rugosa*			0.34
Young	*Miconia foveolata*			0.34
Young	*Xylosma schwaneckeana*			0.17
Mature	*Cordia borinquensis*	2.38	1.24	0.56
Mature	*Cyathea portoricensis*			0.56
Mature	*Sapium laurocerasus*			0.56
Lower montane wet forest				
	Cordia borinquensis		5.07	1.69
	Croton poecilanthus	1.77	3.76	3.39
	Eugenia borinquensis	1.59		0.85
	Henriettea squamulosum	1.48	4.52	24.58
	Miconia pycnoneura		1.69	5.08
	Micropholis garciniifolia	5.29	12.13	7.63
	Psychotria maleolens			5.08
	Tetrazygia urbanii	0.71		
	Byrsonima wadsworthii		1.31	0.72
Upper montane wet forest				
	Cordia borinquensis	0.40	1.43	0.72
	Croton poecilanthus		3.43	1.08
	Eugenia borinquensis			0.72
	Henriettea squamulosum	1.24	1.49	1.80
	Hirtella rugosa		2.36	2.16
	Micropholis garciniifolia	1.87		
	Ocotea moschata			0.36
	Psychotria maleolens			7.19

Empty cells mean absence of the species.

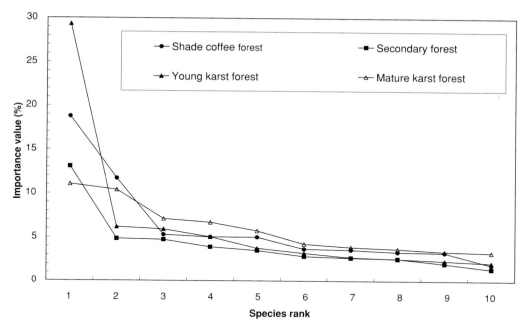

Figure 20.2 Ranked importance-value curves for shade coffee forest, secondary forest, young karst forest and mature karst forest. All data are for 2002.

Spathodea campanulata, Adenanthera pavonina, Senna siamea, A. procera and *Delonix regia* were relatively common alien canopy tree species also regenerating in young karst stands. However, there were 125 native species encountered, about half of which were found in the tree size class. Most of the native species also grew as saplings and almost all were found in the seedling class. The highest seedling, sapling and tree species counts in the study occurred in the young karst forest (Table 20.3). Compared to all forest types studied, this forest had the highest importance value for its top ranked species (Fig. 20.2). The second-ranked species had a low importance value compared with species ranked second in shade coffee and mature karst forest.

There were fewer plots measured in mature karst forest, which explains the fewer species found (82 total species in mature stands versus 161 species in young stands). However, there is still indication that alien species had a lower dominance in mature karst forest than they did in young karst forest (Table 20.3). *Spathodea campanulata* is much less important, less dominant, in mature stands. The native generalist species *G. guidonia* and *A. inermis* assume the dominant positions in the tree class, and the sapling class in the case of *G. guidonia*. Only seven alien species occurred in mature karst forest as compared with seventy-five native species found (Table 20.3). The slope of the ranked importance-value curve

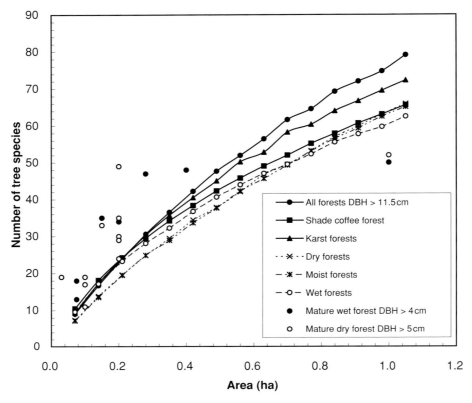

Figure 20.3 Species–area curves for five secondary forest types and for all forest inventory plots combined. Data are for 2002 and trees with diameters at breast height DBH > 11.5 cm. Individual points are for two types of undisturbed mature native forests in Puerto Rico (Lugo *et al.* 2002). Note the change in minimum diameter for these individual data points.

in mature karst was shallow indicating low dominance for the most important species (Fig. 20.2).

Shade-coffee forest

Guarea guidonia was the dominant species in shade coffee followed by *S. campanulata*. The importance value of these two species was responsible for the steep slope of the ranked importance-value curve for this forest type (Fig. 20.2). *Inga laurina* is a common sapling component that is only regenerating well in younger stands. *Inga vera* is also common, and minor, but more successful at regenerating in more mature stands. *Erythrina poeppigiana* is also widespread in shade-coffee and in some young moist and karst stands, but not as a sapling or seedlings. *Erythrina berteriana* is more successful at regenerating, but only in recently abandoned shade coffee.

Table 20.5 *Twenty-two-year trends in the number of species and the importance value of alien species in three forest types on Puerto Rico*

Parameter	1980	1990	2002
Shade-coffee forest			
Number of tree species	101	87	62
Alien species (% of total)	19.8	20.7	19.3
Importance value of alien species (%)	25.2	28.0	27.8
Secondary forest			
Number of tree species	207	215	223
Alien species (% of total)	13	13.5	19.7
Importance value of alien species (%)	20.3	19.3	28.1
Urban forest			
Number of tree species			33
Alien species (% of total)			21.0
Importance value of alien species (%)			53.5

Secondary forests include moist- and wet-forest life zones combined. Data for the urban forest in 2002 are for comparison. Empty cells mean absence of data.

Urban forest

The top-ranked three species of urban forests had higher importance values than the corresponding species in dry, moist, and wet forests (Fig. 20.1). Values are similar to those measured in the trees of shade coffee but lower value than those of young karst (Fig. 20.2). It appears that species richness is less in the urban forest than in other forested areas, but the sampling area has to increase before reaching such a conclusion. Alien species attain their highest importance and abundance in urban forests (Table 20.3). *Albizia procera*, an alien species, was the second-most important urban tree species that invades vacant barren and institutional lands with highly compacted soils.

Species–area curves

None of the five forest types in Fig. 20.3 saturated the species–area curve at the 1-ha level of sampling. Karst forests had the steepest curve and highest number of species at one hectare. Moist and dry forests had a slower rate of tree species accumulation in area sampled. Shade-coffee forest was initially intermediate between the extremes, but reached a similar number of tree species as moist and dry forests at the 1-ha sample. Wet forests started with a steep rate of species accumulation in area sampled but ended with the lowest number of species at the 1-ha sample.

Table 20.6 *Alien species found in mature stands of various forest types throughout Puerto Rico*

Species	Moist forest	Wet forest	Karst forest	Shade coffee forest	Trees	Saplings	Seedlings
Cananga odorata				1			1
Carapa guianensis	1			1			1
Ceratonia siliqua	1			1			1
Cordia oblique				1			1
Erythrina poeppigiana				1	1		
Eucalyptus robusta		1			1		
Ixora ferrea		1			1		
Mangifera indica		1	1		1		
Parkia biglandulosa		1					1
Pinus sp.			1				1
Spathodea campanulata	1	1	1	1	1	1	1
Syzygium jambos	1	1	1	1	1	1	1
Syzygium malaccense		1					1

1 means the species was found, empty cells mean it was not.

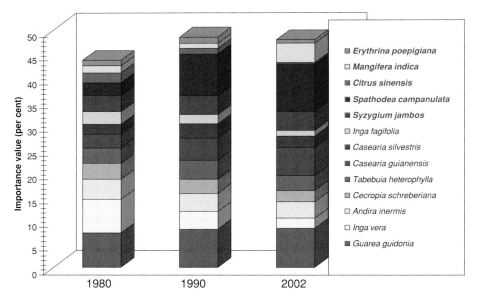

Figure 20.4 Temporal variation in the importance value of the dominant species in inventoried forests of Puerto Rico. The first five species on the top of the list are alien species.

Temporal change

Over the 22-year interval between the first and third forest inventory, the percentage of alien species remained relatively constant in shade coffee (about 20 per cent) but increased in all moist and wet secondary forests from 13 to 20 per cent (Table 20.5). These alien species accounted for 28 per cent of the importance value of forest stands, a slight increase from 1980 values. In shade coffee, *G. guidonia* had a large increase in importance value while *I. vera* decreased by 2002. In the secondary forest, *S. campanulata* increased its importance value.

The aggregation of all inventory data shows that alien species increased in their share of the importance value island-wide. However, when compared with urban forests, both the proportion of alien species in the tree flora and their share of the importance values were much lower than in the urban. The trend by species (Fig. 20.4) shows both increases and decreases of both native and alien species that appear more dependent on the history of land use. For example, the alien *Citrus sinensis* and the native *I. vera*, two species common in actively managed coffee shade, decreased, while the alien *Mangifera indica*, a favoured fruit tree, and the native *G. guidonia*, which thrives in abandoned shade-grown coffee and agricultural land, increased in importance value.

CHAPTER TWENTY ONE

The dynamics of a tropical dry forest in India: climate, fire, elephants and the evolution of life-history strategies

R. SUKUMAR, H. S. SURESH, H. S. DATTARAJA,
S. SRINIDHI AND C. NATH

Indian Institute of Science, Bangalore

Introduction

Tropical dry forests constitute over 40% of all tropical forests (Murphy & Lugo 1986), yet their dynamics have been poorly studied relative to tropical moist forests (Bullock *et al.* 1995). Major ecological factors influencing the dynamics of tropical dry forests include high variability in climate, herbivory by wild mammals and domestic livestock (Skarpe 1991; Sukumar *et al.* 1998), natural and human-induced fires (Swain 1992; Goldammer 1993) and human extraction of a variety of products (Chopra 1993; Narendran *et al.* 2001). There has been much thinking on the issue of stability of tropical forests; some of this follows from or is related to the broader issue of the stability–diversity debate (e.g. Connell 1978; Pimm 1984; Lawton 1994; Johnson *et al.* 1996) or that of the turnover rate of tropical forests (Philips *et al.* 1994; Sheil 1995). These have been discussed mostly in the context of tropical moist forests.

It is being increasingly recognized that environmental variability is a major influence in shaping the structure, functioning and evolution of communities. In particular, we can expect that environmental variability would influence the evolution of life-history traits of species that constitute a particular community (e.g. Murphy 1968; Gadgil & Bossert 1970; Stearns 1977, 1992; Boyce & Daley 1980). Interannual variation in climate (e.g. precipitation) is usually taken as the most important measure of environmental variability that shapes life-history traits in a species. At the same time, the role of disturbances (e.g. hurricanes, fire) in eliciting short-term ecological responses should also be considered. In principle, a recurring disturbance that operates over longer time scales could be a source of environmental variability; for instance, fires (both natural and anthropogenic) are known to have occurred in tropical regions during the Pleistocene and the Holocene (see review by Schule 1990) and would have introduced

Biotic Interactions in the Tropics: Their Role in the Maintenance of Species Diversity, ed. D. F. R. P. Burslem, M. A. Pinard and S. E. Hartley. Published by Cambridge University Press. © Cambridge University Press 2005.

ecological stress into systems already subject to climate-related stress such as drought. It could be expected that plant species associated with such communities would show life-history adaptations to these factors.

From this perspective we think it would be extremely useful to compare the structure, life-history traits and the dynamics of tropical forests along environmental and disturbance gradients; ideally this should span the entire range of 'forest types' from the seemingly simple woodland-savanna vegetation through dry forest and seasonal moist forest to the complex aseasonal rain forest. Are there predictable patterns of life-history traits such as the nature of sapling recruitment, growth rates of juvenile versus adult stems, mortality rates in relation to stem size, and reproductive structures across an environmental gradient? How are these traits eventually reflected in the structure of forests across the gradient?

We have been studying the structure and dynamics of a tropical dry forest in southern India since 1988 (Sukumar *et al.* 1992, 1998, 2004; John *et al.* 2002). This is part of an international network of large-scale (50-ha) plots coordinated by the Center for Tropical Forest Science (see Condit 1995) that provides an opportunity for comparisons across forests. One such comparison between two tropical moist forests, one in Panama (Hubbell & Foster 1983) and the other in Malaysia (Manokaran *et al.* 1992), was reported by Condit *et al.* (1999). In this chapter, we first describe the observed patterns in the structure and dynamics of the tropical dry-forest plot in southern India, and then go on to make some comparisons with the Panamanian and Malaysian plots from published information (Condit *et al.* 1999) to make possible inferences about the role of environmental variability and disturbance from fire in shaping life-history traits of tropical tree species. Rather than describe in detail the population structure and changes of individual species, we have chosen to characterize the dry-forest community in terms of overall species population, size distribution of stems, and the guild of canopy versus understorey species before making comparisons with the other forest sites. More details of forest structure and the short-term dynamics of individual species are either available (Sukumar *et al.* 1992, 1998; John *et al.* 2002) or will be presented elsewhere.

Material and methods
The study area
Our tropical dry-forest plot was set up during 1988–89 in the Mudumalai Wildlife Sanctuary of Tamilnadu State in southern India. Mudumalai is located to the east of the Nilgiri hills, which are part of the Western Ghats chain of mountains in peninsular India. The natural vegetation of Mudumalai varies along a rainfall gradient from tropical dry thorn forest in the east (*c.* 800 mm annual precipitation), through dry deciduous forest over much of the reserve to moist deciduous forest in the southwest (*c.* 1800 mm annual precipitation). Small patches of

semi-evergreen forest are also seen in the reserve, as are swampy grasslands within the tract of moist forests. The dry season extends for a period of about 5 months (December–April) during which little or no rain occurs.

The 50-ha permanent plot (11° 35′ 41″ to 11° 35′ 57″ N, 76° 31′ 50″ to 76° 32′ 22″ E) is centrally located at Mudumalai, within Compartment 17, close to the transition between the dry and moist deciduous forests (Sukumar *et al.* 1992, 2004). *Lagerstroemia microcarpa* is at present the most abundant tree in the plot, followed by *Terminalia crenulata, Anogeissus latifolia* and *Tectona grandis*. The main understorey trees are *Cassia fistula, Xeromphis spinosa* and *Kydia calycina* (the most abundant woody plant at the time the plot was set up), while the most common shrub is *Helicteres isora*. The reserve has sizeable populations of the larger mammals characteristic of peninsular India, including the Asian elephant (*Elephas maximus*), which makes a major impact on the vegetation. Densities of the larger herbivorous mammals are about 2 individuals km^{-2} for elephant, 25 km^{-2} for axis deer (*Axis axis*), 7 km^{-2} for sambar (*Cervus unicolor*) and 6 km^{-2} for gaur (*Bos gaurus*) (Varman & Sukumar 1995, and unpublished data for 1996–2000). The reserve is bordered by human settlements to the south and the east. In addition, several swampy grasslands in the western sector are under cultivation. To the north and the west the reserve is bordered by other protected areas. Direct human influences include some fuel-wood removal by villagers and grazing of livestock, particularly in the eastern part of the reserve; however, there is practically no direct human impact on the 50-ha plot. Human-induced ground fires are also common during the dry months of January–April. The 50-ha plot is generally free from direct human disturbances with the exception of fires that may spread here from other parts of the reserve. The forests of Mudumalai have a history of logging going back to at least the early part of the nineteenth century (Ranganathan 1939). Selective logging of Compartment 17, where the plot is located, was last carried out during 1968.

Field methods

The basic field methods used in the study have been described in an earlier publication (Sukumar *et al.* 1992). These will be briefly recapitulated here. The plot was surveyed with a theodolite and grid into quadrats of 20 m × 20 m after making appropriate corrections for slope. All woody plants >1 cm diameter at breast height (dbh) were tagged with numbers during the period May 1988 to May 1989. The plants were identified, measured for dbh, tagged and mapped to the nearest 0.5 m. Each 20 m × 20 m quadrat was further divided into sub-quadrats of 10 m × 10 m with the aid of ropes before mapping stem locations.

As most of the stems shed their leaves during January–April, the censuses during subsequent years began in June by which time they had flushed leaf, enabling greater accuracy in identification of species. The annual censuses recorded mortality of stems and new recruits (stems growing to >1 cm dbh). In practice, a

stem was considered dead if the above-ground portion was burnt and no coppices had reached 1 cm dbh. However, a stem was considered as living if it had
been merely broken below 1.3 m but was still alive. A recruit was considered as
the appearance of a new stem >1 cm dbh; while these included previously tagged
stems that might have burnt completely but coppiced subsequently and grown
to 1 cm dbh, they excluded broken stems that had been recorded as alive in the
previous census. For every recruit an attempt was made to determine whether it
had originated from a seed (sexual reproduction) or through vegetative means
(clonal propagation or coppicing from the root stock of a 'dead' stem).

At the end of the dry season the extent of burning, if any, was mapped at a
10 m × 10 m resolution. During the study period fires occurred during the dry
seasons of 1989, 1991, 1992, 1994 and 1996. Every four years the stems were also
measured for calculating growth rates. Stem sizes were measured again during
1992, 1996 and 2000.

Analyses

Because the Mudumalai plot is censused annually, the recruitment rates and
size-dependent mortality rates are expressed as arithmetic means of the yearly
rates. The annual size-based mortality rates are accurate only for the years 1989,
1993 and 1997 for which all stems had been measured during the previous year
(1988, 1992 and 1996, respectively, when complete censuses were carried out).
During other years, a certain proportion of stems would have moved into a
higher size class at the time of death because of growth. Growth rates of stems
in various size classes are expressed as mean annual increments (in mm yr^{-1})
computed from the 4-yr census intervals.

Using the observed rates of recruitment, growth and mortality in various
size classes, we used a deterministic, matrix-projection model (Caswell 1989) to
simulate the trends in population size and basal area of the four major canopy
trees, *Lagerstroemia microcarpa*, *Tectona grandis*, *Anogeissus latifolia* and *Terminalia
crenulata*, that constituted 44% of all individuals (woody species >1 cm dbh) and
67% of basal area in the plot during the 1988–89 census. Details of the model
and the results are being published elsewhere but its basic features are described
here.

For each of the four species, individuals were categorized into size classes for
analysis of growth and mortality, and for population projections. For each of the
approximately 4-yr intervals of stem measurement (1988–92, 1992–1996, 1996–
2000) the annual growth rates were computed for individuals in each of the size
classes. Up to 2% of outliers showing extremely fast growth or shrinkage were
removed from the data in order to minimize possible errors in field recording of
measurements. The mean size-class specific growth rates calculated across the
various size classes were smoothed and best-fit functions used for incorporating growth into the model. Annual mortality rates were likewise calculated for

various size classes of a species and the rates smoothed before fitting suitable functions. Average mortality rates were calculated for the 12 years of the study as well as for years with fire (1989, 1991, 1992, 1994 and 1996) and without fire. Recruitment was incorporated as the number of stems of size 1–3 cm dbh (growth from <1 cm to >3 cm dbh was considered unlikely to occur within a year) of a species appearing each year on average for the period 1988–2000 as well as during years with fire and without fire. Because most of the recruits were vegetative sprouts, we did not relate these to the adult population size of the species.

Beginning with the population recorded in 1988, the model projected the number of individuals in each of the size classes surviving and moving into one of the higher size classes (through growth), remaining in the same size class or moving into a lower size class (through shrinkage of stem). For the projections we used a 0.5-cm dbh class interval for the faster-growing *L. microcarpa* and *T. grandis*, but a 0.25-cm class interval for the relatively slow-growing *A. latifolia* and *T. crenulata*. Apart from an 'average scenario' incorporating the mean growth, mortality and recruitment rates observed during 1988–2000, we also ran other scenarios including lower and higher fire frequencies; one of these was a 'worst case' scenario that used the lowest recruitment and stem growth rates and the highest mortality rates observed during the study period. Only the projected basal areas are reported here.

Comparison with two moist-forest plots

We compare the Mudumalai dry forest plot with the moist-forest plots at Barro Colorado Island (BCI) and Pasoh Reserve, basing the comparison entirely on the published work of Condit *et al.* (1999) for the latter two plots. These three Forest Dynamics Plots represent a gradient of environmental variability. For our purposes we use interannual climatic variability as a simple measure of environmental variability. The rainfall data for the period 1990–2002 indicate that Mudumalai has the highest rainfall variability (annual mean = 1118 mm; coefficient of variation [CV] = 26.1%) and strong seasonality (dry season of five months; a dry month defined as one with <100 mm rainfall), BCI moderate variability (mean = 2642 mm; CV = 18.8%) and seasonality (four-month dry season), and Pasoh the lowest variability (mean = 1788 mm; CV = c. 15%) and relative lack of seasonality (no dry month). In addition, the Mudumalai forest is subject to dry-season fire, a stress factor absent at the other two sites.

Results
Changes in population size, species richness and size distributions

The total population (all stems >1.0 cm dbh) of individuals in the Mudumalai plot declined noticeably over the period 1988–1996 and then increased between 1996 and 2000. The first census (in 1988–89) of the plot recorded 25 554 individuals from 72 species (excluding the one species of bamboo – a 'tree grass'). The numbers fell steeply to 17 654 individuals from 70 species by 1992 and at a lower rate to 15 346 individuals (65 species) by the 1996 census. During this period the years 1993 and 1995 recorded small increases in population size over the previous years whereas all other annual censuses registered declines. After 1996 the population increased steadily to 18 177 individuals (75 species) by the year 2000. Of the original number of species recorded during 1988–89, two species have been lost, while three new species have appeared and persisted until the year 2000. Five species have been ephemeral during this period. Four species have disappeared during at least one of the yearly censuses but have later appeared. Correspondingly, the population sizes of most species also declined overall during the period 1988–1996 (62 of the 72 species recorded in 1988 declined by 1996), but many of them then increased during 1996–2000 (29 of the 65 species recorded in 1996 increased by 2000, while 25 species still declined).

Declines have been disproportionately higher in the smaller size classes for practically all species. Thus, the total change between 1988 and 1996 was −13.1% for stems >10 cm dbh as compared with −79.0% for stems 1–10 cm dbh. The increase in population during 1996–2000 is also characterized by a disproportionate increase in the smaller stems, as can be expected because of recruitment of seedlings and young saplings into >1 cm dbh size and their subsequent growth (147% increase in stems 1–10 cm dbh and −3.9% decline in stems >10 cm dbh).

Population change has varied considerably among the common canopy trees, understorey trees and shrubs (Table 21.1). *Kydia calycina* (understorey tree), the most abundant species in the plot during 1988 (5175 individuals), underwent a precipitous decline through to 1996 (247 individuals); its population size was 190 individuals in the year 2000. The shrub *Helicteres isora* also declined steeply from a population of 2569 individuals in 1988 to 307 individuals by 1996, but then increased to 548 individuals in 2000, mainly through a doubling during the last year. The decline of both these species could be largely attributed to herbivory by elephant, though fire also played a role in the death of smaller stems. *Lagerstroemia microcarpa* (canopy tree), the second most abundant species in 1988 (and at present the most abundant), maintained an almost identical population size in 2000 in spite of registering a decline of about 10% during the intermediate period. The canopy tree *Tectona grandis* declined by 16.7% between 1988 and 1996, but since then increased by 5.7% by 2000. Two other abundant canopy species, *Terminalia crenulata* (−8.5%) and *Anogeissus latifolia* (−5.8%) registered modest overall declines between 1988 and 2000. The population of *Cassia fistula* (understorey) in the plot showed the highest interannual fluctuations amongst the commoner species, registering an annual increase as high as 50% (1992–93) and decreases up to −37% (1995–96).

Table 21.1 *Population changes in 10 most abundant species*
(1988 census) during the period 1988–2000

Species (life form)	Population size during full census year			
	1988	1992	1996	2000
Kydia calycina (U)	5175	1096	247	172
Lagerstroemia microcarpa (C)	3982	3817	3567	3981
Terminalia crenulata (C)	2771	2678	2604	2536
Helicteres isora (S)	2569	808	307	548
Anogeissus latifolia (C)	2281	2232	2187	2149
Tectona grandis (C)	2139	1857	1782	1884
Cassia fistula (U)	1884	1092	1214	3200
Xeromphis spinosa (U)	770	725	589	519
Emblica officinalis (U)	577	500	456	422
Grewia tiliifolia (C)	540	459	396	375

C = canopy tree, U = understorey tree, S = shrub

Recruitment rates and patterns

The overall rate of recruitment defined as appearance of stems >1.0 cm dbh during a census, averaged 4.4% over the 12-yr period of the study. Mean recruitment rates were significantly higher during years without dry-season fire (6.0%) than during years with dry-season fire (2.2%). In particular, the recruitment rate was highest during the census interval 1999–2000, when the plot registered a 12.6% rate of recruitment, twice as high as the next highest rate recorded (6.1% in each of the periods 1996–97 and 1997–98). During the three full census periods, the recruitment rates were lowest during 1988–92 and highest during 1996–2000. However, the average rate obscures the fact that only a few species showed strong recruitment, while at the other extreme several moderately abundant and abundant species did not show any recruitment over the 12-year period. In particular, the abundant understorey tree *Cassia fistula*, with a mean recruitment rate of 25% per annum, contributed over 50% of all recruits recorded during the study period.

Recruitment was almost exclusively (>99%) through vegetative means. Very often stems that had died above ground in previous years through fire coppiced from the root; likewise stems broken to < 1.3 m in height also coppiced. In one sense, the disappearance of a young stem through burning does not necessarily imply death, but rather survival as a root or stem at ground level. An examination of seedlings showed that these too sprouted from roots or rhizomes. It is of course possible that a certain proportion of these may have originally sprouted from seeds (i.e. sexually) but repeatedly burnt to ground level. Because of the need to avoid disturbance to the soil and young plants in the plot, it was not possible to distinguish clonal propagation from vegetative coppicing; the former could be occurring in some species but has to be studied out of the plot.

Growth rates and patterns

We examined growth rates for two full census intervals, 1988–92 and 1992–96, across stem sizes for all canopy trees and all understorey trees (plus the few shrubs in the plot). Among canopy tree species the annual growth increments were highest in the smallest stems (1–5 cm dbh) and in the large trees (>60 cm dbh) during the period 1988–92 (Fig. 21.1a). The smallest stems similarly increased in diameter at the highest rate during the subsequent period (1992–96) but growth of the large trees came to a virtual halt; in fact, the largest stems (over about 80 cm dbh) in the plot shrank slightly on average. Overall, the growth rates across all stem sizes of canopy species were also higher during the first census interval than the second interval. Among understorey species the smallest stems actually grew at higher rates during the second census interval, although a depression in growth could be seen among stems >5 cm dbh (Fig. 21.1b).

Mortality rates and patterns

The average annual mortality rate of all individuals >1.0 cm dbh during 1988–2000 was 7.0% (range 1.9–13.1%). Mortality rates varied substantially across size classes with the average rates being progressively lower in the larger size classes (Fig. 21.2). Thus, the average mortality rate of 23.5% in the 1–5 cm size class decreases to about 4% in the 10–20 cm size class and thereafter to under 0.6% in stems >30 cm dbh. During 1990, however, there was little difference in mortality rates between the 1–5 cm and 5–10 cm size classes, and during 1993 the mortality rate was highest in the 5–10 cm dbh class. These were years without dry-season fires, which would have killed a high proportion of smaller stems. Therefore mortality rates were relatively high in medium-sized stems, largely because of elephant utilization of *Kydia calycina*. Interannual variability in mortality rates was generally higher in smaller individuals and decreased with size, except for the very large trees (demographic rates of very large trees are not reliable because of small sample sizes). In general, there was a declining trend in mortality rates of most size classes from the first full census interval (1988–92) to the subsequent two full census intervals (1992–96 and 1996–2000).

Causes of mortality were broadly classified as due to fire, elephant herbivory or other causes. For 'other' causes, it was possible to identify the agent in only a few cases, as in tree fall due to strong wind or a storm. Dry-season fires occurred in the plot during five (1989, 1991, 1992, 1994 and 1996) of the 12 years of the study reported here. Of these the fires were widespread (>80% area burnt) across the plot during the years 1991, 1992 and 1996. Fire-related mortality rates were obviously higher during these years. During the non-fire years a small proportion

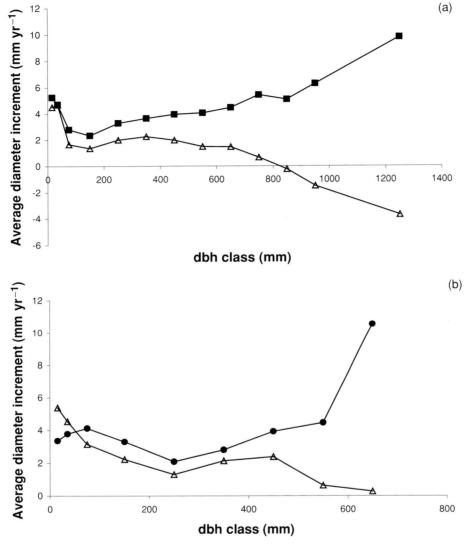

Figure 21.1 Growth rates (mean increment in stem diameter in mm yr^{-1}) of (a) canopy species during the period 1988–92 (solid square) and 1992–96 (open triangle), and (b) understorey species during 1988–92 (solid circle) and 1992–96 (open triangle) in the Mudumalai Forest Dynamics Plot.

of deaths was attributed to weakening of stems from fire during the preceding year. Over the 12 years of the study, fire was the leading agent of mortality during four years, while elephant browsing was the leading cause in five years (of these the mortality rates due to fire and to elephants were identical during 1988–89). In later years, the mortality due to elephants declined because of sharp reductions in the population of their favoured food-plant species.

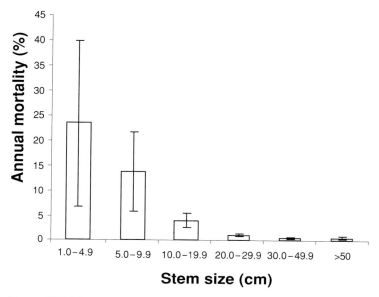

Figure 21.2 Mean size-class specific overall mortality rates during 1988–2000 in the Mudumalai plot. The error bars represent one standard deviation.

When causes of mortality are examined across size classes, it is clear that fire was the main agent of mortality in the 1–5 cm dbh class, elephant her-bivory in the 5–10 cm dbh class and 'other causes' in stems above these sizes (Fig. 21.3). The influence of fire is greatest on stems of small size and declines rapidly with size. Elephant herbivory was restricted to a few species; debarking and pushing over of trees generally occurred in the size range of 5–10 cm dbh. Elephant-related mortality was lower for very small stems, but it is possible that elephants do trample and uproot seedlings and saplings <1 cm dbh. Elephants rarely push over very large trees, but large trees of many species are debarked, which probably weakens their ability to resist fire and/or disease attack. The mortality function for 'other causes' was flatter than the ones for fire and ele-phant. This category captures a multitude of causes related to direct physical damage by treefalls or windstorms, and attacks by termites, diseases and pests, which are more likely to attack stems weakened by physiological and ecological stress. Mortality rates due to these causes were lower than those due to fire and elephant for small- and medium-sized stems respectively, but they accounted for most of the mortality in large trees.

Changes in size-class distributions and basal area

When changes in basal area of the different species are considered, a very dif-ferent picture emerges. Thirty-one of the 72 species recorded in the first census actually increased in basal area by 1996 even though most of them declined in

Table 21.2 *Number of individuals in various size classes during the full census years in the Mudumalai plot*

Size class (cm)	Number of individuals			
	1988	1992	1996	2000
1.0–4.9	5039	1891	1247	4594
5.0–9.9	5478	1720	1017	991
10.0–19.9	6162	5065	4166	3549
20.0–29.9	3986	3868	3662	3528
30.0–49.9	3576	3701	3774	3895
>50	1313	1408	1487	1613
Total	25 554	17 653	15 353	18 170

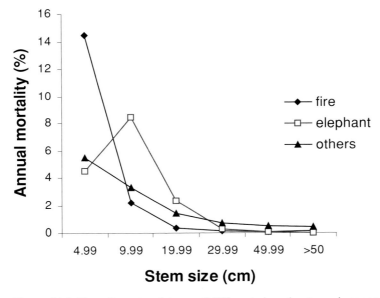

Figure 21.3 Mortality rate of stems of different sizes due to various causes during 1988–2000 in the Mudumalai plot.

absolute numbers (all stems >1 cm dbh). During 1996–2000, 38 of the 65 species recorded in 1996 increased in basal area.

Even though the total number of stems declined in the plot during 1988–96, this decline was not uniform across size classes (Table 21.2). Stems over 30 cm dbh actually increased during this period. Thus, stems in the 30–50 cm size class registered a 4.5% increase and those over 50 cm a 12% increase during this 8-yr period. In the absence of fire during 1996–2000, saplings <5 cm dbh also increased in total numbers. As the larger stems also contribute to basal area

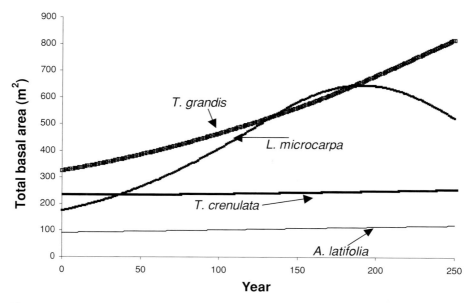

Figure 21.4 Projected trends in the basal area of the four major canopy trees (*Lagerstroemia microcarpa; Tectona grandis; Terminalia crenulata* and *Anogeissus latifolia*) in the Mudumalai plot.

in a geometrically increasing fashion, the basal area of all stems in the plot, which averaged 24.4 m^2 ha^{-1} in 1988, increased slightly to 24.8 m^2 ha^{-1} in 1992 and remained at that level in 1996. During 1996–2000 there was an increase in stems across most size classes with a corresponding increase in basal area to 25.9 m^2 ha^{-1}.

Projection of canopy-tree dynamics

All four canopy species increase in basal area over a 200-yr period when the mean rates of recruitment, growth and mortality during the period 1988–2000 are used in the model (Fig. 21.4). These demographic variables incorporate fire during five of the twelve years of observation. The increase in basal area is modest for the slow-growing species *Terminalia crenulata* and *Anogeissus latifolia* and much higher for the other two canopy trees. Even if one were to discount the steep increase in basal area of *Lagerstroemia microcarpa*, the total basal area of the four major canopy trees in the plot is still projected to increase overall during the next two centuries. The results are broadly similar even when fire frequencies are increased from the observed frequencies. Only under the worst-case scenario with sharply enhanced mortality and lowered recruitment and growth do the basal areas of three species (*Tectona grandis, T. crenulata* and *A. latifolia*) decline while that of *L. microcarpa* still continues to increase.

Discussion

Dynamics and life-history patterns of a tropical dry forest

The 50-ha forest plot at Mudumalai, typical of the tropical dry deciduous forests of the Western Ghats of southern India, shows considerable short-term instability in the face of interannual variation in climate, occurrence of fire and herbivory by large mammals such as elephants. The main features of structure and dynamics of this forest type can be summarized as follows:

(a)   The species richness of woody plants is very low compared with tropical moist forests and perhaps even lower than other tropical dry forests such as the dry dipterocarp forests of Thailand (Rundel & Boonpragob 1995). Indeed, our larger data set from Mudumalai itself indicates that the (alpha) plant diversity of the dry deciduous forest is lower than that of dry thorn forests which have even lower rainfall; this is the consequence of higher fuel loads in the dry deciduous forest resulting in more intense dry-season fires (unpublished data). However, the dry deciduous forest plot has shown a remarkable ability to maintain species numbers in spite of the dynamic nature of the forest.

(b)   Recruitment in the dry deciduous forest is almost entirely through resprouting from below-ground organs or from broken stems and not directly from germination of seeds. Rates of recruitment vary enormously amongst species, with only a few species showing strong recruitment during the decadal time scale of the study. At the other extreme, certain common canopy species showed virtually no recruitment during this period. Overall rates of recruitment were, however, comparable to other tropical forests, even though the absolute number of recruits was low.

(c)   The growth rates across stem size showed a distinct pattern of high growth during the young sapling stage followed by a decrease. In stems of canopy species >20 cm dbh there was considerable variation between two time periods (1988–92 and 1992–96) for which growth rates have been calculated; in the first census period these stems increased in growth rate with size whereas in the second census period they remained relatively steady or declined in growth. The understorey species showed relative constancy in growth rates with size, although the growth rate of larger stems was higher during the first census interval whereas that of the smaller stems was higher during the second.

(d)   Mortality rates are highly variable across stem sizes with very high rates being recorded in the smaller stems and a general decline with increase in stem size. Fire and elephant are the two most important agents of mortality, the former killing the smaller stems of a wide range of species and the latter destroying medium-sized stems of a few favoured forage species. Mortality rates of large trees (>30 cm dbh) were well below 1% per annum, one of the lowest rates recorded in tropical forests globally.

(e)   The Mudumalai forest shows high turnover and high instability in the sub-
      canopy strata of small understorey trees and shrubs but remarkable stability
      in the canopy layer. The results from the matrix-projection model indicate
      that the most common canopy trees would continue to survive and grow
      over the time scale of several decades or even a couple of centuries in the
      face of environmental stresses such as observed climate variability and fre-
      quent fire. We recognize that it is unlikely that basal area of the canopy
      trees would continue to increase in the projected manner over a period
      of two centuries; this would mean a doubling of the basal area of these
      four species during this period. There are obviously factors, not well under-
      stood at present, that could regulate the dynamics of such forests over time
      scales longer than the decadal time period of our observations. These could
      include deterministic factors such as density-dependent regulation (John
      et al. 2002) or stochastic factors such as severe climatic events (e.g. pro-
      longed drought, which could reduce growth and increase mortality across
      stems of all sizes, or windstorms that could sharply increase mortality of
      the large trees). However, it is reasonable to assume on the basis of the
      existing observations and model inferences that the basal area and stand-
      ing biomass of this forest type would be maintained or even increase in the
      coming decades.

   Dry tropical forests subject to periodic disturbances from fire and other factors
may have adapted through a regime of strong, episodic (vegetative) reproduction,
high juvenile growth rates and very low adult mortality, enabling populations
to persist for long periods of time. Such a system may exemplify the 'storage
effect' and the 'lottery competitive system' (Chesson & Warner 1981; Warner &
Chesson 1985). The frequency and intensity of ground fires are therefore key
determinants of the dynamics of tropical dry deciduous forests such as those in
south and Southeast Asia (Goldammer 1993; Stott 1986; Puyravaud *et al.* 1995).

Comparison of the dynamics of a dry forest with two moist forests

We can compare the Mudumalai dry forest plot (high environmental variability
and stress) with the plots at Barro Colorado Island (seasonal moist forest with
moderate environmental variability) and Pasoh Reserve (aseasonal rain forest
with low environmental variability).

(a)   The dry-forest plot at Mudumalai has far fewer species and individuals
      >1 cm dbh (72 species, 25 500 individuals in 1988–89) than the moist-forest
      plots at BCI (305 species, 235 000 individuals in 1982) and Pasoh (815 species,
      335 000 individuals in 1987). In spite of this tremendous difference in
      species and population density the Mudumalai plot had a greater number
      of large trees (5042 individuals >30 cm dbh in 1996; only the main stem
      measured in multiple-stem individuals) than BCI (4113 individuals during
      1995) and Pasoh (3738 individuals during 1995) (Table 21.3). The basal area

Table 21.3 *Size-class distribution of trees (only main stems of multiple-stem individuals) in the Mudumalai, BCI and Pasoh plots*

	Number of individuals		
Size (dbh cm)	Mudumalai (1996)	BCI (1995)	Pasoh (1995)
1–4.9	1247	177 850	244 834
5–9.9	1017	29 751	44 805
10–19.9	4166	13 635	20 092
20–29.9	3662	3687	5047
30–39.9	2348	1777	1889
40–49.9	1426	941	808
50–99.9	1457	1232	985
100+	30	163	56

Data for BCI and Pasoh based on Condit *et al*. (1999).

of all woody stems >1 cm dbh is lower at Mudumalai (25.8 $m^2\,ha^{-1}$), but the difference with BCI (31.8 $m^2\,ha^{-1}$) and Pasoh (30.3 $m^2\,ha^{-1}$) is much less than would be indicated by the differences in population size. When the basal area of only stems >10 cm dbh is considered, there is much less difference between Mudumalai (24.6 $m^2\,ha^{-1}$) and BCI (27.8 $m^2\,ha^{-1}$) and practically no difference with Pasoh (24.4 $m^2\,ha^{-1}$). At this time we cannot convert the stem-size distribution or basal area to standing biomass because the regression functions used for calculating biomass may vary from one forest type to another (for instance, the trees at Pasoh have relatively greater ratios of height to stem diameter than BCI).

(b) In terms of absolute number of recruits the Mudumalai plot had few recruits relative to BCI and Pasoh, but the rate of recruitment was higher than at either of the other sites (Table 21.4). Recruitment at Mudumalai was highly variable (range 1.2–12.6% annually). Pasoh had the lowest rate of recruitment (although the number of recruits during 1990 is an underestimate; Condit *et al*. 1999).

(c) The growth rate of small saplings (1–2 cm dbh) of both canopy species and understorey species at Mudumalai is 5–10-fold greater than those observed at BCI and Pasoh (see Condit *et al*. 1999). Whereas saplings grew at only 0.64 mm yr^{-1} at BCI and at 0.59 mm yr^{-1} at Pasoh on average during all the census periods, those at Mudumalai grew at >3 mm yr^{-1} and up to 5.3 mm yr^{-1} during the two census periods for which we calculated the rates. This difference between the dry and moist forest sites also seems to be true for larger saplings up to c. 5 cm dbh. However, stems greater than c. 5 cm dbh at BCI and Pasoh grew as fast as or faster than those at Mudumalai.

Table 21.4 *Comparison of mean recruitment rates (stems >1 cm dbh) at Mudumalai, BCI and Pasoh*

Location	Interval	Recruitment rate ($\pm$ SE)
Mudumalai	1988–1992	2.09 $\pm$ 0.51
	1992–1996	3.80 $\pm$ 0.78
	1996–2000	7.35 $\pm$ 2.06
BCI	1982–1985	4.43 $\pm$ 0.05
	1985–1990	3.34 $\pm$ 0.03
	1990–1995	2.09 $\pm$ 0.03
Pasoh	1987–1990	0.60 $\pm$ 0.02
	1990–1995	2.37 $\pm$ 0.02

Data of BCI and Pasoh based on Condit *et al.* (1999). The recruitment rate for Pasoh is an underestimate because recruits were undercounted during 1990.

Table 21.5 *Comparison of mortality rates of woody plants at Mudumalai, BCI and Pasoh*

Location	Period	Stem-size class (cm)	Mortality rate (mean)
Mudumalai	1988–2000	>1 cm	6.98
		>10 cm	2.80
		>30 cm	0.58
BCI	1982–1995	>1 cm	2.64
		>10 cm	2.26
		>30 cm	*
Pasoh	1987–1995	>1 cm	1.46
		>10 cm	1.48
		>30 cm	*

* Actual mortality rates could not be obtained, but as interpreted from Condit *et al.* (1999) the rates were considerably higher (>1%) than at Mudumalai.

(d) There are striking differences between the dry- and the moist-forest sites in their mortality rates as a function of stem size (Table 21.5). Mortality rates of canopy trees at BCI and Pasoh decline slightly from 1 cm to 5 cm dbh and are relatively constant thereafter through to larger size classes, or rise marginally (see Condit *et al.* 1999). One exception was during a drought period at BCI (1982–85), when mortality rates rose sharply in larger trees. In the understorey trees at both sites, the mortality rates were relatively constant across size classes. In contrast, at Mudumalai there was a sharp decline in mortality rates with stem size in all species. While the mortality rates of smaller-sized stems (1–20 cm dbh) at Mudumalai were higher than corresponding stems at BCI and Pasoh, the mortality rates of larger trees (>20 cm dbh) were significantly lower at Mudumalai than at the wet sites. Indeed the mortality rates of large trees (>30 cm dbh) are consistently in the range of 0.5% to 1.0% per annum, i.e. less than half the rates recorded at the moist sites.

Life-history variation in relation to environmental variability and stress

The patterns of life-history variation in woody plants across these three sites, experiencing considerable differences in the magnitude of environmental variability and stress, can be summarized as follows:

(a) The trade-off between regenerating through resprouting (vegetative) and seeding (sexual) clearly favours the former in habitats subject to environmental stress through rainfall seasonality and fire (Bellingham & Sparrow 2000). Recruitment in the dry-forest plot is almost entirely through vegetative sprouts from roots or coppices from stems burnt to ground level; thus a burnt stem survives below ground. An examination of seedlings showed that these too sprout from roots or rhizomes. While these may be root suckers (a stem originating from a root bud at some distance from the parent stem, as defined by Hoffman 1998), it is possible that some may have originally sprouted from seeds (i.e. sexually) but repeatedly burnt to ground level. Seedlings may require favourable periods (absence of fires, release from competition by grasses etc.) of several years before they are able to emerge to a size at which they can escape from ground fires. In such plants the strategy may be to store the small quantity of photosynthetic products of seedlings in underground organs rather than attempting to use them in above-ground vegetative growth immediately. The resources stored below ground could then be mobilized during favourable periods for rapid growth. In the relatively open dry forests the relative lack of light limitation would favour rapid growth, while in a rain forest such rapid growth would be possible only in canopy gaps. On the other hand, reproduction through seeding would be the favoured strategy in the more stable, moist tropical habitats. Genetic recombination through sexual reproduction would also be a necessity for plants in their co-evolutionary battle with the high diversity of pests and pathogens characteristic of the moist tropical forests.

(b) Overall recruitment rates can be expected to be higher in habitats subject to frequent disturbance, if recruitment has to compensate for high mortality of small individuals. At the same time, recruitment may also be highly variable depending on favourable conditions not just across the population of woody stems but also across species within the community.

(c) Where frequent fires kill the saplings of woody plants, we can expect that recruitment opportunities would be limited to perhaps a few 'windows' of several years without fire. Saplings should thus be in a position to grow rapidly and establish themselves in the fire-immune size class during a short favourable period. The observed growth rates in the three plots support this expectation, with the saplings at Mudumalai showing much higher growth rates than the other two plots, and saplings at BCI showing higher growth rate than at Pasoh.

(d) Perhaps the most striking aspect of life-history evolution is that of very low adult-mortality rates in the highly seasonal habitat subject to various stresses such as fire and herbivory. The mortality rates of the large stems at Mudumalai are among the lowest recorded for large trees in tropical forests globally. The low death rates of large trees permit the persistence of the population over time scales of several centuries. Thus, it is possible that the long-lived trees could have higher lifetime reproductive success irrespective of short-term recruitment rates.

(e) Although the turnover of stems in the understorey of a tropical dry forest is very high, the canopy layer may be remarkably stable. This is exemplified by the constancy or increase in basal area projected for the four common canopy trees at Mudumalai. In contrast, both canopy and understorey species in a moist tropical forest show relative stability. In this respect the Pasoh forest seems more stable than BCI (Condit *et al.* 1999) as can be perhaps expected from the greater climatic stability in the former site.

Management strategies for tropical forests have to take into account differences in life-history traits that represent evolutionary and ecological adaptations to regional environmental variability. This would include the management of fire, logging and the harvest of various non-timber forest products.

Acknowledgements

This study has been funded by the Ministry of Environment and Forests, Government of India. We thank the Tamilnadu Forest Department for research permissions. We also thank N. V. Joshi, several field assistants at Mudumalai and the Center for Tropical Science for their help at various stages of the work. Three anonymous referees made useful comments that improved the structure of this paper.

References

Bellingham, P. J. & Sparrow, A. D. 2000. Resprouting as a life history strategy in woody plant communities. *Oikos* **89**: 409–416.

Boyce, M. S. & Daley, D. J. 1980. Population tracking of fluctuating environments and natural selection for tracking ability. *American Naturalist* **115**: 480–491.

Bullock, S. H., Mooney, H. A. & Medina, E. (eds.) 1995. *Seasonally Dry Tropical Forests*. Cambridge: Cambridge University Press.

Caswell, H. 1989. *Matrix Population Models: Construction, Analysis and Interpretation*. Sunderland, MA: Sinauer Associates Inc.

Chesson, P. L. & Warner, R. R. 1981. Environmental variability promotes coexistence in lottery competitive systems. *American Naturalist* **117**: 923–943.

Chopra, K. 1993. The value of non-timber forest products: an estimation for tropical deciduous forests in India. *Economic Botany* **47**: 251–257.

Condit, R. 1995. Research in large, long-term tropical forest plots. *Trends in Ecology and Evolution* **10**: 18–22.

Condit, R., Ashton, P. S., Manokaran, N., LaFrankie, J. V., Hubbell, S. P. & Foster, R. B. 1999. Dynamics of the forest communities at Pasoh and Barro Colorado: comparing two 50-ha plots. *Philosophical Transactions of the Royal Society of London B* **354**: 1739–1748.

Connell, J. H. 1978. Diversity in tropical rainforests and coral reefs. *Science* **199**: 1302–1310.

Gadgil, M. & Bossert, W. H. 1970. Life historical consequences of natural selection. *American Naturalist* **104**: 1–24.

Goldammer, J. G. 1993. Fire management. In Pancel, L. (ed.) *Tropical Forestry Handbook*. Vol. 2. Berlin: Springer Verlag, pp. 1221–1268.

Hoffmann, A. W. 1998. Post-burn reproduction of woody plants in a neotropical savanna: the relative importance of sexual and vegetative reproduction. *Journal of Applied Ecology* **35**: 422–433.

Hubbell, S. P. & Foster, R. B. 1983. Diversity of canopy trees in a neotropical forest and implications for conservation. In Sutton, S. L., T. C. Whitmore & A. C. Chadwick (eds.) *Tropical Rain Forest: Ecology and Management*. Oxford: Blackwell Scientific Publications, pp. 25–41.

John, R., Dattaraja, H. S., Suresh, H. S. & Sukumar, R. 2002. Density-dependence in common tree species in a tropical dry forest in Mudumalai, southern India. *Journal of Vegetation Science* **13**: 45–56.

Johnson, K. H., Vogt, K. A., Clark, H. J., Schmitz, O. J. & Vogt, D. J. 1996. Biodiversity and the productivity and stability of ecosystems. *Trends in Ecology and Evolution* **11**: 372–377.

Lawton, J. H. 1994. What do species do in ecosytems? *Oikos* **71**: 367–374.

Manokaran, N., LaFrankie, J. V., Kochummen, K. M. *et al.* 1992. *Stand Table and Distribution of species in the 50-ha Research Plot at Pasoh Forest Reserve*. Research Data No. 1. Kepong: Forest Research Institute of Malaysia.

Murphy, G. I. 1968. Pattern in life history and the environment. *The American Naturalist* **102**: 391–403.

Murphy, P. G. & Lugo, A. E. 1986. Ecology of tropical dry forest. *Annual Review of Ecology and Systematics* **17**: 67–88.

Narendran, K., Murthy, I. K., Suresh, H. S., Dattaraja, H. S., Ravindranath, N. H. & Sukumar, R. 2001. Non-timber forest product extraction, utilization and valuation: a case study from the Nilgiri Biosphere Reserve, southern India. *Economic Botany* **55**: 528–538.

Philips, O. L., Hall, P., Gentry, A. H., Sawyer, S. A. & Vasquez, R. 1994. Dynamics and species richness of tropical rain forests. *Proceedings of the National Academy of Sciences* **91**: 2805–2809.

Pimm, S. L. 1984. The complexity and stability of ecosystems. *Nature* **307**: 321–326.

Puyravaud, J. P., Sridhar, D., Gaulier, A., Aravajy, S. & Ramalingam, S. 1995. Impact of fire on a dry deciduous forest in the Bandipur National Park, southern India: preliminary assessment and implications for management. *Current Science* **68**: 745–751.

Ranganathan, C. R. 1939. *Working Plan for the Nilgiri Forest Division*. Madras: Government Press.

Rundel, W. P. & Boonpragob, K. 1995. Dry forest ecosystems of Thailand. In S. P. Bullock, H. A. Mooney & E. Medina (eds.) *Seasonally Dry Tropical Forests*. Cambridge: Cambridge University Press, pp. 93–123.

Schule, W. 1990. Landscapes and climate in prehistory: interactions of wildlife, man and fire. In Goldammer, J. (ed.) *Fire and the Tropical Biota*. New York: Springer-Verlag, pp. 273–319.

Sheil, D. 1995. Evaluating turnover in tropical forests. *Science* **268**: 894.

Skarpe, C. 1991. Impact of grazing in savanna ecosystems. *Ambio* **20**: 351–356.

Stott, P. 1986. The spatial pattern of dry season fires in the savanna forests of Thailand. *Journal of Biogeography* **13**: 345–358.

Stearns, S. C. 1977. The evolution of life history traits: a critique of the theory and a review of the data. *Annual Review of Ecology and Systematics* **8**: 145–171.

   1992. *The Evolution of Life Histories*. New York: Oxford University Press.

Sukumar, R., Dattaraja, H. S., Suresh, H. S. *et al.* 1992. Long-term monitoring of vegetation in a tropical deciduous forest in Mudumalai, southern India. *Current Science* **62**: 608–616.

Sukumar, R. Suresh, H. S., Dattaraja, H. S. & Joshi, N. V. 1998. Dynamics of a tropical deciduous forest: population changes (1988 through 1993) in a 50-ha plot at Mudumalai, southern India. In F. Dallmeier & J. A. Comiskey (eds.) *Forest Biodiversity Research, Monitoring and Modeling: Conceptual Background and Old World Case Studies*. Man and the Biosphere Series. Paris: UNESCO; New York: The Parthenon Publishing Group, pp. 495–506.

Sukumar, R., Suresh, H. S., Dattaraja, H. S., John, R. & Joshi, N. V. (2004). Mudumalai Forest Dynamics Plot, Mudumalai Wildlife Sanctuary, India. In Losos, E. C. & Leigh, E. G. Jr (eds.) *Tropical Forest Diversity and Dynamism: Findings from a Large-scale Plot Network*. Chicago: University of Chicago Press.

Swaine, M. D. 1992. Characteristics of dry forest in West Africa and the influence of fire. *Journal of Vegetation Science* **3**: 365–374.

Varman, K. S. & Sukumar, R. 1995. The line transect method for estimating densities of large mammals in a tropical deciduous forest: an evaluation of models and field experiments. *Journal of Biosciences* **20**: 273–287.

Warner, R. R. & Chesson, P. L. 1985. Coexistence mediated by recruitment fluctuations: a field guide to the storage effect. *American Naturalist* **125**: 769–787.

CHAPTER TWENTY TWO

Changes in plant communities associated with timber management in natural forests in the moist tropics

MICHELLE A. PINARD

University of Aberdeen

Introduction

Disturbance is integral to a forest's ecology, and selective logging is a form of disturbance. Whether disturbance is natural or anthropogenic, it is likely that the biotic interactions within the system are influenced by it. This volume attests to the importance of biotic interactions to tropical forest function and diversity. And while there are still many unknowns in how interactions change with perturbation, particularly when the impacts are as variable as those associated with timber management, an understanding of these changes is important to inform the development of management interventions that are ecologically sustainable, i.e. where the manipulations stay within the limits of natural disturbance patterns for the forests (e.g. ecological forestry, *sensu* Seymour & Hunter 1999) and where ecosystem functions are maintained (Forest Stewardship Council (FSC) 2000).

Generally, approaches to the study of impacts of logging and other silvicultural interventions are based on comparisons, for example control versus impact designs where unlogged, old-growth forests are used as the control (Johns 1983; Lambert 1992; Silva *et al.* 1995), or various treatments are compared, for example different logging systems (Johns *et al.* 1996; Pinard & Putz 1996), different harvest intensities (Kasenene 1987; Panfil & Gullison 1998; van der Hout 1999), or different stand-improvement treatments (Fox & Chai 1982; de Graaf *et al.* 1999; Gerwing 2001). But as with other forms of anthropogenic impacts, detecting change is difficult because of the complexity and variability (temporal and spatial) of natural systems (Underwood 1991, 1993). Bayesian procedures have been applied to some datasets to increase the power of detecting change (e.g. Crome *et al.* 1996). Also, models have been developed (e.g. Liu & Ashton 1999; Lobo Lopez 2003) and mechanistic research conducted (e.g. Kennard 1998; Howlett & Davidson 2003) to examine a range of ecological questions. In this chapter I draw from this literature to describe the patterns in tropical moist-forest responses to selective logging, vine cutting and, to a lesser extent, stand treatment. I discuss

Biotic Interactions in the Tropics: Their Role in the Maintenance of Species Diversity, ed. D. F. R. P. Burslem, M. A. Pinard and S. E. Hartley. Published by Cambridge University Press. © Cambridge University Press 2005.

the factors associated with disrupting succession and shifting the forest to a less diverse, low-biomass state. Finally, I discuss the implications for silviculture in these forests.

I focus on the plant community rather than the range of biotic interactions that are relevant to the topic for several reasons. Firstly, silvicultural manipulations are directed primarily at the plant community, therefore the plant responses to interventions are more frequently studied than are those of animals or microbes. The literature relevant to silvicultural impacts on forest plant communities has been summarized (Putz *et al.* 2000; Putz *et al.* 2001), but not reviewed in detail. Secondly, several recent and thorough reviews have been published that cover logging impacts on animals (Greiser Johns 1997; Fimbel *et al.* 2001) and on biodiversity conservation (Putz *et al.* 2000). Thirdly, the most significant impacts of logging on animals often tend to be indirect, associated with hunting or fragmentation (see Laurance & Bierregaard 1997; Bennett & Robinson 2000; Robinson & Bennett 2000; Laurance, this volume, for recent reviews).

Any attempt to draw general patterns across the moist tropics from forests that are selectively logged is difficult because the forests and logging systems are extremely variable. However, the variety and variability also provide richness of experience and insights into the diversity of responses to disturbance. This review is put forward, not to represent a comprehensive review of the literature, but rather as an effort to identify patterns and to generate discussion that could inform the development of silviculture for natural forests in the tropics.

Timber management in natural forests in the tropics

Natural-forest management in tropical forests usually focuses on harvesting as the only silvicultural intervention. Although several silvicultural systems have been developed for tropical forests (see Dawkins & Philip 1998 for review), more-complete systems that focus on regeneration and tending, as well as harvesting, are atypical, even in forests where current practice is considered consistent with sustainable practices, as evidenced by awards of FSC certification (e.g. Fredericksen *et al.* 2003; Freitas 2004).

Logging in natural moist forests in the tropics is typically selective, where a relatively small number of trees, above a threshold size or minimum felling diameter, are felled and extracted. The logs are usually skidded out of the forest using bulldozers, rubber-tyred skidders, farm tractors or draft animals.

Logging impacts are variable depending on harvest intensity, levels of pre-harvest planning and control, soil type and terrain, forest stature and structure, among other things. Across the moist tropics, one finds examples of very low-intensity logging, for example, where a single species is being harvested (e.g. *Swietenia macrophylla* (big leaf mahogany) in Bolivia or *Chlorocardium rodiei* (greenheart) in Guyana), and examples of very high-intensity logging (e.g. dipterocarp

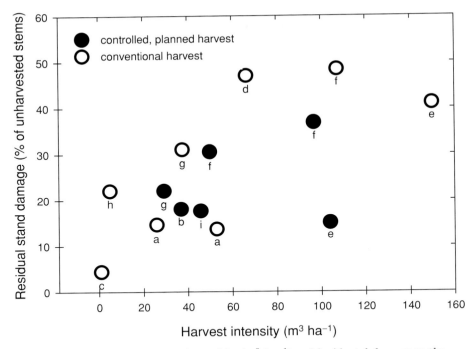

Figure 22.1 Variability in harvest intensities (m³ ha⁻¹) and incidental damage to the residual stand (percentage of unharvested trees) across the tropics. Open symbols are from areas where logging was planned and controlled; closed symbols are from areas where logging was done in a conventional way (typical for the region, generally with little planning and control over operations). Modified from Putz *et al.* 2000 and Webb 1997. (a = Adam 2003; b = Crome *et al.* 1992; c = Gullison & Hardner 1993; d = Kammesheidt 1998; e = Pinard & Putz 1996; f = Sist *et al.* 1998; g = van der Hout 1999; h = Verissimo *et al.* 1995; i = Webb 1997).

forest in Borneo). As harvest intensity increases, the collateral damage to the residual stand also increases. However, there is tremendous variability around this relationship, some of which relates to the level of planning, control and monitoring of operations (Fig. 22.1).

Because of the selective nature of harvesting, forests are repeatedly logged after intervals of a few years to several decades. Where logging entries are scheduled so as to achieve sustained yield of timber, this approach to forest management is referred to as a polycyclic silvicultural system. Where properly designed, several harvesting events (or cuts) are scheduled within a rotation (nominally the time it takes for a tree to grow to maturity). In theory, cutting cycle length is related to forest productivity (i.e. stand growth and nutrient budgets), as is the harvest intensity, although harvesting can be done more strategically, as a treatment to promote the regeneration of certain groups or guilds of trees.

In some (few) forests, management is more intensive, and other silvicultural interventions are used to promote regeneration and growth of the commercial stock (see Lamprecht 1993 or Wadsworth 1997). Woody vines are sometimes cut either post-logging or at intervals through a cutting cycle, to liberate crop trees and to improve stem form. Vine-cutting is also used prior to logging to reduce incidental damage to the residual forest (vine connections between tree crowns can mean that felling one tree will result in the falling of several of its neighbours). Liberation thinning, or the selective removal (through girdling or felling) of trees overtopping crop trees, is used to promote growth. More-intensive refinement treatments are aimed at eliminating undesirable trees, vines and shrubs to promote site dominance by commercial species. Where the target commercial species fail to regenerate, soil treatments such as scarification or controlled burning might be used to prepare a seed bed. As more forests gain certificates for sustainable management, more silvicultural interventions to promote regeneration and for tending crops are likely to be used. Heavy refinement treatments may not be readily adopted because of the likely negative impacts on biodiversity (Putz *et al.* 2000), but less-intensive treatments, such as targeted vine-cutting and liberation may be more compatible with conservation objectives.

Weed responses

Silvicultural weeds, often large herbs, herbaceous climbers and woody vines, are cited by many authors as a problem for natural regeneration of trees following selective logging. Across the tropics, the identity and ubiquitity of the aggressive weedy species varies. For example, in some areas, exotic species are the main problem (e.g. *Chromolaena odorata* in Kerala, India (Chandrashekara & Ramakrishnan 1994); *Lantana camara* in Queensland, Australia (Horne & Hickey 1991)), in others they are native, large-stature herbs (e.g. Maratanceae, Zingiberaceae in Ghana (Hawthorne 1993)), grasses, climbers (e.g. *Merremia* (Neil 1984), *Dinochloa* (climbing bamboo; Liew 1973) or palms (Dekker & de Graaf 2003). Although sprouting stems of woody vines (Putz 1991), bamboos (Liew 1973) and root-sprouting species (Dickinson *et al.* 2000) can become a problem for tree development, weeds are often new recruits establishing in areas with soil and canopy disturbance (Bacon 1982; Neil 1984; Fox 1976; Hawthorne 1994; Kennard 1998; Schnitzer *et al.* 2000).

Soil disturbance associated with ground-based log extraction differs from the small, localized mineral-soil exposure associated with uprooting of trees, animal excavations or natural meandering of streams, typical of protected, old-growth forests (Douglas *et al.* 1992). Where farm tractors, skidders or bulldozers are used to drag logs out of the forest to a loading area, soil disturbance is inevitable. Where harvesting is planned and controlled, the area affected by extraction activities is generally relatively small (<10% of the area; Johns *et al.* 1996;

Pinard & Putz 1996; Adam 2003). The types of soil disturbance associated with ground-based extraction include exposed mineral soil, compacted soil, exposed subsoil where topsoils have been displaced, ditches and mounds.

In areas with disturbed soil, the soil seed bank provides much of the seed for early recovery (Howlett & Davidson 2003). In many sites, exposed mineral soil allows the establishment of short-lived tree species (Williams-Linera 1990) and many herbs and vines considered to be weeds (Liew 1973). Plant recruitment is reduced, at least temporarily, where topsoils are displaced (Howlett & Davidson 2003). Where soils are compacted, as well as scraped of topsoil, seed wash (Pinard et al. 1996), nutrient limitation (Nussbaum et al. 1995), and root restriction may limit woody-plant establishment (Woodward 1995). In some areas, however, the commercial tree species benefit from soil disturbance, either through reduced competition with herbs (Fredericksen & Mostacedo 2000) or greater establishment success for tree species requiring exposed mineral soil (Dickinson et al. 2000).

Disturbance to the canopy is also thought to be an important factor in the development of weed problems in managed forest. Both natural gaps and felling gaps vary tremendously in size, with gap size dependent on the size of the gap maker, the degree to which the canopy is interconnected with woody vines, forest stature and structure, and soil stability, among other things. Few data are available that allow a comparison of natural and felling gap sizes, but those available suggest that felling gaps can be larger, especially if felling and skidding is uncontrolled (Johns et al. 1996; Webb 1997; Dickinson et al. 2000). The difference in size is probably related to the difference in size and composition of the gap makers: natural mortality is often random whereas selective logging focuses on the larger trees within the forest (Dickinson et al. 2000).

Larger gaps may be more prone to weed invasions than smaller gaps (Sist & Nguyen-Thé 1992; Strushaker et al. 1996), but in many instances, interactions between gap size and other factors such as soil disturbance, fire and herbivory may be more important than gap size per se. In dipterocarp forest, soil disturbance was a better predictor of germination from the soil seed bank than was gap size (Kennedy & Swaine 1992). Larger gaps become vulnerable to ignition more rapidly after rainfall events than do smaller gaps (Holdsworth & Uhl 1997), and the association between forest fire and weeds has been documented across the tropics (Woods 1989; Swaine 1992; Hawthorne 1994; Pinard et al. 1999a). In Kibale Forest in Uganda, gap size interacts with damage to tree regeneration by elephants (Kasenene 2001). Here, larger gaps are more likely to develop into herbaceous vine tangles, and are visited more frequently by elephants than are smaller gaps (Strushaker et al. 1996). Herbivores also can limit tree establishment on skid trails (Howlett & Davidson 2003), although in natural gaps in Panama, herbivory did not explain differences for pioneer-tree niche-partitioning whereas seed attributes did (Pearson et al. 2003).

In some managed forests, weedy, early successional vegetation appears to be short-lived. In subtropical rain forest in Australia, 25 years after selective logging, weeds were not a problem ((King and Chapman 1983, cited by Horne and Hickey 1991). In Kerala, India, the herb component of gap vegetation declined over the first decade after logging (Chandrashekara & Ramakrishnan 1994). In *terra firma* forest near Manaus, understorey plant abundance declined with time since logging, and after 11 yr was similar to that in unlogged forest (Costa & Magnusson 2003).

Vine responses

Vine densities vary with forest type and conditions. Within tropical moist forests, evidence from the neotropics suggests that they are more abundant in gap and building phases than in mature phases (Vidal *et al.* 1997), and more abundant and diverse in younger (20 and 40 yr) relative to older (70 yr, 100 yr, old growth) forests (Dewalt *et al.* 2000). An increase in vine densities following selective logging has been observed by many, and attributed to sprouting of stems that fall out of the tree crowns (Appanah & Putz 1984; Schnitzer *et al.* 2003) and new recruitment from seed in gaps created by felling and log extraction (Kennard 1998; Schnitzer *et al.* 2000).

Vine densities are also influenced by silvicultural interventions other than logging. Results from vine-cutting experiments indicate that resprouting from cut stems can increase vine densities (Appanah & Putz 1984; Parren & Bongers 2001; Gerwing & Vidal 2002), but over time, net densities are lower (5–10 yr: Aiken 2002; Gerwing & Uhl 2002; Schnitzer *et al.* 2003) or similar to pre-intervention densities (Gardette 1996). Because species of vines differ in their propensity to resprout, logging and vine cutting may alter the composition of the woody vine flora (Vidal *et al.* 1997; Parren & Bongers 2001; Gerwing & Uhl 2002; Gerwing & Vidal 2002). Long-term studies of vine-cutting impacts are few, but in Pasoh Forest Reserve, 40 yr after logging and vine-cutting, the woody vine flora differed from that in non-intervention forest with fewer individuals of animal-dispersed species and more individuals of species that sprout at the base of the stem (Gardette 1996).

Implications of an increased vine density are several and have been reviewed by several authors (Putz 1991; Schnitzer & Bongers 2002; Parren 2003). Vines compete with trees, reducing their growth (Gerwing 2001; Pérez-Salicrup 2001), their stem quality (Putz *et al.* 1984; Putz 1991) and their survival (Schnitzer *et al.* 2000), and they can slow or stall succession in gaps (Putz 1984; Schnitzer *et al.* 2000; Plumptre 2001). Relatively high numbers of vines in gaps may influence the composition of trees regenerating in gaps, favouring pioneer species over more slow-growing guilds (Schnitzer *et al.* 2000).

Where cutting cycles are long enough to allow the balance between trees and vines to return to pre-intervention levels, there may be few long-term impacts of rises and falls in vine densities. However, if cutting cycles are short, or if gaps are created in area with elevated vine densities, succession to closed forest may be stalled for long periods.

Vine growth and survival is constrained by the availability of supports (Putz 1984) and silvicultural interventions like logging and thinning are likely to influence the availability, size and distribution of supports. Breaks in the forest canopy resulting from felling, skidding or road-building will restrict, at least temporarily, the provision of climbing supports or trellises. Where trees (>20 cm diameter at breast height, DBH) are isolated from their neighbours, they are difficult to ascend by twining (Putz 1985), unless root climbers or the roots of hemiepiphytes facilitate their ascent by providing a trellis (Pinard & Putz 1994; Putz 1994). Where climbing bamboos may do well in logged forests it is because they are able to climb where others cannot, their stems are rigid, allowing the free-standing culms greater reach (Yap *et al.* 1995).

Tree-community responses

Conservationists, ecologists and silviculturalists have variously expressed concern that selective logging and other silvicultural interventions will result in a shift in forest composition, either towards early successional and light-demanding tree species, or towards slow-growing, non-commercial tree species. I summarize the reasoning behind these concerns here. Generally, the expectation for a shift towards light-demanding species is based on one or more of the following observations or assumptions: (1) logging creates gaps, and gaps favour the regeneration of gap-dependent species; (2) gaps created by logging are more frequent, more numerous and larger than those created by treefalls, therefore gap-dependent species will be favoured in logged forest; (3) soil disturbance associated with log extraction should aid the establishment of early successional, light-demanding species; and (4) cutting cycles are generally too short to allow succession to proceed to a point where mature forest species replace the early successional species that colonize sites disturbed by the harvest. The first two of these reasons are based on a view that tropical tree species are specialized in relation to their requirements for light; the evidence for niche-partitioning along light gradients is reviewed by Dalling and Burslem (Chapter 3, this volume) and Brokaw and Busing (2000).

Reasons for expecting a shift towards slow-growing, non-commercial, shade-tolerant species are these. (1) The selective removal of commercial species will result in an over-representation of non-commercial stems (i.e. slow-growing, shade-tolerant) in the residual forest, and this imbalance will be carried over to seed production, deposition and seedling recruitment. Thus, over time, the composition will shift to favour these species. (2) The largest stems within a population are not necessarily the oldest chronologically; they may represent phenotypes that are superior for commercial production (fast-growing, straight bole), and the selective removal of these phenotypically strong individuals may promote dysgenic selection, leaving behind suppressed, poor-quality individuals that may not be able to respond to canopy-opening in order to develop into canopy trees.

Shifts in composition associated with logging

Evidence for shifts in composition of the tree component of the forest is mixed (Table 22.1). Given the variability across sites in terms of forest types and histories, and variability in the intensity and intention of the interventions, this is not surprising. Also, sampling designs and scales are important in comparisons of species composition, as species richness relates to the number of stems sampled as well as the area sampled (Denslow 1995; Cannon *et al.* 1998); the heterogeneity in space and time in logged forests also poses a difficult sampling problem. In the next section I review the evidence for shifts in composition in the tree community, and discuss the support for the assumptions put forward in the previous section.

It is indisputable that selective logging creates canopy gaps, but a shift in composition towards gap-dependent species is not a necessary outcome (e.g. Crome *et al.* 1992). Many authors have documented that small gaps created by selective logging are insufficient to allow the regeneration of commercially favoured light-demanding species (e.g. Snook 1996; Dickinson *et al.* 2000; Fredericksen & Mostacedo 2000; Hall *et al.* 2003). Succession in relatively small felling gaps is likely to follow that in natural gaps, where pre-existing plants have the advantage over new recruitment (Brown & Jennings 1998). The disturbance stimulates regeneration and recruitment but pioneer and light-demanding species are not necessarily favoured (Hubbell *et al.* 1999).

To the extent that felling and extraction activities create new habitats, by exposing mineral soil and creating relatively large gaps, pioneer species recruitment may increase species richness (8 yr post-logging, Cannon *et al.* 1998; 3 yr post-logging, Hawthorne 1993) in logged relative to unlogged forest. If the regenerative capacity of the non-pioneer species is maintained, either through deliberate efforts of seed-tree retention or by circumstance, the change in species richness should be short-lived. Indeed, several authors report that over time, these species decline in abundance (Chandrashekara & Ramakrishnan 1994; Salick *et al.* 1995). When harvest intensity is high, however, recruitment of canopy species can be limited by low densities of potential seed trees (Meijer 1970; Plumptre 1995). Reductions in the density of large trees within populations may have negative implications for pollen transfer, seed set or seed quality (Ghazoul, Chapter 10, this volume). Recruitment limitation due to lack of seed trees is likely to be more of a problem for species with irregular size distributions than for species with size distributions represented by reverse-J curves (Snook 1996;

Table 22.1 *Brief descriptions of studies published that examine changes in plant communities related to selective logging or other silvicultural interventions in tropical forests. Only studies with observations beyond 5 yr post-logging were included.*

Location	Site and context	Findings	Citation(s)
Latin America Brazil	*Terra firme* rain forest in the Tapajós Forest; logged in 1979, removing 75 m³ ha⁻¹, or 16 trees ha⁻¹, using a minimum felling diameter (MFD) of 45 cm dbh; skid trails opened by bulldozer, logs extracted by wheeled skidder; permanent plots established in 1981, re-measured in 1987 and 1992.	In 1992, 13 yr post-logging, the same 22 species comprised 60% of basal area as those in 1981, but some species changed ranks, with light-demanding species increasing their relative abundance (in 1981, light-demanding species comprised 27% of the basal area, whereas in 1992, they comprised 42%). *Rinorea*, a small-stature shade-tolerant species shifted from 1st to 5th rank while the rank importance of the light-demanding and pioneer species *Bixa*, *Cecropia* and *Inga* increased.	Silva *et al.* 1995
Brazil	*Terra firme* rain forest in the Tapajos Forest; plots were established in 1981, area logged in 1982, and re-measurements made in 1983, 1987, 1989; two logging treatments imposed: MFD of 45 cm, MFD of 55 cm.	At 7 yr post-logging, tree species diversity (stems >5 cm dbh) was not affected by logging and was similarly low across treatments and over time; the areas species richness dropped with logging but increased over time; the areas with heavier exploitation (MFD of 45 cm) had higher richness than the lighter logging (MFD of 55 cm); species richness in the unlogged forest declined over time.	de Carvalho *et al.* 1999
Mexico	Semi-deciduous forest in the Maya Forest region of Quintana Roo; compared natural gaps felling gaps and closed canopy forest in six logging areas between 4 and 10 yr post-logging; logging was highly selective, primarily for *Swietenia macrophylla*; rubber-tyred skidders used to extract logs.	Advance regeneration of shade-tolerant species dominated natural gaps whereas felling gaps had more tree establishment from seed. Soil disturbance from the skidder in felling gaps promoted regeneration of shade-intolerant species, and also root-sprouting of both shade-intolerant and shade-tolerant species, although less of the latter. Results suggest that the maintenance of timber species requires both large gaps with understorey and soil disturbance, as well as relatively small gaps with only minimal understorey damage.	Dickinson *et al.* 2000
Venezuela	Carapo Forest Reserve, moist deciduous lowland forest located on alluvial plain; compared unlogged stands with stands logged 5, 8 and 19 yr prior to the study; logging was selective (in early years only five species), intensity estimated to be 89.9, 54 and 55.5 m³ ha⁻¹ in the 5-, 8- and 19- yr-old stands, respectively.	Commercial species changed ranks (basal area), with harvested species declining but no clear pattern emerging relative to shade tolerance among other species; logged forest had greater densities of regenerating stems (<10 cm dbh), although *Bombacopsis* and *Swietenia* were absent as saplings.	Kammesheidt 1998

Location	Description	Results	Reference
Suriname	Tonka research site in Kabo creek region of Suriname, tropical rain forest; experiment started in 1978, with three harvest intensities (15, 23 and 46 m^3 ha^{-1}), and three silvicultural treatments (control, light refinement (basal area (BA) reduced to 18 m^2), heavy refinement (BA reduced to 14 m^2) and unlogged control. Sampled vegetation in 3–10 m height level, focussed on eight species, commercial and non-commercial, light-demanding and shade-tolerant species.	Across all treatments, shade-tolerant species were more abundant than light-demanding species, with relative densities similar in treated and unlogged forest; natural heterogeneity was greater than treatment effects. Liana and palm density was not consistently related to the intensity of the intervention; original site conditions seemed more important than intervention type.	Dekker & de Graaf 2003
Nicaragua	Atlantic lowland tropical rain forest; primary forest logged in 1984, liberation thinning in 1992, sample plots established in 1990 and re-measured in 1993.	One year post-logging, plant density and species richness increased as early successional vegetation established; the greater the logging damage, the greater the densities of vines and secondary species. The increase in species richness and density slowly disappeared over time (9 yr); liberation did not affect basic physiognomy of the forest.	Salick et al. 1995
Asia Kalimantan, Indonesia	Lowland dipterocarp forest, intermixed with swamp forest; compared plots in forest selectively logged (at relatively high intensity, total basal area removal was 43%, 40% area disturbed to some degree) and in unlogged patches of forest within the matrix, avoiding the swamp forest which was not logged.	Eight years post-logging, tree species richness was higher but stem densities lower than in unlogged forest, perhaps because of increase in habitat heterogeneity or because logging reduced the dominance of some of the commercial species.	Cannon et al. 1998
Peninsular Malaysia	Pasoh Forest Reserve, lowland dipterocarp forest, compared primary forest with forest logged in 1955 under the Malayan Uniform System, with girdling and climber cutting treatments; measurements made in 1984 in primary and 1989 in treated forest (40–45 yr post-treatment).	Treated forest had 44% higher density of trees (>10 cm DBH), 157% more commercial stems (as the treatment intended), greater tree species richness and diversity than untreated forest; canopy species were proportionally more abundant; total basal area was similar and ranking of top five families was similar.	Manokaran 1996
India Kerala	Humid tropical forest in Western Ghats of Kerala State; comparison of undisturbed, natural forest with natural canopy gaps and logging gaps 1, 5 and 10 yr post-logging; no information given about logging.	Gap type and age influenced plant community; biomass in ground vegetation increased one year after gap formation and then declined with time; no differences in biomass were found between natural and felling gaps at any time period; plant diversity was higher in felling gaps than natural gaps at all time periods, although tree-sapling diversity tended to follow the opposite pattern; Chromolaena odorata, an exotic shrub, was found only in felling gaps, and its presence declined over time; Nilgirianthus ciliatus, a shrub with monocarpic flowering, influenced the composition of regeneration in any gap in which it established, by casting dense shade until death.	Chandrashekara & Ramakrishnan 1994

(cont.)

Table 22.1 (*cont.*)

Location	Site and context	Findings	Citation(s)
Australia Queensland	Rainforest (Notophyll vine forest) on Windsor Tableland; comparison of unlogged with area selectively logged in 1985 at average intensity of 37 m³ ha⁻¹ (6.6 trees ha⁻¹); logs extracted using bulldozers and skidders; measurements made pre- and post-logging.	No change in proportional representation of major overstorey tree species; reduction in numbers of large-diameter trees.	Crome *et al.* 1992
East Africa Uganda	Budongo Forest Reserve, moist, semi-deciduous forest; analysis of data from 60-y-period, when forest was selectively harvested and treated with arboricides to non-commercial species (particularly the shade-tolerant *Cynometra* that develops into monodominant, 'climax' forest) to open the forest and promote the growth of *Khaya* and *Entandrophragma* and to favour mixed forest over *Cynometra* forest.	Species distribution was more strongly influenced by geographical position than by history of silvicultural interventions, with greater species richness in the west and in logged compartments; arboricide treatments were effective in increasing area of mixed forest, and decreasing that of *Cynometra* forest; pioneer species occurred at relatively high abundance where logging and poisoning treatments were heavy, despite some of these species being killed by arboricides.	Plumptre 1996
Uganda	Kibale Forest Reserve, tropical moist forest; a comparison of plant recruitment into gaps in four sites where selective logging and stand treatments of different intensities occurred 30 yr prior to study; measured canopy gap size, soil seed bank composition, seed rain, herbivory and plant regeneration.	Soil seed banks in both gap and forest soils were dominated by non-commercial species, and herbaceous species outnumbered tree species in all sites; new recruitment into gaps was limited by existing vegetation, and tree-seedling mortality was higher in large gaps in heavily logged forest than in gaps in lightly logged and unlogged forest; saplings in large gaps experienced a higher incidence of browsing damage than did those in smaller gaps.	Kasenene 2001
West Africa Ghana	Bia South Game Production Reserve, transitional between moist evergreen and moist semi-deciduous forest; selectively logged 3 to 6 yr prior to study, intensity not given.	Logging disturbance created opportunities for the recruitment of light-demanding species that were absent from unlogged patches; regeneration on skid trails tended to have more non-pioneer light-demanding species and larger gaps more pioneers, proposed explanation being distance to mother trees was less on skid trails; the weed *Eupatorium* was dominant in the heavily disturbed areas, roads and loading bays; no shift in species composition recorded.	Hawthorne 1993
Ghana	Six Forest Management Units from moist semi-deciduous forest and moist evergreen forest; harvested selectively with one or two cutting cycles, with intensity per cycle varying from 0 to 209 m³ km⁻².	Despite the variability in harvest intensities, the proportion of stems (5–29 cm DBH) of commercial species by guild (pioneers, non-pioneer light-demanding species, non-pioneer shade bearers) was similar across management units.	Adam 2003

Jennings *et al.* 2001) or where residual stand damage has been high (Fox 1976; Woods 1989).

Harvesting can also influence species composition by reducing the relative dominance of commercial species (Cannon *et al.* 1998). Kammesheidt (1998) reported a decline in the rank (basal area) of commercial species following selective logging that did not appear related to shade tolerance. In the extreme, high grading can eliminate species from large tracts of forest, as so often has been documented for high-value species such as *Swietenia*, *Khaya* and *Entandrophragma*. Where logging intensities and disturbance is moderate, it may be that guild ratios are stable despite the loss of individual species (Adam 2003).

Shifts in composition in response to refinement treatments

When interventions (logging and refinement) were implemented with an intention to shift composition in favour of the commercial species, some successes have been recorded. In Pasoh Forest in Malaysia, in forest where the Malayan Uniform System was used for logging and refinement, at 40 yr post-treatment, treated forest had greater species richness and diversity than unlogged forest, with canopy species becoming proportionally more abundant, as was intended by the intervention (Manokaran 1996). In Budongo Forest in Uganda in an area where selective logging and refinement treatments were used, the composition of the forest changed over time as intended by the refinement, with a reduction in the amount of monodominant, *Cynometra* forest, for an increase in more mixed species forest (Plumptre 1996). The natural heterogeneity in forest composition, however, was much greater than any that could be attributed to silvicultural interventions, reinforcing the idea that it is difficult to detect change in natural systems.

In another area with refinement treatments, the CELOS experiments at the Tonka research site in Suriname, no shift towards earlier successional species was recorded at 20 yr post-treatment. In a comparison of saplings of eight tree species and other vegetation in the 3–10-m height stratum of the forest, Dekker and de Graaf (2003) found that the relative densities of shade-tolerant and light-demanding species were similar in unlogged and treated forest; in this experiment, treatments included three harvest intensities and three intensities of refinement. Similar to Budongo, the heterogeneity across sites was much greater than that attributable to treatments. Differences in pre-treatment vegetation were important to understanding post-treatment liana and palm densities. Increasing intensity of intervention resulted in an increase in liana and palm density in two of the three plots, but where there were initially high densities, no change in density was recorded (Dekker & de Graaf 2003).

In some forests it may be difficult to shift the composition of the tree community away from the dominant species. For example, in semi-deciduous forest in Lomerio, Bolivia conditions in felling gaps were manipulated to encourage the recruitment of light-demanding species: gaps were enlarged, soils scarified, competing vegetation slashed. After three years the composition of the tree regeneration in the gaps was almost identical in rank abundance to similar-sized regeneration in the forest understorey (Pinard *et al.* 1998; D. Kennard, unpublished data). The manipulations were ineffective in shifting the relative abundance of species represented in recruitment. It may be, as suggested by Pitman and others (2001), that species that are common at a site have certain attributes that allow them to become dominants, and that some ecological processes allow them to maintain a competitive edge. In the experimental area in Bolivia, irregular seed production limits recruitment for many species (Kennard *et al.* 2001).

The tropical forestry literature includes many references to the difficulties in inducing regeneration (Dawkins 1958; Wyatt-Smith 1987; Catinot 1997; Dawkins & Philip 1998) and observations of lack of tree recruitment following logging (Pinard *et al.* 1998; Chapman *et al.* 2002) or tending treatments (Dolanc *et al.* 2003; Pariona *et al.* 2003). However, for tropical tree species that characteristically have unpredictable recruitment and clumped distributions, failure to foster regeneration within a narrow time frame defined by management objectives is not surprising.

Variability in response among forests

Variability among forests in terms of their responses to disturbance is related to many factors. The forest's previous disturbance history influences its composition and thus its response to subsequent disturbance (Sheil & Burslem 2003). Different floras will also respond differently. For example, pioneer tree and weedy herbaceous species differ in terms of dispersal capacity, persistence in the soil seed bank and seed-germination requirements, and these variables influence recruitment success in logged forest. The proportion of tree species that are able to sprout and survive stem breakage differs among forests, and this characteristic has been shown to be important to their persistence post-logging (Dickinson *et al.* 2000).

Variability within forests is also expected because of the influence of time of gap creation in relation to fruiting phenologies, environmental conditions at the time of disturbance, and neighbourhood effects in terms of composition (Liu & Ashton 1999; Dekker & de Graaf 2003), forest structure and microclimate (Pinard *et al.* 2000), and proximity to areas of secondary succession (Janzen 1983). Site conditions, for example soil nutrient levels, may influence rates of succession (Sheil & Burslem 2003) and competitive interactions among species (Dalling & Burslem, Chapter 3, this volume).

Differences in the animal communities among forests, not only the composition and abundance of the herbivores, pollinators, seed dispersers and seed predators, but also the coincidence of temporal cycles among species (Fragoso,

Chapter 12, this volume) and between animal species and logging disturbance, are likely to affect responses.

Factors associated with disrupted succession

Thresholds in environmental conditions and feedbacks between biotic and abiotic factors are common in degraded systems (Suding *et al.* 2004). In the literature on logging impacts in tropical forests, a variety of factors are implicated for disrupting succession, or causing a shift from forest cover to a less diverse, low-biomass system. Soil damage through compaction or loss of topsoil can restrict forest recovery in affected areas (Pinard *et al.* 1996; Pinard *et al.* 2000; Nussbaum *et al.* 1995), although on certain soils and with certain flora, soil disturbance is a prerequisite for regenerating commercial tree species (Dickinson *et al.* 2000). Canopy gap size and area are associated with fuel dry-down rates, and vulnerability to ignition (Holdsworth & Uhl 1997). The scale and patterns of disturbance are important, not only because of cumulative and neighbourhood effects in terms of microclimate, but also because of the implications for seed dispersal into disturbed areas (Hawthorne 1993) and between patches (Sheil & Burslem 2003).

Fire can have devastating effects in logged forest. Even relatively low-intensity surfaces fires can damage or kill a large proportion of advance regeneration (Woods 1989; Pinard *et al.* 1999a) and encourage the establishment of vegetation that perpetuates fire.

Altered patterns of herbivory and seed predation can act as a threshold in influencing forest composition. Interactions between gap size, herbaceous vegetation, elephant activity, seed predation by rodents and poor tree recruitment described in Kibale Forest (Struhsaker *et al.* 1996; Kasenene 2001) suggest a disturbance threshold that when exceeded causes the system to shift to an alternative state, one where herb tangles persist.

Forest fragmentation may also reduce ecological resilience in managed forest and may bring unanticipated outcomes. For example, fragmentation can influence reproductive synchrony among dipterocarp species in West Kalimantan by altering tree densities and climatic conditions; seed-predator distributions and densities are also affected with implications for seed mortality and tree recruitment (Curran *et al.* 1999).

As patches of non-forest land or plantation may have important ecological consequences for processes in the managed forest matrix, so do patches of unlogged or protected forest. These areas may contribute to resilience by providing propagules, serving as refuge and a source of genetic diversity.

Implications for silviculture and research

From this literature review, several themes with relevance to silviculture in these forests emerge. Firstly, the intensity, extent and type of disturbance generated by interventions needs to match the requirements for the species in the forest. In

Ugandan forests, Kasenene (2001) argues that it is important to minimize stand damage during harvesting and to keep felling gaps small if forests are to regenerate post-logging without additional rehabilitation treatments (e.g. enrichment planting and protection). In other forests, silvicultural intensification is needed to maintain the complement of light-demanding species that occur (Fredericksen & Putz 2003). Community responses to interventions must be considered with the view that natural-forest communities are dynamic and a product of their history.

Retention of adequate numbers and quality of seed trees is important yet is unlikely to be sufficient to ensure recruitment post-disturbance. Current research directed at identifying the factors that limit tree recruitment, and the relative importance of dispersal limitation or fruiting events, is relevant to management. Timber management has tended to be more successful in the tropics in the low-diversity forests (Dawkins & Philip 1998). Whether this is related to the high levels of seed production or abundant advance regeneration (Saenz & Guariguata 2001; Guariguata & Saenz 2002), research in these forests may provide insights for management of more-diverse forests. Recruitment from the seed bank and sprouting are important but tree recruitment post-logging is often limited; therefore protection of advance regeneration is critical for the recovery of closed forest conditions.

Forests and tree species differ in their vulnerability to decline with selective logging. Efforts to characterize species have tended to focus on regeneration requirements for light, relative abundance or rarity, growth rates and dispersal modes (Martini *et al.* 1994; Pinard *et al.* 1999b; Freitas 2004). Persistence of seedlings in shaded understorey, and capacity to sprout following stem breakage, appear important, as does the vulnerability of a species to factors beyond the stand or management unit (e.g. fluctuations in seed-predator densities, fragmentation and isolation, vulnerability to fire). Research exploring the forms and levels of ecological redundancy among tree species may provide insights into forest resilience to disturbance.

Variability in time and space, in both abiotic and biotic factors that influence forest dynamics and successional processes, makes it a challenge to predict forest responses to disturbances. As lists of species with commercial potential grow longer, managers will have more flexibility in managing diverse forests for timber. Some tree species are easier to manage than others, and it may be pragmatic to focus exploitation on the species and forests that are the most resilient to harvesting.

Most of the research from which these observations and tentative conclusions are drawn comes from relatively short-term observations and from forests that have gone through only one or two cutting cycles, so our understanding and experience is still incomplete. Long-term, spatially explicit research is a priority, as is focused, mechanistic research.

Acknowledgements

I thank Joberto Veloso de Freitas for stimulating discussions and his assistance in tracking references, and Martin Barker and two anonymous reviewers for their suggestions to improve the manuscript.

References

Adam, K. A. (2003) Tree selection and selective logging: ecological and silvicultural considerations for natural forest management in Ghana. Unpublished Ph.D. thesis, University of Aberdeen.

Aiken, R. (2002) Liana responses to different treatments nine years after selective logging in Sabah Malaysia. Unpublished BSc thesis, Department of Agriculture & Forestry, University of Aberdeen.

Appanah, S. & Putz, F. E. (1984) Climber abundance in virgin dipterocarp forest and the effect of pre-felling climber cutting on logging damage. *The Malaysian Forester*, **47**, 335–342.

Bacon, P. S. (1982) The weedy species of *Merremia* (Convolvulaceae) occurring in the Solomon Islands and a description of a new species. *Botanical Journal of the Linnean Society*, **84**, 257–264.

Bennett, E. I. & Robinson, J. G. (2000) *Hunting of Wildlife in Tropical Forests: Implications for Biodiversity and Forest Peoples*. Biodiversity Series, Impact Studies, Paper no. 76. Washington DC: World Bank Environment Department.

Brokaw, N. & Busing, R. T. (2000) Niche versus chance and tree diversity in forest gaps. *Trends in Ecology and Evolution*, **15**, 183–188.

Brown, N. D., & Jennings, S. (1998) Gap-size niche differentiation by tropical rainforest trees: a testable hypothesis or a broken-down bandwagon? In: *Dynamics of Tropical Communities* (ed. D. M. Newbery, H. H. T. Prins & N. D. Brown). Oxford: Blackwell Science Ltd, pp. 79–94.

Cannon, C. H., Peart, D. R. & Leighton, M. (1998) Tree species diversity in commercially logged Bornean rainforest. *Science*, **281**, 1366–1368.

Catinot, R. (1997) *The Sustainable Management of Tropical Rainforests*. Paris: Scytale Publishing.

Chandrashekara, U. M. & Ramakrishnan, P. S. (1994) Successional patterns and gap phase dynamics of a humid tropical forest of the Western Ghats of Kerala, India: ground vegetation, biomass, productivity and nutrient cycling. *Forest Ecology and Management*, **70**, 23–40.

Chapman, C. A., Chapman, L. J., Zanne, A. & Burgess, M. A. (2002) Does weeding promote regeneration of an indigenous tree community in felled pine plantations in Uganda. *Restoration Ecology*, **10**, 408–415.

Costa, F. R. C. & Magnusson, W. E. (2003) Effects of selective logging on the diversity and abundance of understorey plants in central Amazonian forest. *Biotropica*, **35**, 103–114.

Crome, F. H. J., Moore, L. A. & Richards, G. C. (1992) A study of logging damage in upland rainforest in North Queensland. *Forest Ecology and Management*, **49**, 1–29.

Crome, F. H. J., Thomas, M. R., & Moore, L. A. (1996) A novel Bayesian approach to assessing impacts of rain forest logging. *Ecological Applications*, **6**, 1104–1123.

Curran, L. M., Caniago, I., Paoli, G. D. *et al.* (1999) Impact of El Niño and logging on canopy tree recruitment in Borneo. *Science*, **286**, 2184–2188.

Dawkins, H. C. (1958) *The Management of Natural Tropical High Forest with Special Reference to Uganda*. Paper No. 34. Oxford: Imperial Forestry Institute.

Dawkins, H. C. & Philip, M. S. (1998) *Tropical Moist Forest Silviculture and Management. A History of Success and Failure.* Wallingford, UK: CAB International Publishing.

de Carvalho, J. O. P., Carmo Alves Lopes, J. & Silva, J. N. M. (1999) Dinâmica da diversidade de espécies em uma floresta de terra firme na Amazônia Brasileira relacionada à intensidade de exploração. 1999. In *Simpóosio Silvicultura Na Amazônia Oriental: Contribuições do Projecto EMBRAPA/DFID*, Bélem, PA. Pará, Brasil: EMBRAPA.

de Graaf, N. R., Poels, R. L. H. & van Rompaey, R. S. A. R. (1999) Effect of silvicultural treatment on growth and mortality of rainforest in Surinam over long periods. *Forest Ecology and Management*, **124**, 123–135.

Dekker, M. & de Graaf, N. R. (2003) Pioneer and climax tree regeneration following selective logging with silviculture in Suriname. *Forest Ecology and Management*, **172**, 183–190.

Denslow, J. S. (1995) Disturbance and diversity in tropical rain forests: the density effect. *Ecological Applications*, **5**, 962–968.

Dewalt, S. J., Schnitzer, S. A. & Denslow, J. S. (2000) Density and diversity of lianas along a chronosequence in a central Panamanian lowland forest. *Journal of Tropical Ecology*, **16**, 1–19.

Dickinson, M. B., Whigham, D. F. & Hermann, S. M. (2000) Tree regeneration in felling and natural treefall disturbances in a semideciduous tropical forest in Mexico. *Forest Ecology and Management*, **134**, 137–151.

Dolanc, C. R., Gorchov, D. L. & Cornejo, F. (2003) The effects of silvicultural thinning on trees regenerating in strip clear-cuts in the Peruvian Amazon. *Forest Ecology and Management*, **182**, 103–116.

Douglas, I., Greer, T., Wong, W. M., Bidin, K., Sinun, W. & Spencer, T. (1992) Controls of sediment discharge in undisturbed and logged tropical rain forest streams. *Fifth International Symposium on River Sedimentation*, Karlsruhe.

Fimbel, R. A., Grajal, A. & Robinson, J. G., eds. (2001) *The Cutting Edge. Conserving Wildlife in Logged Tropical Forests.* New York: Columbia University Press, p. 808.

Forest Stewardship Council (FSC) (2000) *Principles and Criteria for Forest Stewardship.* http://www.fscoax.org/principal.htm. Accessed regularly between 2002 and 2004.

Fox, J. E. D. (1976) Constraints on the natural regeneration of tropical moist forest. *Forest Ecology and Management*, **1**, 37–65.

Fox, J. E. D. & Chai, D. (1982) Refinement of a regenerating stand of the *Parashorea tomentella/Eusideroxylon zwageri* type of lowland dipterocarp forest in Sabah – a problem in silvicultural management. *Malaysian Forester*, **45**, 133–183.

Fredericksen, T. S. & Mostacedo, B. (2000) Regeneration of timber species following selective logging in a Bolivian tropical dry forest. *Forest Ecology and Management*, **131**, 47–55.

Fredericksen, T. S. & Putz, F. E. (2003) Silvicultural intensification for tropical forest conservation. *Biodiversity and Conservation*, **12**, 1445–1453.

Fredericksen, T. S., Putz, F. E., Pattie, P., Pariona, W. & Pea-Claros, M. (2003) Sustainable forestry in Bolivia. Beyond planned logging. *Journal of Forestry*, **101**, 37–40.

Freitas, J. V. de (2004) Improving tree selection for felling and retention in natural forest in Amazonia through spatial control and targeted seed tree retention: a case study of a forest management project in Amazonas State, Brazil. Unpublished Ph.D. thesis, University of Aberdeen.

Gardette, E. (1996) The effect of selective timber logging on the diversity of woody climbers at Pasoh. In *Conservation, Management and Development of Forest Resources: Proceedings of the Malaysia–United Kingdom Programme*

Workshop, 21–24 October, 1996, Kuala Lumpur, Malaysia (ed. S. S. Lee, D. Y. May, I. D. Gauld & J. Bishop). Kepong, Malaysia: Forest Research Institute Malaysia, pp. 115–126.

Gerwing, J. J. (2001) Testing liana cutting and controlled burning as silvicultural treatments for a logged forest in the eastern Amazon. *Journal of Applied Ecology*, **38**, 1264–1276.

Gerwing, J. J. & Vidal, E. (2002) Changes in liana abundance and species diversity eight years after liana cutting and logging in an eastern Amazonian forest. *Conservation Biology*, **16**, 544–548.

Gerwing, J. J. & Uhl, C. (2002) Pre-logging liana cutting reduces liana regeneration in logging gaps in the eastern Brazilian Amazon. *Ecological Applications*, **12**, 1642–1651.

Grieser Johns, A. (1997) *Timber Production and Biodiversity Conservation in Tropical Rain Forests*. Cambridge: Cambridge University Press.

Guariguata, M. R. & Saenz, G. P. (2002) Post-logging acorn production and oak regeneration in a tropical montane forest Costa Rica. *Forest Ecology and Management*, **167**, 285–293.

Gullison, R. E. & Hardner, J. J. (1993) The effects of road design and harvest intensity on forest damage caused by selective logging: empirical results and a simulation model from the Bosque Chimanes, Bolivia. *Forest Ecology and Management*, **59**, 1–14.

Hall, J. S., Medjibe, V., Berlyn, G. & Ashton, P. M. S. (2003) Seedling growth of three co-occurring *Entandrophragma* species (Melliaceae) under simulated light environments: implications for forest management in central Africa. *Forest Ecology and Management*, **179**, 135–144.

Hawthorne, W. D. (1993) *Forest Regeneration After Logging*. ODA Forestry Series No. 3. London: Overseas Development Agency, p. 52.

(1994) *Fire Damage and Forest Regeneration in Ghana*. ODA Forestry Series No. 4. London: Overseas Development Agency.

Holdsworth, A. R. & Uhl, C. (1997) Fire in Amazonian selectively logged rain forest and the potential for fire reduction. *Ecological Applications*, **7**, 713–725.

Horne, R. & Hickey, J. (1991) Ecological sensitivity of Australian rainforests to selective logging. *Australian Journal of Ecology*, **16**, 119–129.

Howlett, B. E. & Davidson, D. W. (2003) Effects of seed availability, site conditions, and herbivory on pioneer recruitment after logging in Sabah, Malaysia. *Forest Ecology and Management*, **184**, 369–383.

Hubbell, S. P., Foster, R. B., O'Brien, S. T. *et al.* (1999) Light-gap disturbances, recruitment limitation, and tree diversity in a neotropical forest. *Science*, **283**, 554–557.

Janzen, D. H. (1983) No park is an island: increase in interference from outside as park size decreases. *Oikos*, **41**, 402–410.

Jennings, S. B., Brown, N. D., Boshier, D. H., Whitmore, T. C. & Lopes, J. D. A. (2001) Ecology provides a pragmatic solution to the maintenance of genetic diversity in sustainably managed tropical rain forests. *Forest Ecology and Management*, **154**, 1–10.

Johns, A. D. (1983) Ecological effects of selective logging in a West Malaysian rain forest. Unpublished Ph.D. thesis, University of Cambridge.

Johns, J. S., Barreto, P. & Uhl, C. (1996) Logging damage during planned and unplanned logging operations in the eastern Amazon. *Forest Ecology and Management*, **89**, 59–77.

Kammesheidt, L. (1998) Stand structure and spatial pattern of commercial species in logged and unlogged Venezuelan forest. *Forest Ecology and Management*, **109**, 163–174.

Kasenene, J. M. (1987) The influence of mechanized selective logging, felling intensity and gap-size on the regeneration of a tropical moist forest in the Kibale Forest Reserve, Uganda. Unpublished Ph.D. thesis, Michigan State University.

Kasenene, J. (2001) Lost logging: problems of tree regeneration in forest gaps in Kibale Forest, Uganda. In *African Rain Forest Ecology and Conservation, An Interdisciplinary Perspective* (ed. W. Weber, L. J. T. White, A. Vedder & L. Naughton-Treves). New Haven, CT: Yale University Press, pp. 1480–1490.

Kennard, D. (1998) Biomechanical properties of tree saplings and free-standing lianas as indicators of susceptibility to logging damage. *Forest Ecology and Management*, **102**, 179–191.

Kennard, D., Gould, K. & Putz, F. E. (2002) Effect of disturbance intensity on regeneration mechanisms in a tropical dry forest. *Forest Ecology and Management*, **162**, 197–208.

Kennedy, D. N. & Swaine, M. D. (1992) Germination and growth of colonizing species in artificial gaps of different sizes in dipterocarp rain forest. *Philosophical Transactions of the Royal Society of London B*, **335**, 357–366.

King, G. C. & Chapman, W. S. (1983) Floristic composition and structure of a rainforest area, 25 years after logging. *Australian Journal of Ecology*, **8**, 415–423.

Lambert, F. R. (1992) The consequences of selective logging for Bornean lowland forest birds. In *Tropical Rain Forest: Disturbance and Recovery* (ed. A. G. Marshall & M. D. Swaine). London: The Royal Society, pp. 443–457.

Lamprecht, H. (1993) Silviculture in the tropical natural forests. In *Tropical Forestry Handbook*, Vol. 2 (ed. L. Pancel). Berlin: Springer, pp. 727–810.

Laurance, W. F. & Bierregaard, R. O. Jr (1997) *Tropical Forest Remnants. Ecology, Management, and Conservation of Fragmented Communities.* Chicago: University of Chicago Press, p. 615.

Liew, T. C. (1973) Occurrence of seeds in virgin forest top soil with particular reference to secondary species in Sabah. *The Malaysian Forester*, **36**, 185–193.

Liu, J. & Ashton, P. S. (1999) Simulating effects of landscape context and timber harvest on tree species diversity. *Ecological Applications*, **9**, 186–201.

Lobo Lopez, E. (2003) Spatial considerations for retention of seed trees in selectively logged forest in Ghana. Unpublished B.Sc. thesis, University of Aberdeen.

Manokaran, N. (1996) Effect, 34 years later, of selective logging in the lowland dipterocarp forest at Pasoh, Peninsular Malaysia, and implications on present-day logging in the hill forests. In *Conservation, Management and Development of Forest Resources* (ed. S. S. Lee, D. Y. May, I. D. Gauld & J. Bishop). Kepong, Malaysia: Forest Research Institute of Malaysia, pp. 41–60.

Martini, A. M. Z., Rosa, N. de A. & Uhl, C. (1994) An attempt to predict which Amazonian tree species may be threatened by logging activities. *Environmental Conservation*, **21**: 152–162.

Meijer, W. (1970) Regeneration of tropical lowland forest in Sabah, Malaysia forty years after logging. *The Malayan Forester*, **33**, 204–229.

Neil, P. E. (1984) Climber problems in Solomon Islands forestry. *Commonwealth Forestry Review*, **63**, 27–35.

Nussbaum, R., Anderson, J. & Spencer, T. (1995) Effects of selective logging on soil characteristics and growth of planted dipterocarp seedlings in Sabah. In *Ecology, Conservation, and Management of Southeast Asian Rainforests* (ed. R. B. Primack & T. E. Lovejoy). New Haven, Yale University Press, pp. 105–115.

Panfil, S. N. & R. E. Gullison. (1998) Short term impacts of experimental timber harvest intensity on forest structure and composition in the Chimanes Forest Bolivia. *Forest Ecology and Management*, **102**, 235–243.

Pariona, W., Fredericksen, T. S. & Licona, J. C. (2003) Natural regeneration and liberation of timber species in logging gaps in two Bolivian tropical forests. *Forest Ecology and Management*, **181**, 313–322.

Parren, M. P. E. (2003) Lianas and logging in West Africa. Ph.D. thesis, Wageningen University.

Parren, M. & Bongers, F. (2001) Does climber cutting reduce felling damage in southern Cameroon? *Forest Ecology and Management*, **141**, 175–188.

Pearson, T. R. H., Burslem, D. F. R. P., Goeriz, R. E., & Dalling, J. W. (2003) Interactions of gap size and herbivory on establishment, growth and survival of three species of neotropical pioneer trees. *Journal of Ecology*, **91**, 785–796.

Pérez-Salicrup, D. R. (2001) Effect of liana cutting on tree regeneration in a liana forest in Amazonian Bolivia. *Ecology*, **82**, 389–396.

Pérez-Salicrup, D. R. & Stork, V. L. (2001) Lianas and trees in a liana forest of Amazonian Bolivia. *Biotropica*, **33**, 34–47.

Pinard, M. A., Barker, M. G., & Tay, J. (2000) Soil disturbance resulting from bulldozer-yarding of logs and post-logging forest recovery on skid trails in Sabah, Malaysia. *Forest Ecology and Management*, **130**, 213–225.

Pinard, M. A., Howlett, B. & Davidson, D. (1996) Site conditions limit pioneer tree recruitment after logging of dipterocarp forests in Sabah, Malaysia. *Biotropica*, **28**, 2–12.

Pinard, M. A., Licona, J. C. & Putz, F. E. (1999a) Tree mortality and vine proliferation following a wildfire in a subhumid tropical forest in eastern Bolivia. *Forest Ecology and Management*, **116**, 247–252.

Pinard, M. A. & Putz, F. E. (1994) Vine infestation of large remnant trees in logged forest in Sabah, Malaysia: biomechanical

facilitation in vine succession. *Journal of Tropical Forest Science*, **3**, 302–309.

(1996) Retaining forest biomass by reducing logging damage. *Biotropica*, **28**, 278–295.

Pinard, M. A., Putz, F. E., Rumiz, D. & Guzman, R. (1999b) Ecological characterization of tree species to guide forest management decisions – an exercise in classification of species in the semideciduous forests of Lomerio, Bolivia. *Forest Ecology and Management*, **113**, 201–213.

Pinard, M. A., Veizaga, R. & Stanley, S. A. (1998) Mejora de la regeneración de especies arbóreas en los claros de corta de un bosque estacionalmente seco en Bolivia. *Las Memorias del Simposio Internacional sobre Posibilidades para el Manejo Forestal Sostenible en America Tropical*, Santa Cruz, Bolivia: IUFRO / BOLFOR, pp. 157–160.

Pitman, N. C. A., Terborgh, J. W., Silman, M. R. *et al.* (2001) Dominance and distribution of tree species in upper Amazonian terra firme forests. *Ecology*, **82**, 2102–2117.

Plumptre, A. J. (1995) The importance of 'seed trees' for the natural regeneration of selectively logged forest. *Commonwealth Forestry Review*, **74**, 253–258.

(1996) Changes following 60 years of selective timber harvesting in the Budongo Forest Reserve, Uganda. *Forest Ecology and Management*, **89**, 101–113.

(2001) The effects of habitat change due to selective logging on the fauna of forests in Africa. In *African Rain Forest Ecology and Conservation. An Interdisciplinary Perspective* (ed. W. Weber, L. J. T. White, A. Vedder & L. Naughton-Treves). New Haven: Yale University Press, pp. 463–479.

Putz, F. E. (1984) The natural history of lianas on Barro Colorado Island, Panama. *Ecology*, **65**, 1713–1725.

(1985) Woody vines and forest management in Malaysia. *Commonwealth Forestry Review*, **64**, 359–365.

(1991) Silvicultural effects of lianas. In *The Biology of Vines* (ed. F. E. Putz & H. A. Mooney). New York: Cambridge University Press, pp. 493–501.

(1994) Relay ascension of big trees by vines in Rock Creek Park, District of Columbia. *Castanea*, **59**, 167–169.

Putz, F. E., Lee, H. S., & Goh, R. (1984) Effects of post-felling silvicultural treatments on woody vines in Sarawak. *The Malaysian Forester*, **47**, 214–226.

Putz, F. E., Redford, K. H., Robinson, J. G., Fimbel, R. & Blate, G. M. (2000) *Biodiversity Conservation in the Context of Tropical Forest Management*. Biodiversity Series: Impact Studies. Washington DC: The World Bank, p. 80.

Putz, F. E., Sirot, L. K., & Pinard, M. A. (2001) Tropical forest management and wildlife: silvicultural effects on forest structure, fruit production, and locomotion of arboreal animals. In *The Cutting Edge* (ed. R. A. Fimbel, A. Grajal & J. G. Robinson). New York: Columbia University Press, pp. 11–34.

Robinson, J. G. & Bennett, E. L. (2000) *Hunting for Sustainability in Tropical Forests*. New York: Columbia University Press, p. 582.

Saenz, G. P. & Guariguata, M. R. (2001) Demographic response of tree juveniles to reduced-inpact logging in a Costa Rican montane forest. *Forest Ecology and Management*, **140**, 75–84.

Salick, J., Mejia, A. & Anderson, T. (1995) Non-timber forest products integrated with natural forest management, Rio San Juan, Nicaragua. *Ecological Applications*, **5**, 878–895.

Schnitzer, S. A. & Bongers, F. (2002) The ecology of lianas and their role in forests. *Trends in Ecology and Evolution*, **17**, 223–230.

Schnitzer, S. A., Dalling, J. W. & Carson, W. P. (2000) The impact of lianas on tree regeneration in tropical forest canopy gaps: evidence for an alternative pathway of gap-phase regeneration. *Journal of Ecology*, **88**, 655–666.

Schnitzer, S. A., Parren, M. P. E. & Bongers, F. (2003) Recruitment of lianas into logging gaps and the effects of pre-harvest climber cutting in a lowland forest in Cameroon. *Forest Ecology and Management*, **190**, 87–98.

Seymour, R. S. & Hunter, M. L. (1999) Principles of ecological forestry. In *Maintaining Biodiversity in Forest Ecosystems* (ed. M. L. Hunter). Cambridge: Cambridge University Press, pp. 22–61.

Sheil, D. & Burslem, D. F. R. P. (2003) Disturbing hypotheses in tropical forests. *Trends in Ecology and Evolution*, **18**, 18–26.

Silva, J. N. M., Carvalho, J. O. P. de, Lopes, J. de C. A. *et al.* (1995) Growth and yield of a tropical rain forest in the Brazilian Amazon 13 years after logging. *Forest Ecology and Management*, **71**, 267–274.

Sist, P. & Nguyen-Thé, N. (1992) Logging damage and the subsequent dynamics of a dipterocarp forest in East Kalimantan (1990–1996). *Forest Ecology and Management*, **165**, 85–103.

Sist, P., Nolan, T., Bretault, J.-G. & Dykstra, D. (1998) Logging intensity versus sustainability in Indonesia. *Forest Ecology and Management*, **108**, 251–260.

Snook, L. (1996) Catastrophic disturbance, logging and the ecology of mahogany (*Swietenia macrophylla* King): grounds for listing a major tropical timber species in CITES. *Botanical Journal of Linnean Society*, **122**, 35–46.

Strushaker, T., Lwanga, J., & Kasenene, J. (1996) Elephants, selective logging and forest regeneration in the Kibale Forest, Uganda. *Journal of Tropical Ecology*, **12**, 45–64.

Suding, K. N., Gross, K. L. & Houseman, G. R. (2004) Alternative states and positive feedbacks in restoration ecology. *Trends in Ecology and Evolution*, **19**, 46–53.

Swaine, M. D. (1992) Characteristics of dry forest in West Africa and the influence of fire. *Journal of Vegetation Science*, **3**, 365–374.

Underwood, A. J. (1991) Beyond BACI: experimental designs for detecting human environmental impacts on temporal variations in natural populations. *Australian Journal of Marine and Freshwater Resources*, **42**, 569–587.

— (1993) The mechanics of spatially replicated sampling programmes to detect environmental impacts in a variable world. *Australian Journal of Ecology*, **18**, 99–116.

van der Hout, P. (1999) *Reduced Impact Logging in the Tropical Rain Forest of Guyana. Ecological, Economic and Silvicultural Consequences.* Georgetown, Guyana: Tropenbos-Guyana Series 6, Tropenbos-Guyana Programme, p. 335.

Verissimo, A., Barreto, P., Tarifa, R. & Uhl, C. (1995) Extraction of a high-value natural resource in Amazonia: the case of mahogany. *Forest Ecology and Management*, **72**, 39–60.

Vidal, E., Johns, J., Gerwing, J. J., Barreto, P. & Uhl, C. (1997) Vine management for reduced-impact logging in eastern Amazonia. *Forest Ecology and Management*, **98**, 105–114.

Wadsworth, F. H. (1997) *Forest Production For Tropical America.* United States Department of Agriculture. Forest Service. Agriculture Handbook 710.

Webb, E. L. (1997) Canopy removal and residual stand damage during controlled selective logging in lowland swamp forest of northeast Costa Rica. *Forest Ecology and Management*, **95**, 117–129.

Williams-Linera, G. (1990) Origin and early development of forest edge vegetation in Panamá. *Biotropica*, **22**, 235–241.

Woods, P. (1989) Effects of logging, drought, and fire on structure and compostion of tropical forests in Sabah, Malaysia. *Biotropica*, **21**, 290–298.

Woodward, C. (1995) Soil compaction and topsoil removal effects on soil properties and seedling growth in Amazonian Ecuador. *Forest Ecology and Management*, **82**, 197–209.

Wyatt-Smith, J. (1987) Problems and prospects for natural management of tropical moist forests. In *Natural Management of Tropical Moist Forests: Silvicultural and Management Prospects of Sustained Utilization* (ed. F. Mergen & J. R. Vincent). New Haven: Yale University School of Forestry and Environmental Studies, pp. 5–22.

Yap, S. W., Chak, C. V., Majuakim, L., Anuar, M. & Putz, F. E. (1995) Climbing bamboo (*Dinochloa* spp.) in Deramakot Forest Reserve, Sabah: biomechanical characteristics, modes of ascent and abundance in a logged-over forest. *Journal of Tropical Forest Science*, **8**, 196–202.

Index